Latin American Economic Outlook 2022

TOWARDS A GREEN AND JUST TRANSITION

This work is published under the responsibility of the Secretary-General of the OECD, the President of the Corporación Andina de Fomento (CAF) and the President of the European Commission. The opinions expressed and arguments employed herein do not necessarily reflect the official views of the Member countries of the OECD, the members of its Development Centre, those of Corporación Andina de Fomento (CAF), or of the European Union, or the Member States of the United Nations.

This document, as well as any data and map included herein, are without prejudice to the status of or sovereignty over any territory, to the delimitation of international frontiers and boundaries and to the name of any territory, city or area.

The names of countries and territories and maps used in this joint publication follow the practice of the OECD.

Please cite this publication as:
OECD et al. (2022), *Latin American Economic Outlook 2022: Towards a Green and Just Transition*, OECD Publishing, Paris, https://doi.org/10.1787/3d5554fc-en.

ISBN 978-92-64-79606-5 (print)
ISBN 978-92-64-95712-1 (pdf)
ISBN 978-92-64-79835-9 (HTML)
ISBN 978-92-64-73266-7 (epub)

Latin American Economic Outlook
ISSN 2072-5159 (print)
ISSN 2072-5140 (online)

CAF Reference Number: CAF-513i-2022

ECLAC Reference Number: LC/PUB.2022/14

European Union
ISBN 978-92-78-43290-4 (PDF)
Catalogue number: OA-05-22-350-EN-N (PDF)

Photo credits: Cover design by Aida Buendía and Mélodie Ly Descours (OECD Development Centre).

Corrigenda to publications may be found on line at: www.oecd.org/about/publishing/corrigenda.htm.

Foreword

Since first launched in November 2007, the *Latin American Economic Outlook* (LEO) has systematically analysed critical aspects related to sustainable and inclusive development in Latin America and the Caribbean (LAC) region. Each year, the report identifies key trends and associated opportunities and challenges. Across diverse topics, it compares LAC's performance with that of other regions, analyses main development challenges and puts forward policy recommendations, experiences and good practices.

The LEO benefits from the expertise and inputs of multiple co-authors. Since 2011, it has been published in conjunction with the United Nations Economic Commission for Latin America and the Caribbean. In 2013, the Development Bank of Latin America joined the team of authors and, the European Commission joined as a main partner from the LEO 2018.

In a context of global economic and social crises, this 15th LEO, *Towards a green and just transition*, aims to analyse the challenges and opportunities of a green and just transition in LAC, ultimately to provide policy recommendations to ensure the well-being of citizens and the region's vast and rich ecosystems. The report explores policy actions to systemically advance the transition through the promotion of a new energy and production matrix and the creation of quality formal jobs. In turn, it stresses the need to pursue this transition as a way to accelerate the new social contract and reduce social disparities. Finally, the publication includes recommendations on how the international green agenda could benefit LAC via new partnerships and new tools to finance the green transition.

Acknowledgements

Partners of this report are the United Nations Economic Commission for Latin America and the Caribbean (ECLAC), the Development Bank of Latin America (CAF), the European Union (EU) and the Development Centre of the Organisation for Economic Co-operation and Development (OECD). This report is supported under Pillar 1 of the European Union Regional Facility for Development in Transition for Latin America and the Caribbean (LAC), a European Union-led initiative, jointly implemented with the OECD and its Development Centre and the ECLAC.

The contribution of the OECD Development Centre to this report was led and managed by Sebastián Nieto-Parra, Head of the Latin America and the Caribbean Unit at the OECD Development Centre. Co-ordination was led by Luis Cecchi and Olivia Cuq, Policy Analysts at the Latin America and the Caribbean Unit of the OECD Development Centre, under the guidance of Ragnheiður Elín Árnadóttir, Director of the OECD Development Centre and Federico Bonaglia, Deputy Director of the OECD Development Centre. The ECLAC's contribution was led by Sebastián Rovira, Economic Affairs Officer at the Innovation and New Technologies Unit, with the support of Nunzia Saporito, research assistant at the Innovation and New Technologies Unit, under the guidance of José Manuel Salazar-Xirinachs, Executive Secretary of ECLAC and Mario Cimoli, former Acting Executive Secretary of ECLAC. The contribution from CAF was led by Adriana Arreaza, Director of Macroeconomic Studies. The European Commission contribution was led by Diana Montero Melis, Deputy Head of Unit at the Directorate-General for International Cooperation and Development of the European Commission (INTPA) and Sergio Martin Moreno, ex Programme Manager in the Latin America and the Caribbean Directorate, under the guidance of Jorge de la Caballería, their Head of Unit.

The report benefited from the research, drafting and fruitful collaboration among various authors across these organisations, including: Aimee Aguilar Jaber (OECD), Adriana Arreaza (CAF), Paul Baldwin (OECD), Pablo Brassiolo (CAF), Cristina Cabutto (OECD), Adriana Caicedo (OECD), Luis Cecchi (OECD), Simone Cecchini (ECLAC), Olivia Cuq (OECD), Rita Da Costa (OECD), Ricardo Estrada (CAF), Lianne Guerra (OECD), Laura Gutiérrez Cadena (OECD), Oswaldo López (CAF), Thomas Manfredi (OECD), Mariana Mirabile (OECD), Nathalia Montoya González (OECD), Sergio Martin Moreno (European Commission), Sebastián Nieto Parra (OECD), René Orozco (OECD), Juan Ortegón Ocampo (OECD), Juana Angela Ospina (Shantalla), Juliana Juana María Otalvaro Mendez (OECD), Camila Ramirez (OECD), Mariana Rodríguez Pico (OECD), Sebastián Rovira (ECLAC), Vicente Ruiz (OECD), Nunzia Saporito (ECLAC), David Schmid (OECD), Daniel Titelman (ECLAC), Juan Vázquez Zamora (OECD) and Juan Nicolás Velandia (OECD). Agustina Vierheller (OECD) provided invaluable administrative support throughout the elaboration of the report.

A group of experts and colleagues have been particularly active and supportive during the production process, providing views, inputs or boxes, comments and strategic orientation to the report. We would like to highlight the support of Sergio Ampudiana Giorgana (OECD), Geraldine Ang (OECD), Karine Badr (OECD), Froukje Boele (OECD), Antoine Bonnet (OECD), Peter Borkey (OECD), Felipe Bosch (European Commission), Enrico Botta (OECD), Pablo Brassiolo (CAF), Álvaro Calderón (ECLAC), Amy Cano Prentice (OECD), Emma Cantera (OECD), Gabriela Casanova Rangel (Universidad del Rosario), Anthony Caubin (AFD), Mathilde Closset (ECLAC), Angie Contreras Sanabria (DNP), Ignacio Corlazzoli (CAF), Filipe Da Silva (ECLAC), Lylah Davies (OECD), Nawal Djaffar (OECD), John Dulac (OECD), Jane Ellis (OECD), Manuel Escudero (OECD), Ricardo Estrada (CAF), Chiara Falduto (OECD), Jimmy Ferrer (ECLAC), Jason Gagnon (OECD), Camilo Gamba Gamba (OECD), Catherine Gamper (OECD), Marina Gil Sevilla (ECLAC), Nathalie Girouard (OECD), Daniel Gómez Gaviria (DNP), Nicolas Gottman (European Commission), Tomas Hos (OECD), Sofía Hurtado del Orbe (OECD), Joerg Husar (IEA), Raphaël Jachnik (OECD), Richard Valery

Jaimes Bonilla (Universidad Javeriana), Katia Karousakis (OECD), Maike Kirsch (OECD), Eija Kiiskinen (OECD), Kumi Kitamori (OECD), Izumi Kotani (OECD), Helen Laubenstein (OECD), Bruno Leclerc (AFD), Xavier Leflaive (OECD), Lahra Liberti (OECD), Carina Lindberg (OECD), Marc Litvine (European Commission), Oswaldo López (CAF), Will Macpherson (OECD), Carlos Fernando Maldonado Valera (ECLAC), Natali Maldonado Valera (Universidad del Rosario), Vasiliki Mavroeidi (OECD), Mauricio Mejia Galvan (OECD), Natalia Moreno Rigollot (Telefónica), Jolien Noels (OECD), Georgina Nuñez (ECLAC), Walid Oueslati (OECD), Lorenzo Pavone (OECD), Carolina Pardo (Universidad del Rosario), Anna Piccini (OECD), Alexander Pick (OECD), Rodrigo Pizarro (OECD), Rayen Quiroga (ECLAC), Rafael Camilo Ramirez Correa (DNP), Ji-Yeun Rim (OECD), Laure Rogès (European Commission), Marta Salafranca (Telefónica), Carlos Santiso (OECD), Juan Manuel Santomé Calleja (Eurosocial), Elliot Smith (OECD), Ernesto Soria Morales (OECD), Katarina Svatikova (OECD), Cecilia Tam (OECD), Jonas Teusch (OECD), Elia Trippel (OECD), Daniela Trucco (ECLAC), Melanie Vilarasau Slade (OECD), Sebastian Weber (OECD), Ulrich Weins (European Commission), Dimitra Xynou (OECD), Robert Youngman (OECD), Yumika Yamada (OECD), Germán Zarama (OECD) and Felix Zimmermann (OECD).

The content of the report was enriched by constructive feedback received during the informal consultation with the LAC countries and members of the OECD Development Centre Governing Board, on 14 February 2022; the virtual Experts Meetings on 28 and 29 April 2022; and the hybrid Experts Dialogue that took place on 03 August 2022. In addition to the LAC delegates to the Governing Board of the OECD Development Centre, we are particularly grateful to the experts, academics, private-sector representatives and other public servants who joined us during the Experts' Meetings: Manuel Albaladejo (UNIDO), Verónica Amarante (University of the Republic, Uruguay), Jorge Arbache (CAF), Alberto Arenas de Mesa (ECLAC), Adriana Arreaza (CAF), José Miguel Benavente (CORFO), Carl Bernadac (AFD), Janos Bertok (OECD), Aida Caldera (OECD), Germán Jorge Carmona Paredes (UNAM), Mario Castillo (ECLAC), Monica Castillo (ILO), Eduardo Cerdá (Ministry of Agriculture livestock and fisheries, Argentina), Amaury Chazeau-Guibert (European Comission), Tommaso Ciarli (UNU-MERITT), Diego Coatz (UIA), Ruben Contreras (ECLAC), Sergi Corbalan (FTAO), Andrea Costafreda (OXFAM), Slim Dali (AFD), Carlos de Miguel (ECLAC), Deborah Delgado Pugley (Catholic University of Peru), Andres Espejo (ECLAC), Joao Carlos Ferraz (Rio de Janeiro Federal University, Brazil), Jimy Ferrer (ECLAC), Juan Flores (University of Geneva), Andrés García (EVCO, Colombia), Ricardo Gorini (IRENA), Stephany Griffith-Jones (Central Bank of Chile), Marcela Jaramillo (2050 Pathways Platform), Ernesto Jeger (EU-LAC), Eija Kiiskinen (OECD), Annette Killmer (IDB), Rodolfo Lacy (OECD), Maria Fernanda Lemos (IPCC), Santiago Lorenzo (ECLAC), Julio Maturana (Ministry of Energy, Chile), Roxana Maurizio (ILO), Catalina Misleh (Reborn Electric Motors), El Iza Mohamedou (OECD), Irene Monasterolo (EDHEC Business School), Carola Moreno Valenzuela (Ministry of Finance, Chile), Pablo Núñez (Ministry of Science Technology and Innovation, Argentina), José Antonio Ocampo (Columbia University), Rodrigo Olivares-Caminal (Queen Mary University of London), Jocelyn Olivari (CORFO), Laura Oroz (AECID), María Paz de la Cruz (Chilean hydrogen association), Javier Perez (Banco de España), Horst Pilger (European Commission), Paula Poblete (Ministry of Social Development and Family, Chile), Andrew Powell (IDB), Martin Rivero (SEGIB), Juan Ruiz (BBVA), José Luis Samaniego (ECLAC), José Antonio Sanahuja (Fundación Carolina), Ewout Sandker (EU Delegation, Chile), André Sapir (ULB), Julián Suárez (CAF), Daniel Titelman (ECLAC), Ricardo Torres (State of Querétaro, Mexico), Maria Elena Valenzuela (ECLAC), Ulrich Volz (University of London), Francesco Vona (University of Milan) and Luis Alberto Zuleta (ECLAC). A group of colleagues from the OECD provided insightful inputs, comments and discussions that significantly improved the report: Shardul Agrawala, José Antonio Ardavín, Jens Arnold, Sofia Blamey Andrusco, Frederic Boehm, Monica Brezzi, Jorge Carbonell, Emanuele Ciani, Joseph Cordonnier, Charlotte Dubald, Mayumi Endoh, Manuela Fitzpatrick, Michael Förster, Mills Gary, Fabio

Gehrke, Santiago González, Felipe González Zapata, Håvard Halland, Jean-Jacques Hible, Elizabeth Holbourne, Michael Jelenic, Fatos Koc, Alexandre Kolev, Kamil Kouhen, Juan de Laiglesia, Iris Mantovani, Claire Mc Evoy, Alejandra Meneses, Martin Neil, Ana Novik, Masayuki Omote, Néstor Pelechà Aigües, Nicolás Penagos, Jan Rielaender, Jacob Arturo Rivera Pérez, Camila Saffirio, Katherine Scrivens, Kimiaki Shinozaki, Ana Stringhini, Enes Sunel, Gabriela Villa Aguayo and Martin Wermelinger.

The country notes benefited from constructive inputs, scrutiny and verification by delegations to the OECD from Chile, Colombia, Costa Rica and Mexico, as well as the embassies in France of Argentina, Brazil, Dominican Republic, Ecuador, El Salvador, Guatemala, Panama, Paraguay, Peru and Uruguay.

The OECD Development Centre would also like to express its sincere gratitude to the *Agence française de développement*, the *Departamento Nacional de Planeación* (DNP) of Colombia, the European Union, the Spanish Ministry of Foreign Affairs, the Swiss Agency for Development and Cooperation, Telefónica and *Universidad del Rosario* (Colombia) for their support to the *Latin American Economic Outlook*.

Finally, many thanks go to the Publications and Communications Unit of the OECD Development Centre, in particular, Aida Buendía, Mélodie Descours, Delphine Grandrieux, Elizabeth Nash, Irit Perry and Henri-Bernard Solignac-Lecomte, for their steadfast patience and expedient work on the production of this report and associated materials. The authors also sincerely appreciate the editing and proofreading activities undertaken by Jessica Hutchings and Marilyn Smith, and the translation and Spanish editing services provided by Alejandro Barranco, Julia Gregory, Alexander Summerfield and Liliana Tafur.

Table of contents

Figures

Tables

Boxes

Acronyms and abbreviations

AAs Association Agreements
ACTO Amazon Cooperation Treaty Organization
AECID Spanish Agency for International Development Cooperation (*Agencia Española de Cooperación Internacional para el Desarrollo*)
AFD French Development Agency (*Agence Française de Développement*)
AGESIC National Digital Government Agency of Uruguay (*Agencia de Gobierno Electrónico y Sociedad de la Información y del Conocimiento*)
AICCA Andes Adaptation to the Impact of Climate Change on Water Resources Project (*Proyecto Adaptación a los Impactos del Cambio Climático en Recursos Hídricos en los Andes*)
AILAC Independent Association of Latin America and the Caribbean (*Asociación Independiente de Latinoamérica y el Caribe*)
ALBA Bolivarian Alliance for the Peoples of Our America (*Alianza Bolivariana para los Pueblos de Nuestra América*)
ALMP Active Labour Market Policies
AOSIS Alliance of Small Island States
BAU Business-as-usual
CABEI Central American Bank for Economic Integration (*Banco Centroamericano de Integración Económica*)
CAF Development Bank of Latin America (*Banco de Desarrollo de América Latina*)
CAIT Climate Analysis Indicators Tool
CAT Climate Action Tracker
CAPEX Capital Expenditures
CARICOM Caribbean Community (*Comunidad del Caribe*)
CBAM Carbon Border Adjustment Mechanism
CBD Convention on Biological Diversity
CBI Climate Bonds Initiative
CCS Carbon Capture and Storage
CCUS Carbon Capture, Use and Storage
Cedefop European Centre for the Development of Vocational Training
CEPII Center for Research and Expertise on the World Economy (*Centre d'Études Prospectives et d'Informations Internationales*)
CfRN Coalition for Rainforest Nations
CIEL Center for International Environmental Law
CIT Corporate Income Tax
CLEG Combined List of Environmental Goods
CO_2(e) Carbon dioxide (equivalent)
CoG Centre of Government
CONPES National Council for Economic and Social Policy of Colombia (*Consejo Nacional de Política Económica y Social*)
COP Colombian Peso
COP United Nations Climate Change Conference
COVID-19 Coronavirus
CSOs Civil Society Organisations
CVF Climate Vulnerable Forum
DAC Development Assistance Committee

DDPLAC Deep Decarbonisation Pathways for Latin America and the Caribbean
DFID United Kingdom Department for International Development
DNP National Planning Department of Colombia (*Departamento Nacional de Planeación*)
DNS Debt-for-nature Swap
EBA Ecosystem-based approach
ECB European Central Bank
EC-JRC European Commission Joint Research Centre
ECLAC United Nations Economic Commission for Latin America and the Caribbean
EEA European Environment Agency
EESI Environmental and Energy Study Institute
EFSD+ European Fund for Sustainable Development Plus
EIB European Investment Bank
EIG Environmental Integrity Group
EITC Earned Income Tax Credit
Embrapa Brazilian Agricultural Research Corporation (*Empresa Brasileira de Pesquisa Agropecuária*)
EMDEs Emerging Markets and Developing Economies
EPR Extended Producer Responsibility
ESG Environmental Social and Governance
ESMAP Energy Sector Management Assistance Program
ETS Emission Trading System
EU European Union
EUR Euro
EVs Electric Vehicles
FAO Food and Agriculture Organisation
FEBRABAN Brazilian Federation of Banks (*Federação Brasileira de Bancos*)
FDI Foreign Direct Investment
FTAO Fair Trade Advocacy Office
G20 Group of Twenty
GCF Green Climate Fund
GDP Gross Domestic Product
GEF Green Environment Facility
GERD Gross Domestic Expenditure in Research and Development
GGGI Global Green Growth Institute
GGO Global Gender Office
GHG Greenhouse Gases
GIZ German Agency for International Cooperation (*Deutsche Gesellschaft für Internationale Zusammenarbeit*)
GMI Global Methane Initiative
GPS Global Positioning System
GSS Green, Social and Sustainability bonds
GSSE General Services Support Estimate
GSSS Green, Social, Sustainable and Sustainability-linked bonds
HDI Human Development Index
HLPF High-level Political Forum on Sustainable Development
HS Harmonised System

ICMA	International Capital Market Association
ICZM	Integrated Coastal Zone Management
IDB	Inter-American Development Bank
IEA	International Energy Agency
IEEE	Spanish Institute for Strategic Studies (*Instituto Español de Estudios Estratégicos*)
IFC	International Finance Corporation
IFL	Intact Forest Landscapes
IFPRI	International Food Policy Research Institute
IFS	Institute for Fiscal Studies
IICA	Inter-American Institute for Cooperation on Agriculture
ILACC	Latin America and the Caribbean Carbon Market Initiative
ILO	International Labour Organization
IMF	International Monetary Fund
INED	French Institute for Demographic Studies (*Institut national d'études démographiques*)
INTPA	Directorate-General for International Partnerships
IDEA	International Institute for Democracy and Electoral Assistance
IOGP	International Association of Oil & Gas Producers
IPAC	OECD International Programme for Action on Climate
IPCC	Intergovernmental Panel on Climate Change
IPC-IG	International Policy Centre for Inclusive Growth
IRENA	International Renewable Energy Agency
IRP	International Resource Panel
ITF	International Transport Forum
IUCN	International Union for Conservation of Nature
LAC	Latin America and the Caribbean
LDCs	Least Developed Countries
LEO	Latin American Economic Outlook
LMDCs	Like-minded Developing Countries
LNG	Liquefied Natural Gas
LUCF	Land use, land-use change, and forestry
MAPA	Brazilian Ministry of Agriculture, Livestock and Food Supply (*Ministério da Agricultura, Pecuária e Abastecimento*)
MCTI	Ministry of Science and Technology of Brazil (*Ministério da Ciência, Tecnologia e Inovações*)
M&E	Monitoring and Evaluation
MENA	Middle East and North Africa
MERCOSUR	Southern Common Market (*Mercado Común del Sur*)
MIC	Middle-income Country
MINAE	Ministry of Environment and Energy of Costa Rica (*Ministerio de Ambiente y Energía*)
MINAM	Ministry of Environment of Peru (*Ministerio del Ambiente*)
MINURVI	Ministers and High-level Authorities of the Housing and Urban Development Sector in Latin America and the Caribbean (*Ministros y Autoridades Máximas de la Vivienda y el Urbanismo de América Latina y el Caribe*)
MNE	Multinational Enterprise

MoU	Memorandum of Understanding
MSMEs	Micro, Small and Medium-sized Enterprises
MSP	Maritime Spatial Planning
NAMA	Nationally Appropriate Mitigation Actions
NAP	National Adaptation Plan
NCP	National Contact Point
NDB	National Development Bank
NDC	Nationally Determined Contribution
NDICI	Neighbourhood, Development and International Cooperation Instrument
NDP	National Development Plan
NGO	Non-governmental Organisation
NIT	Negative Income Tax
OCHA	United Nations Office for the Coordination of Humanitarian Affairs
ODA	Official Development Assistance
OECD	Organisation for Economic Co-operation and Development
OPEC	Organisation of the Petroleum Exporting Countries
PCAB	Partnership for the Conservation of Amazon Biodiversity
PIT	Personal Income Tax
PPA	Partnership Platform for the Amazon
PV	Photovoltaics
(P) VAT	(Personalised) Value-added Tax
R&D	Research and Development
RBC	Responsible Business Conduct
RELAC	Renewables in Latin America and the Caribbean
REDD+	Reducing Emissions from Deforestation and forest Degradation
RICYT	Network for Science and Technology Indicators – Ibero-American and Inter-American
RST	Resilience and Sustainability Trust
SBFN	Sustainable Banking and Finance Network
SDGs	Sustainable Development Goals
SDR	Social Discount Rates
SDSN	Sustainable Development Solutions Network
SEGIB	Iberoamerican General Secretary (Secretaría General Iberoamericana)
SIDS	Small Islands Developing States
SINAMECC	National Climate Change Metrics System of Costa Rica (*Sistema Nacional de Métrica de Cambio Climático*)
SINARE	National System for the Reduction of Greenhouse Gas Emissions of Brazil (*Sistema Nacional de Redução de Emissões*)
SLB	Sustainability-linked Bond
SMEs	Small and Medium-sized Enterprises
SSC	South-South Co-operation
SSTC	South-South and Triangular Cooperation
STI	Science, technology and innovation
TFP	Total Factor Productivity
TPI	Transmission Protection Instrument
TrC	Triangular Co-operation
UK PACT	United Kingdom Partnering for Accelerated Climate Transitions
ULB	Free University of Brussels (*Université libre de Bruxelles*)

UN	United Nations
UNCCD	United Nations Convention to Combat Desertification
UNCTAD	United Nations Conference on Trade and Development
UNDESA	United Nations Department of Economic and Social Affairs
UNDP	United Nations Development Programme
UNECA	United Nations Economic Commission for Africa
UNECE	United Nations Economic Commission for Europe
UNEP	United Nations Environment Programme
UNESCO	United Nations Educational, Scientific and Cultural Organization
UNFCCC	United Nations Framework Convention on Climate Change
UNIDO	United Nations Industrial Development Organization
UNWTO	United Nations World Tourism Organization
USAID	United States Agency for International Development
USD	United States Dollar
VAT	Value Added Tax
VNR	Voluntary National Reviews
WEF	World Economic Forum
WHO	World Health Organization
WIN	Worldwide Independent Network of Market Research
WIR	World Inequality Report
WMO	World Meteorological Organization
WRI	World Resources Institute
WTO	World Trade Organization
WWF	World Wide Fund for Nature

Editorial

Latin American and Caribbean (LAC) countries are at a critical juncture. Just as the region was looking forward to a rebound in growth and a more sustainable development path after the pandemic, the current inflationary pressures and stark geopolitical tensions are throwing spanners in the regional recovery works. In parallel, LAC faces the challenge of implementing ambitious adaptation and mitigation strategies to address the climate emergency, while at the same time accelerating the pace of social, economic and institutional progress, all within narrowing fiscal space.

This 15th edition of the *Latin American Economic Outlook* (LEO 2022) argues that the green and just transition agenda is a unique opportunity for Latin America and the Caribbean to improve well-being for all, tackle inequalities, tap into new sources of employment and financing, and chart more sustainable and inclusive development pathways.

We need to act now. This ambitious agenda must protect and value the region's unique natural and renewable energy capital. Innovative industrial policies and green, blue and circular economy approaches can transform LAC's energy and production matrices, drive productivity growth and support the development of new economic sectors, while reducing greenhouse gas (GHG) emissions.

A green transition will not be possible, nor would it be desirable, unless it is just. A green and just transition should strengthen social protection systems, promote job formalisation and bridge existing gaps across socioeconomic groups, territories and generations. Active labour market policies are essential to provide people with the necessary skills and help them transition from brown to green industries, and from low-quality to better jobs.

A transition to a low-carbon and inclusive economy requires the mobilisation of vast amounts of financial resources. Fiscal policy, national and multilateral development banks, and the private sector will have to play a stronger and more co-ordinated role to catalyse investment, develop market instruments and regulatory tools, and compensate the most affected and vulnerable populations.

A legitimate and politically viable transition requires a broad consensus on its objectives, and proceeds through reforms that result from a collaborative and inclusive dialogue. This is particularly important in a region where support for green policies is high, but where trust in public institutions has seen a sharp erosion. It is therefore an opportunity to renew the social contract in the region.

New forms of cross-border partnerships and co-operation will be crucial, as national efforts will not suffice. Addressing climate change and de-carbonising economies require co-ordinated global action. Boosting international partnerships, in particular with the European Union as part of the Global Gateway, can help LAC adapt to the "greening" of trade norms and regulations, attract greater investment, access new technologies and decisively contribute to shaping global norms and actions.

This is a challenge we cannot fail to address, an opportunity not to be missed. We stand ready to work together to support the region's efforts, and trust that LEO 2022 provides a solid basis for the ambitious and inclusive policy dialogue ahead of us.

José Manuel Salazar-Xirinachs Executive Secretary ECLAC	Sergio Díaz-Granados Executive President CAF – Development Bank of Latin America	Mathias Cormann Secretary-General OECD	Jutta Urpilainen European Commissioner for International Partnerships

Executive summary

Countries in Latin America and the Caribbean (LAC) face an ambitious agenda to ensure that the green transition is an opportunity to enhance development in the region. The global impacts of the COVID-19 pandemic and of Russia's war against Ukraine have been a reminder of LAC's weak resilience to shocks. This stems from its structural challenges: fragile social protection systems; low productivity; weak institutions; and an environmentally unsustainable development model. A systemic green and just transition could help the region overcome its development "traps" and strengthen its resilience while improving Latin Americans' well-being. LAC is highly exposed to the effects of climate change and, as such, governments should seize the recovery as a strategic opportunity to launch broad and deep transformation.

LAC faces a complex context, on both domestic and international scales

After a strong rebound from the impacts of COVID-19 in 2021, recovery in LAC has slowed. Domestically, this reflects low potential growth and diverse structural challenges. But fallout from Russia's invasion of Ukraine and an economic slowdown in China confirm that LAC is also deeply connected to an increasingly complex international context. Across the LAC region, economies are experiencing increased inflationary pressures and uncertainty, as well as disruptions in trade with key economic partners. Reduced macroeconomic policy space – both monetary and fiscal – makes it harder for LAC economies to sustain international and environmental shocks, as they strive to support the economic recovery and protect the most vulnerable. Estimates suggest that by the end of 2022, 33.7% of the LAC population will be in poverty and 14.9% in extreme poverty, as rising prices have more profound effects on the most vulnerable population. In the first five months of 2022, in selected LAC countries, inflation rates for the extreme poor were 3.6 percentage points higher than total inflation. Fiscal policy should promote growth and social inclusion, based on strategic pillars of environmental and fiscal sustainability, decarbonisation and resilience.

A green and just transition can make LAC societies more resilient to climate change while promoting better development

LAC is one of the regions most vulnerable to climate change; 13 of the 50 countries identified as most affected by the climate emergency are in LAC. Hence, the urgent need for a green and just transition. This high risk is disproportionate in that LAC's share in total greenhouse gas (GHG) emissions (8.1%) is proportional to its share of total global population (8.4%) and slightly higher than its share in total gross domestic product (GDP) (6.4%). If implemented in a systemic way, active mitigation and adaptation policies can reduce the disproportionate consequences of climate change on inequalities across countries, socio-economic groups, territories, generations and gender. A green and just transition must go beyond fighting climate change and put citizens' well-being at its centre.

Transforming the energy and production matrix can help boost productivity, develop new and more sustainable economic sectors, and create more formal jobs

Transformation of the energy matrix is key to promoting greater well-being for citizens and fostering more resilient societies in LAC. The region is endowed with high potential for renewable energy resources; at present, renewables account for 33% of total energy supply in the region, compared to just 13% globally. Investing in renewable technologies can substantially reduce GHG emissions while also providing lower-cost power and, for some LAC countries, reducing reliance on imported fossil fuel products. A successful transition to net zero emissions will be contingent on systemic decarbonisation through electrification across all sectors. In parallel, investments in green hydrogen and other low-carbon alternative fuels, including sustainable biofuels, will be key to decarbonising hard-to-abate sectors. A total of 17 million people still have no access to electricity, especially in rural areas. Ensuring universal access to energy is a crucial element of a green and just transition as it can help overcome inequalities in access to basic public services and stimulate local economic growth. Industrial, circular and blue policies should also be key elements of LAC's sustainable productive transformation. Advancing a green transition can potentially add 10.5% more net jobs in LAC by 2030. To achieve this, it is necessary to secure additional public and private investments that contribute to an increase of 3 percentage points in the value added of green sectors. The transition to new green jobs implies development of active labour market policies and well-targeted social policies to support those workers and households who will be negatively impacted by the transition.

Financing the green transition requires environmentally sustainable fiscal policies and a scaling up of novel financial instruments

The LAC region needs to mobilise substantial resources to finance the green transition. A 2.5°C global warming scenario could cost the region between 1.5% and 5.0% of its GDP by 2050. Since the cost of inaction is high, governments have to gradually phase out subsidies and leverage the potential of environmentally related taxes while also scaling up novel financial instruments such as debt-for-nature swaps, natural disaster clauses, catastrophe bonds, and green, social, sustainability and sustainability-linked (GSSS) bonds. Between 2014 and 2021, LAC's GSSS bonds issuance in international markets reached an accumulated USD 73 billion, of which USD 31 billion came from green bonds alone. Of the total GSSS issuance, 42% came from the corporate sector, 37% from the sovereign sector, 12% from quasi-sovereign, and 5% from supranational issuers. Promoting active roles for sub-national, national and international development finance institutions can contribute to the necessary green financing, by mobilising and catalysing key investments for the transformation to a low-carbon economy. In the LAC context, facilitating participation of the private sector is fundamental. Finally, enhancing and developing regulatory tools, such as sustainability or green bond standards and taxonomies, is crucial.

Advancing the green transition demands institutional mechanisms to foster consensus and build a new sustainable social contract

To make the green transition viable over time, LAC governments should ensure participative and inclusive processes. The multiplicity of actors, sectors and constituencies affected by the green agenda requires a shared platform for negotiation and consensus building. The green agenda can be the binding element of a new sustainable social contract for the region, as 68% of LAC citizens recognise climate change as a very serious threat to their country in the next 20 years, above other regions. Public institutions will have to work strategically to balance economic, social and environmental objectives; move forward the objectives of Nationally Determined Contributions; and ensure consistent and progressive implementation of long-term strategies, such as those set out in National Development Plans. The active participation of the private sector, academia, civil society and local communities is essential in every aspect of the public policy cycle, to consider their voices and develop participatory and transparent policies.

International partnerships are key to harnessing the potential of a green and just transition

Advancing a green transition and achieving the associated low-carbon development targets involves a series of challenges that cannot be addressed exclusively at the national level. Regional and international co-operation is necessary to ensure successful implementation of climate change mitigation and adaptation policies and a broader green agenda. Natural resource endowment and having 50% of the world's biodiversity make many LAC countries key players in international climate negotiations. In the transition to a sustainable model of development, LAC governments will benefit from having a strong convening power and a single voice in multilateral environmental agendas. This will help to better showcase the region's particularities while also better aligning national policies with internationally established environmental goals. LAC governments also have to face how green policies adopted in third countries will impact on trade. Co-operation with international actors, including private and multilateral institutions, will be key to leveraging the newly established international green norms and regulations.

Overview

Towards a green and just transition

Latin America and the Caribbean (LAC) faces an ambitious agenda to leverage the green transition as an opportunity to achieve a more just and sustainable development model. The recovery requires adopting a systemic approach to tackle the challenges and harness opportunities of the green transition to improve citizens' well-being. The green transition is a once-in-a-generation social and economic opportunity. A green transition that is truly just should advance the transformation of LAC's energy and productive matrices, and develop new and more sustainable economic sectors while also promoting quality jobs and supporting workers and households throughout the transition. Making this possible demands the mobilisation of high amounts of resources by rethinking fiscal systems and applying innovative financial solutions; reaching broad consensus across income groups, generations and territories by crafting a new sustainable social contract; and working to forge new active regional and international partnerships.

LAC is experiencing sluggish economic growth with a disproportionate impact on the vulnerable populations

Following a robust economic rebound in 2021, growth in LAC economies will slow in 2022. This is driven by an increasingly adverse global backdrop, fiscal and monetary stimuli rollbacks, and low potential growth. Inflationary pressures are high, and most central banks in the region are reacting by raising policy interest rates. At the international level, the scenario is becoming more complex. Weaker global economic growth is expected due to Russia's war of aggression against Ukraine and the zero-COVID policy of the People's Republic of China (hereafter "China") (OECD, 2022[1]). How the complex international context will affect LAC overall remains uncertain but will be transmitted through terms of trade, as well as through volatility in international financial markets and the possible persistence of inflationary pressures. Modest growth rates also reflect low potential growth, a pre-pandemic structural challenge. Potential gross domestic product (GDP) per-capita growth in LAC remains stagnated at low levels (below 1% since 1980), and below that of advanced economies, hindering convergence (Figure 1).

Figure 1. **Potential GDP per-capita growth in LAC and advanced economies**

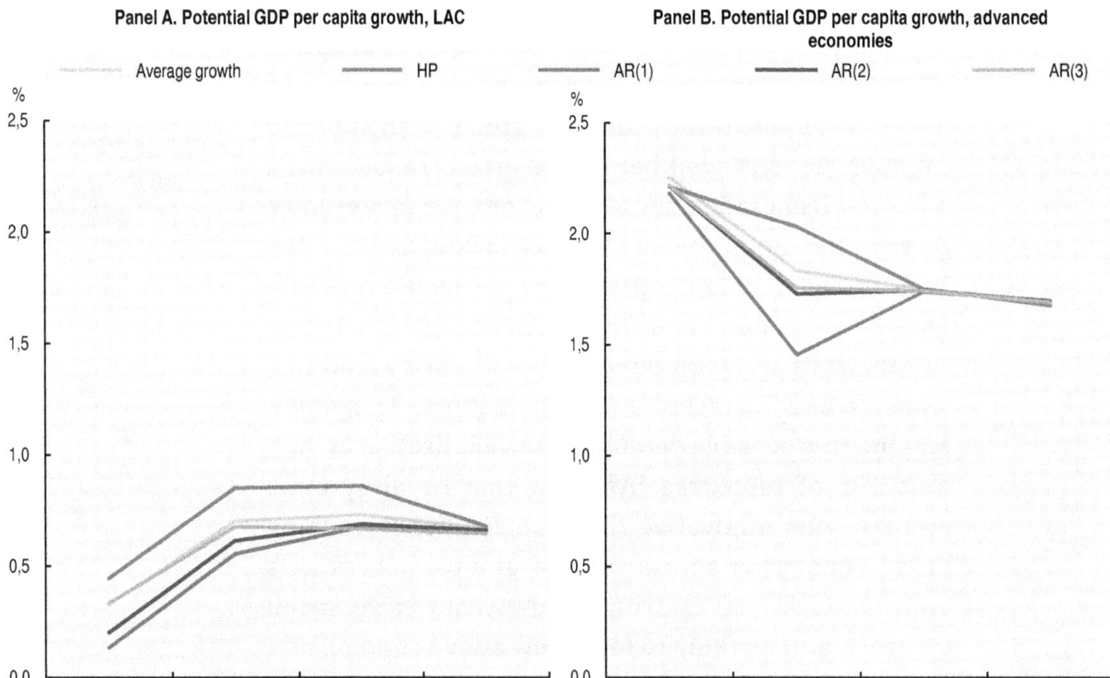

Note: AR stands for autoregressive model, which uses GDP per-capita growth data. The number of lags (1 and 2) was determined by analysing autocorrelation function and by choosing the model that maximised the log-likelihood. AR(1) Refers to an autoregressive model with one lag. To create a smoothed curve (lambda 100), the Hodrick-Prescott (HP) filter was used as an alternative model due to its resilience to short-term shocks. The LAC series refers to the 33 countries covered by the International Monetary Fund's (IMF) World Economic Outlook database, October 2022.

Source: Authors' elaboration based on (IMF, 2022[2]).

StatLink https://stat.link/4ofcsa

With restrictive monetary conditions, the management of fiscal policy is at the core of the LAC recovery. As in other regions, as a response to inflationary pressures, most central banks have reacted appropriately with interest rate increases to anchor expectations. In the case of fiscal policy, LAC economies must balance support for the economic recovery with protecting the most vulnerable households and preserving fiscal sustainability.

The COVID-19 pandemic aggravated social challenges in the region, leading to increased levels of poverty and inequality. These are projected to increase in 2022, due mainly to the economic slowdown and rising inflation, especially in food prices, which affect the most vulnerable. In the first five months of 2022, extremely poor households in LAC faced an average price increase of 3.6 percentage points more than the nationally representative household (Figure 2). Estimates suggest that, by the end of 2022, 33.7% of the population will be in poverty and 14.9% in extreme poverty. As a response to worsening social conditions, governments should complement monetary measures with fiscal policies, including targeted safety net interventions. In parallel, gradually advancing towards universal, comprehensive, resilient and sustainable social protection systems will be essential.

Figure 2. **Impact of inflation on overall population and on the extreme poor in 2022 in selected LAC countries**

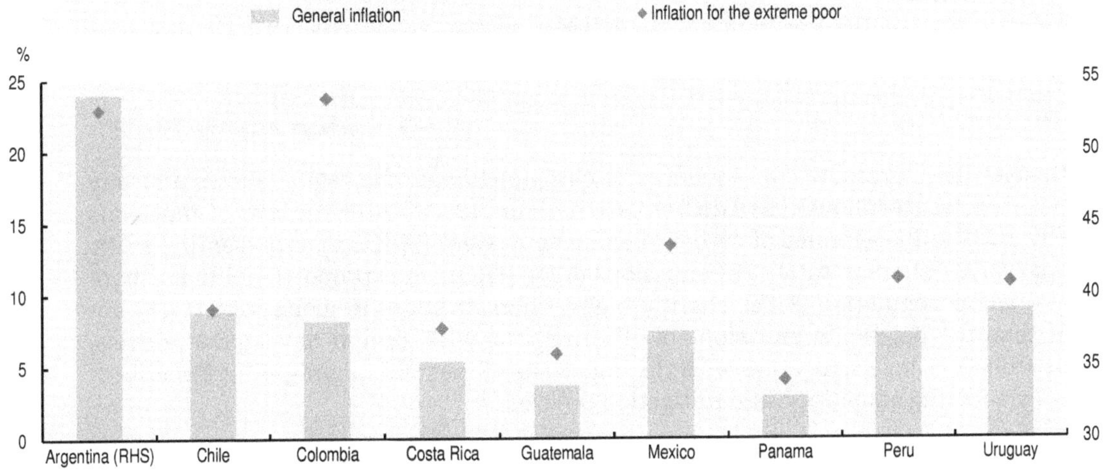

Notes: Year-to-date average of year-over-year growth of national consumer price indexes (CPI) vs. growth of extreme poverty lines 2022. Extreme poverty lines are based on the cost of a basic food basket that covers basic food needs and provides the minimum caloric requirement of the members of a reference household. The Chilean extreme poverty line also includes a share of non-food basic goods and services. For Colombia and Peru, the food and non-alcoholic beverages division of their CPI was used. For Panama, the data cover the districts of Panama and San Miguelito. Argentina is plotted on the right-hand side (RHS) axis.

Source: Authors' elaboration based on data from national statistic offices on CPIs and poverty lines.

StatLink *https://stat.link/navo20*

A green and just transition can help the LAC region improve the development model and reduce its vulnerability to climate change

LAC is disproportionately affected by the consequences of climate change: 13 of the 50 countries most affected by climate change worldwide are in the region. The average quantity of extreme climate-related weather events in LAC increased in most countries between 2001 and 2022 compared to the previous two decades (Figure 3). In total, 17.1% of the 11 933 climate-related extreme weather events registered worldwide between 1970 and 2022 occurred in LAC. Warming temperatures, extreme precipitation events leading to floods, landslides and droughts, sea level rise, coastal erosion, ocean and lake acidification resulting in coral bleaching, and storm surges are expected to increase in frequency and severity, with adverse socio-economic consequences on populations (IPCC, 2022[3]).

Figure 3. **Frequency of climate-related extreme weather events in LAC, 1980-2022**

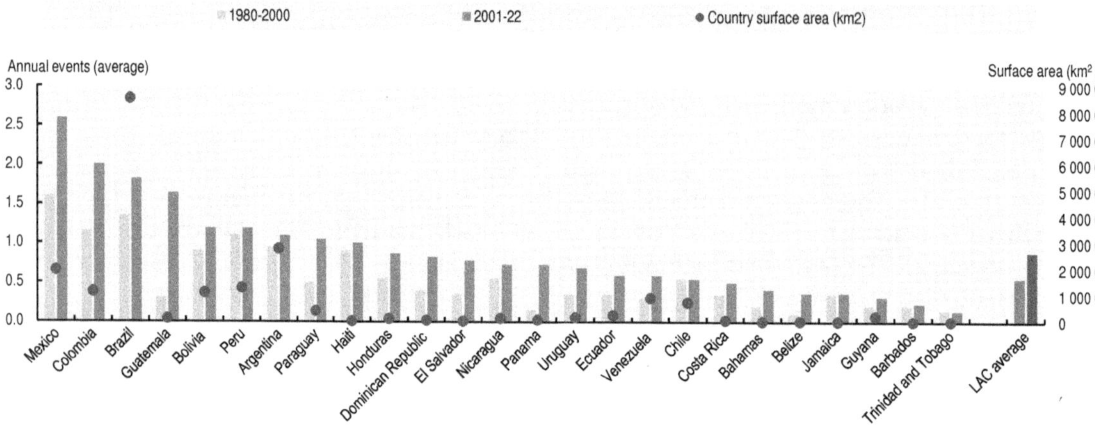

Notes: Based on (Alejos, 2018[4]), extreme weather events were defined as a natural disaster resulting in 100 000 or more people affected, or 1 000 or more deaths, or at least 2% of GDP in estimated economic damages. The following natural disasters were considered: landslides, storms, droughts and floods. The secondary axis refers to the countries' surface area.
Source: Authors' elaboration based on data from (EM-DAT, n.d.[5]); (IDB, 2021[6]); (Alejos, 2021[7]); (FAO, 2018[8]).

StatLink *https://stat.link/x76f0u*

Despite the increasing consequences of climate change, the region shows a steady increase of total greenhouse gas (GHG) emissions. From 1990 to 2019, the level of emissions rose by 1 223 million tonnes of carbon dioxide equivalent (Mt CO_2e), representing a 61% increase. LAC's share in total GHG emissions (8.1%) (Figure 4) is proportional to its share in total world population (8.4%), slightly higher than its share in global GDP (6.4%) but lower than the per-capita emissions of other regions with similar development levels. These emissions levels, together with the high costs of inaction, highlight the benefits of urgently adopting adaptation and mitigation policies.

Figure 4. **Regional shares of total GHG emissions, 2019**

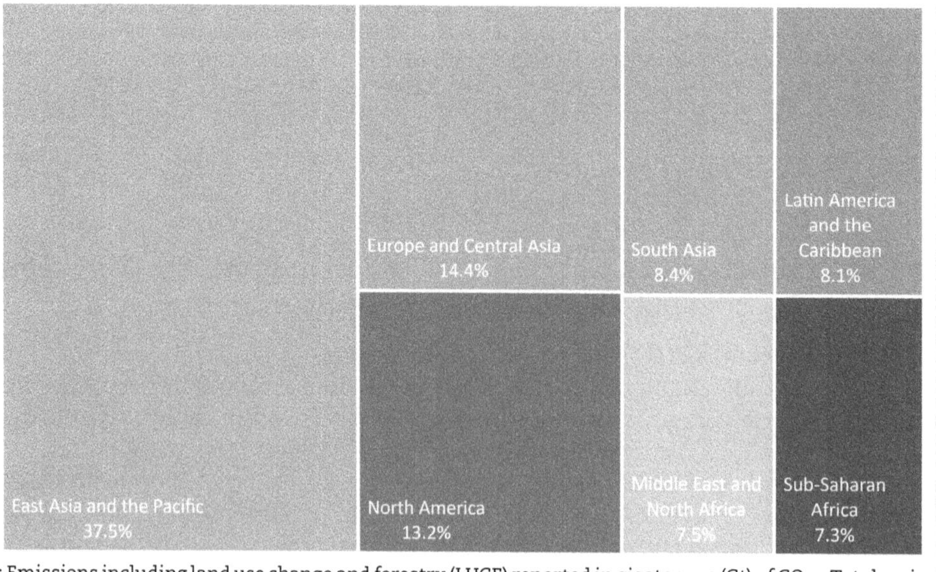

Notes: Emissions including land use change and forestry (LUCF) reported in gigatonnes (Gt) of CO_2e. Total emissions do not include bunker fuels. The Climate Analysis Indicators Tool (CAIT) was used as the data source as it is the most comprehensive dataset on Climate Watch and includes all sectors and gases. Climate Watch Historical GHG Emissions data (previously published through CAIT Climate Data Explorer) are derived from several sources. The use of LUCF or agriculture data is cited as (FAO, 2022[9]). Fuel combustion data is cited as (OECD/IEA, 2021[10]).
Sources: (Climate Watch, 2022[11]); (FAO, 2022[9]); (OECD/IEA, 2021[10]).

StatLink *https://stat.link/1i6y47*

The green transition should not, however, focus exclusively on fighting climate change. The recovery context is a timely opportunity to combine economic and social measures with green policies, advancing a just transition that could help achieve greater levels of well-being. A systemic approach can help policy makers reprioritise climate action towards improving systems' functioning and accelerate the transition to systems that are net zero emissions by design. If focused on effectively transforming the systems that underpin LAC's economy and society, the green transition has potential to improve significantly every aspect of Latin Americans' lives.

Towards a transformational change of the energy and productive matrices to reduce GHG emissions and promote quality jobs

A more sustainable and diversified energy matrix will support LAC in reducing emissions, harnessing the potential of its vast renewable energy resources and driving universal access to energy. The region is endowed with high potential renewable energy resources; at present, they represent 33% of total energy supply compared to 13% globally (Figure 5). Over the last two decades, many LAC countries have made substantial progress in building renewable energy markets and diversifying their energy mix. In 2020, renewable energy accounted for 61% (952 TWh) of regional electricity generation, broken down as 75% from hydroelectricity and 25% from solar, wind, biomass and geothermal. Significant variations exist across the region, however. While Brazil generates 84% of its electric power from renewables, Jamaica relies on imported oil derivatives for 87% of its electricity generation.

Figure 5. **World and LAC total energy supply matrix, 2020**

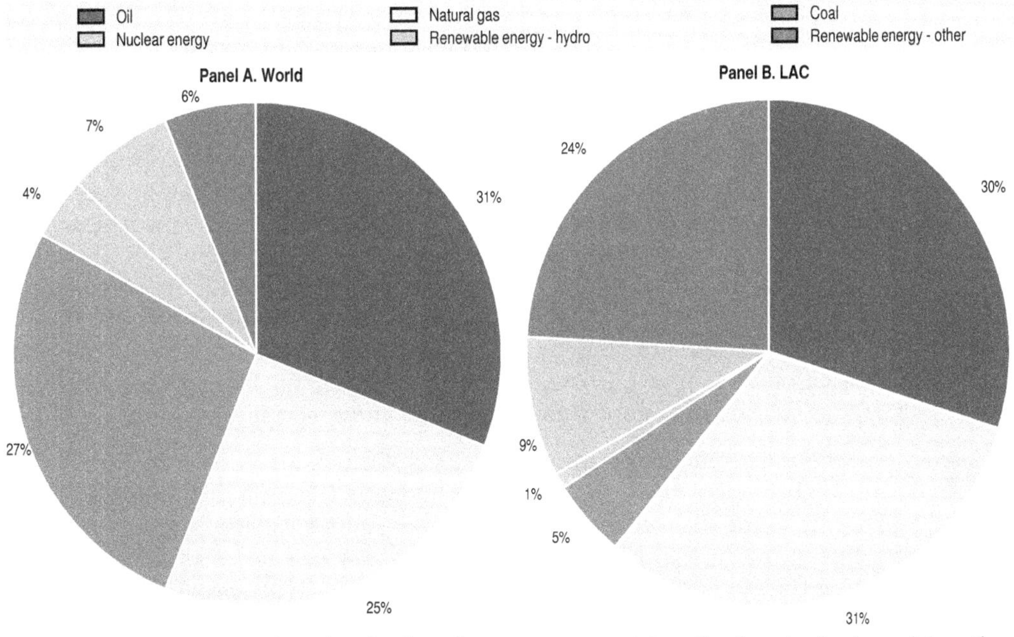

Notes: Total energy supply consists of production + imports – exports – international marine bunkers – international aviation bunkers +/- stock changes. "Renewable energy – other" includes biofuels, solar, wind and geothermal energy.
Source: Authors' elaboration based on (Sistema de Informacion energetica de Latinoamerica y el Caribe (SieLAC), 2020[12]).

StatLink https://stat.link/uop61q

Looking ahead, LAC countries producing electricity from renewables, given their abundance of low-cost renewable energy and relatively clean electricity matrices, can position themselves as green hydrogen industrial hubs. This could help support decarbonisation of hard-to-abate sectors such as heavy industries and transportation – for which no viable alternatives to fossil fuels currently exist. Under certain conditions, natural gas can be considered as a transitional activity towards a net-zero economy. The repurposing existing oil and gas infrastructure could help create a hydrogen industry (e.g. using pipelines for transport or depleted oil and gas reservoirs for carbon capture utilisation and storage projects). Moreover, sustainable hydrogen can promote vertical and horizontal linkages along its value chain, creating value added and promoting innovative industries. The region also has a strategic position to supply key minerals for the energy transition. In 2017, 61% of global lithium reserves, 39% of global copper, and 32% of global nickel and silver reserves were in LAC. Achieving universal access to electricity is crucial. Across LAC, a total of 17 million people have no access to electricity, especially in rural areas and among poorer households and indigenous and Afro-descendant populations.

Industrial, circular and blue policies can transform LAC's production structure and are key components of a green and just transition. Renewed industrial policies are needed to encourage and attract investments in green innovation. So far, the region's gross domestic expenditure in research and development (GERD) has been only 0.3% of GDP in 2018 (vs. 2% of GDP in the OECD) and remains highly government-driven (56.5% of the total). The transition to a circular economy is expected to have net positive effects on GDP growth and employment while reducing GHG emissions. Net effects expected for Chile, Colombia, Mexico and Peru are increased GDP (from 0.82% in Chile to 2.03% in Peru) and job creation (from 1.1% in Chile and Colombia to 1.9% in Peru). The blue economy can also contribute to LAC development, but its potential remains unexploited. In 2018, the total GDP contribution of ocean services was estimated at USD 25 billion for LAC and USD 7 billion for Caribbean countries alone. In terms of employment, fishery and aquaculture employed more than 2.5 million people.

A green transition is a good opportunity to create quality jobs for LAC citizens. While some jobs in brown sectors will most likely be lost as countries move towards a net zero carbon model, if effective policies are put in place, many others jobs can be created in green sectors by 2030 (Figure 6). These include policies to favour green investments as well as active labour market policies (ALMPs) to facilitate the transition from brown to green sectors and from informal to formal jobs. Net job creation will indeed depend on the magnitude of investments. In a high-impact scenario in which additional public and private investments contribute to an increase of 3 percentage points in the value added of green sectors (compared to the business-as-usual scenario), the green transition could add 10.5% more net jobs of total employment in brown and green sectors.

Labour market and social protection policies play a crucial role, both in stimulating the creation of high-quality new jobs and in cushioning the downside consequences of the transition towards cleaner economies. Well-designed ALMPs, such as training programmes, hiring incentives or placement services, are crucial to promoting green jobs and boosting the skills of those workers who will lose their actual jobs. Life-long learning will be central, although currently, only 15% of LAC workers receive some form of training, compared to 56% at the OECD. Well-targeted social policies, such as income support measures or conditional cash transfer programmes, can play a positive role in minimising the temporary income losses of families with workers negatively impacted by green policies.

Figure 6. **Job creation in green sectors in LAC, 2020-30**

Change in employment in green sectors in LAC, under various green policy scenarios compared to the BAU, as percentage of 2020 baseline employment in green industries

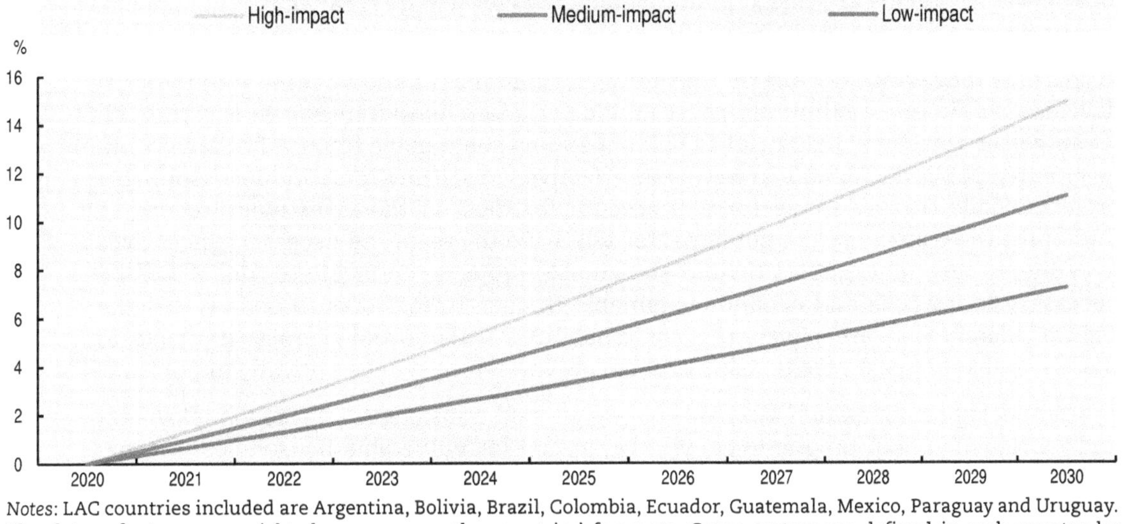

Notes: LAC countries included are Argentina, Bolivia, Brazil, Colombia, Ecuador, Guatemala, Mexico, Paraguay and Uruguay. The data refer to an unweighted average over the countries' forecasts. Green sectors are defined in each country by first identifying the number of green tasks that workers perform in their occupations and then by examining the top ten industries across which those jobs are distributed. The baseline scenario assumes that, in each green sector, value added and employment will follow the same dynamic as in the past ten years. The counterfactual scenarios are defined according to the impact of a green policy that aims to boost investment in fixed and human capital, with a positive impact on value-added growth in each green sector. The high-impact scenario assumes that the value added in each sector will increase by 3 percentage points per year, adjusting to the new equilibrium. The medium-impact scenario assumes that the value added will increase by 2 percentage points per year, while the low-impact scenario assumes it will increase by 1 percentage point per year. In all forecasts, total factor productivity will increase by 1 percentage point due to lower climate damages and new technology-induced change. Employment change is forecast using the estimated short-term elasticity to the value added, applying a panel dynamic model, defined by each sector and country, in the last ten years.

Sources: Authors' estimates based on Labour Force Surveys, National Accounts data by industries, (Vona et al., 2018[13]) and (Hardy, Keister and Lewandowski, 2018[14]).

StatLink https://stat.link/hlceqd

Financing the transition requires environmentally sustainable fiscal policies and strategies to mobilise resources from both public and private sectors

The LAC region faces the challenge of financing the green transition under a tight fiscal space. A 2.5°C global warming scenario could cost the region between 1.5% and 5.0% of its GDP by 2050. Since the cost of inaction is high, countries must develop environmentally sustainable fiscal policies. These include infrastructure plans with adaptation and mitigation goals. These policies also entail focusing on more and better spending on clean energy and energy efficiency, and on phasing-out fossil fuel subsidies and support measures, particularly to the most affluent population.

In addition, the region needs to mobilise new ways to generate additional revenues, such as environmentally-related taxes, emissions trading systems (ETS) and scaling up debt tools. On average in LAC, environment-related tax revenues amounted to only 1% of GDP in 2020, just half of the estimated OECD average of 2% of GDP. Scaling up tools of debt, such as green, social, sustainable and sustainability-Linked (GSSS) bonds, debt-for-nature swaps, catastrophe bonds, and natural disaster clauses, can also help raise additional revenues to ensure flows of resources target climate action. The GSSS market reached an accumulated USD 73 billion from 2014 to September 2021, of which green bond issuance accounted for USD 31 billion alone followed by social with USD 17 billion (Figure 7, Panel A).

The role of the corporate sector has become increasingly significant in the region. Over the period December 2014 to September 2021, corporates led total GSSS bond volumes with a share of 42% of the total LAC GSSS bond issuance, while sovereigns represented 37%, quasi-sovereigns 12% and supranational issuers 5% (Figure 7, Panel B).

Financial strategies will have to enhance private and public sector resource mobilisation, in part by engaging key actors including subnational, national and international development finance institutions (DFIs). Enhancing green fiscal frameworks (e.g. through green golden rules) will be crucial as will expanding sustainable finance frameworks to ensure that public and private investments effectively reach environmentally sustainable projects. Since the private sector will account for most of the investment needed to undertake the transition, the public sector will have to create the necessary incentives to redirect these investments toward sustainable projects. To facilitate this, it will be necessary to improve and expand sustainable finance frameworks to ensure that the right regulatory tools are in place (e.g. sustainability standards and green, sustainable, or transition taxonomies). Mechanisms to avoid greenwashing will be critically important.

Developing compensation mechanisms (e.g. in-kind transfers, ALMPs, self-employment and entrepreneurship programmes) will be crucial for vulnerable households affected negatively by climate reform. Well-targeted cash transfers and in-kind transfers will continue to be essential, together with compensation policies to support the relocation and retraining of workers, promote decent work in rural areas, develop new business models, and provide support for displaced workers.

Figure 7. **Total LAC GSSS bond issuance in international markets, by type of instrument and issuer, December 2014 to September 2021**

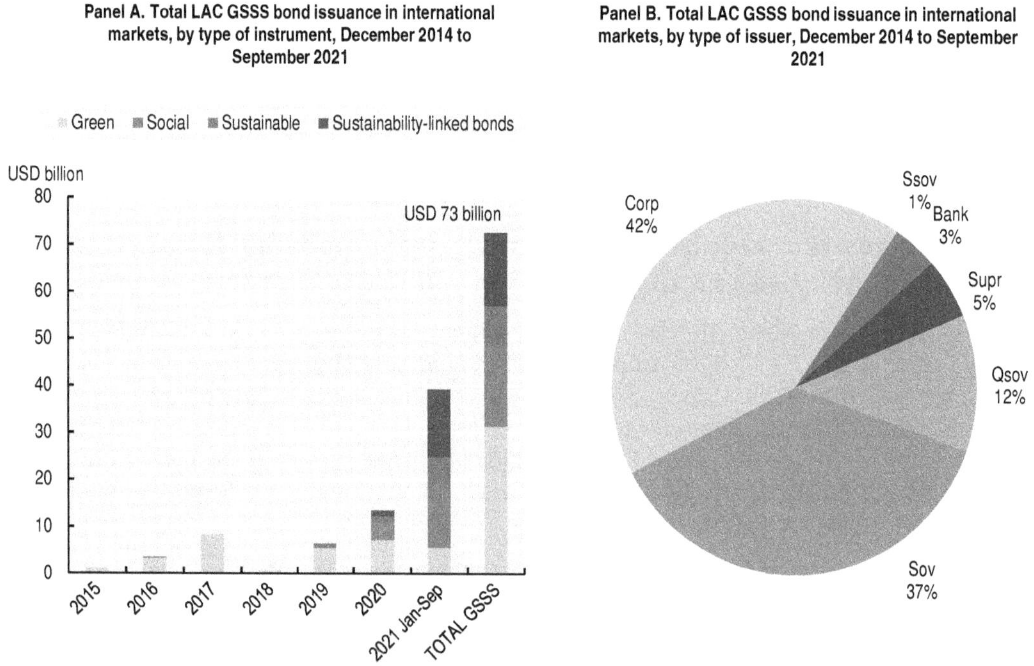

Note: Panel B: Sov = sovereign. Corp = corporate. Ssov = sub-sovereign (states, cities and provinces). Supr = supranational. Qsov = quasi-sovereign. Quasi-sovereign issuers are defined as companies with full or partial government ownership or control. Supranational issuers are defined as entities formed by two or more central governments to promote economic development for the member countries. The "bank" category refers to commercial banks. Other non-bank financial institutions are included in corporates.

Source: (Núñez, Velloso and Da Silva, 2022[15]).

StatLink https://stat.link/5wluj3

Advancing the green transition will require institutional mechanisms to foster consensus and build a long-term vision underpinning a new social contract

LAC citizens are broadly concerned about the seriousness of climate change - and relatively more so than other regions globally. This suggests that the recovery could represent a "critical juncture" for advancing towards a new, sustainable social contract. On average, 68% of citizens in LAC recognise that climate change is a very serious threat to their country in the next 20 years (Figure 8). In sharp contrast to some countries, such as the United States, concern about climate change in LAC is consistent across the political spectrum (Evans and Zeichmeister, 2018[16]). The importance LAC citizens place on the green agenda could make the green transition the cohesive element of a wider social contract for the region.

Figure 8. **Share of citizens who agree that climate change is a very serious threat to the country in 20 years, 2019**

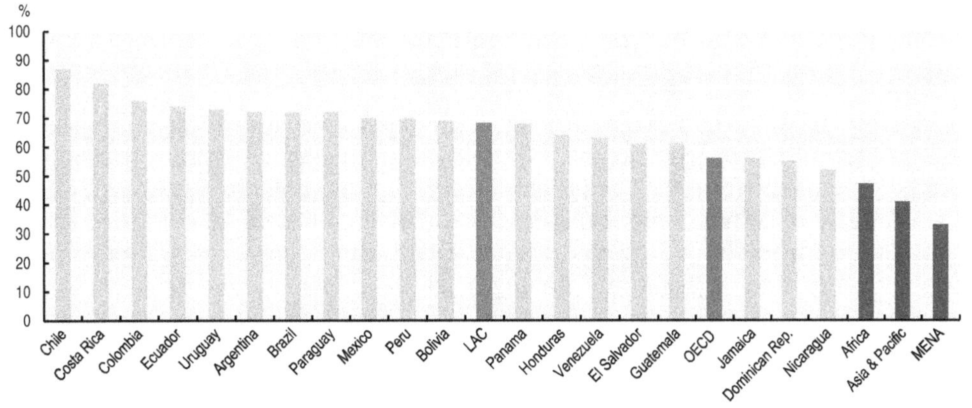

Notes: Question for Figure 8: "Do you think that climate change is a very serious threat, a somewhat serious threat, or not a threat at all to the people in this country in the next 20 years? If you don't know, please just say so".
Source: Authors' elaboration based on (Lloyd's Register Foundation, 2020[17]).
StatLink https://stat.link/i85nup

In practice, as the green transition may involve a shift of resources among economic sectors and political constituencies that could trigger the opposition of some interest groups, to build consensus it will be important to establish inclusive and shared platforms for reaching a negotiated stance. Encouraging the participation of citizens, civil society groups, women, indigenous and local communities throughout the policy-making process can promote greater local ownership and generate more inclusive policies that appropriately consider local needs. Policy makers should also bring on board the private sector by raising awareness of responsible business conduct (RBC) practices and establishing stronger integrity policies to avert the risk of environmental policy capture by powerful groups. In addition, adapting the strategy to specific socio-political contexts is vital, as is devising empowering and empathic communication strategies about the proposed green reform agenda. In turn, strategies must include specific compensation mechanisms for vulnerable groups that may be negatively affected in the short term.

As the green transition affects virtually every domain of public policy, policy makers should work more strategically and achieve better co-ordination across sectors and levels of government to ensure a coherent green agenda. An integrated approach will be needed to balance economic, social and environmental trade-offs while also leveraging policy spillovers among these fields.

Linking policy objectives with long-term plans is also key to ensure consistent implementation over time, beyond short-term political cycles. Governments need to articulate a long-term vision to align their actions. This can be done through frameworks such as national development plans (NDPs), nationally determined contributions (NDCs) coupled with climate strategies, and defined policies and regulations to underpin their pledges. NDCs establish concrete targets and policies, setting the basis for the contributions of various stakeholders in national efforts to achieve the long-term goals of the Paris Agreement. While most LAC countries have already submitted an update to their NDCs, Costa Rica's 2020 update is among the few that are rated 2°C compatible (CAT, 2020[18]). Argentina, Brazil, Colombia, Costa Rica and Panama have set only unconditional targets. In contrast, the majority of LAC countries also set conditional targets, meaning that implementation of these commitments depends on the delivery of international financial and technical support. This highlights the critical importance of collective action and co-operation at both national and international levels for achieving the goals of the Paris Agreement.

LAC should profit from an international green agenda that proposes new partnerships for the region's development as well as new tools to foster the transition

Considering the global nature of the increasingly urgent need to reduce CO_2 emissions, all countries are called to participate in individual and collective efforts. Decoupling economic development from CO_2 emissions has proven possible, as illustrated by the recent trajectories of certain regions (Figure 9). At the same time, LAC countries should play a predominant role in this global agenda by sharing experiences of sustainable development with other regions and having a leading voice at climate negotiations. Climate change has shown that continuing a path of exponential growth of CO_2 emissions is no longer an option. LAC countries can reach higher human development index (HDI) levels while meeting low emissions targets.

Figure 9. CO_2 **emissions per capita versus HDI, 1995-2019**

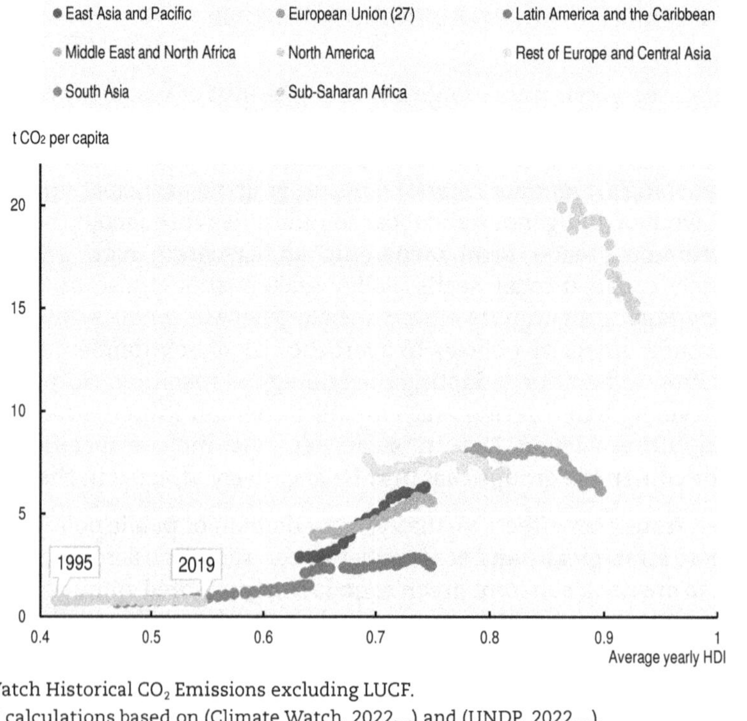

Note: Climate Watch Historical CO_2 Emissions excluding LUCF.
Source: Authors' calculations based on (Climate Watch, 2022[11]) and (UNDP, 2022[19]).

StatLink https://stat.link/djn2w1

To date, as each country negotiates within multiple climate-related international coalitions, the LAC region has lacked a unified voice in the international arena (Figure 10). This is the result of both the existing fragmentation of LAC regional integration processes and subregional economic ties. Preventing further fragmentation of environmental policies and the politicisation of environmental instruments is crucial to harness the full potential of the green transition. In many ways, LAC's fragmented voice in climate negotiations constitutes a missed opportunity, especially considering that the region hosts 50% of the planet's biodiversity. Future efforts will need to prioritise enhanced policy dialogue and a regional environmental agenda.

Figure 10. **LAC countries' participation in climate-related international coalitions**

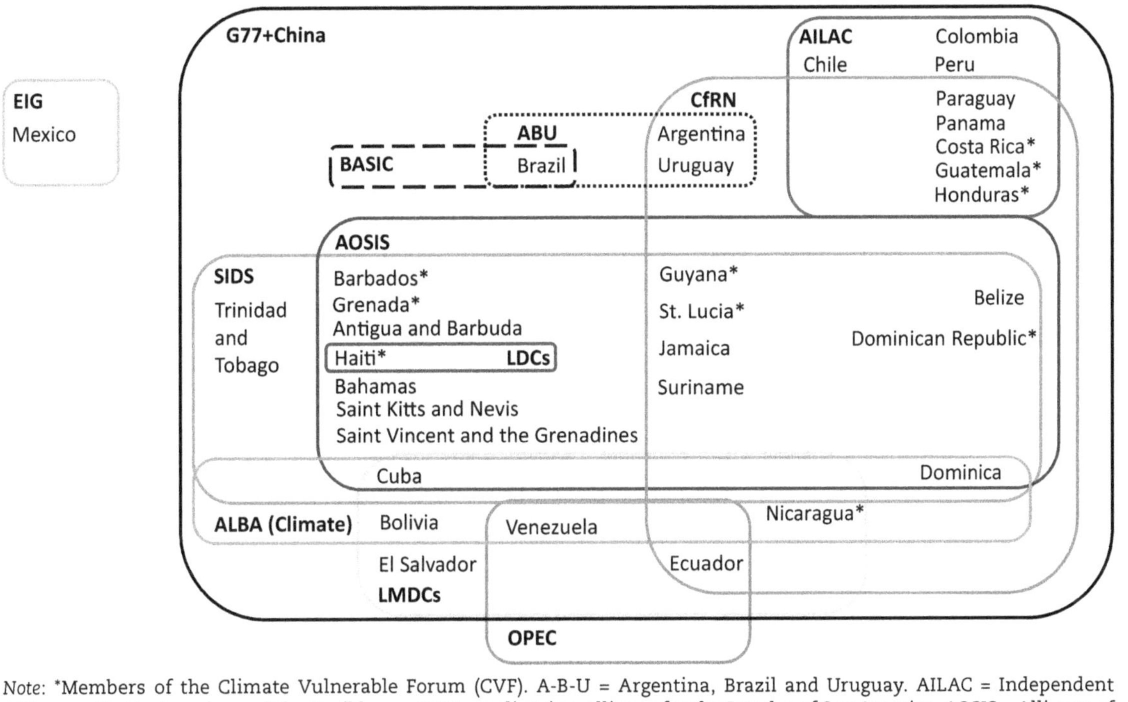

Note: *Members of the Climate Vulnerable Forum (CVF). A-B-U = Argentina, Brazil and Uruguay. AILAC = Independent Alliance of Latin America and the Caribbean. ALBA = Bolivarian Alliance for the Peoples of Our America. AOSIS = Alliance of Small Island States. CfRN = Coalition for Rainforest Nations. EIG = Environmental Integrity Group. LDCs = least-developed countries. LMDCs = like-minded developing countries. OPEC = Organisation of the Petroleum Exporting Countries. SIDS = small island developing states. Figure is a non-exhaustive representation of coalitions in the region; some coalitions relate to the environment as part of a broader agenda.

Source: Authors' elaboration based on (Delgado Pugley, 2021[20]); (Klöck et al., 2020[21]) and (Watts and Depledge, 2018[22]).

Trade is one of the channels through which the green transition will impact the LAC region. It represents a challenge in that, over the last two decades, LAC has consistently posted a deficit in its trade in environmental goods (environmental specific services, environmental sole-purpose products, adapted goods and environmental technologies). Three-quarters of the region's imports of environmental goods come from China, the United States and the European Union, while intraregional imports account for just 5% of total expenditure. Moreover, regional export capacity is highly concentrated: between 2018 and 2020, just one country (Mexico) accounted for 84% of regional exports of environmental goods.

The European Green Deal may have implications for LAC countries, potentially affecting trade between the two regions. In particular, as the EU Green Deal increases demands for traceability, transparency, compliance and due diligence, as well as low-carbon, organic and sustainable production and reinforcement of the circular economy, LAC countries will need to adapt to these new international environmental standards and regulations. LAC countries that trade with Europe have the opportunity to align national climate change mitigation plans so as to use "the new rules of the game" to execute a productive transition.

Indeed, the trade channel holds opportunities as well. The transition to a circular economy requires LAC countries to design specific public policies for the entire life cycle of products, including production, consumption, waste management and recycling. Public and private investments and co-operation are critical for increasing capacity building, innovation and technology transfer. Transitioning to the circular economy also depends on the co-ordination of LAC's national and international efforts. From reducing tariffs and non-tariff barriers to enhancing the granularity of international trade classifications, harmonising standards for circular economy goods could help firms, countries and regional actors adopt sustainable practices.

References

Alejos, L. (2021), *What are the fiscal risks from extreme weather events and how can we deal with them?*, Inter-American Development Bank, Washington, DC, https://blogs.iadb.org/gestion-fiscal/en/what-are-the-fiscal-risks-from-extreme-weather-events-and-how-can-we-deal-with-them/#:~:text=It%20is%20estimated%20that%20the,income%20countries (Figure 2). [7]

Alejos, L. (2018), *Three Essays in Public Finance in Developing Countries*, University of Michigan, Ann Arbor, MI, https://deepblue.lib.umich.edu/bitstream/handle/2027.42/147524/lalejos_1.pdf?sequence=1&isAllowed=y. [4]

CAT (2020), *Climate Target Update Tracker: Costa Rica*, Climate Action Tracker, Climate Analytics/NewClimate Institute, Berlin, https://climateactiontracker.org/climate-target-update-tracker/costa-rica/. [18]

Climate Watch (2022), *Historical GHG Emissions*, World Resources Institute, Washington, DC, https://www.climatewatchdata.org/ghg-emissions. [11]

Delgado Pugley, D. (2021), *América Latina frente a la COP26: Posiciones y perspectivas*, Fundación Carolina, Madrid, https://doi.org/10.33960/issn-e.1885-9119.DT58. [20]

EM-DAT (n.d.), *EM-DAT Public*, Emergency Events Database, Brussels, https://www.emdat.be/database. [5]

Evans, C. and E. Zeichmeister (2018), *Education and Risk Assessments Predict Climate Change Concerns in Latin America and the Caribbean*, Latin American Public Opinion Project, Vanderbilt University, Nashville, https://www.vanderbilt.edu/lapop/insights/IO929en.pdf. [16]

FAO (2022), *FAOSTAT Emissions*, Food and Agriculture Organization, Rome, https://www.fao.org/food-agriculture-statistics/data-release/data-release-detail/en/c/1304919/. [9]

FAO (2018), *FAOSTAT Surface Area 1961-2018*, Food and Agriculture Organization, Rome, https://www.fao.org/faostat/en/#data. [8]

Hardy, W., R. Keister and P. Lewandowski (2018), "Educational upgrading, structural change and the task composition of jobs in Europe", *Economics Of Transition*, Vol. 26, https://onlinelibrary.wiley.com/doi/full/10.1111/ecot.12145. [14]

IDB (2021), *Fiscal Policy and Climate Change: Recent Experiences of Finance Ministries in Latin America and the Caribbean*, Inter-American Development Bank, Washington, DC, https://publications.iadb.org/publications/english/document/Fiscal-Policy-and-Climate-Change-Recent-Experiences-of-Finance-Ministries-in-Latin-America-and-the-Caribbean.pdf. [6]

IMF (2022), *World Economic Outlook, April 2022: Wars Set Back the Global Economy*, International Monetary Fund, Washington, DC, http://www.imf.org/en/Publications/WEO/Issues/2022/04/19/world-economic-outlook-april-2022. [2]

IPCC (2022), *Climate Change 2022: Impacts, Adaptation and Vulnerability. Working Group II Contribution to the IPCC Sixth Assessment Report*, Cambridge University Press, https://www.ipcc.ch/report/sixth-assessment-report-working-group-ii/. [3]

Klöck, C. et al. (2020), *Coalitions in the Climate Change Negotiations*, Routledge, London, https://doi.org/10.4324/9780429316258. [21]

Lloyd's Register Foundation (2020), *World Risk Poll*, powered by Gallup, Lloyd's Register Foundation, London, https://wrp.lrfoundation.org.uk/explore-the-poll/. [17]

Núñez, G., H. Velloso and F. Da Silva (2022), *Corporate governance in Latin America and the Caribbean: Using ESG debt instruments to finance sustainable investment projects*, Economic Commission for Latin America and the Caribbean, Santiago, https://repositorio.cepal.org/handle/11362/47778. [15]

OECD (2022), *OECD Economic Outlook, Volume 2022 Issue 1*, OECD Publishing, Paris, https://doi.org/10.1787/62d0ca31-en. [1]

OECD/IEA (2021), *GHG Emissions from Fuel Combustion*, OECD Publishing/International Energy Agency, Paris, https://www.oecd-ilibrary.org/energy/data/iea-co2-emissions-from-fuel-combustion-statistics_co2-data-en. [10]

Sistema de Informacion energetica de Latinoamerica y el Caribe (SieLAC) (2020), *Estadística Energética [database]*, https://sielac.olade.org/default.aspx. [12]

UNDP (2022), *Human Development Report Data Center*, https://hdr.undp.org/data-center/documentation-and-downloads. [19]

Vona, F. et al. (2018), "Environmental Regulation and Green Skills: An Empirical Exploration", *Journal of the Association of Environmental and Resource Economists*, Vol. 5/4, pp. 713-753, https://doi.org/10.1086/698859. [13]

Watts, J. and J. Depledge (2018), "Latin America in the climate change negotiations: Exploring the AILAC and ALBA coalitions", *WIREs Climate Change*, Wiley Interdisciplinary Reviews, Wiley-Blackwell, Hoboken, NJ, https://doi.org/10.1002/wcc.533. [22]

Chapter 1

Addressing the structural macro context to drive the green transition

The COVID-19 recovery in Latin America and the Caribbean (LAC) is slowing down, reflecting low potential growth and an increasingly complex international context driven by Russia's war against Ukraine and an economic slowdown in China. The socio-economic consequences of COVID-19 linger, with poverty and extreme poverty still high. With macroeconomic policy space being reduced, most LAC countries will face the multidimensional challenge of balancing recovery stimulus, financing the green transition, and protecting the most vulnerable, particularly from the impact of inflationary pressures. After analysing the global context, this chapter presents the economic performance and critical factors affecting the pace and shape of the recovery in LAC. The chapter then analyses the weight of climate change on fiscal accounts and explores some options to mobilise further resources to promote the green transition. Before concluding, the chapter discusses the deteriorated post-COVID-19 social conditions, particularly poverty and inequality, and the need to strengthen social protection systems.

Addressing the macroeconomic challenges of LAC to ensure a green and just transition

Growth in LAC economies will slow down in 2022 driven by an increasingly adverse global backdrop, fiscal and monetary stimuli rollbacks, and low potential growth

The extreme poor face a higher price increase than the average household

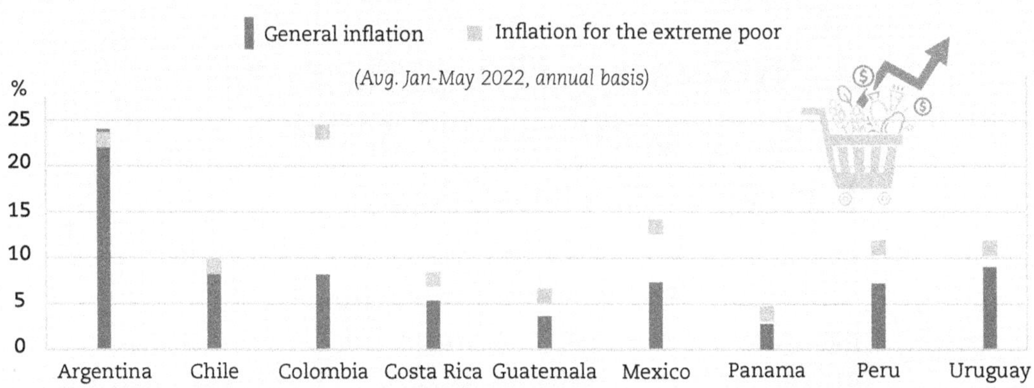

Fiscal space is limited, and the level and cost of public debt have been rising

Public debt-to-tax ratios increased from 2013-19 and rose significantly in 2020

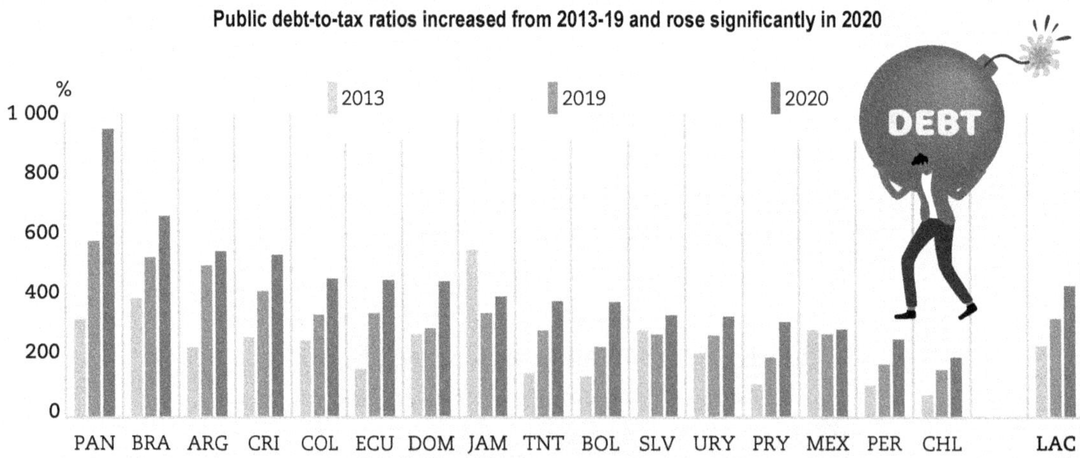

Poverty and extreme poverty are expected to increase in 2022, due to low economic growth and high inflationary pressures

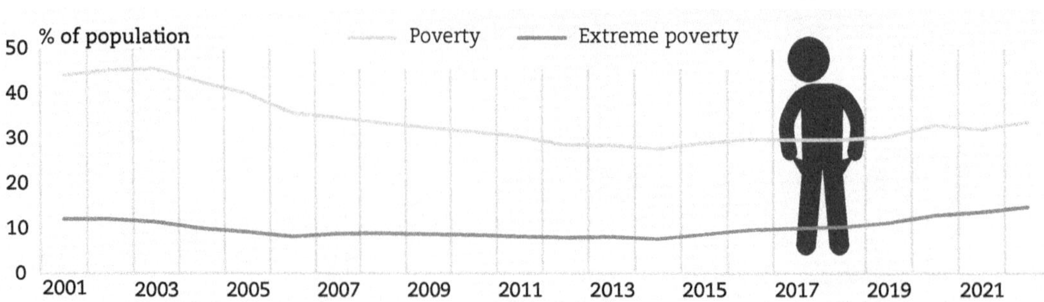

Introduction

Following a robust economic rebound in 2021, growth in LAC economies will slow down in 2022. This is driven by an increasingly adverse global backdrop, fiscal and monetary stimuli rollbacks, and low potential growth. Inflationary pressures are mounting, and most central banks in the region are raising policy rates.

Social challenges from the pandemic remain, with an expected increase in poverty in 2022. Even though there was a decrease in total poverty levels between 2020 and 2021, these are projected to increase in 2022 due to rising inflation, especially in food prices. Extreme poverty did not recede in 2021 and is expected to increase in 2022. It is estimated that, by 2022, 33.7% of the LAC population will be in poverty and 14.9% in extreme poverty (ECLAC, 2022[1]). This has translated into downward mobility in the socio-economic stratum. The incidence of poverty is heterogeneous not only among countries in the region but also among population groups. For instance, women aged 25 to 39 have higher poverty rates than men of the same age in all countries. Inequality in income distribution has also increased in most countries, with current high inflation posing a risk of a further increase (ECLAC, 2022[2]).

With restrictive monetary conditions, fiscal policy management is at the recovery's core. As in other regions, inflation rates have increased substantially; therefore, most central banks have responded appropriately with interest rate increases. Since the end of 2021, many of the region's economies have started to withdraw some fiscal stimulus, and tax revenues have increased thanks to the improvement in economic conditions, thus narrowing primary fiscal deficits. LAC economies must support economic conditions and fiscal sustainability, while protecting the most vulnerable through strengthening social protection systems. In the future, climate change and the green transition can weigh heavily on fiscal accounts as natural disasters, the phasing out of fossil fuels from the energy matrix, or stranded assets can diminish revenues. Therefore, the region will need to mobilise resources to compensate for shortfalls and to invest more, better and greener to reduce the adverse effects of climate change and finance the green transition. A green transition goes beyond fighting climate change. It also aims to advance a more sustainable and inclusive model of production and consumption that creates new, quality, green jobs, generates the conditions for workers to successfully navigate the transition, and supports firms to adopt more sustainable production schemes and citizens to change their consumption habits (Chapter 2).

Global economic prospects are weakening stemming from the invasion of Ukraine by the Russian Federation (hereafter "Russia") and the lingering effects of the coronavirus (COVID-19) pandemic and strategies to contain it. The impact of Russia's invasion of Ukraine will vary across regions, depending on their commercial and financial exposure to Russia or Ukraine. Russia's invasion of Ukraine has also pushed up commodity prices, fuelling inflation. Disruptions in global supply chains, high freight costs, and demand and supply imbalances have contributed to the build-up of inflationary pressures not seen in decades and that go beyond food and energy. Similarly, the zero-COVID policy of the People's Republic of China (hereafter "China") continues to weigh on the global outlook and trade flows (OECD, 2022[3]; OECD, 2022[4]).

This chapter first examines the global context, focusing on the consequences of Russia's war against Ukraine and the growing inflationary pressures. It then presents the economic performance in LAC, highlighting the region's heterogeneity, external accounts, growing inflationary pressures, and tight fiscal space, especially to finance the green transition and face the adverse effects of climate change. Last, the chapter examines the remaining social consequences of the COVID-19 crisis, focusing on poverty and inequality, the nexus with inflation, and the importance of strengthening universal, comprehensive, sustainable and resilient social protection systems.

An increasingly challenging global context

After the strong recession generated by the COVID-19 pandemic in 2020, the global economy grew at a robust pace in 2021, supported by the advance of vaccination programmes and the massive fiscal and monetary stimulus packages applied by most countries. The economic rebound of 2021 was widespread, with an estimated global growth of 5.8%, much higher than the contraction of 3.4% reported in 2020 (IMF, 2022[5]; OECD, 2022[3]; OECD, 2022[4]).

In 2022, global economic growth has slowed due to deteriorating global conditions, fuelled by the war in Ukraine and the lingering effects of COVID-19, mostly China's zero-COVID policy. Russia's war against Ukraine impacted global recovery. Before the outbreak of the war, global growth was projected to return to rates similar to those prevailing before the pandemic, and inflation was seen as a temporary phenomenon. China's zero-COVID policy has similarly affected the global outlook by creating bottlenecks in international trade and adding to inflationary pressures. Estimates suggest that for 2022 global growth could slow to around 3.0%, and 2.2% in 2023 (OECD, 2022[3]; OECD, 2022[4]).

The impact of the war will vary across regions, depending on their ties to Russia and Ukraine, their main trading partners, their export basket, and their financial exposure. In Emerging Asia, the impact will likely be less acute, as the economic ties with the countries involved in the war are not very strong. Africa also has relatively small investment ties with Russia and Ukraine. Still, Russia's large-scale aggression against Ukraine will significantly affect its major trade and investment partners, along with the inflationary pressures already causing a food crisis. In LAC, the war can have diverse and significant impacts, although mostly indirectly through higher commodity prices, slow global growth, disruptions in trade, and financial volatility.

For the global economy, Russia's invasion of Ukraine and the slowdown in China have at least three relevant channels of transmission

The first channel is an increase in energy and food commodity prices. Considering the importance of Russia in the global energy commodity market, Russia's war against Ukraine pushed up prices amid pre-existing supply and demand imbalances (Figure 1.1).

In the case of Russia's war against Ukraine, geopolitical risk and uncertainty from the application of potential larger-scale sanctions have fuelled volatility in global energy markets (Box 1.1). As a result of uncertainty, prices of oil, gas, coal and industrial metals soared in March 2022, and fluctuated around higher levels over the next few months. Energy prices have remained elevated, but a weaker demand from China has eased some of the pressures on metal prices (OECD, 2022[4]).

Beyond energy, the price of food raw materials has further risen due to the disruption of essential trading channels in the cereals and fertilisers segments. Ukraine and Russia contribute 30% of the wheat sold globally and are relevant corn, oats and sunflower producers. Belarus (Ukraine's border country) and Russia are important exporters of potassium and phosphorus worldwide, minerals that are critical inputs to produce fertilisers used in multiple crops. The agreements that allowed some agricultural exports from Ukraine have helped to ease food price pressures (OECD, 2022[4]).

The increase in commodity prices will weaken the post-pandemic economic recovery by accelerating inflation. Energy and food inflation directly affects the purchasing power of households, limiting private consumption spending, trade, and global growth as well as generating social tensions.

Figure 1.1. **Commodity prices**

December 2019 Index = 100

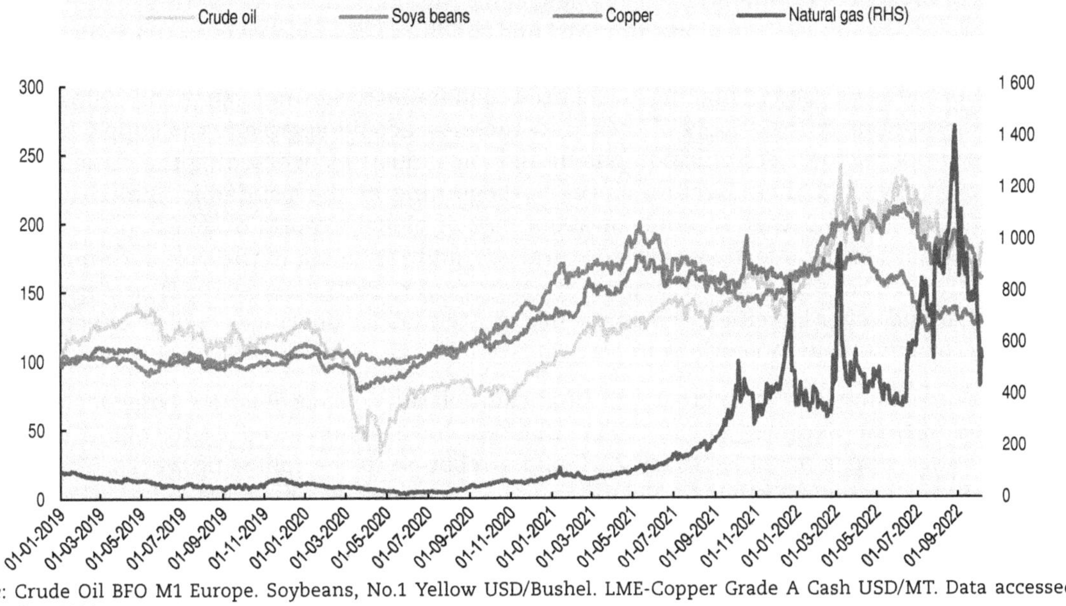

Note: Crude Oil BFO M1 Europe. Soybeans, No.1 Yellow USD/Bushel. LME-Copper Grade A Cash USD/MT. Data accessed 7 October 2022.

Source: Thomson Reuters Datastream.

StatLink https://stat.link/8dmqka

Box 1.1. **Could a total embargo on Russian oil shipments lead to a global economic crisis?**

At the time of writing, the oil price had risen more than 20% since the Russian invasion of Ukraine, primarily driven by expectations of a drop in Russian crude availabilities rather than effective supply disruptions linked to sanctions. The current economic scenario is reminiscent of the oil crisis of 1973-74 when the Organization of the Petroleum Exporting Countries (OPEC) decided to suspend the sale of crude oil to the United States for its military support to Israel during the Arab-Israeli war. The price of oil quadrupled, causing inflation to rise sharply and central banks to raise interest rates sharply, giving way to a major global recession.

Although a Russian oil crash may have profound inflationary impacts, with severe effects on domestic demand on a global scale, it seems unlikely it will generate a global recession as the one observed almost half a century ago. This is due to the more efficient use of oil that the advanced economies of the West have developed, which translates into less dependence on this raw material.

The global industry has a much lower dependence on oil than in the 1970s. In 1973, for example, the world used about one barrel of oil to produce USD 1 000 of gross domestic product (GDP) (2015 prices) (Ruhl and Erker, 2021[6]). In 2019, before the COVID-19 pandemic, the intensity of oil use had fallen to 0.43 barrels by the same magnitude of global GDP (-56%). Additionally, today there is a more diversified matrix of energy sources, where oil generates about one-third of the world's energy compared to 53% at the beginning of the 1970s, having ceded space to biofuels and nuclear reactors.

Another no less important element is the development of the fracking industry in the United States in recent years, which has allowed it to improve its oil trade balance. Thus, while harming consumer spending, an oil price shock such as the current one also benefits domestic producers. An energy matrix less dependent on oil and a broader range of producers makes the global economy less vulnerable to abrupt disruptions in crude oil supply and energy price shocks.

The second channel is the disruption of world trade. While, in general, the share of Russia and Ukraine in global trade and production is relatively small, they are critical suppliers of inputs to certain industrial value chains. Russia is one of the world's leading producers of palladium (26% of global imports) and rhodium (7% of global imports), which are inputs for producing catalytic converters for automobiles. Ukraine supplies more than 90% of the neon used in making lasers used to manufacture American micro-chips. Possible disruptions in the supply of these raw materials could aggravate the supply of semiconductors for the electronic equipment and car industries, worsening the critical shortages observed in these sectors since the beginning of the pandemic. Similarly, Russia and Ukraine together account for about 30% of global exports of wheat, 15% of corn, 20% of both mineral fertilisers and natural gas, and 11% of oil. In the case of China, trade disruptions have arisen as a result of the impact of the strict zero-COVID strategy. In Shanghai and other big cities, the policy has created labour shortages, which affects transport capacity, slows operations in ports and reduces air traffic (OECD, 2022[3]).

The third channel is an increase in financial volatility. Since Russia's large-scale aggression against Ukraine, global capital markets have experienced high volatility, with an initial plunge on 24 February 2022 and a rebound in the following weeks. The Chicago Board Options Exchange Volatility Index, a proxy of the standard market volatility in international capital markets, reached its highest peak of 2022 in March and has remained relatively high, although still considerably below the volatility seen in 2020 due to the COVID-19 outbreak. Overall, the impacts of Russia's war against Ukraine and China's economic slowdown on global capital markets have been more moderate than the pandemic's. However, as central banks have responded to above target inflation rates, financial conditions have tightened and capital outflows from emerging-market economies have intensified (OECD, 2022[4]).

Central banks are reacting to growing inflationary pressures not seen in decades

One of the main economic policy challenges of Russia's war against Ukraine is soaring commodity prices, further fuelling inflation. Disruptions in global supply chains, high freight costs, and demand and supply imbalances contributed to the build-up of inflationary pressures not seen in decades. China's zero-COVID policy can add to inflationary pressures via producer prices (OECD, 2022[3]; OECD, 2022[4]).

Central banks in major economies initially assessed this rise as transitory. However, it has persisted much longer than originally anticipated by authorities, leading to an unexpected global inflationary scenario. In the United States, inflation reached 8.6% year on year in March 2022 – a 40-year high. For some advanced economies, May 2022 had the highest level of inflation. This was observed in the Eurozone, with inflation reaching around 8.0%, Canada (around 7.7%) and the United Kingdom (above 9.0%), where headline and core price growth has already far exceeded the respective official inflation targets.

The main risk of this prolonged deviation of inflation from the target is the de-anchoring of medium- and long-term inflation expectations. Central banks in developed economies are accelerating the speed of monetary policy normalisation by reducing asset purchases and increasing interest rates from historical minimum levels.

The first major central bank to move forward with interest rate normalisation was the Bank of England, raising its benchmark rate from 0.10% to 0.25% in December 2021 and it has since then increased it up to 1.75%. In March 2022, the US Federal Reserve began its cycle of interest rate hikes, applying the first increase of the Fed Funds Rate in four years, from 0.50% to 0.75%. Raising interest rates has continued. In June and September 2022, the US Federal Reserve approved hikes of 0.75 basis points up to 2.37, the largest since 1994 (OECD, 2022[4]).

The European Central Bank (ECB) has also been taking measures towards monetary normalisation, mostly as this region is more directly exposed to Russia's war against Ukraine. In March 2022, the ECB announced that the pandemic emergency purchase program would come to an end and that it would gradually reduce its debt purchase until the end of June. In July 2022, the ECB raised its key interest rate by half a percentage point, the first increase in over a decade. Similarly, it implemented the Transmission Protection Instrument (TPI) to ensure that the monetary policy stance is transmitted smoothly across all euro area countries.

Although many emerging economies, particularly in Latin America, have advanced in the withdrawal of monetary COVID-19 stimuli since last year, emerging markets will face challenges as interest rate hikes continue in advanced economies. Higher interest rates may pose risks to highly indebted households and companies, compromising the banking sector (IMF, 2022[5]). Moreover, following the pandemic-related sharp increase in public debt, higher interest rates could threaten sustainability, particularly in countries experiencing lower growth.

The recovery in Latin America is slowing down, reflecting low potential growth

LAC had a strong rebound in 2021. GDP growth in LAC rebounded to above 6% in 2021, driven by fiscal stimulus, more favourable external conditions and the acceleration of the region's vaccination campaigns, which allowed the economies to reopen. This happened after the region registered one of the most severe output contractions (6.9%) globally in 2020, causing poverty and inequality to rise. However, the recovery was uneven. While several countries regained pre-pandemic GDP levels, decreasing tourism flows and limited fiscal space for stimulus constrained a fuller comeback in the Caribbean and Mexico (ECLAC, 2022[1]).

In 2022, a slowdown in LAC growth is expected as external conditions deteriorate, reopening effects dissipate, and local authorities roll back fiscal and monetary stimuli. The rapid convergence to modest expansion rates reflects low potential growth. Projected economic growth in the region will be insufficient to revert the increases in poverty and inequality accentuated by the pandemic. Inflationary pressures have prompted central banks to increase interest rates since 2021, and the tightening cycle will continue while inflation rates remain above central banks' targets. Fiscal deficits will narrow in 2022 as governments withdraw spending and activity stabilises. However, debt levels will remain high, and further consolidation efforts might be needed in the medium term to recover fiscal space and debt sustainability. Risks to this outlook stem from a steeper and faster-than-anticipated tightening of financial conditions, new waves of the pandemic, longer-lasting disruptions in global supply chains, political uncertainty in the region, and the repercussions of both Russia's invasion of Ukraine and China's economic slowdown.

Potential growth, a pre-pandemic structural challenge in LAC, remains stagnated at low levels. Moreover, regardless of how potential growth is calculated, it has weakened. Potential GDP per capita growth has been below 1% since 1980, slightly increasing following the commodity boom (between 2003 and 2013). Since then, per-capita potential output growth has stagnated. Furthermore, potential GDP per capita growth remains below advanced economies, hindering convergence. Over the last decades, the per-capita GDP gap between LAC and advanced economies has narrowed, but gains stopped after 2015 as potential regional output faltered (Figure 1.2).

Figure 1.2. **Potential GDP per capita growth in LAC and advanced economies, estimated under different methods**

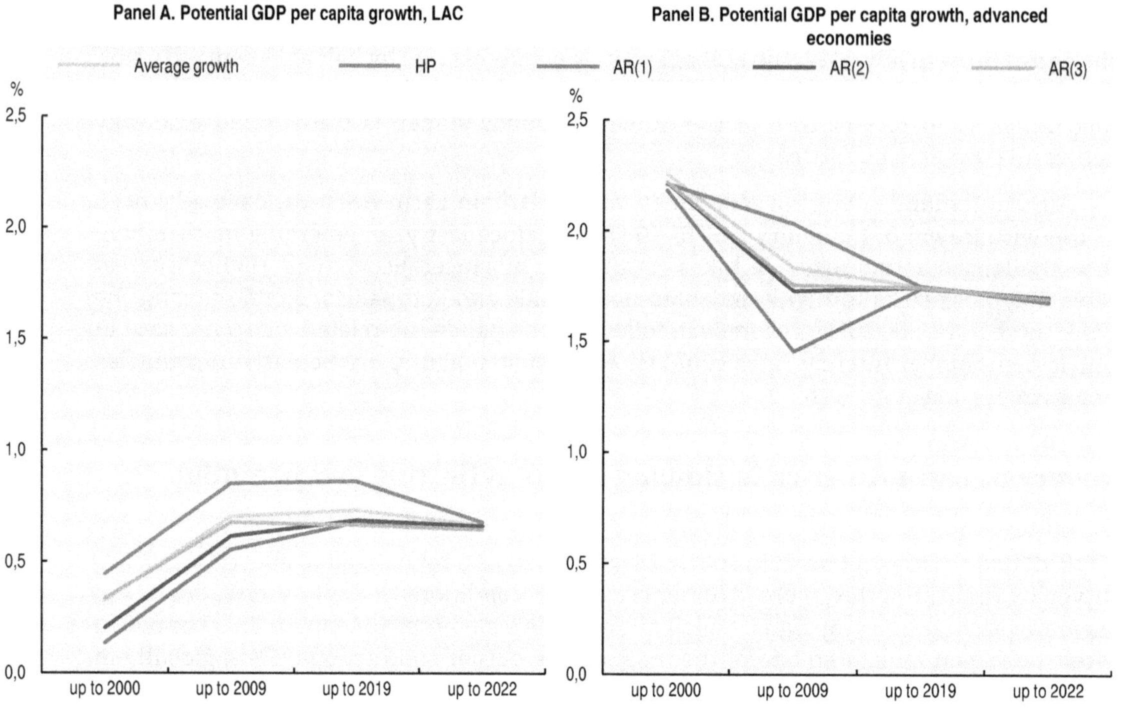

Notes: AR = autoregressive model, which uses GDP per capita growth data. The number of lags (1 and 2) was determined by analysing the autocorrelation function and choosing the model that maximised the log-likelihood. HP = the Hodrick-Prescott filter, which was used as an alternative model due to its resilience to short-term shocks to create a smoothed curve (lambda 100). The LAC series refers to the 33 countries covered by the IMF's World Economic Outlook database, October 2022.

Source: Authors' calculations based on (IMF, 2022[5]).

StatLink *https://stat.link/y94d8a*

Overall effects on LAC of Russia's war against Ukraine and China's slowdown remain uncertain and are transmitted through three main channels

The first channel is the effect of higher commodity prices on external accounts. The reduced exposure to Russia and Ukraine limits the direct impact on trade in the LAC region, which does not exceed 1% of the region's total trade. Only Ecuador (4% of total trade) and Paraguay (8% of total trade) exhibit a more significant exposure to trade with Russia. From the point of view of investment flows, Russia's involvement in the region is shallow, except for participation in some energy projects in Brazil and Mexico. Therefore, the main impact of the crisis on the LAC region is through the terms of trade because of increases in the prices of energy and some agricultural raw materials.

The outcome depends on whether countries are net exporters or importers of energy and food. Net exporters in South America benefit from more favourable terms of trade, improving current account balances, and generating additional fiscal revenues that may stimulate demand and, consequently, growth and employment. Nonetheless, the improvement in the terms of trade may not be significant because of the accumulated increases in imported input prices due to the disruptions in global supply chains. Central American and Caribbean countries experience the opposite effects (Figure 1.3, Panel A). Similarly, the economic slowdown experienced by China and globally affects the trade channel, particularly in economies such as Brazil, Chile, Peru and Uruguay, where

China is a crucial trading partner. Negatively affected countries, such as Chile, Panama, Paraguay and Peru, have sufficient international reserves to face the transitory negative shock (Figure 1.3, Panel B) and maintain access to global financial markets at relatively low costs.

Figure 1.3. **Current account deficits and international reserves**

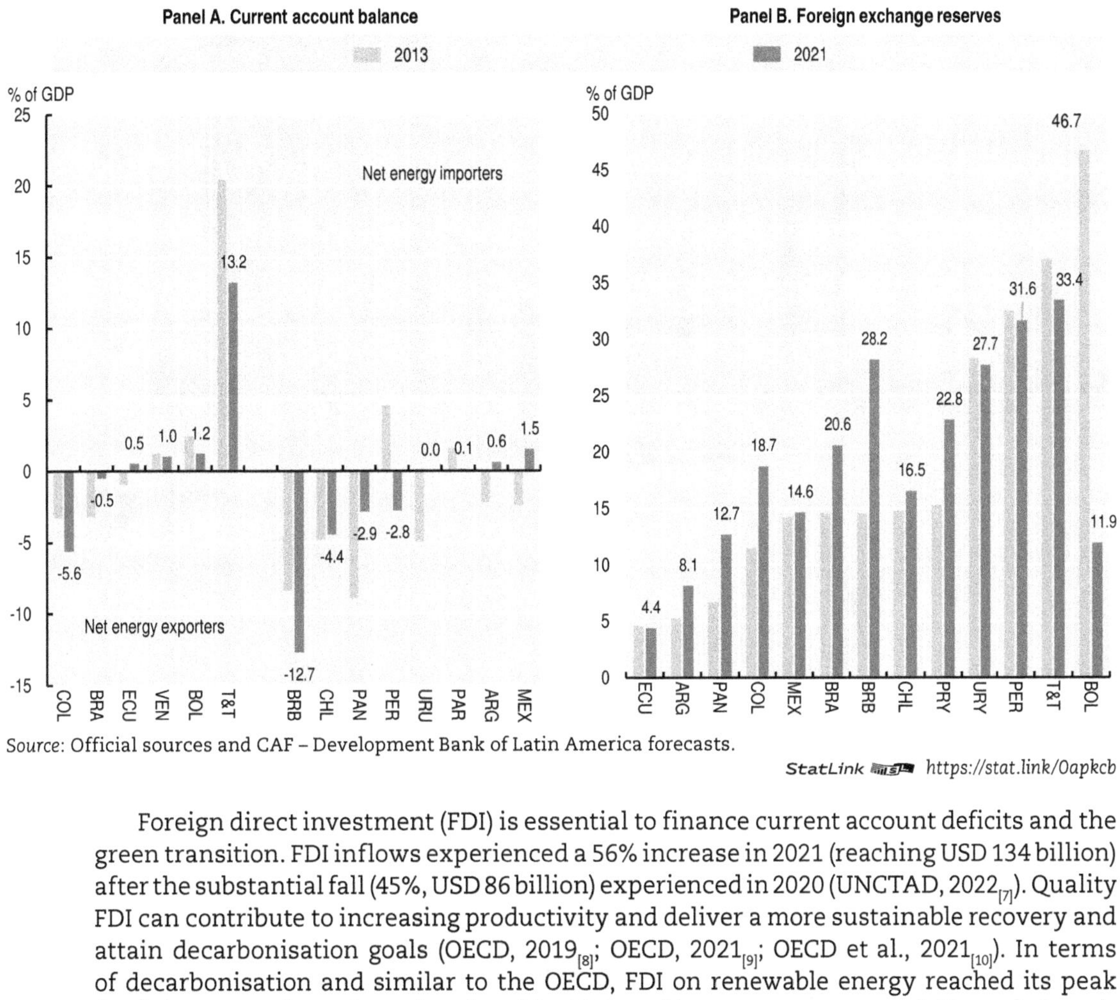

Source: Official sources and CAF – Development Bank of Latin America forecasts.

StatLink https://stat.link/0apkcb

Foreign direct investment (FDI) is essential to finance current account deficits and the green transition. FDI inflows experienced a 56% increase in 2021 (reaching USD 134 billion) after the substantial fall (45%, USD 86 billion) experienced in 2020 (UNCTAD, 2022[7]). Quality FDI can contribute to increasing productivity and deliver a more sustainable recovery and attain decarbonisation goals (OECD, 2019[8]; OECD, 2021[9]; OECD et al., 2021[10]). In terms of decarbonisation and similar to the OECD, FDI on renewable energy reached its peak (both in USD and number of projects) in 2019 and has not yet recovered. Nevertheless, it remains above the pre-2019 levels. In LAC, FDI on renewables remains above the levels invested in oil coal and gas (both in USD and number of projects) (Figure 1.4).

Figure 1.4. **LAC and OECD FDI investment in renewable energy and in coal, oil and gas, LAC**

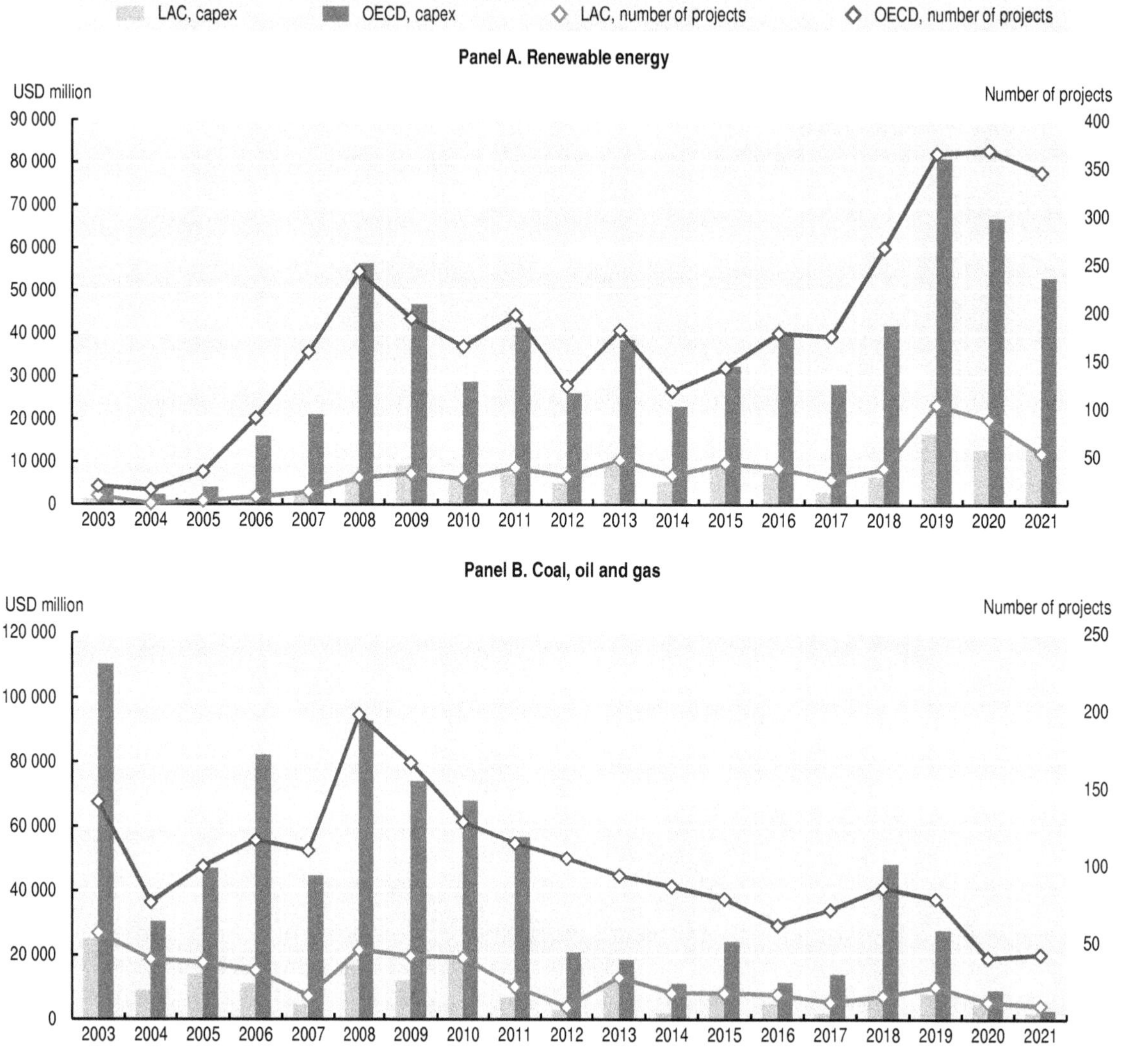

Notes: Countries included are Argentina, Belize, Bolivia, Brazil, Chile, Colombia, Costa Rica, Cuba, Dominican Republic, Ecuador, El Salvador, Guatemala, Guyana, Haiti, Honduras, Jamaica, Mexico, Nicaragua, Paraguay, Peru, Saint Lucia, Suriname, Trinidad and Tobago, Uruguay, and Venezuela.

Source: Financial Times' fDi Markets database.

StatLink https://stat.link/vkugip

The second channel of transmission is through international financial markets. In addition to the shock in markets generated by the Russian war against Ukraine, other factors including the normalisation of monetary policy in advanced economies, and specific domestic aspects within LAC countries affected capital markets' behaviour. Between March 2022 and October 2022 risk premia increased, although they remain below the levels seen during the COVID-19 pandemic. Similarly, during the same period, most local currencies have depreciated (Figure 1.5), with the exception of some economies such as Mexico where it has appreciated. Exchange rate depreciations in most LAC countries follow a trend that precedes the war, and recently is a combination of domestic and external factors, including increases in inflation rates.

Figure 1.5. **Nominal effective exchange rate and sovereign bond spreads**

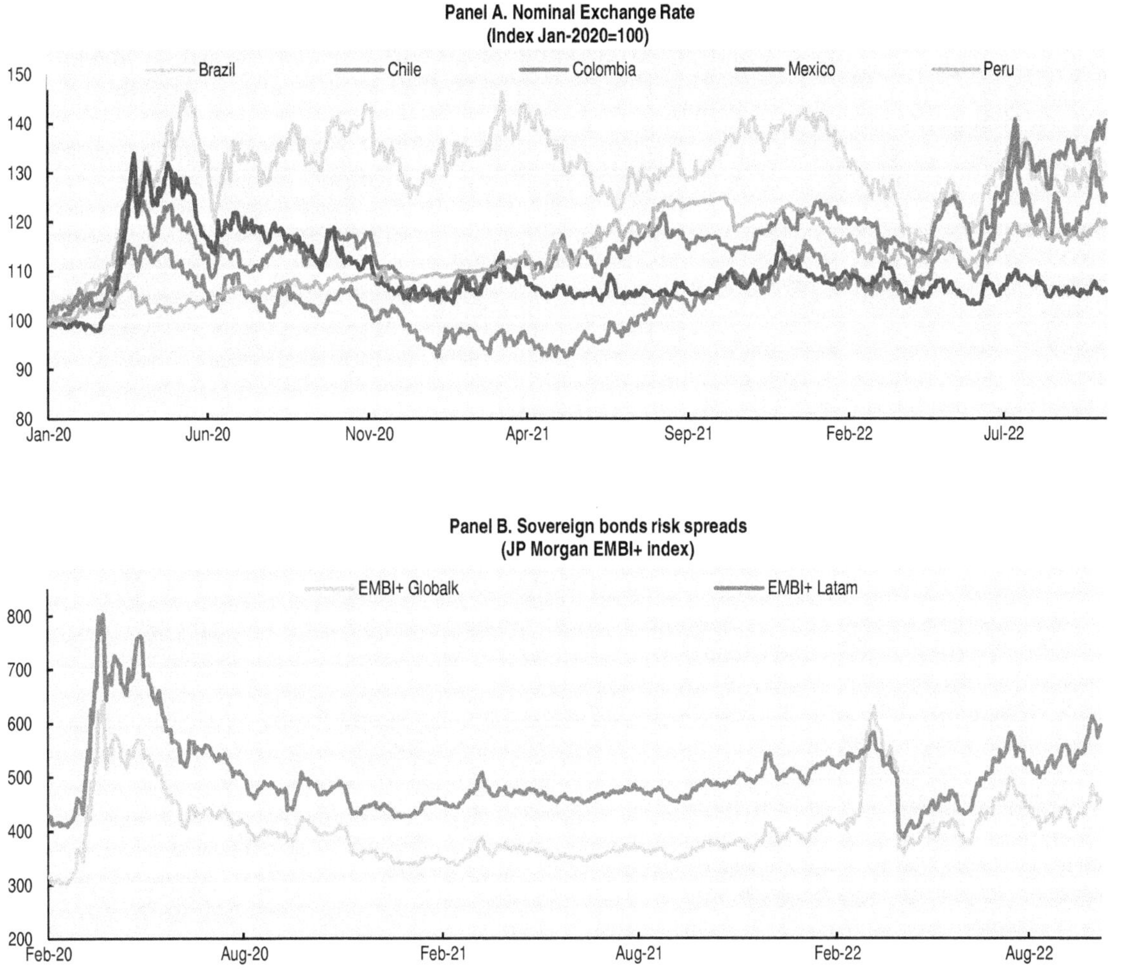

Note: Data until 7 October 2022.
Source: Thomson Reuters Datastream.

StatLink https://stat.link/pg0cmo

The third channel is the intensification of inflationary pressures. First, the pandemic and its recovery disrupted trade and caused supply bottlenecks, input shortages and increases in transport costs and commodity prices (IMF, 2022[11]). In addition, the war has exacerbated the rise in commodity prices in most LAC countries. As a result, inflation in 2022 has been above the policy target in LAC economies (Figure 1.6, Panel A). In Chile for instance, inflation has risen to a 30-year high (OECD, 2022[12]). Before Russia's war against Ukraine, the price rise (Figure 1.6, Panel B) had already prompted central banks across LAC to increase interest rates to anchor expectations. As inflationary pressures mount due to the increase in commodity prices, and the US Federal Reserve hikes rates, this process will be difficult to reverse. The marked rise in energy and food prices reduces the purchasing power of households, particularly the more vulnerable, still reeling from the effects of the pandemic. This does nothing to reverse the increase in poverty and inequality in the region, already hampered by the projected weaker economic growth.

Monetary policy should consider and align with climate goals and policies. Different climate policies have distinct implications for the price system, for instance a fixed carbon price can affect price fluctuations (Chen et al., 2021[13]). Similarly, incorporating environmental, social and governance (ESG) factors into central bank mandates will be critical to efficiently safeguarding price and financial stability given the high-level impact that climate risks could have on the traditional core responsibilities of this institution. Central banks in LAC can share and learn from experiences with other central banks. The ECB aims to monitor the financial system, including private banks, so that the risks from climate change are included (ECB, 2021[14]).

Figure 1.6. **Inflation and policy rates in inflation targeting in selected LAC economies**

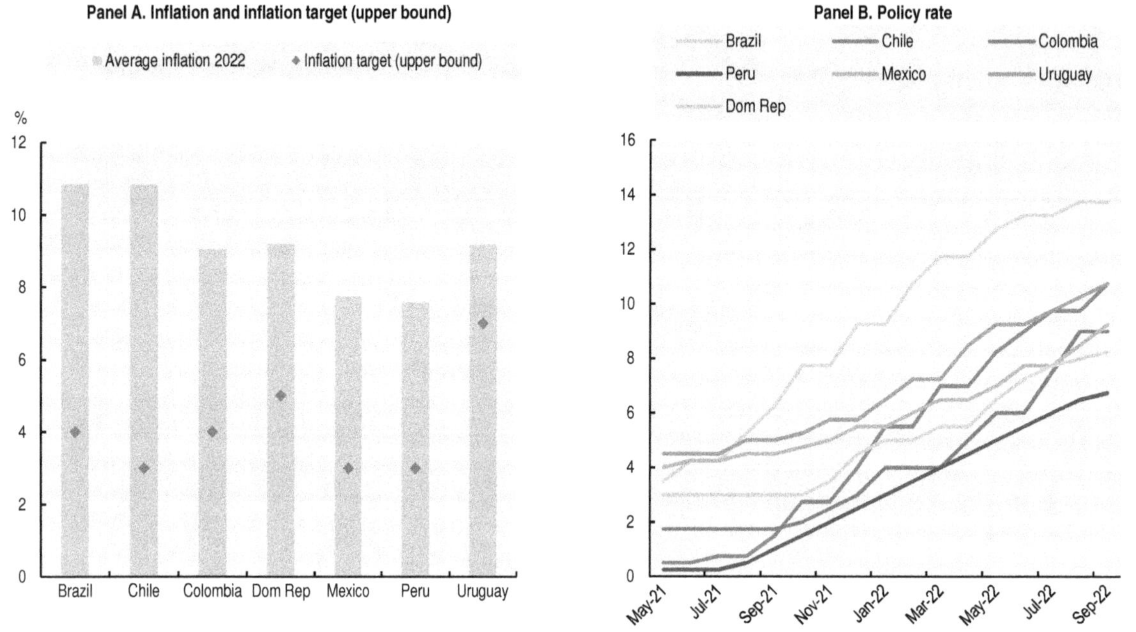

Note: Data are from January to August 2022 (Panel A) and from January 2021 to September 2022 (Panel B).
Source: Thomson Reuters Datastream and official sources.

StatLink https://stat.link/kafhyg

Low fiscal space to navigate the challenging context

To cope with the COVID-19 crisis, public expenditure in LAC reached a historic high of 13.6% of GDP in 2020 at the central government level, an increase of 2.3 percentage points compared to 2019. This level was even higher than that seen during the last economic crisis in 2008, when the increase was 1.1 percentage points (ECLAC, 2022[2]). Social spending is the main component of total public spending (ECLAC, 2022[2]). The countries that stand out for the largest increases in 2020 (with respect to 2019) are Brazil and El Salvador (5.3 percentage points of GDP) and Argentina, Barbados and the Dominican Republic (between 4.2 and 4.6 percentage points of GDP) (ECLAC, 2022[2]).

In 2021, LAC economies started to withdraw the fiscal stimulus and revenues increased with the economic recovery; therefore, primary fiscal deficits for the central government have been narrowing. On average, deficits reduced to 4.2% of GDP in 2021, an improvement of 2.7 percentage points from 2020. Tax revenues increased in 2021 thanks to strong GDP growth, while spending fell as LAC economies reduced emergency transfers (ECLAC, 2022[15]). This came after a drop in tax revenues of 0.8 percentage points in 2020 relative to 2019. The COVID-19 pandemic resulted in historic falls in both nominal tax revenues and nominal GDP, with taxes falling more sharply than nominal GDP. There is still some fiscal stimulus related to the pandemic, as current expenditure remains above 2019 levels (IDB, 2022[16]; OECD et al., 2022[17]).

Even though the 2021 recovery helped relieve some of the pressure on fiscal accounts, structurally tight fiscal space in LAC still needs to be addressed. Public debt-to-tax ratios, a proxy indicator of countries' financial capacity to pay for public debt, were higher in 2019 than in 2013 and increased significantly in 2020 (Figure 1.7).

Tax revenues remain low in LAC. In 2020, the average tax-to-GDP ratio was 21.9% of GDP compared to 33.5% in the OECD. As a result, the gap between the LAC and OECD tax-to-GDP ratios widened to 11.6% of GDP in 2020 from 10.7% in 2019 (OECD et al., 2022[17]). The gap is mainly explained by the region's low revenues from income taxes and social security contributions relative to the OECD average. The combined share of taxes on income and profits (especially personal income taxes) and social security contributions was much lower in the LAC region than in the OECD (44.1% versus 60.0% in 2019, on average).

Figure 1.7. **Gross public debt-to-tax ratio in selected Latin American countries**

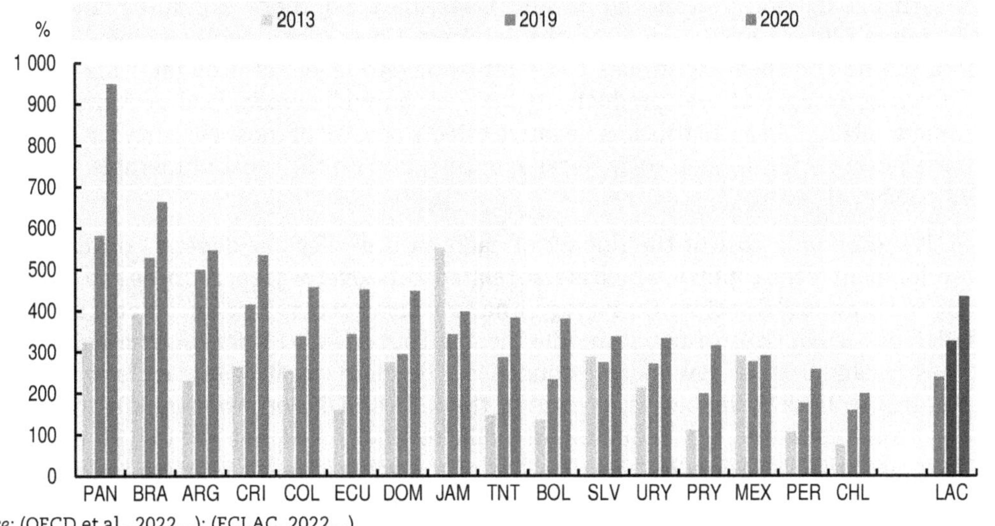

Source: (OECD et al., 2022[17]); (ECLAC, 2022[18]).

StatLink https://stat.link/cr9yxn

After a substantial increase in 2020, LAC debt declined in 2021. Debt stood at 53.7% of GDP in 2021, below the 56.5% registered in 2020 (ECLAC, 2022[15]). The economic recovery and rising inflation helped reduce debt, despite fiscal deficits. Notwithstanding, volatility in the financial markets, rising debt costs, and the need to finance the green transition highlight the necessity for adequate, proper fiscal frameworks (including fiscal rules) and globally co-ordinated debt management under key guidelines (Chapter 4) (OECD et al., 2021[10]; Arreaza et al., 2022[19]).

Going forward, LAC economies must support economic conditions and fiscal sustainability, particularly in the face of rising food inflation, financing the green transition, while protecting the most vulnerable through strengthening social protection systems. The composition of the fiscal consolidation, the timing and the balance between capital and current expenditure will play important roles in the shape and inclusiveness of the recovery. If public investment is safeguarded, relative to current expenditure, it can neutralise the contractionary effects of fiscal adjustment in the short run and might stimulate growth in the medium term (Ardanaz et al., 2021[20]). In the short term, the focus should remain in protecting the most vulnerable against rising inflation.

With a tight fiscal space and with high debt levels, countries in the region will need to face the growing negative effects of climate change and finance the green transition. Fiscal efforts should develop comprehensive frameworks that combine decarbonisation and resilience strategies with the promotion of growth and social inclusion (D'Arcangelo et al., 2022[21]).

Natural disasters weigh heavily on fiscal sustainability

Climate change is already challenging fiscal sustainability when natural disasters strike. The frequency of natural disasters in the region has increased over the last decades. An average of 17 hurricanes per year and 23 category-5 hurricanes in total were recorded between 2000 and 2019, mostly affecting Caribbean and Central American countries (OCHA, 2020[22]).

Other natural disasters, such as floods, wildfires and droughts, are common in LAC. They have severely affected the whole region, with flooding and wildfires being the most common events in South America's Southern Cone. Drought affects the largest number of people in the region, with crop yield reductions in 2018 of between 50% and 75% in eastern El Salvador, central and eastern Guatemala, southern Honduras, and parts of Nicaragua (OCHA, 2020[23]). In 2022, the Southern Cone, traditionally a global pantry for both grains and meat, continues to be hit by severe levels of drought. This has led to a decrease in agricultural productivity, as well as widespread food security concerns (Amaya, 2022[24]). In turn, this has negative effects on GDP of most countries in the region (Banerjee et al., 2021[25]), on their fiscal accounts, and on the most vulnerable (Chapter 2) (Bárcena et al., 2020[26]).

The economic cost of the impact of each natural disaster depends on the level of development of the country, which is also related to its level of preparedness and capacity to respond to natural disasters. On average, a natural disaster results in an increase in the fiscal deficit of 0.3% of GDP for upper-middle-income countries, 0.8% for lower-middle-income countries and 0.9% for low-income countries. The main negative impact is concentrated in a drop in fiscal revenues as a result of the fall in GDP. For lower-middle-income and low-income countries, this reduction in public revenues is equivalent to 0.8% and 1.1% of GDP, respectively (Alejos, 2021[27]). The size of the government's reaction in the event of a disaster depends not only on the severity of the disaster but also on the government's rate of compliance in meeting its liabilities – that is, its ability and willingness to meet or exceed its *ex-ante* commitments to shoulder specific disaster-related costs (OECD/The World Bank, 2019[28]).

The effects of fiscal deficits can be specifically harmful to economies with tight fiscal space and can leave long-term scars on fiscal accounts. Natural disasters can lead to dramatic increases in public debt, the abandonment or postponement of new investment projects, and the procyclicality of fiscal policy, especially for countries that do not have adequate insurance mechanisms for natural disaster risks (such as catastrophe bonds or disaster insurance) (Chapter 4) (IDB, 2021[29]). It is also important to consider that similar negative fiscal effects can be caused by the concentration of multiple non-extreme events in a short period, especially when there are conditions of high exposure and vulnerability in the country.

Some Caribbean economies present the highest levels of indebtedness: in 2020, nearly three-quarters of small states with unsustainable debt levels were in the Caribbean. Natural disasters, alongside poor economic performance, insufficient fiscal restraint and high financing costs in capital markets, are the main reasons for these high debt levels. The cost of debt service for these economies greatly reduces their fiscal space and undermines their ability to react to further shocks and to fund the necessary public services and public investment to drive their development process (OECD et al., 2021[10]).

LAC will need to diversify its tax revenue to compensate for the coming fall in public revenues from hydrocarbons in main producing economies

As the world transitions to clean energy sources, demand for non-renewable resources will decline, entailing a drop in public revenues in a group of LAC countries that export hydrocarbons. As alternative technologies become cheaper, and measures to address climate change and implement the Paris Agreement are put in place, demand for oil is expected to fall (IDB, 2021[29]). Before Russia's war against Ukraine, it was estimated that, in scenarios that meet the Paris Agreement targets, oil production in LAC needed to fall by 60% by 2035, which would entail losing about USD 3 trillion in tax revenue (Solano-Rodríguez et al., 2019[30]; Vogt-Schilb, Reyes-Tagle and Edwards, 2021[31]). Similarly, the role that natural gas plays in the region's economy will be progressively reduced, leaving half the reserves untapped and reducing associated tax revenues by up to 80% (Welsby et al., 2021[32]).

This phasing out among major hydrocarbon producers will have significant negative effects on fiscal revenues and on foreign exchange. Hydrocarbons are a major generator of foreign exchange, with its exports accounting for one-third or more of total exports in several countries. Concerning fiscal revenues, the exploration and production of oil and gas account, on average, for around 3% of GDP (Figure 1.8, Panel A). This can reach above 15% of total revenues in Bolivia, Mexico, and Trinidad and Tobago and 24.2% in Ecuador (Titelman et al., 2022[33]). These revenues can excede, on average, 3% of GDP and in some cases more than 7%, as in Ecuador and Trinidad and Tobago (Figure 1.8, Panel B) (OECD et al., 2022[17]). If the decline in hydrocarbon revenues is not offset by an increase of those from other sources or economic diversification, these countries will experience large fiscal deficits (Titelman et al., 2022[33]).

Figure 1.8. **General government fiscal revenues from non-renewable natural resources in selected LAC economies**

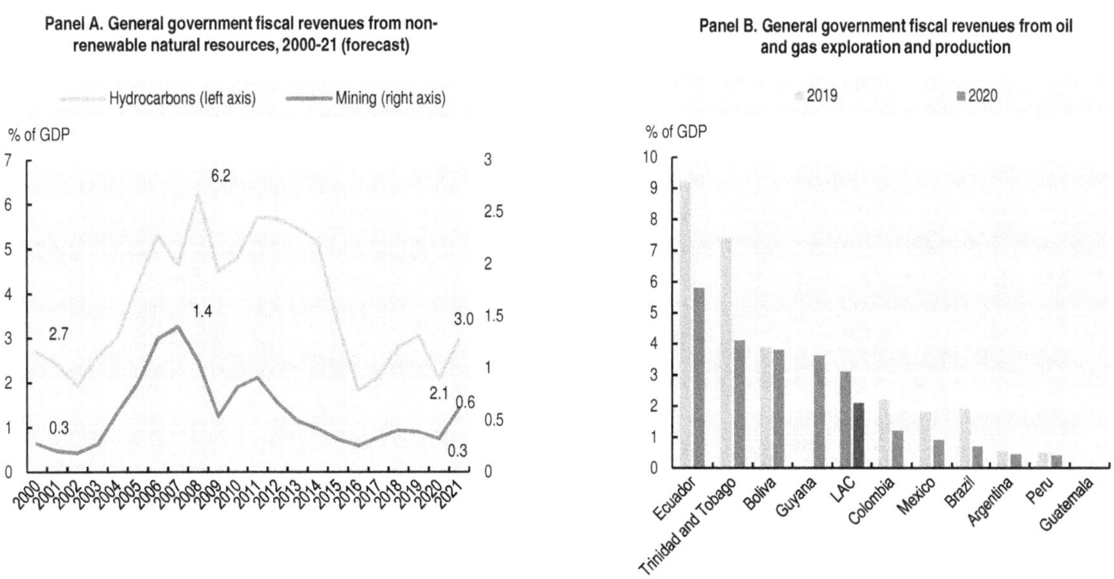

Note: For Panel A the series refers to the average of the countries presented in Panel B.
Source: (OECD et al., 2022[17]).

StatLink https://stat.link/lf1jsp

Stranded assets can be an additional cost of inaction against climate change

As climate policies progress in the region and the phasing out of fossil fuels makes headway, hydrocarbon reserves and infrastructure are at risk of becoming stranded assets, resulting in financial losses for LAC economies.

Many economies in the region continue to develop new oil and gas projects that risk ending up as stranded assets. For instance, Argentina, Brazil and Mexico have ambitious plans to increase hydrocarbon production; others, such as Guyana, have plans to start exploiting it on a scale transformative for their economies (IEA, 2017[34]). If all planned fossil fuel power plants were built, there would be a 150% increase in the region's "committed emissions" (IDB, 2021[29]). In addition, ten coal-based power plants were inaugurated in the last decade, illustrating how, between 2009 and 2016, countries were still choosing coal-based thermal power over other cleaner options, such as solar, wind or hydro (Bermúdez, 2020[35]).

Some types of stranded assets will represent higher costs. For instance, coal technologies are expected to make up a small percentage of the total capacity stranded in the LAC region. Still, the cost associated with the stranding of these assets is the highest. This is because they are more capital-intensive than gas and oil plants and are assumed to have longer lifetimes (60 vs. 45 years) and thus a lower depreciation rate (Binsted et al., 2020[36]).

Economies in the region that are more advanced in their transition to low-carbon energy production are already closing coal-based plants. For instance, in Chile, since the Decarbonisation Plan was announced in 2019, 6 coal-based thermoelectric plants have closed. Another 5 will close by the end of 2024, and the remaining 17 will close before 2040 (Parra, 2021[37]).

Spending decisions should consider the effects of climate change and the green transition

LAC must close its investment gap by allocating more resources in infrastructure to increase resilience and achieve its decarbonisation goals. The region already falls behind in terms of investment. In 2021, it invested around 19.5% of GDP, below the 22% invested in advanced economies and the 39% invested in Emerging and Developing Asia (IMF, 2022[5]). To achieve the United Nations Sustainable Development Goals (including resilience and decarbonisation goals), the region will need to increase its investment in infrastructure by about 5% of GDP (Galindo, Hoffman and Vogt-Schilb, 2022[38]). Despite the need for high investment, the region must continue to balance between capital and current spending, using the momentum to finance the green transition.

LAC must invest more to become more resilient against the negative effects of climate change. Climate change increases the frequency of natural disasters and causes changes in precipitation and temperature, along with coastal flooding. The current infrastructure of the region can be vulnerable to these events. New energy, transport, water or telecommunications projects must envisage the risk that these events pose. Moreover, existing infrastructure must be made resilient to these events, while new infrastructure must be created to reduce their effects (e.g. flood control dams, floodwalls or water dams to preserve and redistribute water to zones experiencing droughts). It is estimated that, for every USD 1 invested in making infrastructure and economies more resilient, USD 4 is avoided in impact costs (Galindo, Hoffman and Vogt-Schilb, 2022[38]). Similarly, the region could explore the role of nature-based solutions (NbS) in limiting and managing the current and future impacts of climate change. NbS are measures that protect, sustainably manage or restore nature, with the goal of maintaining or enhancing ecosystem services to address a variety of social, environmental and economic challenges (OECD, 2020[39]).

Although the investment required to achieve resilience is high, spending reallocation and early planning can help reduce costs. Climate strategies are important for anticipating long-term objectives in government planning, managing risks appropriately and greening public spending. To achieve carbon neutrality and climate resilience in 2050 a multi-sectoral climate strategy is needed, that aligns all sectoral strategies and incorporates decarbonisation and resilience criteria into public investment and budget systems (Galindo, Hoffman and Vogt-Schilb, 2022[38]).

Alongside the correct infrastructure, financial frameworks that manage climate-related risks and increase financial resilience are needed. LAC economies must develop frameworks that can manage national risks while promoting global climate financial resilience. As such, frameworks that identify risks and better understand them in terms of their components (hazards, exposure and vulnerability) and sources can play a key role, mitigating financial losses through risk reduction.

Despite best efforts, risk will always remain. Thus, coherent and integrated multipronged government financial strategies must be put in place (OECD, forthcoming[40]). Examples to minimise risk include improved building codes, better territorial and watershed planning, analysis of the budgetary impact of risk, and financial preparedness, including the use of insurance and reinsurance financial instruments (Galindo, Hoffman and Vogt-Schilb, 2022[38]). It is also necessary to scale up investment towards financial instruments such as green, social, sustainable and sustainability-linked bonds (GSSS Bonds) within internationally aligned frameworks that include green standards and sustainable taxonomies (Chapter 4). These will allow for increased and more effective redirection of spending towards sustainable projects that contribute to climate change mitigation and adaptation. Additionally, as some adaptation measures are costly, governments must carefully evaluate the *ex-post* impacts and the probability of disasters occurring. Post-disaster loss assessment can provide qualitative and quantitative information to help identify the strengths and weaknesses of risk assessment.

If fiscal policy is to be effective, it must be through co-ordinated efforts

For any fiscal policy or reform to be effective, it must co-ordinate tax, spending and debt policies, considering the socio-economic and political context through well-defined sequencing of actions. It also needs to be backed by a broad consensus, built through national dialogue and clear communication (Chapter 5). The political economy of fiscal policy is more important than ever (Nieto-Parra, Orozco and Mora, 2021[41]).

Moreover, a fiscal sustainability framework focused on strengthening public revenues will be required to ensure the viability of a growing green expenditure trajectory. In the short term, the region will have to focus on reducing tax evasion (6.1% of GDP in 2018) and on reviewing tax expenditures (ECLAC, 2021[42]). It will also be key for countries to align their tax codes with new international best practices and to strengthen tax frameworks applied to the extractive sector (Titelman, 2022[43]). In the medium term, the region will have to focus on generating more progressive and greener fiscal pacts, with the aim of increasing tax collection and strengthening direct income and property revenues. It will also need a fully-fledged redesign of hydrocarbon taxation and subsidy policies (Chapter 4) (Titelman et al., 2022[33]). Additionally, to guarantee a just transition it will require public investments that attract green private investment (crowding-in); generate direct tax incentives towards renewable energies and decarbonisation, digital inclusion, and research and development; and lay the foundations for universal social protection systems (Chapter 4) (Titelman, 2022[43]). In terms of green tax incentives, special attention must be given to their design in relation to policy objectives. If misused, they can reduce revenue-raising capacity; create economic distortions; erode the principle of equity;

increase administrative and compliance costs; and potentially trigger harmful tax competition and windfall gains to investors for projects that would have taken place in the absence of the incentive (Celani, Dressler and Wermelinger, 2022[44]).

Social conditions remain worse than before COVID-19

Poverty and extreme poverty in LAC remain above pre-pandemic levels. By 2022, ECLAC estimated that 33% of the population would be in poverty, and 14.5% would live in extreme poverty as a consequence of limited economic performance and increasing inflation (ECLAC, 2022[1]). Poverty rates in 2022 are the highest since 2008, before the global financial crisis. After the strong increase in 2020 due to the COVID-19 crisis, poverty slightly receded in 2021 due to the strong rebound (Figure 1.9). However, low economic growth and rising inflation in 2022 reversed these small gains. Extreme poverty has steadily increased every year since 2014 – even in 2021, despite the strong rebound (ECLAC, 2021[45]; ECLAC, 2022[1]). The rise of poverty and extreme poverty is a result of massive losses of jobs and livelihoods, low potential growth, high inflation rates, lack of sufficiently strong social protection systems and, in some countries, reduction in emergency income transfers that was not offset by the expected increase in employment income (ECLAC, 2021[45]; OECD et al., 2021[10]; OECD, 2022[12]).

Since the pandemic, there has also been a generalised impoverishment of much of the population in Latin America, which has translated into downward mobility. This means that, since 2020, the low and lower middle-income strata's share of the population has increased, to the detriment of that of the upper middle- and high-income strata. With the recovery measures and economic upturn in 2021, there was a clear improvement in the middle classes, which regained some of their income. Still, the situation for low-income and low-middle-income classes worsened, and extreme poverty increased (ECLAC, 2022[2]).

Figure 1.9. **Poverty and extreme poverty rate evolution**

Percentage of the population

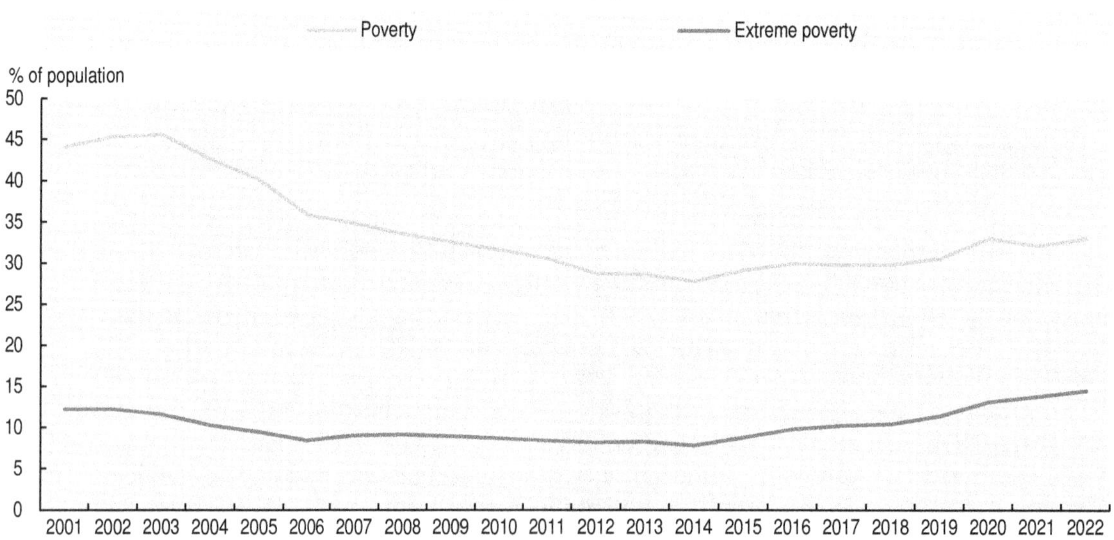

Source: (OECD et al., 2021[10]; ECLAC, 2021[45]; ECLAC, 2022[2]; ECLAC, 2022[1]).

StatLink https://stat.link/ipegru

Emergency cash transfers are key to offsetting the negative effects of the pandemic and external shocks on employment income and, therefore, to addressing poverty. However, in most countries, these are not enough to curb the increase in poverty. For instance, compared to the previous years in most countries in the region, the increase in the total income of households that received transfers in 2020 was smaller than the fall in their labour income. The notable exception was Brazil, where the fall in labour income represented a loss in the total income of about 4%, while transfers were equivalent to an increase in the total income of 7% (ECLAC, 2022[2]).

Poverty is heterogeneous not only among countries in the region but also among population groups. Across all countries, the trend shows that women aged 25 to 39 have higher poverty rates than men of the same age. Likewise, poverty rates can be 1.3 to 1.8 times higher for people under age 15 than for the next age group (ages 15 to 39). Additionally, the largest gaps are found in countries with low poverty rates, such as Brazil, Chile, the Dominican Republic and Uruguay. In countries where the incidence of poverty is higher, the gap between age groups tends to narrow (ECLAC, 2022[2]).

Informality remains a key challenge to tackling poverty and extreme poverty. In LAC, both informal and mixed households account for two-thirds of the total population. On average, almost half (45.3%) of people in LAC countries live in a household that depends solely on informal employment, 21.7% live in households with formal and informal workers (mixed households), and the remaining 33.1% live in completely formal households. In Bolivia, Honduras and Nicaragua, more than 60% of households rely entirely on informal employment, making them especially vulnerable to shocks such as the COVID-19 crisis (Figure 1.10) (OECD, forthcoming[46]).

Figure 1.10. **Distribution of overall population, by degree of informality of households, 2019 or latest**

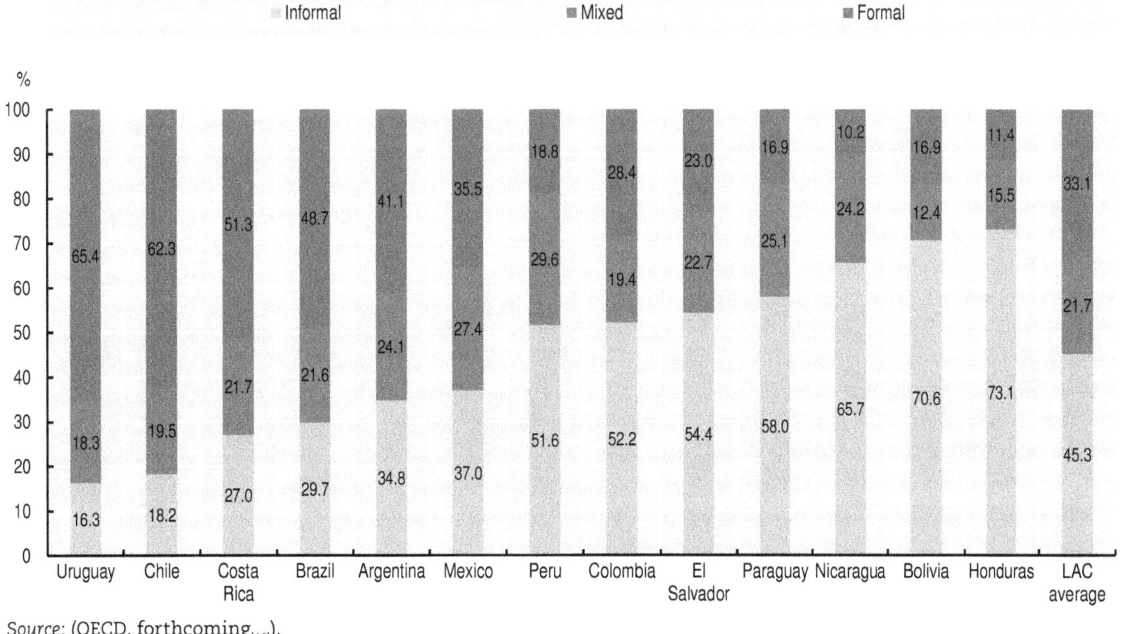

Source: (OECD, forthcoming[46]).

StatLink https://stat.link/r6o9ix

Having a perspective on household informality is key to bringing about new policy insights, as the informality status of working members within a household has implications for its dependent members, and indicates the various vulnerabilities to which they are exposed. Such insight is also key to analysing the opportunities that the green agenda will bring to create new high-quality formal jobs in LAC (Chapter 3) (OECD et al., 2021[10]; OECD, forthcoming[46]). This is also the case for analysis that links informality with the structural axes of the social inequality matrix in the region, including a territorial perspective of informality, as this is not a phenomenon evenly distributed within countries (Abramo, 2022[47]; Espejo, 2022[48]). It is crucial that strategies to reduce informality, including those connected to a green transition, take its diverse expressions into account.

Although the slow trend of inequality reduction was reversed in 2020, cash transfers prevented a larger increase than occurred

Inequality in income distribution has also increased in most countries in the LAC region. In 2020, the deterioration in distribution affected the poorest sectors the most, halting the trend of falling inequality that had been slowing down since 2002 and had lost pace since 2010. The Gini coefficient for the Latin American average went from 0.54 in 2002 to 0.46 in 2020, with very slight reductions from 2010 onwards. Countries demonstrating among the highest inequality coefficients in 2020 included Brazil, Colombia and Panama, with averages above 0.50. The lowest coefficients occurred in countries including Argentina, the Dominican Republic and Uruguay, with an index of 0.40. A comparison of the situation in 2017 with that prevailing in 2019 and 2020 shows that inequality, as measured by the Gini coefficient, increased in nine countries and decreased in six (ECLAC, 2022[2]). The distributional deterioration affected the poorest segments of the population the most (ECLAC, 2022[2]).

To better explain the evolution of changes in inequality among countries, it is necessary to refer to the evolution of average household income. Although there was a decrease in total average, the determining differential factor is the way in which the losses were distributed. In countries where inequality increased, the better-off quintiles lost less than the poorest. In this sense, the fall in income for the poorest quintile was on average 3.2 times the reduction in total income for the richest quintile. Thus, the wage income of the poorest quintile collapsed on average by 39.4%, which represents 5.1 times the decline in wage income experienced by the richest quintile (-7.8%). In contrast, in countries where inequality decreased, the richest quintile contracted the most. (ECLAC, 2022[2]). Emergency cash transfers also contributed to the reduction of inequality. The transfers implemented by states specifically to respond to the drop in income caused by the COVID-19 pandemic were essential in preventing a further increase in inequality. The Gini coefficient would have increased on average by 4% without them, compared to the actual increase of 1% in Bolivia, Costa Rica, Ecuador, Paraguay, Peru and the Dominican Republic. Similarly, in these countries the Atkinson index would have increased to 13.8% without them, compared to its actual growth of 5.1%. (ECLAC, 2022[2]). In other economies, such as Chile, inequality actually dropped from 0.45 in 2020 to 0.39 in 2021 thanks to the import support policies (OECD, 2022[12]).

Strengthening social protection systems will be key to protecting the population

The implementation of social protection measures in response to the pandemic have been central to the welfare of the population but has also highlighted areas for improvement. During the last 15 years, countries in Latin America and the Caribbean have expanded the coverage of both contributory (financed by wages) and non-contributory (financed by taxes) social protection schemes (OECD et al., 2021[10]; ECLAC, 2022[2]). While significant progress has been made in building social protection systems, many informal workers remain often excluded from it (OECD et al., 2021[10]; OECD/ILO, 2019[49]; ECLAC, 2022[2]). More than half of the workers in the region do not participate in any contributory social security system against risks related to illness, unemployment, and old age (ILO, 2018[50]). On average, in the LAC region, more than 60% of economically vulnerable and informal workers do not benefit from labour-based social protection or a social assistance programme (OECD et al., 2021[10]). Despite their lower incomes and greater need for protection, informal workers often fall through the cracks of social protection systems, making many incomes insecure or vulnerable to income poverty affecting their families. Thus, it is fundamental that countries advance towards universal, comprehensive, sustainable and resilient social protection systems (ECLAC, 2022[2]).

Individuals and households in LAC, have a long tradition of informal networks of mutual support to cope with risks and uncertainty, especially in contexts where public options are absent or limited like in rural areas. Informal support is often organised around lifecycle or livelihood risk and vulnerability. Private transfers received from friends, relatives and other households are another element of this form of inter-household informal protection. Around the mid-2010s the share of private transfers in household income varies from 4% in Bolivia and Honduras to around 15% in Costa Rica (OECD/ILO, 2019[49]). However, relying on informal ties or mechanism for protection has several limitations and strong, public social protection systems are essential as part of a social contract to advance towards the exercise of economic and social rights. In the absence of universal access to social protection and social security policies, studies suggest that informal risk-sharing mechanisms are close to efficiency when it protects from idiosyncratic shocks linked to individuals, households, or lifecycle events, such as illness or death. They may fall short when it comes to broader shocks that affect a wider geographical area, such as a neighbourhood or community, which is likely the case for health environmental risks and the broad changes brought by green agendas. This can particularly hurt poorer households, which are already financially constrained (Watson, 2016[51]). It is therefore crucial to strengthen social protection systems that are universal, comprehensive, sustainable and resilient, and which can progressively expand coverage to informal workers, ensuring a green and just transition for all (OECD/The World Bank, 2020[52]; ITF, forthcoming[53]; OECD, 2021[54]; ECLAC, 2021[59]).

Regarding the contributory scheme, some Latin American countries have extended its coverage to informal economy workers. Reasons for success include several measures, such as combining the support for the formalisation of enterprises with access to social protection schemes; extending statutory coverage to previously uncovered workers; adapting benefits, contributions and administrative procedures to reflect the needs of informal workers; and subsidising contributions for those with very low incomes. In addition, several countries expanded the fiscal space needed to scale up social protection programmes financed through general government revenues. These efforts have significantly contributed to building social protection floors that guarantee universal health coverage and at least basic income security throughout the lifecycle, for instance through tax-financed pensions, disability benefits, child benefits, maternity benefits or employment guarantee schemes (OECD et al., 2021[10]) (Chapter 3).

Regarding the non-contributory scheme, according to official information, between 1 March 2020 and 31 October 2021, 33 countries in LAC adopted 468 non-contributory emergency measures and other support measures. These included three types: 1) monetary transfers; 2) in-kind transfers (including the provision of food, medical and education materials, as well as support for labour and productive inclusion); and 3) securing and facilitating access to basic services (water, energy, telephone and Internet) (ECLAC, 2022[2]). There were also measures aimed at containing and reducing household expenses, through tax relief, price setting and control for basic goods, and rents and payment facilities (Brooks, Jambeck and Mozo-Reyes, 2020[55]). While most of these measures were put in place in 2020, given the prolongation and depth of the economic and social consequences of the pandemic, it has been necessary to extend some measures or implement new ones (ECLAC, 2021[45]).

Going forward, it will be necessary to strengthen social protection systems (not only by increasing coverage but also by improving co-ordination and interoperability), make them more flexible against different types of shocks, ensure that they have positive long-term effects and improve their functioning while guaranteeing that they are properly funded.

Social protection schemes or cash transfers should have long-term development goals. Along with guaranteeing adequate levels of income, these goals could be promoting just, green education, labour or formalisation outcomes. For instance, there is evidence that targeted cash transfers, especially conditional, can spur investment in child schooling (OECD et al., 2021[10]; OECD, 2019[56]). Similarly, payments for environmental services can encourage a long-term change in behaviour to prevent future ecosystem degradation (Porras and Asquith, 2018[57]). A good example is Peru. Between 2014 and 2018, the Ministry of Environment and the Deutsche Gesellschaft für Internationale Zusammenarbeit (GIZ) developed conditional cash transfers to stimulate community protection of tropical forests in the Amazon region. To date, 188 indigenous communities have been covered, and over 1 800 000 hectares (ha) of tropical forest have been placed under protection. The conditional cash transfers are benefiting indigenous families, which have been able to improve their income levels and livelihoods without endangering their forests. Sustainable development skills are also being developed, as beneficiaries are advised on how to use the conditional cash transfer to initiate and manage projects and to monitor forest conversion (deforestation) (GIZ, 2014[58]).

Governments must also improve the flexibility of non-contributory programmes with strategies that provide protection against different types of shocks, for example natural disasters in some cases exacerbated by climate change. Such programmes should complement existing strategies and programmes that focus on structural poverty with others that ensure income support in the face of systemic (or idiosyncratic) shocks. Social protection must therefore articulate and co-ordinate with disaster management, with a view to increasing social and institutional resilience in order to cope with the impacts of increasing disasters and allow for a transformative recovery that is equitable (ECLAC, 2021[59]). Hence, responsive social assistance programmes should be promoted so that countries can adapt quickly to contingencies and find flexible ways to respond to the needs of individuals and households affected by shocks. These are also key in preventing shocks from systematically translating into persistently higher levels of poverty and inequality (Stampini et al., 2021[60]).

Countries must address the most pressing challenges affecting the functioning of their social protection systems. Efforts include, among others: enhancing information systems and digital platforms to better identify potential recipients and participants; increasing levels of coverage and updated information; improving the institutional framework, considering the various territorial levels working towards a unified social registry, and improving the level of interoperability; improving electronic payment systems; and making income support programmes more sustainable (Stampini et al., 2021[60]; Berner and Van Hemelryck, 2021[61]; Alvarez et al., 2021[62]). Additionally, the COVID-19 crisis highlighted the need to improve co-ordination and interoperability among the labour, health and education systems to enhance countries' overall social protection systems (Cabutto, Nieto-Parra and Vázquez-Zamora, 2021[63]; IPCC, 2022[64]). It is important to build on the advances in sectoral co-ordination that were consolidated during the pandemic, since they positively influenced the unequal distribution of the social determinants of health, such as poverty and unemployment (IPCC, 2022[64]).

In addition, it is important to take into account a comprehensive view of households when designing social protection systems. For instance, households with only informal workers face different vulnerabilities or in a different magnitude than mixed households, or those with household members working in the formal economy. These differences present an opportunity to design differentiated public policies that address specific needs to effectively mitigate the vulnerabilities and negative consequences of informality on individuals' and households' well-being.

Finally, it is essential that social protection systems have funding sources that will ensure their financial sustainability. This can be achieved, in part, by increasing tax revenues through the reduction of generalised subsidies and tax exemptions. For example, energy subsidies have negative externalities (Chapter 4). Among other measures to ensure their financial sustainability are: reducing tax evasion (ECLAC, 2017[65]); creating reserve funds, insurance and catastrophe bonds (Chapter 4); and developing regional risk-sharing mechanisms. The social protection system becomes more resilient and responsive, contributing to the fight against climate change and supporting a just transition to net zero emissions societies (Stampini et al., 2021[60]). This highlights the importance of comprehensive social protection policies that stimulate return on investment, a lever of structural change. It must be channelled in the right direction through public policy instruments – including taxation, financial policy, technology policy and regulatory policy – to increase relative returns for the benefit of the sectors that drive recovery (ECLAC, 2022[66]). However, a fiscal compact for equality and sustainability first requires a social compact to make this possible. The new social contract needs far-reaching political agreement and consensus and a different balance among the state, the market, society and the environment (OECD et al., 2021[10]; ECLAC, 2022[67]).

Innovative tax mechanisms could help increase revenues, promote formalisation and reduce inequalities

Innovative income support policies could be worth exploring to increase progressivity and formalisation in tax systems, although further research is needed and so far it has only been used in advanced economies. In addition to promoting further progressivity in personal income taxes, possible useful policies in LAC include negative income tax (NIT), earned income tax credit (EITC) or personalised value-added tax (P VAT). NIT or EITC programmes generate fewer distortions or disincentives to formalisation of employment than traditional non-contributory welfare programmes (Pessino et al., 2021[68]). They also avoid generating more fiscal pressure while addressing the equity impact of indirect taxes and reducing poverty, in much the same way as cash transfer programmes. NIT guarantees the revenue of a traditional cash transfer – independent of whether a person is unemployed, employed in informality or employed in the formal sector. As such, when a person secures a formal job, income tax benefits are not eliminated (as in traditional welfare programmes); rather, the worker's net income increases and, as they earn a higher wage, the refundable tax credit gradually reduces to the point they start paying. That the person continues to receive government aid along with a formal salary ensures that formal job wages are higher, more attractive and more affordable. It should be noted that NIT is targeted to a specific population, not all individuals (Pessino et al., 2021[68]).There is evidence that EITC programmes have many positive effects, including increased labour force participation and poverty reduction, as they reward employment. The difference between EITC and NIT is that, for low-income wages, transfers increase proportionately to salary; once the salary is high enough, they gradually reduce to zero. Evidence has also shown that EITC helped to reduce informality in the United States, as it can be complemented by other measures, such as refundable tax credits (as long as a person is formally employed) or demand-side subsidies to firms to incentivise formal employment demand (Pessino et al., 2021[68]). In general, the proposed initiatives have so far been developed and evaluated only in developed countries. Before moving forward with such measures, further research and policies that establish the building blocks are needed, for instance the improvement of the registration/identification of informal workers.

P VAT might be an option to explore to increase fiscal resources while addressing the equity impact of indirect taxes and informality. This strategy consists in applying the VAT to all products and services at a standardised rate and implementing a tax refund based on the incidence of VAT on consumption among the poorest deciles. Using these targeted instruments also addresses the high degree of informality, a feature of most developing countries. As informality makes individuals in the lower deciles invisible, this population rarely benefits from the provision of transfers and public social services. The P VAT is useful in overcoming this situation by including individuals in the informal sector (Barreix et al., 2022[69]).

Higher inflation rates hit low-income households the hardest

One of the most concerning effects of the Russian invasion of Ukraine on LAC economies is the rise in global commodity prices. Inflation rates are putting pressure on households' real incomes and savings, and middle- and low-income households are the most vulnerable (Gill and Nagle, 2022[70]). It is estimated that a 1% rise in inflation increases the share of low-income households by around 7%, while it reduces the share of high-income households by only 1% (Nuguer and Powell, 2020[71]). Consequently, inflation could also exacerbate inequality, aggravating existing social tensions in one of the most unequal regions in the world (Jaramillo and O'Brien, 2022[72]).

There are three main channels through which inflation affects the poorest particularly hard. The first channel is the unprotected nature of their assets, which undermines their purchasing power during shocks. This results from poor access to financial products and financial assets other than cash, such as interest-bearing accounts, that could protect their assets against inflation. The second channel is their income composition. In the formal sector, it depends on wages, pensions and social benefits, which are mostly rigid in the short term and respond more slowly to price volatility than the non-wage income of richer households (Ha, Kose and Ohnsorge, 2019[73]). As 45.3% of the LAC population lives in a household that relies exclusively on informal employment, their incomes are often not indexed and are not protected by social protection systems (Nuguer and Powell, 2020[71]). The third channel is the composition of each household's consumption basket relative to patterns of price spikes. In emerging economies, 50% of lower-income people's spending is on food vs. 20% for the rich (Gill and Nagle, 2022[70]). These channels explain why the most vulnerable are suffering the most from current food prices, which are at historic highs, as is the FAO's Food Price Index. The situation in the agricultural sector is pushing up prices further, as agricultural production is exposed to rising input costs, particularly from fertilisers and fuels (FAO, 2022[74]).

Recognising the heterogeneity in LAC, it is noteworthy that about one-fifth of poor households in developing countries are net food producers; thus, this segment might be positively affected by higher food prices (Gill and Nagle, 2022[70]). However, the recent increase in fertiliser prices has been so steep that it has outpaced the gains from higher food prices, even for net food-producing households (FAO, 2022[75]).

Inflation is affecting the extreme poor on a larger scale than for the national representative household in LAC countries and is pushing millions back into these conditions. There is evidence that the extreme poor face an average price increase 3.55 percentage points higher than the total households in selected LAC countries (Figure 1.11). In Argentina, although year-on-year Consumer Price Index (CPI) growth was higher for the general population in the first four months of 2022, in April, year-on-year inflation was 1.41 percentage points higher for the extreme poor. Prices are driving more people below the poverty line, pushing millions into food and energy insecurity. It is estimated that in LAC, if prices rise by 2 percentage points above a baseline scenario, extreme poverty would increase by 1.1 percentage points compared to 2021 levels, which translates into 7.8 million people joining the 86.4 million who are food insecure. This adds to the reversed trend of poverty reduction that started in 2020 compared to 2019, with poverty and extreme poverty increasing by 2.5 and 1.7 percentage points, respectively (ECLAC, 2022[1]).

Figure 1.11. **Year-on-year inflation for the representative consumption basket vs. inflation for the extreme poverty basket, year-to-date average 2022, selected LAC countries**

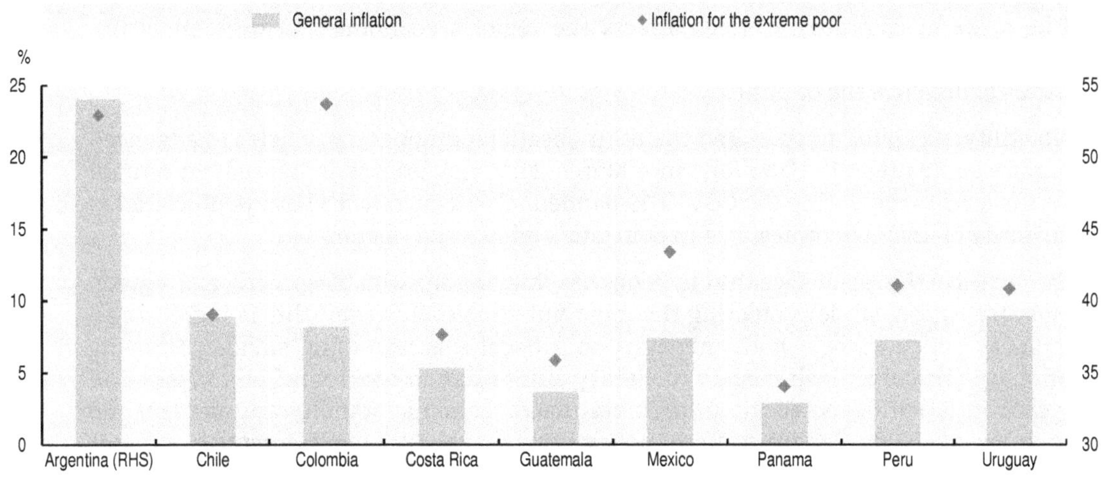

Notes: Year-to-date average of year-over-year growth of national CPIs vs. growth of extreme poverty lines 2022. Extreme poverty lines are based on the cost of a basic food basket that covers basic food needs and provides the minimum caloric requirement of the members of a reference household. The Chilean extreme poverty line also includes a share of non-food basic goods and services. For Colombia and Peru, the food and non-alcoholic beverages division of their CPI was used. For Panama, the data covers the districts of Panama and San Miguelito. Argentina is plotted on the right hand side (RHS) axis.
Source: Local sources, OECD construction based on data from national statistic offices on CPIs and poverty lines.

StatLink https://stat.link/ywso1e

Policy action is needed to reduce the negative effects of inflation rates on the most vulnerable. In the short run, non-contributory social protection policies, such as cash transfers, school feeding programmes, and food and in-kind transfers, could mitigate negative impacts on poor households, as they did for millions of Latin Americans during the pandemic (Jaramillo and O'Brien, 2022[72]). In the long term, governments should put in place reforms to tackle the structural channels through which poor households' assets are more vulnerable to inflation. Promoting access to financial products and increasing labour formalisation would protect the value of poor households' assets and shield them through social protection systems (Ha, Kose and Ohnsorge, 2019[73]). In addition, a more granular measurement of the asymmetric impacts of inflation on various income groups would better guide the design of social protection policies (Gill and Nagle, 2022[70]).

Food, energy and fertiliser security must be addressed. In the short term, governments should keep markets open, avoid trade restrictions and use instruments such reducing value-added taxes on basic consumption baskets. In the long term, greater regional trade integration could generate positive effects on food security, and regional co-ordination in fertiliser production could help achieve a long-term goal of reducing dependence on fossil or mineral fertilisers (ECLAC, 2022[1]). Renewable energy mandates have already generated long-term benefits and have the potential to mitigate the regressive effects of fossil energy price peaks (World Bank Group, 2022[76]). Building a favourable regional ecosystem for green transformation could enable countries to develop more inclusive green energy matrices (ECLAC, 2022[1]).

Key policy messages

Overall, the LAC region is expected to undergo a strong economic slowdown in 2022 due to remaining structural challenges, such as low potential growth and a complicated international context. At the international level, Russia's war against Ukraine and China's economic slowdown are making the already challenging scenario of low potential growth and high social vulnerabilities in the LAC region even more challenging. In particular, main indirect channels of Russia's war against Ukraine relate to lower global GDP growth, disruptions in trade, and higher volatility in both commodity prices and capital markets.

The spike in commodity prices affects the region's economies in different ways, with net exporters in South America benefiting while Central American and Caribbean countries experience the opposite.

Volatility in capital markets and the normalisation of monetary policies in advanced economies could affect LAC. So far, since March 2022, risk premia increased but remain below the levels seen during the COVID-19 pandemic, and currencies have depreciated in the majority of LAC economies, due to domestic and external factors.

Countries in the region face the challenge of achieving sustainable growth, and a green and just transition, while protecting the most vulnerable. However, this is increasingly difficult, as policy space – both monetary and fiscal – is becoming smaller. Growing inflationary pressures have pushed monetary authorities to control inflation rates and to anchor inflation expectations to promote macroeconomic stability and prevent the negative social effects. Regarding fiscal policy, greater spending requirements to support economic recovery face smaller fiscal capacity as a result of higher debt burdens after the COVID-19 pandemic and higher debt service costs. Going forward, climate change and the green transition can weigh heavily on fiscal accounts. Therefore, the region will need to mobilise resources to compensate shortfalls and invest more, better and greener to reduce the adverse effects of climate change and finance the green transition.

Although overall poverty levels declined in 2021 compared to 2020, the economic slowdown of 2022 and rising inflation have erased the gains. As inflation most affects the vulnerable, and there is a declining trend in the number and coverage of social protection programmes implemented to address the COVID-19 crises, it is critical that countries strengthen universal, comprehensive, sustainable and resilient social protection systems.

Box 1.2. Key policy messages

There is an intensification of inflationary pressures that does not appear to be temporary. Central banks across LAC have been increasing policy rates to anchor expectations. In some LAC countries, further increases in interest rates will be necessary to promote macroeconomic stability and limit the negative impact of price increases on the most vulnerable populations.

Unchecked inflation generates poverty traps and takes a particular toll on the most vulnerable, who see their purchasing power quickly erode. With macroeconomic policy space being reduced, most LAC countries will face the multi-dimensional challenge of balancing recovery stimulus, financing the green transition, and protecting the most vulnerable, particularly from the impact of inflationary pressures. The composition of the fiscal consolidation that some economies will undertake plays a key role in the strength and inclusiveness of the recovery.

Box 1.2. **Key policy messages** *(cont.)*

Fiscal policy will continue to be at the core of the recovery. To be effective, it must take into account the current complex context through well-defined sequencing of actions and be backed by a broad consensus built through national dialogue and clear communication.

A set of tax policy options are available in LAC that could help increase revenues without compromising the recovery. These include measures to reduce tax evasion and avoidance, increase the progressivity of personal income taxes and policies to improve tax compliance, strengthen tax administration and eliminate inefficient tax expenditures.

A green and just transition will require mobilising further revenues to increase investment that aims to reduce climate-related risks (physical and transition risks), and helps to advance a more sustainable and inclusive model of production and consumption that creates new quality green jobs.

Social challenges from the COVID-19 pandemic remain. Households living in poverty and extreme poverty, but also non-poor, low-income and lower-middle-income households, face persisting high inequality and high levels of vulnerability. Inflation exacerbates this, as increases in food, energy and fertiliser prices affect the most vulnerable through various structural channels, such as the nature of their assets and the composition of their income and consumption basket.

To counteract the regressive effects of inflation in LAC, governments should complement monetary measures with fiscal policies, such as social protection policies. These could protect poor households, as they did for millions of Latin Americans during COVID-19. Longer-term policies to protect the value of poor households' assets include promoting access to financial products and increasing labour formalisation, which would shield them through social protection systems.

Strengthening universal, comprehensive, resilient and sustainable social protection systems will be a key determinant in containing the social crisis.

In response to the episode of rising inflation and to ensure food and energy security, the region has the opportunity to strengthen domestic trade and to foster an integrated regional ecosystem to boost fertiliser security and inclusive energy matrices.

References

Abramo, L. (2022), *Policies to address the challenges of existing and new forms of informality in Latin America*, United Nations Economic Commission for Latin America and the Caribbean, Santiago, https://repositorio.cepal.org/bitstream/handle/11362/47774/1/S2100648_en.pdf. [47]

Alejos, L. (2021), *¿Cuáles son los riesgos fiscales de los eventos climáticos extremos y cómo enfrentarlos?*, Inter-American Development Bank, Washington, DC, https://blogs.iadb.org/gestion-fiscal/es/cuales-son-los-riesgos-fiscales-de-los-eventos-climaticos-extremos-y-como-enfrentarlos/. [27]

Alvarez et al., (2021), *Pension and healthcare systems in Latin America: Challenges posed by aging, technological change, and informality*, http://scioteca.caf.com/handle/123456789/1730. [62]

Amaya, L. (2022), *Drought threatens fertile fields of South America's Southern Cone*, Swiss Broadcasting Corporation, Bern, https://www.swissinfo.ch/spa/d%C3%ADa-agua-cono-sur--panor%C3%A1mica- la-sequ%C3%ADa-amenaza-los-f%C3%A9rtiles-campos-del-cono-sur-de-am%C3%A9rica/47453458. [24]

Ardanaz, M. et al. (2021), *The Output Effects of Fiscal Consolidations: Does Spending Composition Matter?*, Inter-American Development Bank, Washington, DC, https://publications.iadb.org/en/output-effects-fiscal-consolidations-does-spending-composition-matter. [20]

Arreaza et al. (2022), *Reglas fiscales para la recuperación en América Latina: experiencias y principales lecciones*, http://cafscioteca.azurewebsites.net/handle/123456789/1905. [19]

Banerjee, O. et al. (2021), *Climate Change Impacts on Agriculture in Latin America and the Caribbean: An Application of the Integrated Economic-Environmental Modeling (IEEM) Platform*, Inter-American Development Bank, Washington, DC, https://publications.iadb.org/publications/english/document/Climate-Change-Impacts-on-Agriculture-in-Latin-America-and-the-Caribbean-An-Application-of-the-Integrated-Economic-Environmental-Modeling-IEEM-Platform.pdf. [25]

Bárcena, A. et al. (2020), *La emergencia del cambio climático en América Latina y el Caribe: Seguimos esperando la catástrofe o pasamos a la acción?*, United Nations Economic Commission for Latin America and the Caribbean, Santiago, http://www.cepal.org/sites/default/files/events/files/19-00711_lbc_160_emergencia-cambio-climatico_web.pdf. [26]

Barreix, A. et al. (2022), *Revisiting Personalized VAT: A Tool for Fiscal Consolidation with Equity*, Inter-American Development Bank, Washington, DC, https://doi.org/10.18235/0004147. [69]

Bermúdez, A. (2020), *Stranded power? Mapping Latin America's thermoelectric energy*, China Dialogue, London, https://chinadialogue.net/en/energy/stranded-power-mapping-latin-americas-thermoelectric-energy/. [35]

Berner, H. and T. Van Hemelryck (2021), *Social information systems and registries of recipients of non-contributory social protection in Latin America in response to COVID-19*, United Nations Economic Commission for Latin America and the Caribbean, Santiago, https://repositorio.cepal.org/bitstream/handle/11362/46868/1/S2100029_en.pdf. [61]

Binsted et al. (2020), "Stranded asset implications of the Paris Agreement in Latin America and the Caribbean", *Environmental Research Letters*, Vol. 15, No.4, https://iopscience.iop.org/article/10.1088/1748-9326/ab506d/pdf. [36]

Brooks, A., J. Jambeck and E. Mozo-Reyes (2020), "Plastic Waste Management and Leakage in Latin America and the Caribbean", *Technical Note*, No. 02058, Inter-American Development Bank, Washington, DC, https://doi.org/10.18235/0002873. [55]

Cabutto, C., S. Nieto-Parra and J. Vázquez-Zamora (2021), *A post-pandemic social contract for Latin America: the why, the what, the how*, VOX.Lacea, Latin American and Caribbean Economic Association, Bogotá, http://vox.lacea.org/?q=blog/social_contract_latam. [63]

Celani, A., L. Dressler and M. Wermelinger (2022), *Building an Investment Tax Incentives database: Methodology and initial findings for 36 developing countries*, https://doi.org/10.1787/62e075a9-en. [44]

Chen et al. (2021), "Engaging central banks in climate change? The mix of monetary and climate policy.", *Energy Economics*, Vol. 103, http://www.sciencedirect.com/science/article/pii/S0140988321004096. [13]

D'Arcangelo, F. et al. (2022), "A framework to decarbonise the economy", *OECD Economic Policy Papers*, No. 31, OECD Publishing, Paris, https://doi.org/10.1787/4e4d973d-en. [21]

ECB (2021), *Climate change and the ECB*, European Central Bank, Frankfurt, Germany, http://www.ecb.europa.eu/ecb/climate/html/index.en.html. [14]

ECLAC (2022), *Building a New Future: Transformative Recovery with Equality and Sustainability*, United Nations Economic Commission for Latin America and the Caribbean, Santiago, http://www.cepal.org/en/publications/46226-building-new-future-transformative-recovery-equality-and-sustainability-summary. [67]

ECLAC (2022), *CEPALSTAT*, United Nations Economic Commission for Latin America and the Caribbean, Santiago, https://statistics.cepal.org/portal/databank/index.html?lang=es&indicator_id=2207&area_id=. [18]

ECLAC (2022), *Fiscal Panorama of Latin America and the Caribbean 2022: Fiscal policy challenges for sustainable and inclusive development*, United Nations Economic Commission for Latin America and the Caribbean, Santiago, http://www.cepal.org/es/publicaciones/47920-panorama-fiscal-america-latina-caribe-2022-desafios-la-politica-fiscal-un. [15]

ECLAC (2022), *How to Finance Sustainable Development: Recovery from the Effects of COVID-19 in Latin America and the Caribbean*, United Nations Economic Commission for Latin America and the Caribbean, Santiago, https://repositorio.cepal.org/bitstream/handle/11362/47721/1/S2100877_en.pdf. [66]

ECLAC (2022), *Repercussions in Latin America and the Caribbean of the war in Ukraine: how should the region face this new crisis?*, United Nations Economic Commission for Latin America and the Caribbean, Santiago, http://www.cepal.org/en/publications/47913-repercussions-latin-america-and-caribbean-war-ukraine-how-should-region-face-new. [1]

ECLAC (2022), *Social Panorama of Latin America 2021*, United Nations Economic Commission for Latin America and the Caribbean, Santiago, http://www.cepal.org/en/publications/47719-social-panorama-latin-america-2021. [2]

ECLAC (2021), *Disasters and inequality in a protracted crisis: Towards universal, comprehensive, resilient and sustainable social protection systems in Latin America and the Caribbean*, United Nations Economic Commission for Latin America and the Caribbean, Santiago, http://www.cepal.org/en/publications/47376-disasters-and-inequality-protracted-crisis-towards-universal-comprehensive. [59]

ECLAC (2021), *Fiscal Panorama of Latin America and the Caribbean 2021: Fiscal policy challenges for transformative recovery post-COVID-19*, United Nations Economic Commission for Latin America and the Caribbean, Santiago, http://www.cepal.org/en/publications/46809-fiscal-panorama-latin-america-and-caribbean-2021-fiscal-policy-challenges. [42]

ECLAC (2021), *Social Panorama of Latin America 2020*, United Nations Economic Commission for Latin America and the Caribbean, Santiago, http://www.cepal.org/en/publications/46688-social-panorama-latin-america-2020. [45]

ECLAC (2017), *Linkages between the social and production spheres: Gaps, pillars and challenges*, United Nations Economic Commission for Latin America and the Caribbean, Santiago, http://www.cepal.org/en/publications/42269-linkages-between-social-and-production-spheres-gaps-pillars-and-challenges. [65]

Espejo, A. (2022), *Informalidad laboral en América Latina: propuesta metodológica para su identificación a nivel subnacional*, United Nations Economic Commission for Latin America and the Caribbean, Santiago, https://repositorio.cepal.org/bitstream/handle/11362/47726/4/S2100889_es.pdf. [48]

FAO (2022), *Food Outlook – Biannual Report on Global Food Markets*, Food and Agriculture Organization of the United Nations, Rome, https://doi.org/10.4060/cb9427en. [75]

FAO (2022), *New UN Food Outlook report: World's most vulnerable are paying more for less food*, Food and Agriculture Organization of the United Nations, Rome, http://www.fao.org/newsroom/detail/new-un-food-outlook-report-world-s-most-vulnerable-are-paying-more-for-less-food/en. [74]

Galindo, L., B. Hoffman and A. Vogt-Schilb (2022), *How Much Will It Cost to Achieve the Climate Goals in Latin America and the Caribbean?*, Inter-American Development Bank, Washington, DC, https://publications.iadb.org/publications/english/document/How-Much-Will-It-Cost-to-Achieve-the-Climate-Goals-in-Latin-America-and-the-Caribbean.pdf. [38]

Gill, I. and P. Nagle (2022), *Inflation could wreak havoc on the world's poor*, World Bank Group, Washington, DC, https://blogs.worldbank.org/voices/inflation-could-wreak-havoc-worlds-poor. [70]

GIZ (2014), *Conditional cash transfers to protect community tropical forests in the Amazon region*, Deutsche Gesellschaft für Internationale Zusammenarbeit, Bonn, Germany, http://www.giz.de/en/worldwide/13301.html. [58]

Ha, J., M. Kose and F. Ohnsorge (2019), *Inflation in Emerging and Developing Economies: Evolution, Drivers, and Policies*, World Bank Group, Washington, DC, https://openknowledge.worldbank.org/handle/10986/30657. [73]

IDB (2022), *From Recovery to Renaissance: Turning Crisis into Opportunity*, Inter-American Development Bank, Washington, DC, https://flagships.iadb.org/en/MacroReport2022/From-Recovery-to-Renaissance-Turning-Crisis-into-Opportunity. [16]

IDB (2021), *Fiscal Policy and Climate Change: Recent Experiences of Finance Ministries in Latin America and the Caribbean*, Inter-American Development Bank, Washington, DC, https://publications.iadb.org/publications/english/document/Fiscal-Policy-and-Climate-Change-Recent-Experiences-of-Finance-Ministries-in-Latin-America-and-the-Caribbean.pdf. [29]

IEA (2017), *Oil 2017: Analysis and Forecasts to 2022*, International Energy Agency, Paris, http://www.iea.org/reports/oil-2017. [34]

ILO (2018), *World Employment and Social Outlook 2018: Greening with jobs*, International Labour Office, Geneva, https://www.ilo.org/wcmsp5/groups/public/---dgreports/---dcomm/---publ/documents/publication/wcms_628654.pdf. [50]

IMF (2022), *How Soaring Shipping Costs Raise Prices Around the World*, International Monetary Fund, Washington, DC, https://blogs.imf.org/2022/03/28/how-soaring-shipping-costs-raise-prices-around-the-world/. [11]

IMF (2022), *World Economic Outlook, April 2022: Wars Set Back the Global Economy*, International Monetary Fund, Washington, DC, http://www.imf.org/en/Publications/WEO/Issues/2022/04/19/world-economic-outlook-april-2022. [5]

IPCC (2022), *IPCC Sixth Assessment Report: Impacts, Adaptation and Vulnerability*, Intergovernmental Panel on Climate Change, Geneva, Switzerland, http://www.ipcc.ch/report/ar6/wg2/. [64]

ITF (forthcoming), *Developing Accessibility Indicators for Latin American Cities*, International Transport Forum, Paris. [53]

Jaramillo, C. and R. O'Brien (2022), *Inflation, a rising threat to the poor and vulnerable in Latin America and the Caribbean*, World Bank Group, Washington, DC, https://blogs.worldbank.org/latinamerica/inflation-rising-threat-poor-and-vulnerable-latin-america-and-caribbean. [72]

Nieto-Parra, S., R. Orozco and S. Mora (2021), *Fiscal policy to drive the recovery in Latin America: the "when" and "how" are key*, VOX.Lacea, Latin American and Caribbean Economic Association, Bogotá, http://lacea.org/vox/?q=blog/fiscal_policy_latam. [41]

Nuguer, V. and A. Powell (2020), *Inclusion in Times of Covid-19*, Inter-American Development Bank, Washington, DC, https://doi.org/10.18235/0002529. [71]

OCHA (2020), *Desastres naturales en América Latina y el Caribe: 2000-2019*, United Nations Office for the Coordination of Humanitarian Affairs, New York, https://reliefweb.int/sites/reliefweb.int/files/resources/OCHA-DESASTRES_NATURALES_ESP%20%281%29.pdf. [23]

OCHA (2020), *Natural Disasters in Latin America and the Caribbean*, https://www.humanitarianresponse.info/en/operations/latin-america-and-caribbean/document/latin-america-and-caribbean-natural-disasaters-2000. [22]

OECD (2022), *OECD Economic Outlook, Interim Report September 2022: Paying the Price of War*, OECD Publishing, Paris, https://doi.org/10.1787/ae8c39ec-en. [4]

OECD (2022), *OECD Economic Outlook, Volume 2022 Issue 1*, OECD Publishing, Paris, https://doi.org/10.1787/62d0ca31-en. [3]

OECD (2022), *OECD Economic Surveys: Chile 2022*, OECD Publishing, Paris, https://doi.org/10.1787/311ec37e-en. [12]

OECD (2021), *Policies for improving the sustainable development impacts of investment*, Consultation Paper March 2021, OECD Publishing, Paris, http://w.oecd.org/daf/inv/investment-policy/FDI-Qualities-Policy-Toolkit-Consultation-Paper-2021.pdf. [9]

OECD (2021), *Transport Strategies for Net-Zero Systems by Design*, OECD Publishing, Paris, https://doi.org/10.1787/0a20f779-en. [54]

OECD (2020), "Nature-based solutions for adapting to water-related climate risks", *OECD Environment Policy Papers*, No. 21, OECD Publishing, Paris, https://doi.org/10.1787/2257873d-en. [39]

OECD (2019), *Can Social Protection Be an Engine for Inclusive Growth?*, Development Centre Studies, OECD Publishing, Paris, https://doi.org/10.1787/9d95b5d0-en. [56]

OECD (2019), *FDI Qualities Indicators: Measuring the sustainable development impacts of investment*, OECD Publishing, Paris, https://www.oecd.org/investment/fdi-qualities-indicators.htm. [8]

OECD (forthcoming), *Building financial resilience to climate impacts: A framework for government action to manage the risks of losses and damages*, OECD Publishing, Paris. [40]

OECD (forthcoming), *Portraying informality in Latin America*, OECD Publishing, Paris. [46]

OECD et al. (2022), *Revenue Statistics in Latin America and the Caribbean 2022*, OECD Publishing, Paris, https://doi.org/10.1787/58a2dc35-en-es. [17]

OECD et al. (2021), *Latin American Economic Outlook 2021: Working Together for a Better Recovery*, OECD Publishing, Paris, https://doi.org/10.1787/5fedabe5-en. [10]

OECD/ILO (2019), *Tackling Vulnerability in the Informal Economy*, Development Centre Studies, OECD Publishing, Paris, https://doi.org/10.1787/939b7bcd-en. [49]

OECD/The World Bank (2020), *Health at a Glance: Latin America and the Caribbean 2020*, OECD Publishing, Paris, https://doi.org/10.1787/6089164f-en. [52]

OECD/The World Bank (2019), *Fiscal Resilience to Natural Disasters: Lessons from Country Experiences*, OECD Publishing, Paris, https://doi.org/10.1787/27a4198a-en. [28]

Parra, F. (2021), *The complex energy transition of Chile's "sacrifice zones"*, China Dialogue, London, https://dialogochino.net/en/climate-energy/42193-the-complex-energy-transition-of-chiles-sacrifice-zones/. [37]

Pessino, C. et al. (2021), *Now it is the time to foster labor formalization in Latin America and the Caribbean*, Inter-American Development Bank, Washington, DC, https://blogs.iadb.org/gestion-fiscal/en/now-it-is-the-time-to-foster-labor-formalization-in-latin-america-and-the-caribbean/. [68]

Porras, I. and N. Asquith (2018), *Ecosystems, poverty alleviation and conditional transfers*, International Institute for Environment and Development, London, https://pubs.iied.org/sites/default/files/pdfs/migrate/16639IIED.pdf. [57]

Ruhl, C. and T. Erker (2021), *Oil intensity: The curiously steady decline of oil in GDP*, School of International and Public Affairs, Columbia University, New York, http://www.energypolicy.columbia.edu/sites/default/files/file-uploads/LongTermOilIntensity_CGEP_Report_083121.pdf. [6]

Solano-Rodríguez, B. et al. (2019), *Implications of climate targets on oil production and fiscal revenues in Latin America and the Caribbean*, Inter-American Development Bank, Washington, DC, https://doi.org/10.18235/0001802. [30]

Stampini, M. et al. (2021), *Adaptive, but not by design: cash transfers in Latin America and the Caribbean before, during and after the COVID-19 Pandemic*, Inter-American Development Bank, Washington, DC, https://doi.org/10.18235/0003795. [60]

Titelman, D. (2022), *Política fiscal para el desarrollo sostenible: cierre de brechas estructurales y adaptación al cambio climático*. [43]

Titelman, D. et al. (2022), *Fiscal Impact Estimates of a Net-Zero Emissions Transit for Major Hydrocarbon Producers in Latin America and the Caribbean*, Task Force on Climate, Development and the International Monetary Fund, http://www.bu.edu/gdp/files/2022/04/TF-WP-004-FIN.pdf. [33]

UNCTAD (2022), *World Investment Report* 2022, United Nations, New York, https://unctad.org/webflyer/world-investment-report-2022. [7]

Vogt-Schilb, A., G. Reyes-Tagle and G. Edwards (2021), *Are Latin America's fossil fuels at risk of becoming stranded assets this decade?*, https://blogs.iadb.org/sostenibilidad/en/are-latin-americas-fossil-fuels-at-risk-of-becoming-stranded-assets-this-decade/. [31]

Watson, C. (2016), *Shock-Responsive Social Protection Systems Research - Working Paper 3: Shock-Responsive Social Protection in the Sahel: Community Perspectives*, Oxford Policy Management, Oxford, https://www.opml.co.uk/files/Publications/a0408-shock-responsive-social-protection-systems/wp3-community-perspectives-sahel-en.pdf?noredirect=1. [51]

Welsby, D. et al. (2021), *High and Dry: Stranded Natural Gas Reserves and Fiscal Revenues in Latin America and the Caribbean*, Inter-American Development Bank, Washington, DC, https://publications.iadb.org/en/high-and-dry-stranded-natural-gas-reserves-and-fiscal-revenues-latin-america-and-caribbean. [32]

World Bank Group (2022), *Commodity Markets Outlook: The Impact of the War in Ukraine on Commodity Markets, April 2022*, World Bank Group, Washington, DC, https://openknowledge.worldbank.org/bitstream/handle/10986/37223/CMO-April-2022.pdf. [76]

Chapter 2

Harnessing the potential of the green transition to build a more inclusive development model

A green transition that places citizens' well-being at its centre could help Latin America and the Caribbean (LAC) move towards a more inclusive and sustainable development model. First, this chapter maps out where the region stands regarding environmental indicators. Second, it argues why the green transition is an opportunity for the post-COVID-19 recovery in the region. Third, it puts forward a multi-dimensional, systemic approach to advancing a green and just transition and focuses, in particular, on a territorial sustainable development model and policies needed to accelerate the transition towards sustainable transport and urban systems. Last, the chapter highlights main policy messages.

LAC needs to advance the green transition while fostering sustainable and inclusive development

LAC is one of the most vulnerable regions to the consequences of climate change

The average number of **extreme climate-related weather events** in LAC increased

between 2001 and 2022 compared to the previous two decades

Climate change is projected to push **5 million more people into poverty** by 2030 in LAC

In the Caribbean, the **annual cost of not dealing with the effects of sea level rise** is estimated to reach

USD 22 billion per year by 2050 (10% of GDP)

and USD 46 billion per year by 2100 (22% of GDP)

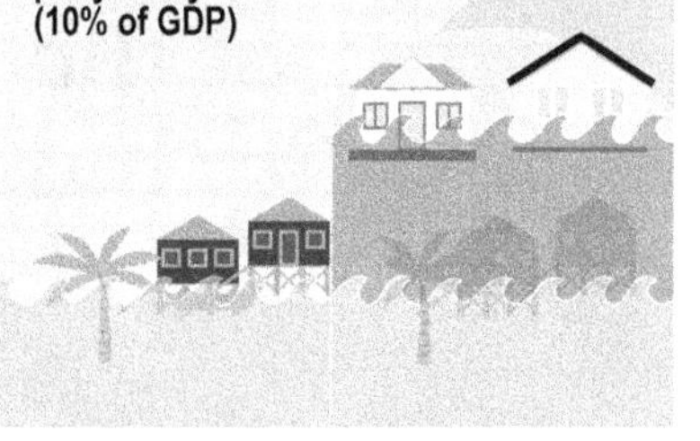

LAC's total greenhouse gas emissions (GHG) have been steadily rising since 1990, reflecting an unsustainable development model in the region

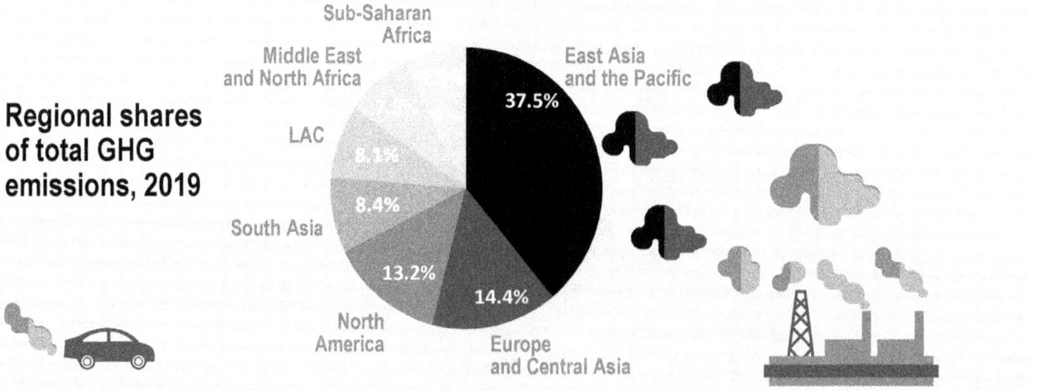

A systemic approach to the green transition can help implement public policies centred on citizens' well-being

Instead of focusing on specific problems, a systemic approach designs systems that produce better social, economic and enviromental results

Systemic transport policies can:

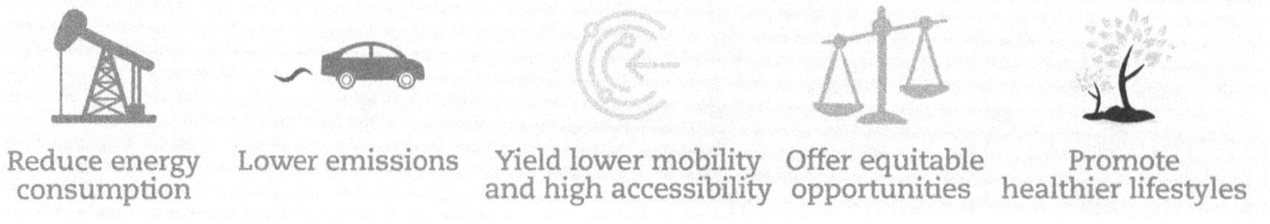

Introduction

Climate change is an urgent concern that demands local and regional actions that support globally agreed targets. At the current rate of global greenhouse gases (GHG) emissions, the Paris Agreement goal of limiting – by 2030 – global warming to a maximum between 1.5°C and 2°C will be difficult to achieve (IPCC, 2018[1]). Despite the LAC region's contribution to global emissions being around 8.1%, the region has proved to be particularly vulnerable to the effects of climate change. The unequal consequences of climate change are being felt at the environmental, economic and social context. Without immediate implementation of polices for mitigation and adaptation,[1] the effects of global warming are expected to intensify in the coming years and to continue affecting disproportionately the most vulnerable countries and their most exposed socio-economic groups.

The recent crisis generated by the coronavirus (COVID-19) pandemic has led to a significant development setback in LAC. Recovery strategies require rethinking health, social and economic strategies and are also an opportunity to address the environmental and climate emergency (OECD et al., 2021[2]). The region is particularly vulnerable to the consequences of climate change and the cost of inaction is high. Hence, it is imperative to minimise risks by increasing resilience, which requires a better understanding of how the region contributes to and is affected by climate change.

LAC is at a critical juncture that poses an opportunity for action. The post-COVID-19 recovery, a complex global scenario and the pre-existing development traps in the region must be seen as a strategic context to make structural changes that help the region move towards a more sustainable, resilient and inclusive development model. A green and just transition represents a unique opportunity to take this leap forward, by focusing on effectively transforming and decarbonising the systems that underpin the economy and society to improve almost every aspect of the lives of Latin American citizens.

We conceive the green transition as a means to foster a more sustainable and just development model in LAC. This model should help close the existing social, economic, institutional and environmental development gaps and avoid generating new ones (OECD et al., 2019[3]). A green transition is not only about fighting climate change. A green and just transition aims to advance a greener and more just model of production and consumption that creates new quality green jobs, generates the conditions for workers to acquire new green and digital skills, and supports firms to adopt more sustainable production schemes, including those in brown sectors, which will be more affected throughout the transition and with a special focus on SMEs. Moreover, a green and just transition should contribute to the eradication of poverty and strengthen social inclusion mechanisms, without concentrating only on compensation schemes.

A green and just transition should adopt a systemic approach that produces better socio-economic and sustainable results and should be co-designed by governments and all members of society, across socio-economic groups, territories, generations and genders. Only a strong consensus on what, why and how the green transition is articulated will allow its proper fulfilment. Given the potentially transformational effect of the green transition, mainstreaming climate mitigation and adaptation policies as cross-cutting issues across levels and agencies of government is key. Failure to integrate the social dimension horizontally and vertically, runs the risk of undermining societal acceptance of environmental policy reforms embedded in the green transition. Integrated approaches would allow social development priorities to be taken fully into account in the transition, helping drastically reduce multi-dimensional inequalities (AFD, 2020[4]).

First, this chapter maps out where the region is standing regarding climate change and environmental degradation. Second, it argues that the green transition should be a priority for LAC and explains how it can respond to structural development challenges. Third, it proposes a systemic approach to guide the policies needed for a green and just transition, giving special relevance to the role of subnational governments. Finally, the chapter provides key policy messages.

A green transition demands a co-ordinated and systemic response to address climate change effects

At current rates, global warming will reach the 1.5°C threshold between 2030 and 2052 (IPCC, 2022[5]). This is a short horizon, particularly for green policies. Even the commitments for the 2°C scenario already require revisions due to the levels of GHG emissions of the four major emitters (the People's Republic of China [hereafter, "China"], India, the United States and the European Union). At major emitters' current rates, the rest of the world will lose its margin of emissions reduction to comply with the global carbon budget of the 2°C threshold (Bárcena et al., 2020[6]). The actions needed to address climate change are challenging from a global perspective. The time frames of mitigation and adaptation policies are longer, given their complexity and transformational effect, and therefore need to be addressed urgently. In this way, international partnerships are crucial to co-ordinate worldwide efforts to achieve a just distribution of emissions reduction (Chapter 6).

Burning of fossil fuels and deforestation are the main causes of climate change. Human activity and its emissions are increasingly causing changes in atmospheric and ocean temperature, precipitation, and wind patterns which have direct consequences for the planet (IPCC, 2021[7]), and for social well-being. The observed negative impacts over the environment make the current development model based on extractive and natural resource-dependent economies unsustainable and calls for a co-ordinated response.

A whole-of-government approach that places environmental issues as a transversal axis of public policy is necessary to mitigate the effects of climate change and develop a more sustainable and inclusive economic model. The post-COVID-19 recovery illustrates the need for such a response. Global carbon dioxide (CO_2) emissions rebounded to their highest level in history in 2021, reaching 36.3 Gt. The increase in global CO_2 emissions of over 2 Gt was the largest in history in absolute terms, more than offsetting the previous year's pandemic-induced decline. Due to the contraction caused by the pandemic, most of the world's economies recorded a reduction in CO_2 emissions of between 5% and 10% in 2020 compared to 2019 (IEA, 2021[8]). The evolution of CO_2 emissions from energy combustion and industrial processes reflects the consequences of promoting a recovery concentrated in setting economies "back on track". The progress gained on GHG emissions trends or clean air during previous years was quickly reversed in 2021 due to a recovery defined by a pre-pandemic economic approach. While productivity indicators might have recovered, it is important to align broader policy efforts towards a sustained transition of emissions reduction. Otherwise, the tendency will continue to be the former carbon-intensive growth path.

Climate change is a global but heterogeneous and asymmetric phenomenon that entails inequalities (Bárcena et al., 2020[6]). It disproportionately affects the most vulnerable, threatening poverty reduction. It is estimated that globally, by 2030, 68 to 135 million people could be pushed into poverty because of climate change (World Bank, 2020[9]).

LAC needs to be part of the global fight against climate change

Given its share in total global GHG emissions compared to countries with similar or higher development levels, LAC is usually associated with better environmental performance. The regional shares of total world GHG emissions are disproportionately distributed, as East Asia and the Pacific had the highest emissions in 2019, contributing 37.5%, while Europe and Central Asia represented 14.4%, North America 13.2%, South Asia 8.4% and the LAC region 8.1%, more than the Middle East and North Africa (MENA) (7.5%) and sub-Saharan Africa (7.3%) (Figure 2.1). LAC's share in total GHG emissions is proportional to its share in total world population (8.4%) and slightly higher than its share in total GDP (6.4%) (Ined, 2022[10]; World Bank, 2022[11]), showing that the region sustains the same carbon-intensive development model as the high emitters. In terms of subregions, the Caribbean contributed 0.4% to total global emissions in 2019, Central America 1.7%, and South America had the biggest share with 6.1%.

Figure 2.1. **Regional shares of total GHG emissions, 2019**

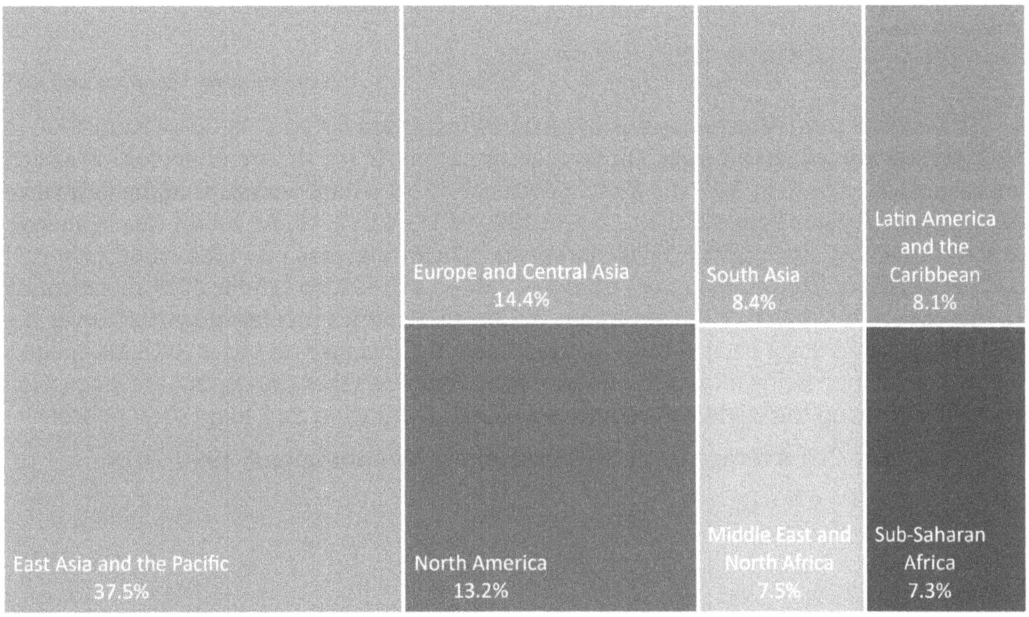

Notes: Emissions including LUCF reported in Gt of CO_2e. Total emissions do not include bunker fuels. The Climate Analysis Indicators Tool (CAIT) was used as the data source. The CAIT dataset is the most comprehensive on Climate Watch and includes all sectors and gases. Climate Watch Historical GHG Emissions data (previously published through CAIT Climate Data Explorer) are derived from several sources. The use of the land use change and forestry (LUCF) or agriculture data is cited as (FAO, 2022[12]). For fuel combustion data, it is cited as (OECD/IEA, 2021[13]).
Sources: (Climate Watch, 2022[14]); (FAO, 2022[12]); (OECD/IEA, 2021[13]).

StatLink https://stat.link/3fpujw

LAC data show a steady increase of total GHG emissions since 1990, reflecting its development model. Between 1990 and 2019, Brazil, Mexico, Argentina, Venezuela and Colombia consistently had higher levels of emissions than other LAC countries (Figure 2.2).[2] This explains the steady increase in South America's emissions, further than the Caribbean and Central America[3] and even OECD countries, which in turn have managed to decrease average total emissions since 2005 (Figure 2.3).[4]

Figure 2.2. **Top 10 countries of total GHG emissions, 1990-2019**

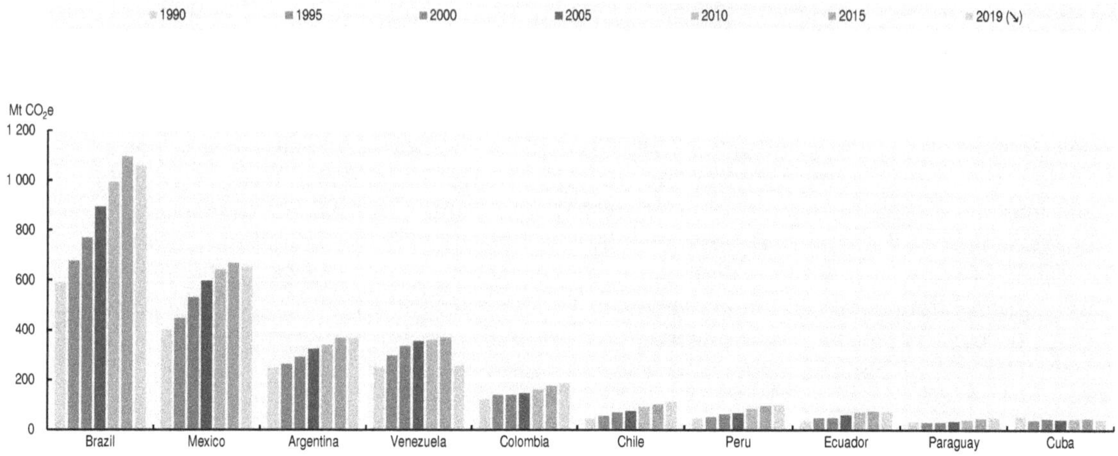

Note: Total GHG emissions excluding LUCF. Top 10 countries are the 10 countries that emit the most in the LAC region when analysing total GHG emissions in 2019.

Sources: (Climate Watch, 2022[14]); (OECD/IEA, 2021[13]); (FAO, 2022[12]).

StatLink https://stat.link/ty5o2i

The level of total GHG emissions in LAC has increased by 1 223 Mt CO_2e from 1990 to 2019, representing a 61% increase. On average, emissions in the three subregions have been increasing with a peak in 2015 and a slight decrease in 2019. Total emissions in the Caribbean subregion increased from 125.8 Mt Co_2e in 1990 to 155.4 in 2019. Although this is a small share of total LAC emissions (5%), it represents a 23.5% increase rate for a small region. If LUCF is considered, the amount rises from 125.1 Mt Co_2e in 1990 to 180.4 in 2019, representing an increase rate of 44.2%. Total emissions in Central America increased by 70.5% over the same time period (Figure 2.3). If LUCF is considered, the increase rate rises to 54.4%. Indeed, the pattern of increasing total GHG emissions in LAC shows that the region is no exception and that reversing the trend will require ambitious mitigation and adaptation policies.

Figure 2.3. **Average total GHG emissions by subregions, 1990-2019**

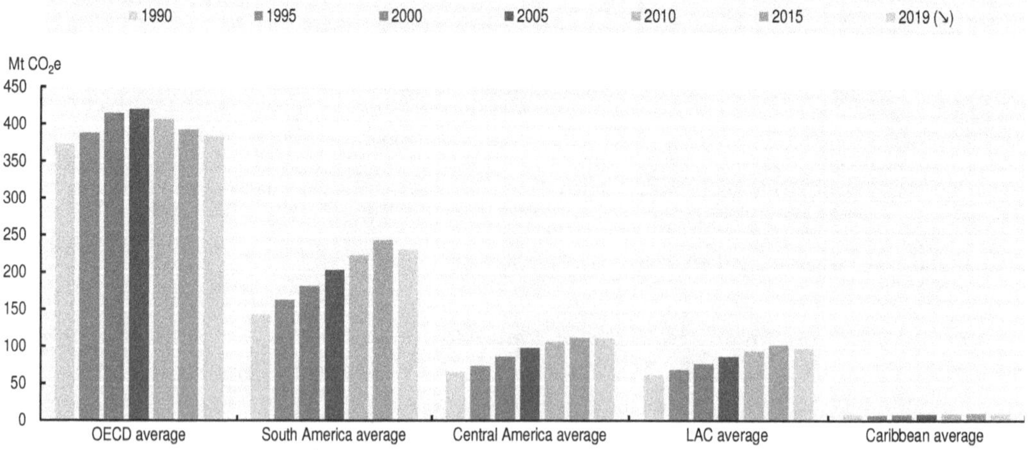

Note: Total GHG emissions excluding LUCF. OECD average is a simple average of the largest set of all OECD member countries, as of May 2022, for which data are available.

Sources: (Climate Watch, 2022[14]); (OECD/IEA, 2021[13]); (FAO, 2022[12]).

In 2019, LAC's average per-capita GHG emissions level equalled the global average (6.3 t CO_2e) and was lower than the OECD average (9.1 t CO_2e). When comparing emissions per capita, the Caribbean countries show the highest levels due to a non-proportional relationship between the size of their population and their high levels of emissions (Figure 2.5).

Grenada, Trinidad and Tobago, and Barbados emitted around 20, 21 and 13 t CO_2e in 2019, respectively. When analysing the top 10 countries with highest total emissions, Venezuela, Argentina and Paraguay show levels of per capita emissions of 9, 8 and 7 t CO_2e, respectively (Figure 2.4). Identifying the origins of these emissions in each country is paramount to help raise awareness of the urgent need to take action in the highest-emitting sectors.

Figure 2.4. **GHG emissions per capita in selected LAC countries**

Top 10 countries of total emissions, 1990-2019

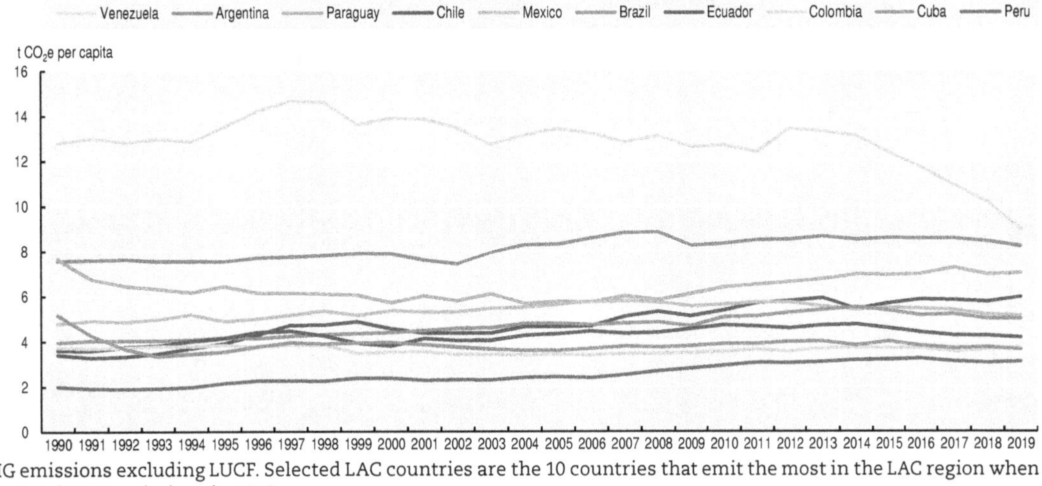

Note: GHG emissions excluding LUCF. Selected LAC countries are the 10 countries that emit the most in the LAC region when analysing total GHG emissions in 2019.

Sources: (Climate Watch, 2022[14]); (OECD/IEA, 2021[13]); (FAO, 2022[12]).

StatLink https://stat.link/qi39sm

Since 1990, GHG emissions per capita in LAC have remained largely constant (between 5.5 and 6.4 t CO_2e without LUCF and between 8.1 and 8.4 including LUCF) while OECD countries have made significant progress in lowering their averages; as such, the gap between the two regions has been closing. The Caribbean subregion has the highest emissions per capita within LAC, and it has managed to slightly decrease them from 8.4 t CO_2e in 1990 to 7.7 t CO_2e in 2019 without LUCF (Figure 2.5)[7] and from 11.5 to 10.7 including LUCF.

Figure 2.5. **Average GHG emissions per capita by subregions, 1990-2019**

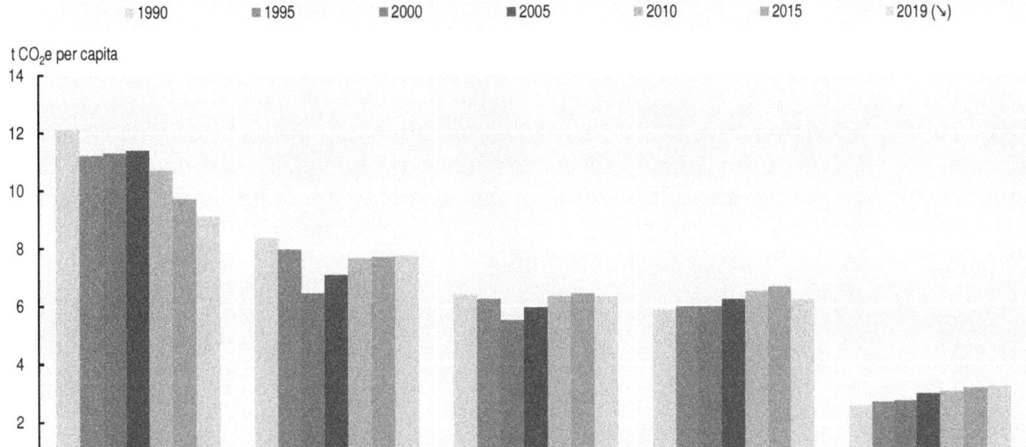

Note: GHG emissions excluding LUCF. OECD average is a simple average of the largest set of all OECD member countries, as of May 2022, for which data are available.

Sources: (Climate Watch, 2022[14]); (OECD/IEA, 2021[13]); (FAO, 2022[12]).

StatLink https://stat.link/df8qaw

Emission levels also evidence a historic, unjust and disproportionate share of responsibility between rich and poor countries and across their various socio-economic groups (Guivarch, Taconet and Méjean, 2021[15]). The double asymmetry explained earlier is illustrated in Figure 2.6. Regionally, North America stands out, with the top 10% of citizens by income emitting 73 t CO_2e per capita. The same quintile shows similar patterns in East Asia and Europe, emitting 30-40 t CO_2e per capita. Even in less-developed regions with lower total emissions, such as Latin America or sub-Saharan Africa, this pattern persists but on a lower scale (Figure 2.6) (Guivarch, Taconet and Méjean, 2021[15]).

Figure 2.6. **GHG emissions by income group and region, 2019**

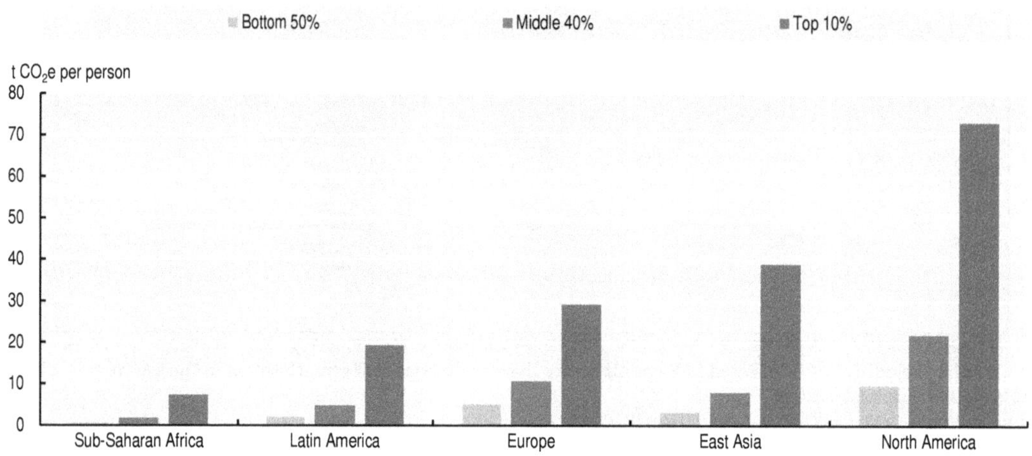

Note: Latin America refers to Argentina, Brazil, Chile, Colombia and Mexico, due to data availability.
Source: (WIR, 2022[16]).

StatLink https://stat.link/sg2k3q

Since 1990, emissions from almost all sectors have grown continuously in LAC, with the greatest increase in the energy sector; 738 Mt CO_2e from 1990 to 2019 (Figure 2.7).[8] In the agricultural sector, GHG emissions increased 100% between 1961 and 2010 (Tubiello et al., 2014[17]), primarily due to the rise in extensive grazing systems in South and Central America. In a smaller, more recent time frame (1990 to 2019) emissions from agriculture increased around 32%. Direct GHG emissions from agriculture are expected to continue rising by 1.1% annually between 2022 and 2031, but the rate of output growth is only around 0.01%, suggesting a persistent carbon intense production (OECD/FAO, 2022[18]). While the industrial and waste sectors produce emissions at a lower scale, their growth rates have been significant (193% and 108%, respectively). To address this constant growth in emissions from almost all sectors, targeted responses are needed in LAC. Programmes and policies should include digitalised and tailored solutions for each sector, from subsistence agriculture to renewable energy and technological intensification for competitiveness and further global integration.

Figure 2.7. **GHG emissions by sector for LAC, 1990-2019**

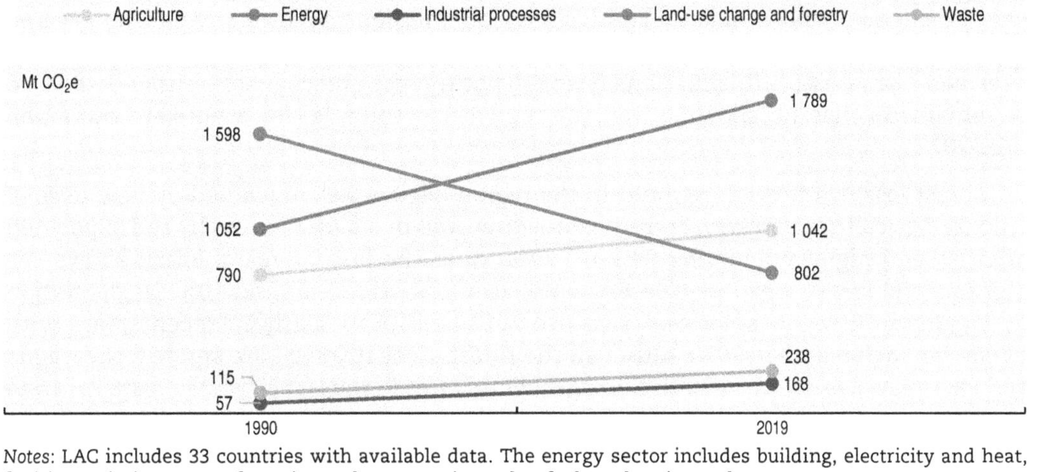

Notes: LAC includes 33 countries with available data. The energy sector includes building, electricity and heat, fugitive emissions, manufacturing and construction, other fuel combustion and transport.

Sources: (Climate Watch, 2022[14]); (FAO, 2022[12]); (OECD/IEA, 2021[13]).

StatLink https://stat.link/h12wto

The structure of emissions in LAC differs from that of the EU and OECD countries. Whereas the energy sector (which includes building, electricity and heat, transport, fugitive emissions, manufacturing and construction, and other fuel combustion) accounts for 83.5% of OECD countries' total emissions and 80% for EU, three sectors represent 88.3% of total emissions in LAC, comprising energy (43.5%), agriculture (25.3%, more than double the OECD) and LUCF (19.5%) (Figure 2.8).[9] Although the energy sector remains the most emission-intensive for all three subregions, each one has its particularities. South America's high emitting sectors are agriculture (28.5%), LUCF (23.8%), and transport (13.4%). The Caribbean differs slightly, with electricity and heat (24.8%) followed by agriculture (15.6%) and LUCF (13.4%, very similar to transport at 11.1%). In Central America, electricity and heat accounts for 23.8% of emissions, while transport accounts for 21.4%, followed by agriculture with 16%. A deeper understanding of the structure of emissions would contribute to a cleaner productive strategy that protects the land and avoids long-term consequences in terms of food security and vulnerability to external shocks (Chapter 3).

Figure 2.8. **LAC's sector shares of total GHG emissions by subregion compared to the European Union and OECD countries, 2019**

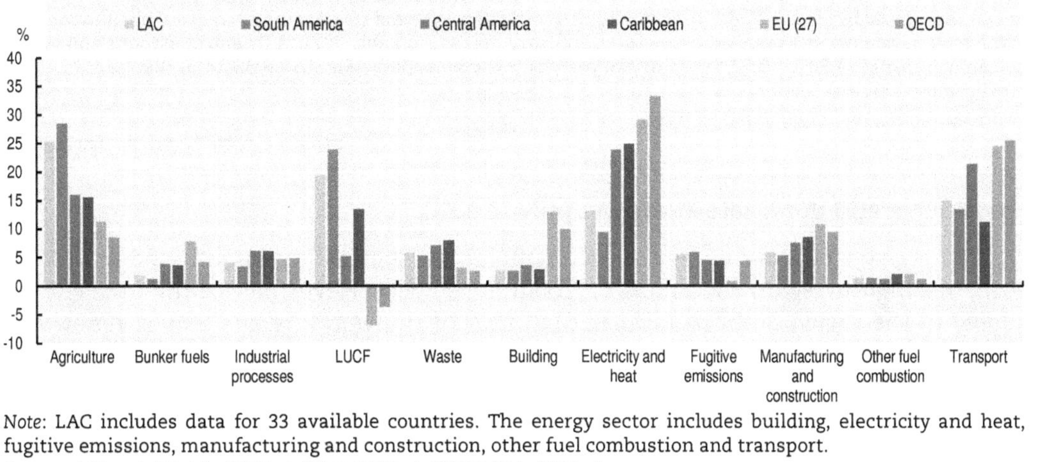

Note: LAC includes data for 33 available countries. The energy sector includes building, electricity and heat, fugitive emissions, manufacturing and construction, other fuel combustion and transport.

Sources: (Climate Watch, 2022[14]); (FAO, 2022[12]); (OECD/IEA, 2021[13]).

StatLink https://stat.link/n5986r

The growth of transport sector emissions contributes the most to the increase of energy-related GHG emissions in LAC (Bárcena et al., 2020[6]), followed by electricity and heat production (where oil is the main source of emissions, followed by natural gas and coal). While figures for all three generators rose between 1990 and 2014, there has since been some progress: GHG emissions from oil, natural gas and coal have either decreased or remained constant, showing that the region has potential for a net-zero transition (IEA, 2021[19]).

Forest loss is a prevailing trend in the region (Figure 2.9), explained by new uses of land for agriculture, forestry, and stockbreeding, and to a lesser extent by the expansion of cities and highway building (ECLAC, 2021[20]). In the last 20 years, Brazil shows the highest total forest area lost (544 690 km²), at a loss rate of around 10%. Deforestation by logging in Brazil has accelerated since 2012, and particularly in recent years with 11 088 km² deforested in 2020. Although the forest areas involved are smaller, Nicaragua and Paraguay also stand out with the highest loss rates in the last 20 years. Costa Rica and Chile stand out as they managed to increase forest cover. Strong government capacity to enforce law in general, and land tenure in particular, can secure property rights and help fight illegal deforestation and unsustainable agricultural and livestock practices.

Figure 2.9. **Change in LAC forest area, 2000-20**

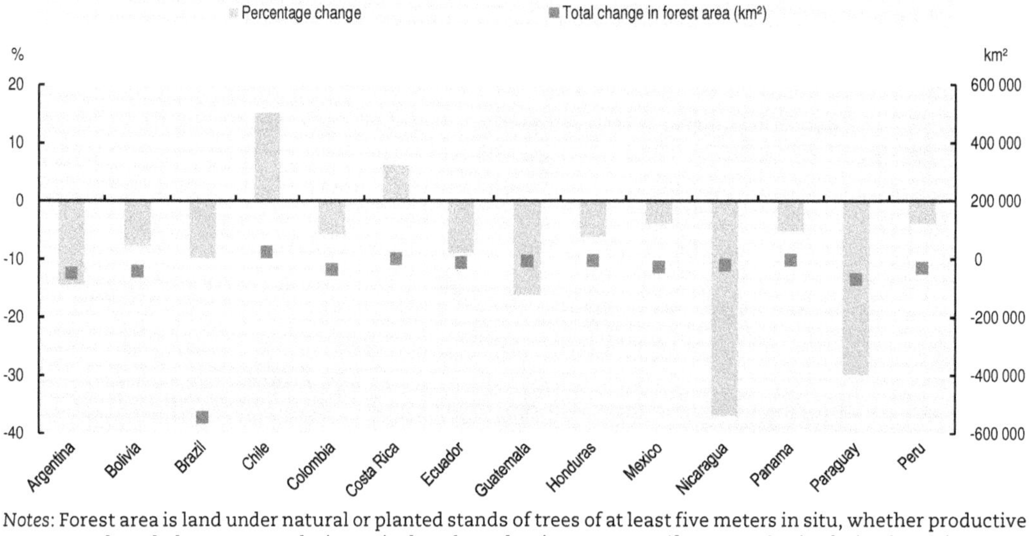

Notes: Forest area is land under natural or planted stands of trees of at least five meters in situ, whether productive or not, and excludes tree stands in agricultural production systems (for example, in fruit plantations and agroforestry systems) and trees in urban parks and gardens (World Bank, 2021[21]). The primary axis shows the percentage change of forest land regarding the year 2000; the secondary axis shows the total change of forest in km² between 2000 and 2020.

Source: Authors' elaboration based on (World Bank, 2021[21]).

StatLink https://stat.link/h0b5xs

Fossil fuels still dominate energy supply in LAC

The region's primary energy supply remains predominantly fossil-based at 66% by 2020, making the region vulnerable to global fuel shocks. Nevertheless, the total energy supply in the region is much cleaner than that of the world, where 33% is renewable compared with 13% at the global level (Figure 2.10). The primary supply of renewables in LAC includes hydroelectric power (9%), biofuels such as firewood and bagasse (18.8%), solar and wind (5.1%) and geothermal (0.9%). The second greatest supply of energy is natural gas (31%), which slightly surpassed the oil share (30%), possibly because of the effects of COVID-19 pandemic, followed by coal (5%), and nuclear energy (1%) (Figure 2.10). The trend in the region is one of absolute and proportional growth of renewable energy sources, mainly hydropower and increasingly solar and wind, followed by biofuels. Between 1970 and 2020, the primary energy supply grew 2.44 times, while the renewable portion grew faster, from 25% in 1971 to 33.6% in 2020, although the decrease in economic activity (-6.8% GDP in 2020) and in the regional energy supply due to the pandemic must be taken into account (Chapter 3).

Figure 2.10. **World and Latin America and the Caribbean total energy supply matrix, 2020**

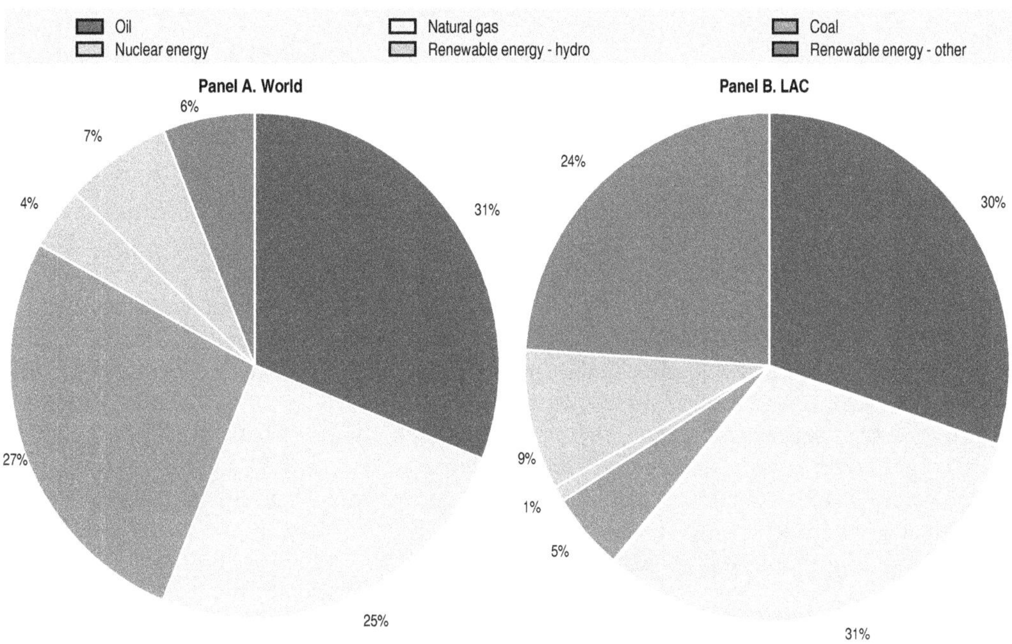

Notes: Total energy supply consists of production + imports – exports – international marine bunkers – international aviation bunkers +/- stock changes. Renewable energy – other includes biofuels, solar, wind, and geothermal energy.
Source: Authors' elaboration based on (Sistema de Informacion energetica de Latinoamerica y el Caribe (SieLAC), 2020[22]).

StatLink https://stat.link/isf1oc

LAC countries have been implementing policies designed to reduce their dependence on fossil fuels. However, emissions generated by oil and gas represent the primary source of pollution across the region, reaching 90% or above of CO_2e in countries such as Costa Rica, El Salvador, Paraguay and Uruguay. Although energy produced by coal has been decreasing across the region, some LAC countries remain highly dependent (e.g. Chile, Colombia, the Dominican Republic and Guatemala) with associated consequences in pollution levels (Tambutti and Gómez, 2020[23]). Nonetheless, in 2019, Chile approved a Coal Phase-Out Plan, aiming to close all coal-fired power plants by 2040 and already 5 of 28 existing plants have been closed (Gobierno de Chile, 2021[24]). The Caribbean is highly dependent on imported fossil energy. Only Trinidad and Tobago, Suriname and recently Guyana have significant domestic energy resources (ECLAC, 2021[25]).

The region has slightly decreased its energy intensity in the last three decades, but mainly because of limited industrial development, the impacts generated by COVID-19 and Russia's invasion of Ukraine. In final energy consumption across the region, the main economic activities are transport (36%), industry (30%) and the residential sector (17%), agriculture, fisheries and mining (6%), and trade and services (5%) (Sistema de Informacion energetica de Latinoamerica y el Caribe (SieLAC), 2020[22]). Governments should progress faster to improve energy intensity in all uses and sectors. The electrification of transport and industry through renewables can offer an alternative to significantly reduce fossil fuel dependency, improve energy security, and greatly increase energy efficiency (e.g. an electric vehicle is 3 to 4 times more efficient than a combustion vehicle, however the electricity must come from renewable sources) (Chapter 3). This should be complemented with the promotion of more efficient public transport systems to ensure a well-being-centred approach. Nuclear energy could also be considered, as it does not produce GHG emissions. Nevertheless, given the radioactive waste management challenges, the potential risk of accidents and the security issues it implies, governments should closely follow the developments of promising innovations such as nuclear fusion in the medium term.

The effects of climate change are severely felt in LAC

LAC is one of the world's most vulnerable regions to the consequences of climate change. Among the 50 countries most affected by climate emergency, 13 are in LAC,[10] based on 2000-2019 data (Germanwatch, 2021[26]). Considering geophysical and climate-related disasters together, the region is the second-most disaster prone in the world (OCHA, 2020[27]). Nearly half the population is assumed to be highly or extremely vulnerable to risks of climate-related impacts (CAF, 2014[28]). The average number of extreme climate-related weather events in LAC increased in most countries between 2001 and 2022 compared to the previous two decades. (Figure 2.11). In total, of the 11 933 climate-related extreme weather events registered worldwide between 1970 and 2022, 17.1% occurred in LAC. Warming temperatures, extreme precipitation events that result in floods, landslides and droughts, sea level rise, coastal erosion, ocean and lake acidification resulting in coral bleaching, and storm surges are expected to increase in frequency and severity, with adverse socio-economic consequences on populations (IPCC, 2022[29]). The region's vulnerability highlights the real and urgent need to address climate change.

Figure 2.11. **Frequency of climate-related extreme weather events in LAC, 1980-2022**

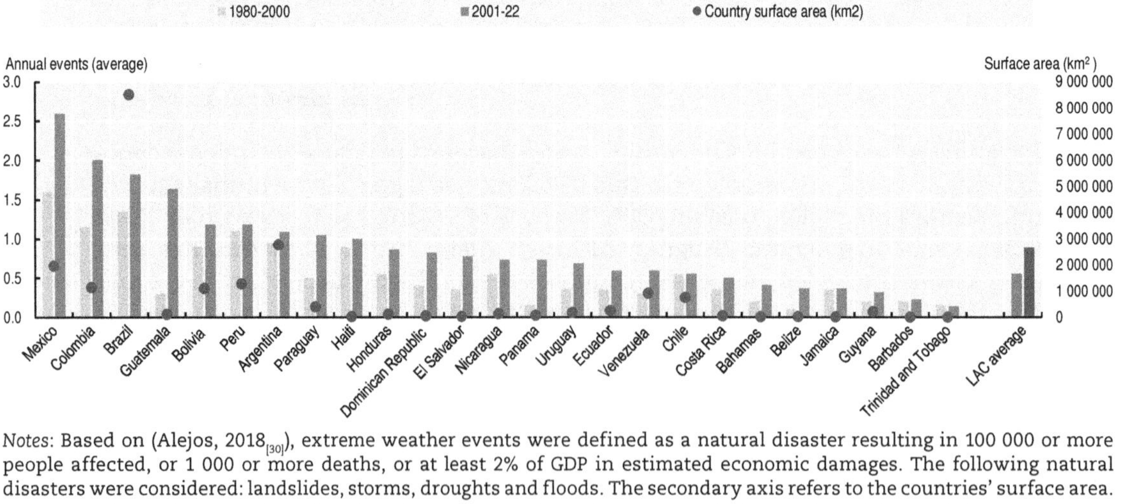

Notes: Based on (Alejos, 2018[30]), extreme weather events were defined as a natural disaster resulting in 100 000 or more people affected, or 1 000 or more deaths, or at least 2% of GDP in estimated economic damages. The following natural disasters were considered: landslides, storms, droughts and floods. The secondary axis refers to the countries' surface area.
Sources: Authors' elaboration based on data from (EM-DAT, 2022[31]); (IDB, 2021[32]); (Alejos, 2021[33]); (FAO, 2018[34]).
StatLink https://stat.link/cfx6s5

The impacts of climate change vary due to differing geographies, capacities to adapt and differing levels of socio-economic vulnerability (IPCC, 2022[29]). Central America and the Caribbean are two of the most vulnerable regions in the world,[11] mainly due to their geographical location and large coastal extensions with high population concentration. Adaptation is therefore among their top development priorities (Bárcena et al., 2020[6]; Germanwatch, 2020[35]; Bleeker et al., 2021[36]). More than half of Caribbean countries are extremely exposed to risks (CAF, 2014[28]) generated by extreme weather events, such as hurricanes and severe storms, increased intensity and frequency of droughts, sea level rise and ocean acidification. In Central America, extreme weather events have increased, on average, by 3% per year over the past 30 years (IPCC, 2022[29]). Vulnerable groups in both regions are the least prepared and the most affected by these events. Lower income households in the Caribbean have a greater chance of suffering longer periods of displacement after a natural disaster, as they might not receive enough financial help to rebuild their houses (Bleeker et al., 2021[36]). In Puerto Rico, after Hurricane Maria (2017), families from upper quintiles were able to rebuild their houses quickly or even leave the island, while poorer families waited months or even years for underfunded relief efforts to help them (McCarthy, 2020[37]). The severe impacts of climate change illustrate the need to continue building resilience and adapt to the current and future effects of climate change.

Climate change is having a direct impact on LAC's biodiversity, declining at twice the rate seen in OECD countries. Chile, Ecuador and Mexico account for the largest falls, but every country in the region is considered to have "high-risk" rates (OECD, 2021[38]). The entire LAC region is home to major ecosystems that are directly under pressure because of climate change and unsustainable development strategies. Retreating glaciers, bleaching coral reefs or loss of ecosystem services undermines the ability of ecosystems to provide a shield against growing climate-related risks and creates additional vulnerabilities (IPCC, 2022[29]). The Amazon rainforest in particular is projected to continue being increasingly threatened by fires and forest degradation.

Average temperatures will continue to rise throughout LAC. The average temperature recorded for the period 1991-2020 is more than 1°C higher than the average for 1901-30 (Figure 2.12). The last 30 years have been the warmest on record, with the sharpest increases in countries located at the region's most northern and southern latitudes. Some locations in Brazil and Paraguay, such as Asuncion, Belo Horizonte, Cuiaba and Curitiba, recorded their highest temperatures ever, and the Caribbean, Central America and Mexico were affected by heatwaves and extreme temperatures. The year 2020 was one of the warmest in the region's history, one of the three warmest in Central America and the Caribbean, and the second-warmest in South America. The largest temperature increases were recorded in the Caribbean, confirming its greater vulnerability to climate change (WMO, 2021[39]).

Figure 2.12. **Latin America and the Caribbean: Temperature anomaly, 1991-2020 compared to 1901-1930**

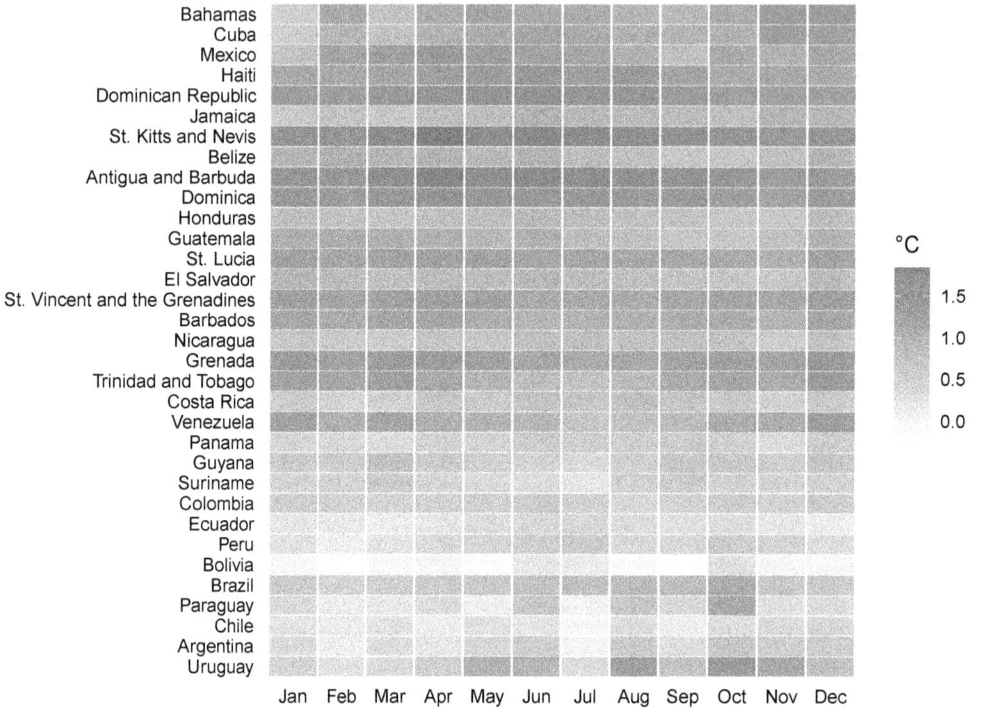

Note: Temperature reported in °C.
Source: Economic Commission for Latin America and the Caribbean (ECLAC) on the basis of (World Bank, 2021[40]).

Global warming has direct impacts on ocean and coastal ecosystems. Global warming is one of the main causes of sea level rise and will increase the intensity and timing of extreme weather events in the Caribbean. More than 50% of Caribbeans live within 1.5 km of the coastline and one-third live in low-elevation zones. Given the region's proximity to the equator, sea level rise generated by higher temperatures will continue to erode coasts, damage ecosystems and lead to land loss, household damage, relocations and business closures (Bleeker et al., 2021[36]).

Temperature rise also directly affects other water sources. Decreasing water availability is another impact of climate change that can particularly affect Central America. By 2100, per-capita water availability in Central America is projected to decrease by 82% and 90%, on average, under low- and a high-emissions scenarios, respectively (ECLAC/DFID, 2010[41]). Water stress in LAC (below 15%) is lower than the OECD (20%) average; however, there is a lot of heterogeneity in the region. In countries including Brazil, Colombia and Peru, the share of withdrawal from freshwater sources is well below 5%, whereas in the Dominican Republic and Mexico, figures are above 25% (OECD, 2021[38]). Even if some of the long-term changes, such as sea level rise, ocean acidification or melting of Arctic ice, will be irreversible, there is still a window of opportunity to avoid the worst consequences, if the right policies are adopted (Chapter 3) (Hickey and Wellenstein, 2021[42]).

The promotion of mitigation and adaptation policies in LAC will require a deeper understanding of its complexities, further funding and a strengthening of their policy coherence. The lack of trustworthy information is a main barrier to better mitigate and adapt to the effects of climate change in LAC is. The governments of the region need to invest in data creation and analysis; raising awareness; financial and technical capacity; co-ordination among relevant policy makers with potentially diverging objectives; and integration of Indigenous and local knowledge systems. Effective climate mitigation and adaptation should include as much detailed information as possible to identify the most exposed areas and their vulnerable groups. Reliable and specified data are also crucial to develop more and better digital risk mitigation tools (CAF, 2014[28]). The transparent design of public policies, as well as the research and development of innovative responses to climate change, depend directly on the availability of this kind of data. LAC countries should continue strengthening the statistical infrastructure and advancing in the construction of digitalised databases, the integration of tools to measure risks, and the design of policies based on solid evidence. These efforts need to be multiplied and better co-ordinated, while also including as many stakeholders as possible (Chapter 5).

Further funding is needed for the implementation of more and better adaptation policies. During 2019 and 2020, funding for mitigation policies in LAC averaged USD 28 billion, while adaptation policies only received USD 4.5 billion (Chapters 4 and 5) (Buchner et al., 2021[43]). While there is marked heterogeneity in the region, overall results are positive and exemplify why the region should deepen its efforts by increasing the funding of adaptation policies. Adaptation policies and strategies designed to address climate risks at the local and national levels can reduce both exposure and vulnerability to climate change impacts. Positive signs of progress show that LAC can implement more and better adaptation policies. Protected areas are the most important policy instrument for biodiversity protection implemented in LAC, namely for ecosystem-based adaptation through conservation and restoration (OECD, 2018[44]). The region has the largest extent of biodiversity protection (8.8 million km^2) in the world (RedParques, 2021[45]). In total, 25% of land and 24% of marine areas are protected, both above the OECD average. Terrestrial protection has increased by 9 percentage points since 2000, while protection of marine areas has more than doubled in some LAC countries (Chapter 3) (OECD, 2021[38]).

Policy coherence among short-, medium- and long-term objectives should also be encouraged, to better preserve the accomplishments and future goals of mitigation and adaptation policies. This can be accomplished by ensuring that short-term actions are consistent with long-term goals and by enforcing existing policies. Addressing the transboundary and long-term effects of policies will help LAC governments to take more informed choices about sustainable development while ensuring the well-being of future generations. Besides a proper funding and access to updated data, governments can strengthen policy coherence with investment in human resources, development of new skills for local bureaucracies, an inclusive governance and strong systems of monitoring and evaluation (M&E) systems (Chapter 5) (IPCC, 2022[29]).

Why should the green transition be a priority for LAC?

In LAC, government's efforts to recover from and offset the impacts of the COVID-19 pandemic have not incorporated the environmental dimension. So far, the opportunity to target recovery spending towards more transformative and green sectors has not been seized. The recovery has prioritised a high demand of non-renewable resources, which underlines the unsustainable nature of the pre-pandemic development pattern (ECLAC, 2022[46]).

The green transition is a structural challenge that LAC will have to contend with sooner or later. The estimated cost of inaction highlights the benefits of urgently adopting policies for adaptation and mitigation, while the extreme consequences of climate change are increasingly being experienced in the region. Given the impact the pandemic had on the region, the post-COVID-19 context is a timely opportunity to combine recovery measures with green policies, and to advance a just transition that could help achieve higher levels of well-being. A new sustainable development model in LAC demands a green transition that is just throughout its process, from its design to its implementation (AFD, 2020[4]). A green transition can also reduce the region's vulnerability to the effects of climate change while opening new, future-oriented market opportunities.

Despite LAC's relatively lower contribution to total GHG emissions, the region is highly vulnerable to the effects of climate change. The last 30 years have been the warmest on record, with the sharpest increases in the countries located at the region's northern and southern latitudes. In 2021, the region experienced several extreme weather events, such as low temperatures and snowfall in southern Brazil and droughts and high temperatures in central Chile. In 2020, for the fifth consecutive year, the Atlantic hurricane season was abnormal (IPCC, 2021[7]).

Climate change and environmental degradation are generating a direct social and economic impact. Many countries in LAC are experiencing a fall in agricultural productivity and tourism, as well as climate-driven migration and high reconstruction costs after natural disasters. In the Dominican Republic, the heavy rains of 2016 generated severe economic losses in crops such as plantain, cassava and sweet potato. During the hurricane season of 2017, an estimated loss of USD 52 million was recorded due to lower touristic activity (OECD/UNCTAD/ECLAC, 2020[48]). Natural disasters create a need to increase expenditure and thus have direct impacts on fiscal deficits and public revenues. On average, a natural disaster results in a reduction of public revenues equivalent to 0.8% and 1.1% of GDP, for lower middle-income and low-income countries respectively (Chapter 1) (Alejos, 2021[33]). In the case of the Caribbean, the annual cost of inaction to deal with the effects of sea level rise, is estimated to reach USD 22 billion per year by 2050 (10% of GDP) and USD 46 billion by 2100 (22% of GDP) (AFD, 2022[47]). In the case of Haiti and Puerto Rico, two of the three most affected countries in the world between 1999 and 2018, annual GDP losses reached 2.38% and 3.76%, respectively (Internal Displacement Monitoring Centre, 2022[49]). If environmental degradation is not addressed soon, LAC governments will have to add the costs of climate change to those of social vulnerability. The link between the dependence on biodiversity and financial security has proved to be very close, and the cost of inaction could have unprecedented economic and social consequences (Chapters 1 and 4) (Bárcena et al., 2020[6]).

A recovery strategy aligned with a green transition is an opportunity to overcome the region's development traps

The crisis generated by COVID-19 has led to a historic economic downturn in LAC. Almost 25% of the jobs lost in 2020 were not recovered in 2021, deepening the social gap that characterises the region (ECLAC/ILO, 2022[50]). The COVID-19 recovery can be a strategic moment to align LAC policy objectives with a green and just transition. Productivity, social vulnerability, institutional and environmental traps – which are closely interlinked – are today the main inhibitors to further inclusive and sustainable growth in the region (OECD et al., 2019[3]).

A recovery based in decarbonising investments and climate mitigation and adaptation policies could promote a more competitive productive matrix, boost job creation and enhance social inclusion (Chapter 3). Most LAC countries adopted measures to respond to the impact of the COVID-19 crisis. For the recovery, further actions are needed to advance towards a more sustainable, resilient and sustainable development model. Advancing sustainable production and consumption patterns and improving the quality and coverage of social services is key for the recovery (ECLAC, 2022[46]).

Combining adaptation strategies with social and economic dimensions through a systemic lens could help overcome the region's development traps. A systemic design of the green transition could be a way to address the unavoidable trade-offs presented by any transformative transition. Confronting climate change is not the only reason for advancing decarbonisation efforts. It can also contribute to stronger public institutions, as governments could increase policy coherence and deliver a more just society based on a new sustainable social contract. If properly designed, a green transition can also help increase the overall well-being of LAC citizens (Chapter 5). A systemic approach with a multi-dimensional perspective could transform the COVID-19 recovery into an opportunity to advance a more sustainable and inclusive development model. Setting green policies at the centre of the recovery ensures countries invest in sustainable long-term economic models, while making the best of the growing international green agenda (Chapter 6) and its emerging market opportunities and various investment and financing initiatives (IPCC, 2021[7]).

Integrating social dimensions into the green transition is key for a better development model

Close interaction between humans and their environment highlights the need to address the challenges of inequality and environmental degradation together (OECD, 2021[51]). If not addressed, the effects of climate change will continue to deepen poverty and inequality in the region. The green transition has the potential to help LAC address all dimensions of inequality, across countries, socio-economic groups, territories, generations and gender.

Both climate change and inequality are pressing issues demanding integrated solutions at the subnational, national, regional and international levels. Climate change exacerbates inequalities within societies and also among countries. Developing countries face a kind of "double asymmetry" in the sense that those who produce the most emissions (the richest countries and social groups) have the greatest capacity to defend themselves against the effects of climate change, while those who produce the least emissions (the poorest countries and social groups) suffer the most and have the least resources to recover (Tambutti and Gómez, 2020[23]; OECD, 2021[51]; ECLAC, 2020[52]). The first asymmetry stems from the fact that the level of emissions reflects consumption capacity and therefore reproduces patterns of income inequality (Figure 2.6). The second asymmetry is derived from unequal distribution of the cost of environmental degradation. The rise in temperature has affected poor countries (Tambutti and Gómez, 2020[23]) and their poorest social groups in particular. After Hurricane Mitch (1998) in Central America, low-income households suffered larger relative loss of assets (31%) than the non-poor (11%) (UNDESA, 2017[53]).

At the international level, the green transition offers a possibility to rebalance the disproportionate burden of climate effects on LAC and developed countries through stronger international co-operation (Chapter 6). If national efforts are not co-ordinated internationally, the speed and effectiveness of mitigation and adaptation policies adopted globally, particularly in the developing world, will continue to prove insufficient (Chapter 6) (IPCC, 2018[1]).

At the regional level, the green transition presents several opportunities to promote better collaboration and integration within LAC. The effects of climate change are also aggravating inequalities among LAC countries. Regional co-operation can help improve data and information generation, water resource management, sustainable production and consumption, and biodiversity management. Regional initiatives based on active co-operation for a more green and just transition could help contain the most vulnerable groups. In LAC, sea level rise, droughts, flooding and wildfires will force people to migrate, exposing them to further vulnerabilities. In 2017, three million Caribbeans were forced to relocate due to the Atlantic hurricane season (Bleeker et al., 2021[36]). The effects of climate change are forecast to drive an estimated 17 million Latin Americans to migrate by 2050

(World Bank, 2021[54]). Relocation programmes co-ordinated at a regional level would help anticipate and contain forced migration due to climate change and reduce unnecessary exposure to risk.

At the national level, a new sustainable development model could help reduce social inequalities across socio-economic groups. The combined impacts of the pandemic and climate change deepen the urgent need for such a model. About 49.2% of the urban population in LAC is either poor or extremely poor (Bárcena et al., 2020[6]). The pandemic increased the number of people living below the poverty line in LAC and climate change is projected to contribute an additional 5 million by 2030 (Hickey and Wellenstein, 2021[42]). Governments should focus on protecting those most in need through targeted social programmes (Chapter 1). As climate change threatens to reverse global health gains of the past 50 years, national healthcare programmes could be an essential element of a more just green transition (Watts et al., 2015[55]; Landrigan et al., 2017[56]).

At the subnational level, environmental degradation is deepening disparities among urban and rural areas and its effects on most vulnerable groups. In some Latin American cities, high rates of socio-economic residential segregation make people living in precarious neighbourhoods more susceptible to climate change effects and other phenomena, such as exposure to heatwaves. This exposure increases their health risks and vulnerability to extreme climate events as a function of their socio-economic status (OECD et al., 2021[2]). High urbanisation rates and unregulated expansion of urban areas have pushed vulnerable groups to locate in high-risk zones with deficient or non-existent infrastructure, such as floodplains and landslide-prone slopes. Those who live near highways or industrial sites are often exposed to high levels of air pollution. In Chile, for example, there is a pronounced difference in exposure to PM2.5 (particulate matter with a diameter of less than 2.5 μm) between Magallanes with the least exposure (5.9 μg/m^3) and Aysén with the greatest exposure (41.9 μg/m^3) (Figure 2.13). A systemic green transition will have to include urban designs that internalise how citizens from poorer areas experience the largest negative impacts of climate change while lacking the capacities necessary to adapt.

Figure 2.13. **Subnational regional disparities in mean annual exposure of the population to outdoor PM2.5, selected LAC countries, 2019**

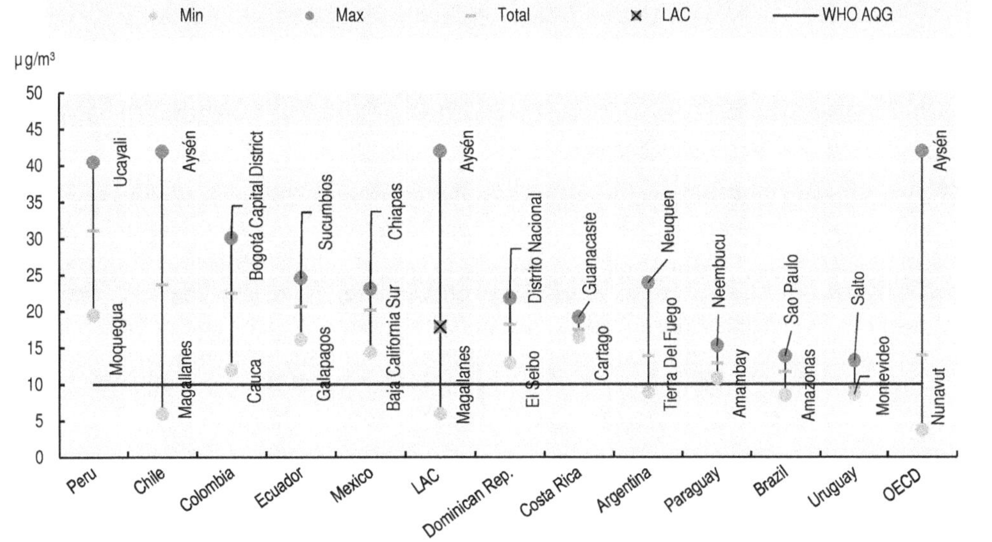

Notes: WHO AQC = World Health Organization Global Air Quality Guidelines. The mean population exposure to outdoor PM2.5 is calculated as the mean annual outdoor PM2.5 concentration weighted by population living in the relevant area, e.g. the concentration level, expressed in µg/m³, to which a typical resident is exposed throughout a year. The country "total" considers the country as a single entity to which each region contributes proportionally. The LAC regional average is calculated by the Organisation for Economic Co-operation and Development (OECD) based on the countries selected.

Sources: (OECD, 2021[38]); (OECD, 2020[61]).

StatLink https://stat.link/sl18ux

The effects of climate change on rural areas are particularly relevant for LAC. Agricultural activities are particularly sensitive to climate change, which is expected to produce changes in structure, yields and crop cycles. Certain crop cycles will probably speed up, which will alter the physical properties of the soil and the supply of water for irrigation, lead to greater evaporation and place greater stress on those crops (Bárcena, et al., 2018[58]). The region's economy relies significantly on agriculture, either through subsistence production (comprising around 15 million smallholder farms in LAC and 20-30% of the workforce in the Caribbean) or through large industries (CAF, 2014[28]). Changing rainfall patterns, droughts and extreme climate events are likely to worsen significantly during the next decade, implying greater vulnerability related to land labour and food and water security (IPCC, 2022[29]). The effects of climate change have direct impacts on rural productivity, which could increase poverty, particularly in Central America (IDB, 2021[57]), as real income is highly dependent on land use. Income declines not only because of reduced availability of fertile land dedicated to agricultural production and livestock but also because heat stress forces lower labour productivity. In most cases, the capacity of farmers to react and adapt to the effects of climate change depends on their wealth. In Peru, temperature rise has forced farmers to sell their livestock, move their crops into fallow land and include children in their farming activities to compensate for income deficits (IFS, 2018[59]).

The gender gap is also deepening in LAC, as women and girls are more vulnerable to the effects of climate change (OECD, 2021[60]). Women are the main food producers in developing countries and have higher dependence on natural resources. When it comes to planted food, the impact of climate change on land and water directly affects their harvests (IUCN/GGO, 2015[62]). This not only reduces the amount of food women can take home but also potential micro-selling initiatives to gain financial independence from their couples. In extreme scenarios, the fall in agricultural production might even cause forced relocations. In the dry corridor of Central America, 62% of poor families depend on corn, beans and sorghum to survive – all products likely to become impossible to grow as temperatures continue to rise (Hickey and Wellenstein, 2021[42]). Women and girls, who also tend to be responsible for water gathering, cleaning and cooking, experience disproportionately the consequences of climate change. While poor water connections and droughts impose greater distances and time to search for water, floods and hurricanes increase the exposure to risk situations (OXFAM, 2018[63]).

The income gap and the uneven distribution of domestic labour deepen the impact of climate change for women, as their lack of access to resources or their greater burden of domestic care activities hampers their capacity to recover quickly after a climate-related natural disaster. As women are likely to be responsible for staying home and taking care of children out of school and injured family members after a hurricane or flooding, they have more chances of losing their jobs or suffer a reduction in their wages. Moreover, women are less likely to be employed in "cash for work" programmes implemented after a disaster to rebuild infrastructure (Bárcena et al., 2020[6]). Additionally, in the context of stressful events, such as climate disasters, the level of domestic violence and street aggression against women rises (IPCC, 2018[1]). Green policies should promote and ensure the development of new skills in future green jobs for women and encourage their involvement throughout the decision-making process for policy to ensure an inclusive response (Chapter 5) (IPCC, 2018[1]).

Climate change also has unequal effects across generations. Nearly 60% of countries had an increase in the number of days people were exposed to very high or extremely high fire danger in 2017-20 compared to 2001-04, and 72% of countries around the world had increased human exposure to wildfires across the same period (Romanello et al., 2021[64]). By 2050, the population over 65 years will double in Latin America, increasing the number of elderly people who are vulnerable to heatwaves and other consequences (CAF, 2020[65]). Exposure to extreme heat poses a health hazard, particularly risky for individuals older than 65 years, populations in urban environments, and people with health conditions. Heat disproportionally affects people who are marginalised or with scarce resources, because they have limited access to cooling mechanisms, fresh water, and health care (Romanello et al., 2021[64]). An active, green education strategy is key to ensure that future generations envision and interact differently with the environment (Vona, 2021[66]), having learned from the experiences of current approaches to production and consumption.

Negligent land use and deforestation will have effects on how much each generation will benefit from a clean and safe environment. Forest land cover in LAC decreased by 8.2% between 2000 and 2020 (World Bank, 2021[21]), meaning that future generations will enjoy less green capital. The region has an important role in preserving forests, being home to 23% of the world's forests including the Amazon, the world's largest rain forest which is shared by eight countries. These ecosystems are key for climate change mitigation and adaptation due to their capacity to absorb CO_2. They also provide environmental services by regulating the water cycle, protect soils, supply resources such as timber, medicines, food and fibres, and provide opportunities for recreation and tourism. More than half of the world's forest area is distributed in just five countries, with Brazil being the world's second more forested. While most LAC countries suffered a significant net loss, Chile and Costa Rica increased their forest area between 2000 and 2020 (Figure 2.9). After a long period of forest loss last century, Costa Rica implemented recovery and reforestation policies and has managed to increase forest cover replacing land previously used for farming and livestock activities (ECLAC, 2021[20]). In Chile, massive exploitation of primary forests and extensive agricultural areas in places with high rainfall were eroding volcanic soils generating problems of stability and water quality. Policies implemented since 2015 focus on the conservation of natural forests, and have encouraged large forest companies to take action in rural development and funding fire prevention which has translated into the recovery of forest land (European Forest Institute, 2019[67]).

Harnessing the green transition through a systemic approach to improve well-being

The wide range of post-COVID-19 recovery strategies have shown several frameworks that can be adopted to build forward better. Each country's conception of recovery prioritised different economic, social or environmental goals and the best strategies to achieve them. How this policy priorities are conceived and aligned, determine the kind of recovery pathway adopted. Each recovery pathway distinguishes itself in the way in which each encompass GHG emission reductions, and integrates these with considerations for wider well-being outcomes (e.g. SDGs) (OECD/IEA, 2021[13]). Globally, how the recovery is designed and implemented will determine how inclusive the transition could be.

The Rebound path disassociates the recovery from the environment and focuses solely on the economic recovery and measures to attain the pre-crises economic indicators such as GDP, job creation, firms' profits or trade rates. Even though this approach might have put certain economies back on track, they have also reversed much of the environmental progress gained during the previous years in terms of GHG emissions or clean air. The Decoupling strategy also conceives recovery in terms of growth and production, but unlike the rebound path, places the mitigation strategy in the centre of its efforts. It aims to decouple the economic growth from the CO_2 emissions associated with it, through initiatives such as energy efficiency or transition to low carbon energy. Nevertheless, this green growth does not conceive the recovery as an opportunity to shift towards policies and consumer patterns that place well-being as their central focus.

A well-being approach advocates for an economic recovery low on emissions and centred in well-being improvements. From this lens, GDP growth may not necessarily be a means to increase current nor future well-being and should not be used as the aggregate measure of progress.[12] LAC countries should especially beware of taking a recovery pathway focused on maximising GDP growth at the expense of environmental degradation and without placing other wider well-being impacts (e.g. health, equity) at the centre of decisions (as would be reflected in a Rebound pathway). Under a wider well-being recovery path, the focus is no longer on greening GDP growth (as in a Decoupling pathway), but on designing systems in ways that increase well-being while reducing emissions and material use. In practice, it means enabling the conditions that make sustainable choices the easiest and cheaper and thus those that people choose. This beyond-GDP systemic vision aims to redesign systems to achieve better results. Through a well-being pathway, policies no longer focus on solving specific problems but on designing systems that do not generate those problems in the first place.

A green transition approach specific to LAC should place citizen's well-being at the centre and account for the particularities of the region. The path towards zero emissions should not be constrained to other region's strategies, but rather it should be the result of a comprehensive analysis of how to make the transition without hampering the region's development opportunities. The "common but differentiated responsibilities" approach and the Glasgow Pact (which commits developed countries to provide USD 100 billion annually to developing countries by 2030), are two important elements to advance global climate justice. Since colonial times, history shows the consequences of not putting Latin American resource preservation and citizens well-being at the centre of decision-making processes. Following the principles of climate justice, LAC should pursue a transition in which the burden of climate change is properly shared by the most developed countries and regions (Figure 2.1). Moreover, a Latin American approach to the green transition should focus on reducing the emissions of citizens who consume and pollute the most, usually those in the upper quintile (Figure 2.6).

A well-being approach for the green transition implies addressing the challenges of placing Latin Americans' well-being at the centre (OECD, 2021[38]). Since 2011, the OECD has been promoting a well-being framework that provides a comprehensive approach to analysing and measuring the determinants of current and future well-being, beyond aggregate traditional measures, such as GDP (OECD, 2021[38]). This approach encompasses multiple dimensions that determine people's current well-being (e.g. income and wealth, work and job quality, housing, health, knowledge and skills, safety and the quality of the environment) and proposes a broader set of indicators to track performance and guide decision making. To analyse the dimensions of current well-being, it measures well-being outcomes by looking at averages, deprivations, and inequalities between groups and between top and bottom performers. Then, it measures the stocks, flows, risk factors and resilience of resources that will determine the well-being of future generations (e.g. natural, human, economic and social capital) (Figure 2.14).

Figure 2.14. **The OECD Well-being Framework**

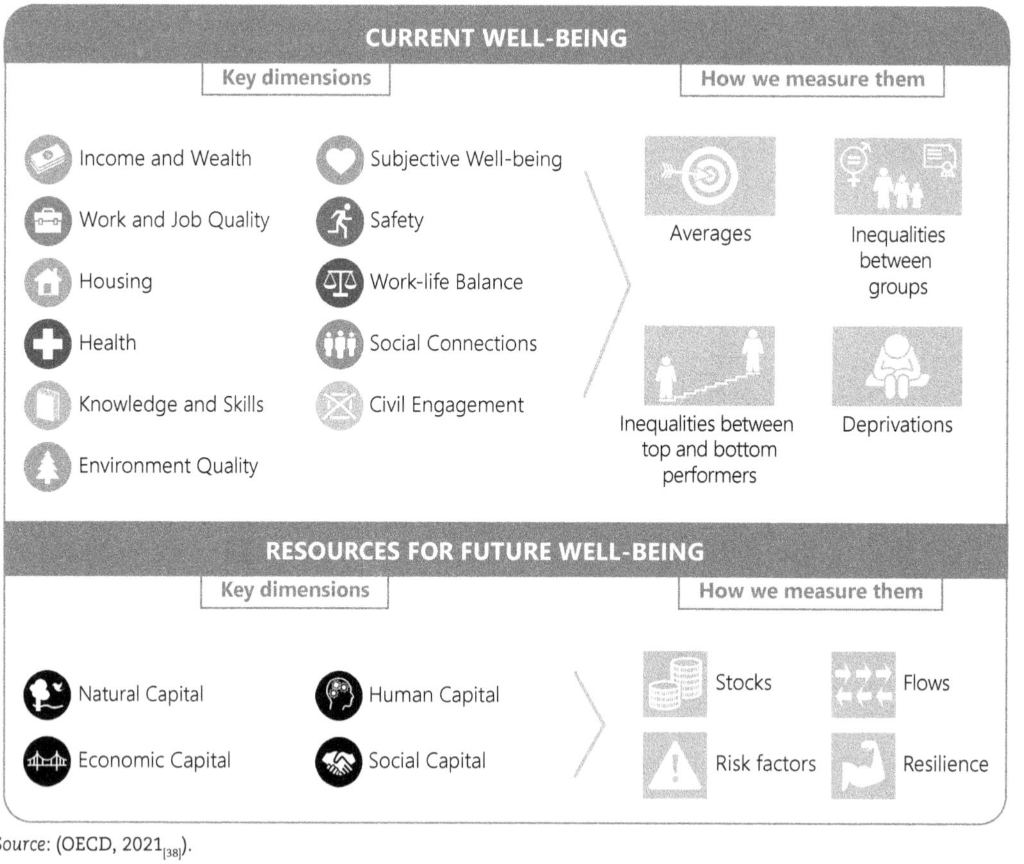

Source: (OECD, 2021[38]).

This well-being approach should be considered for developing national statistics, policies and promoting recognition of multi-dimensional well-being – e.g. measuring all dimensions, beyond income, that affect people's well-being (OECD, 2021[38]). Measuring well-being through inequalities often reveals wide variation between and within countries. In the context of environmental policy, looking beyond averages is particularly relevant as the impacts of environmental degradation are often concentrated among vulnerable groups and households (OECD, 2021[51]). In addition, a well-being approach calls for reframing measurement systems around well-being outcomes (e.g. improved learning outcomes) instead of focusing on intermediate outputs (e.g. increased education coverage) (OECD, 2019[68]).

Delivering systems that can, by their design, improve well-being while reducing energy and material use, and therefore emissions, calls for policy-making that is able to understand and reshape current unsustainable systems. Analytical rather than systemic approaches are usually pursued, leaving governments ill-equipped to design effective climate change mitigation and adaptation action. An analytical approach consists of solving problems by dividing them into parts and trying to optimise those parts. For example, countries make significant efforts to improve the energy efficiency of or to electrify vehicles (a part in the transport system). Such efforts focus on reducing the undesired output (e.g. emissions) per unit of output (e.g. kilometres driven). Evidence shows, however, that the increase in the overall number of vehicles and the kilometres driven (pushed by the functioning of car-dependent systems) offset the emissions reduction obtained via such policies (Lamb et al., 2021[69]).

Complex problems, such as climate change or poverty, are rarely caused by specific parts in a system but instead by the way parts are organised and interrelated. Partially blind to the system that generates the results their policies aim to change, analytical approaches are prone to making assumptions about certain trends or behaviours. For example, analytical thinking led to decades of transport policies that assumed an increase in the number of vehicles in circulation and the congestion generated as inevitable consequences of progress. Transport policies therefore focused on increasing road capacity but ended up exacerbating congestion and directly affecting climate action.

With a systemic lens, climate action is no longer limited to reacting to or anticipating inevitable trends. Rather, it gives the opportunity to modify trends via (re)designing systems that are behind such trends. Furthermore, when policies are designed with a systemic lens, what used to be trade-offs between climate and well-being objectives may become synergies (OECD, 2019[68]).

The OECD supports the adoption of a systemic approach to help policy makers reprioritise climate action towards improving system functioning and accelerating the transition to systems that are net zero by design (OECD, 2021[70]). The OECD has developed the Systems innovation for net-zero process,[13] which builds on systems thinking and consists of three steps: 1) *envision* the outcomes that a well-functioning system achieves; 2) *understand* why the current system is not achieving such outcomes, which mental schemes support such a system and how it could be redesigned to produce better results; and 3) identify policies able to transform or *(re)design* systems (Figure 2.15). This model could serve as a framework for policy makers designing and implementing green transitions, while underpinning the importance of policy coherence across multiple dimensions and time periods (Chapter 5).

Figure 2.15. **The systems innovation for net-zero process for transformative climate action**

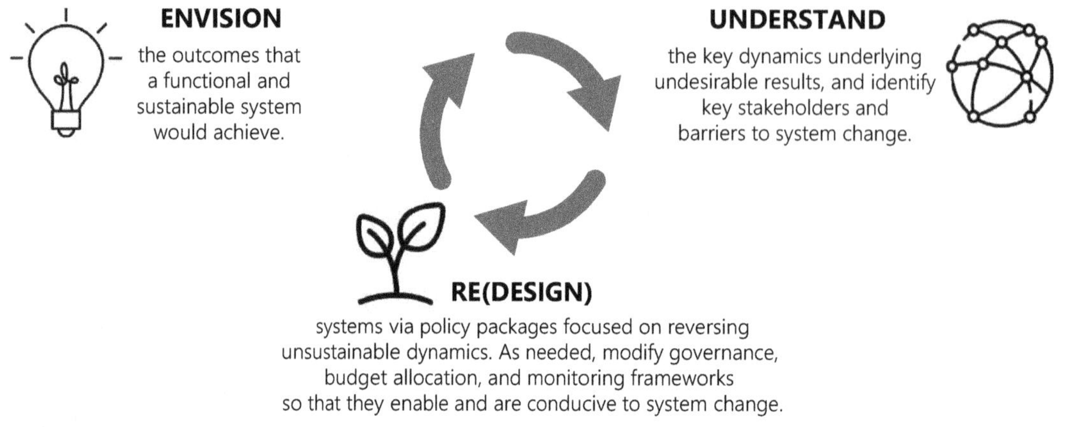

Source: (OECD, 2022[71]).

Thinking in systems for sustainable territorial development

The urbanisation process in LAC has had favourable economic and social consequences, such as greater dynamism in production activities, development of services, increased productivity and harnessing economies of scale (ECLAC/MINURVI/UN-Habitat, 2017[72]). However, the lack of a systemic approach has also given rise to negative impacts such as air pollution and GHG emissions, road congestion and accidents, health problems, and water pollution, all of which are eroding the foundations of economic dynamism (Bárcena et al., 2020[6]).

Air pollution is the world's single largest environmental health risk (UNECE, 2021[73]). Of its components, PM2.5 has the highest health impacts. Although PM2.5 emissions are, on average, lower in LAC cities than in Southeast Asia or North America, levels have been rising faster lately (Florczyk et al., 2019[74]). As Latin American cities grow, their pollution levels are increasing (Gouveia et al., 2019[75]). In general, CO_2 emissions have been growing faster than populations in cities, resulting in increased CO_2 per capita across the region.

The percentage of people living in urban areas in LAC has doubled since 1950 (UNDESA, 2019[76]). South America is the most urbanised subregion in LAC (84.6%), above the high-income countries group (81.9%). Slightly lower in Central America (75.4%) and the Caribbean (72.2%) are still much higher than other developing regions including Sub-Saharan Africa (41.4%) and Southern Asia (36.6%) (Figure 2.16). The Caribbean has experienced the greatest increase in urbanisation, from 36.3% in 1950 to 72.2% in 2020. In particular, the countries with greater urbanisation increases are the Dominican Republic (from 23.7% to 82.5%) and Haiti (from 12.2% to 57.1%) (UNDESA, 2019[76]). The Caribbean's fast urbanisation has had direct impacts on the environment. Most urban, industrial and commercial developments did not follow sustainable urban design. Activities, including tourism and its infrastructure, have thus contributed to destruction of natural habitat, landscape transformation and coastal erosion (Ecosystem Profile, 2019[77]). Many metropolitan areas in LAC, including their residents and infrastructure, that experienced rapid urbanisation now face significant climate-related risks, such as flood plains or slopes prone to landslides. Such risks are projected to increase in the future (IPCC, 2022[29]).

Figure 2.16. **Global urbanisation trends, 1950-2050**

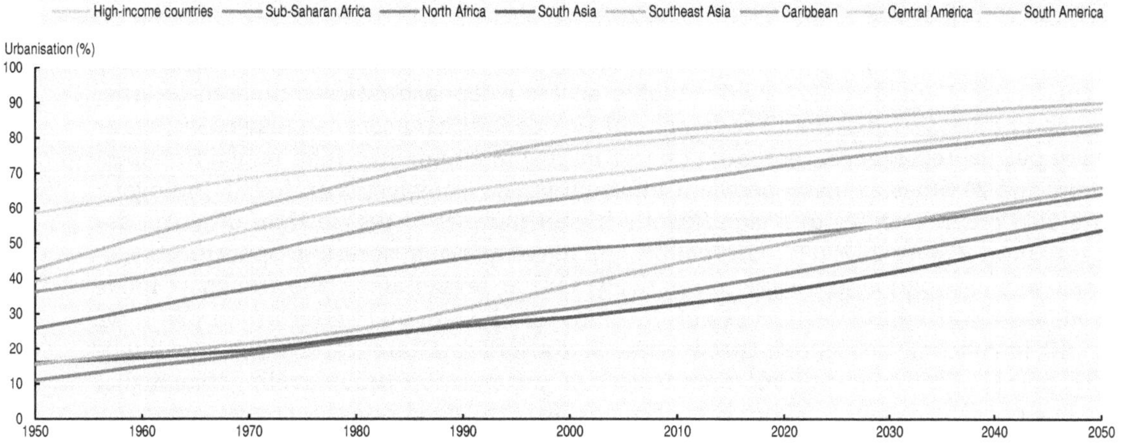

Notes: Available data until 2018; from 2020-50, the values are forecasts. Urbanisation trends refers to annual percentage of population at mid-year residing in urban areas. The country classification by income level is based on 2016 GNI per capita from the World Bank.

Source: Authors' elaboration based on (UNDESA, 2018[78]).

StatLink https://stat.link/10iox2

Car-dependent territories with high sprawl perform poorly in terms of well-being

Transport is one of the main sources of air pollution and high GHG emission levels, mainly due to the large size of many LAC countries. In Latin American cities, private motorised transport is responsible for 75% of CO_2 emissions and produces 82% of PM10 pollutants, both of which are negatively associated with health outcomes (Vasconcelos, 2019[79]). Moreover, the Caribbean, land-based transportation accounts for the second-largest share of fossil-energy consumption. The subregion has one of the highest vehicle motorisation rates in the world, with 201 vehicles per 1 000 inhabitants. In Latin American and Caribbean countries, the urban layout and the availability and quality of public transport influence automobile use and shape car-dependent and sprawled territories (ECLAC, 2020[52]). This concept refers to the dynamic by which people move away from city centres but still commute daily to such centres. To a great extent, this is the result of decades of short-sighted transport and spatial planning policies (or sometimes the lack of these policies). While this section focuses on transport, a systemic approach can be applied to other sectors in LAC countries. Applied to the residential sector, for example, a systemic approach can shed light on the drivers of urban sprawl (why people "choose" to move further away from city centres) and thus inform policy decisions aiming to contain such development.

Car-dependent and sprawled territories perform poorly in terms of present and future well-being of citizens. Undesirable results include traffic congestion,[14] growing inequalities, poverty perpetuation, road fatalities, high emissions, air pollution and reduced capacity to adapt to climate change. Peripheries host most of the low-income settlements (ITF, 2019[80]; IDB/ITF, forthcoming[81]), and after decades of transport investment heavily biased towards highways (ECLAC, 2020[82]; Lardé, 2021[83]; Sanchez et al., 2017[84]), these areas lack proper public transport infrastructure, preventing the most vulnerable groups from easily accessing labour (ITF, 2020[85]). For example, for a similar trip duration, inhabitants of peripheries of Bogota have access to four times less job opportunities by public transport (around 20 000 jobs) than people living in the city centre (IDB/ITF, forthcoming[81]). Moreover, the predilection for highways is directly detrimental to the well-being of citizens since, of all modes of transport, highways have the highest

CO_2 emissions (between 55 and 256 grams per tonne-kilometre), a figure much higher than maritime transport (between 11 and 101), river transport (between 17 and 38) and rail transport (between 2 and 21) (ITF, 2022[87]).

The infrastructure of many cities in Latin America prioritises mobility by car, making it the transport mode that provides higher access to opportunities compared to other transport modes such as public transport. In Mexico City, car users can reach 13 times as many people (representing essential opportunities) in 30 minutes than public transport users, and 20 times as many people if informal public transport is excluded. In Bogota, cars also provide access to a significantly higher number of people than other modes of transport (Figure 2.17) (ITF, forthcoming[86]). In comparison, access to opportunities in cities such as London and Paris is relatively similar regardless of the transport mode used, showing higher effectiveness of the transport system and better urban planning. In LAC, low-income households that are unable to afford cars[15] are "captive users" of public transport (and, most recently, motorcycle use); they spend more time traveling than higher income households able to afford cars, face unsafe travel conditions and may need to pay various single tickets per commute. In Brazilian cities of more than 60 000 inhabitants, the average commuting time in public transport is 36 minutes – more than double the 15 minutes the same trip would take using individual transport (Vasconcelos, 2019[79]).

Figure 2.17. **Absolute accessibility to opportunities by transport mode, 2019**

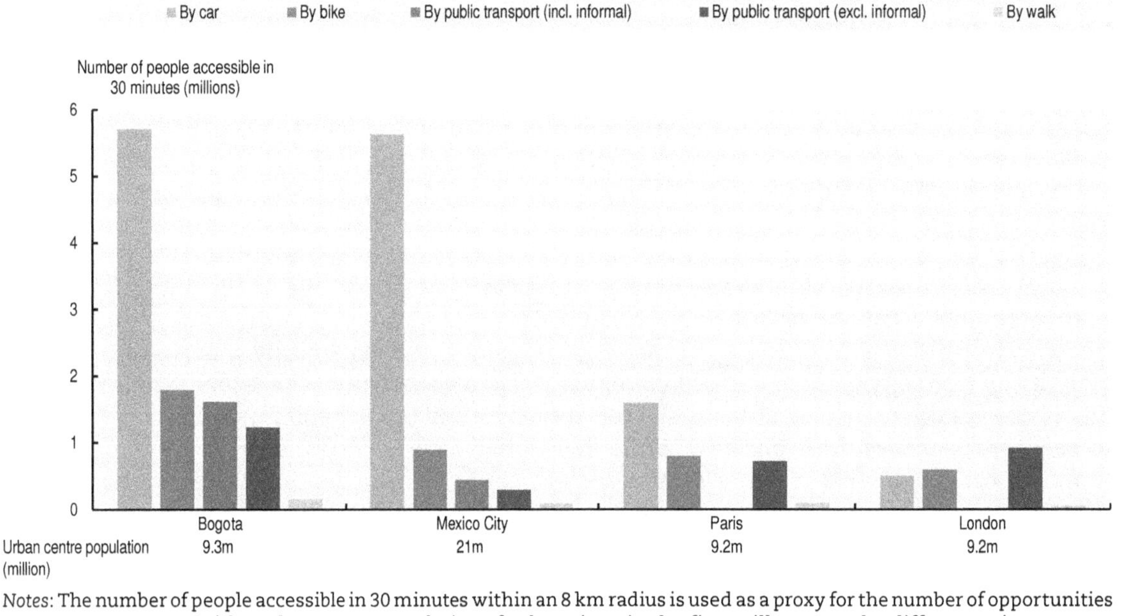

Notes: The number of people accessible in 30 minutes within an 8 km radius is used as a proxy for the number of opportunities a person can access with each transport mode (ITF, forthcoming[86]). The figure illustrates the differences in access to opportunities via different transport modes within cities. The population density and total area vary across cities, which in turn affects the number of people that can be reached in 30 minutes within an 8 km radius. The area of each city refers to the Urban Centre, which is usually larger than the administrative city. It is suggested that the data be interpreted comparing transport modes within not across cities.

Source: (ITF, forthcoming[86]).

StatLink https://stat.link/58wtd0

Car-dependent and territories with high sprawl also perform poorly in terms of environmental sustainability and adaptation, which also has negative impacts on future well-being of citizens. High vehicle ownership results in disposal of used oils, tires and expired vehicles, which pollutes natural and coastal waterways and increases the use of scarce land for landfilling (ECLAC, 2021[25]). Car-dependent and sprawled territories are also very difficult to decarbonise. First, rapidly growing private car and motorcycle use will offset emissions reduction from cleaner vehicles (Lamb et al., 2021[69]). Second, decarbonisation is costly and slow as it implies incentivising the shift of large fleets of vehicles towards cleaner technologies. Third, efforts to decarbonise car-dependent and sprawled cities can lead to trade-offs between climate and other well-being outcomes such as equity, making them politically unattractive. Fourth, trade-offs between climate actions and wider environmental goals may hinder across-government collaboration. Sprawled territories also reduce the capacity of urban agglomerations to adapt to climate change. They are space-intensive, and dedicate most public space to car use,[16] reducing space available for green areas in cities (necessary to cope with heatwaves) and/or leading to built-in development expanding into natural ecosystems that formerly provided ecosystem services (such as water absorption to cope with floods).

Three vicious cycles to revert

As cities expanded, resources for public (especially mass transit) and active transport modes have lagged, especially outside of capital cities. Deregulation of public transport services in the region during the 1980s has also contributed to the erosion of public transport service, as profits were prioritised over maintenance, service expansion and upgrades.[17] In parallel to the prioritisation of infrastructure for car use, single-use development has been the norm in spatial planning. Both low-income social housing and high-income closed neighbourhoods are now concentrated in city outskirts, far from services and activities and often with lower density than in city centres – which drives up car dependency. Such policy prioritisation has led to transport and urban systems organised around car driving and is largely driven by three unsustainable and interconnected dynamics: *induced demand, urban sprawl, and the erosion of active shared modes of transport* (Figure 2.18). These dynamics are at the source of increased car use in the region. Between 2007 and 2014, private vehicle ownership increased by 35% in LAC (CAF, 2016[88]) – a trend that is still observed in the region and in other areas of the world.

Induced demand refers to the phenomenon by which public investment in roads for car use ends up causing more, rather than less, traffic congestion. The dynamic works like this: public investments in road capacity for car use (1) play a major role in fostering urban sprawl. As road capacity for cars (2) increases, so does the catchment area (7). Induced demand occurs when public investment in roads for car use ends up causing more, rather than less, congestion. As mapped out in Figure 2.18 public investment in roads for car use (1) leads to increased road capacity for cars (2). While the intended objective of these investments is to reduce travel time (and thus congestion) (3) they end up having the opposite effect. As travel time (congestion) (3) by car is reduced, the attractiveness of driving and owning a car (4) increases, inducing people to "choose" cars over other modes, and causing traffic volume (5) to increase. As traffic volume (5) goes up, so does congestion (3) leading to public pressure (6) on policymakers to reduce congestion. Most countries have responded to this pressure by investing further in road capacity for cars (1), which restarts the cycle, rather than solves the problem.

Both induced demand and urban sprawl exacerbate the erosion of public transport and active modes, the third vicious cycle at the source of increased car use and emissions. As more and more people are induced to drive cars, and as policymakers respond to that "choice" by further increasing the road capacity for cars (2), traffic volume (5) of motorised vehicles increases as does the space and funding allocated to these modes. Unsurprisingly, in such systems, the attractiveness of public transport and active modes (8) is low: public transport often takes longer and provides less access to places than driving a car and riding a bike is not safe.

Figure 2.18. **Policies designed with an analytical lens have led to car-dependent and sprawled cities**

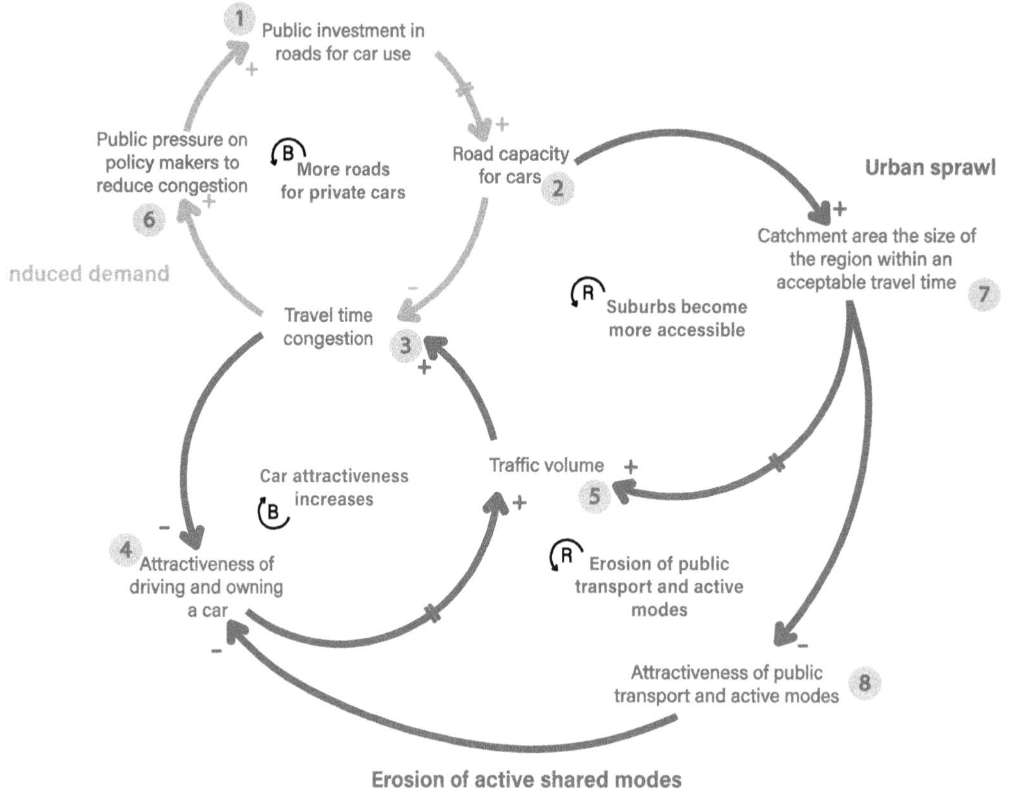

Source: (OECD, 2021[70]).

Taken together, these dynamics lead to territories in which most people need to travel long distances daily and private cars or motorcycles are the most attractive options for the bulk of these trips. As a result, these are the modes most people "choose" as soon as they can afford them. Understanding the dynamics above allows policy makers to see that this "choice" is not really an individual preference but the result of the systems design.

Applying a diet analogy, "unhealthy" transport systems are those in which most people use motorised vehicles (the sugar and fat in the diet analogy) for most of their trips (Figure 2.19). "Healthy" transport systems are those in which people can access places by walking, cycling and using micro- or shared mobility for most trips while high-emitting and space-intensive modes are used less frequently. Such a "diet" is possible thanks to: 1) the proximity between people and places; and 2) public space and investment being allocated to privilege active and shared modes, such that they are the most convenient and people choose them most often. By design, the "healthy system" needs less energy to function and has lower emissions, yields lower mobility but high accessibility,[18] offers more equitable and safe access to opportunities and promotes healthier lifestyles.

Figure 2.19. **From "unhealthy" to "healthy" transport systems**

Note: The icons illustrate the most frequent means of transportation used per type of trip.
Source: (OECD, 2021[70]).

Policies to transform territories and get to better environmental and well-being results

LAC countries need to focus efforts on policies that can transform their transport systems away from car dependency. Two of these policies are the following:

Public space reallocation is a key policy to revert the dynamic of induced demand and has also the potential to contain urban sprawl and the erosion of active and shared modes (Figure 2.18). Rebalancing space use between modes and uses could lead to "disappearing traffic", the opposite phenomenon than the increase in the use of cars and motorcycles experienced in LAC. Barcelona's superblocks are a good example of a wide scale plan for road space reallocation and street redesign, implemented with climate and well-being ambitions in parallel. Barcelona's superblocks *restructure* the city into polygons of 400 m x 400 m with inner roads closed to motorised vehicles. The superblocks become spaces welcoming active modes (walking, cycling) and recreation (Ajuntament de Barcelona, 2014[89]). Superblocks are often referred to as low-cost urbanism; they have demonstrated great potential to transform the urban ecosystem and bring about health, safety, social and environmental benefits in the short run (López, Ortega and Pardo, 2020[90]). Parking policy (e.g. through parking pricing and regulation) is also crucial to public space reallocation aligned with environmental and social goals. In Mexico City, minimum parking requirements that forced developers to build a minimum number of parking spaces in each project were abolished in 2017 and replaced with maximum parking requirements as a way to limit the space allocated to cars (Ciudad de México, 2017[91]). As part of this legal reform, minimum parking requirements for bicycles were also introduced (Guzmán, 2020[92]). While there are good examples in the region, road space reallocation and street redesign need to be implemented at a wide scale and across territories; and be prioritised in climate strategies.

Policies to mainstream shared mobility (including active modes and micro-mobility) are fundamental to reverse the erosion of active and shared transport modes and accelerate the development of multimodal and sustainable transport networks. Support[19] to mainstream shared bicycles and micro-mobility, as well as the expansion of on-demand micro-transit services can significantly increase the attractiveness of sustainable alternatives at a much faster pace than if solely focusing on existing public transport services. The development of digital tools will be key to harnes shared mobility. Global

Positioning System (GPS) technologies and apps today allow people to share vehicles (e.g. bikes) and rides and combine transport modes in cost-effective ways. If used for this purpose, these technologies could allow the shift from a system which requires car ownership to systems in which a multiplicity of transport modes (including shared electric cars) are available for people to choose from and combine according to particular needs. Importantly, increasing the feasibility and attractiveness of these modes will highly depend on the road space reallocation (discussed above) away from private car use. Thus, there is a high potential for change if implementing these two policies (mainstreaming of shared mobility and road space reallocation) in tandem.

The GHG emissions of the region's transport sector relative to GDP are 2.2 times as high as in Europe and 1.3 times as high as in Asia, which means there is ample scope to increase carbon efficiency (ECLAC, 2020[82]). If a better balance were achieved with electric railway transportation, the environmental performance of cargo transport would improve, while at the same time enhancing the competitiveness and flexibility of the sector. Decarbonising the transport sector in the region would also create 4 million new jobs in heavy vehicle operation and maintenance activities and more than 1.5 million in the light vehicle industry (UNEP, 2019[94]).

Implementing transformative policies can also increase the effectiveness and feasibility of other policies (e.g. carbon and road prices), which can help accelerate the transition towards systems that are sustainable by design. Moreover, electrifying transport will be more effective and rapid in a system that is no longer based on private vehicle ownership and use, but rather increases the participation of modes (e.g. micro-mobility, public transport) that are more resource efficient and which, as discussed in (IPCC AR6 WGIII, 2021[93]), have already a higher penetration of electric vehicles.

Infrastructure investments should shift away from hydrocarbon-based transport modes towards transport modes that allow for multimodal distribution. This could also lead to a reduction of negative impacts on ecosystems, a reduction of emissions and a better protection of biodiversity, among others. Transformative policies are especially relevant for small and medium-sized cities in LAC. As these cities are still expanding, early interventions that address the vicious cycles (Figure 2.18) can make these agglomerations sustainable by design, avoiding carbon lock-in and improving their climate resilience (OECD, forthcoming[96]).

The region is well-placed to produce the material basis for electric mobility. Three countries are major car manufacturers: Argentina, Brazil and Mexico. The Brazilian automotive industry accounts for 5% of GDP and employs 500 000 people directly and 1.3 million indirectly. In Mexico, the industry generated 3.7% of GDP and employed 824 000 people directly in 2017. In addition, three countries in the region, Argentina, Chile and Bolivia, have the world's largest reserves of lithium, and there are areas that are very well endowed with solar and wind energy that would allow hydrogen to be generated at very low costs. Chile and Peru also have large reserves of copper, a metal that is more in demand for the manufacture of electric vehicles than those with internal combustion engines (Chapter 3) (ECLAC, 2020[82]).

Recovery pathways from a systemic lens

Focusing COVID-19 recovery on low-carbon strategies that set the basis for a systemic transformation could accelerate the transition towards transport systems based on renewable energies, able to simultaneously reduce income inequalities and emissions. The transport and residential sectors are determinant for the region's current and future energy consumption, emissions and well-being (IDB, 2016[95]); yet there is little focus on them on recovery packages. Principally, recovery packages have been mainly focused on recovering employment and compensating the loss of household revenues (OECD et al., 2021[2]). Although these policies are fundamental to offset the COVID-19 crisis, moving forward requires holistic recovery packages that address the structural causes of social and economic vulnerability as well as the already-existing climate challenges.

The following tables categorise transport and residential sector policies found in different LAC country recovery plans[20] according to the three recovery pathways: Rebound, Decoupling and Wider well-being. The tables assess whether the transport and residential sector policies align with the Wider well-being pathway or rather are locking countries into less effective development pathways (Table 2.1 and Table 2.2).

Table 2.1. **Rebound, Decoupling and Wider well-being pathways for the transport sector**

	Transport	Selected recovery policies implemented
Rebound: fostering car-dependency	The focus is on boosting economic growth, jobs and disposable incomes by maintaining and reinforcing car-dependent systems. The policy rationale maintains a "traditional vision" of mobility – e.g. physical movement and speed – as central performance indicators for the sector. It also reflects the belief of a positive correlation and virtuous cycle between transport volumes and GDP as the ultimate goal for the economy. Because the focus is on mobility, the role of proximity between people and places is ignored and mobility by car – a space- and carbon-intensive mode – is privileged.	Chile: Containment of paraffin, petrol and benzine prices. Panama: Infrastructure projects focused on highway improvement and enlargement. Colombia: Project *Concluir* to finish the construction of 400 km across 27 road projects in 23 departments. Construction of 21 road sections for legality and reactivation in 18 departments.
Decoupling: promoting clean car-dependency	The mind-set is still around "supporting mobility for economic growth". As such, the aim is not to transform transport systems but rather to decarbonise existing (e.g. car-dependent and mobility-intensive) ones. Mitigation efforts concentrate, to a great extent, on improving parts (e.g. vehicles and fuels), while maintaining current systems. Efforts focus on improving the energy efficiency and reducing the carbon intensity of the vehicle fleet (especially private cars) and the fuels they use, fostering improvement of vehicle technologies. As significant travel reduction and modal shift will not be main drivers of mitigation in this pathway, "shift" and "avoid" type policies and measures have a smaller role than "improve" actions.	Uruguay: Electric mobility solutions with major focus on electric private cars. Panama: Definition of the National Strategy of Electro Mobility.
Wider well-being: shifting away from car-dependency while promoting clean vehicles	Wider well-being shifts the focus from mobility to sustainable accessibility, opening the door to envision systems that significantly reduce emissions not only by reducing mobility, but also by increasing "proximity" and "access". It places emphasis on reversing and shifting away from car-dependency while simultaneously improving the vehicles that are still needed (e.g. public buses). It prioritises the use of space for space efficient and low/zero carbon modes (walking, cycling, micro-mobility and public transport) while focusing important action on reallocation of road space and, in cases where policies have historically prioritised building infrastructure, redesigning streets for car use. Incentives for electric vehicles (EVs, including charging infrastructure) is provided and planned with the aim of embedding electrification in the wider aim of shifting away from a system based on privately owned cars with low occupancy.	Uruguay: Programme to increase the use of public and active travel (MOVÉS). Argentina: Strengthen the metropolitan transport agency, extend Ecobici's capacity, and reallocate road space in Avenida del Libertador (Buenos Aires). Mexico: Permanent cycling path Insurgentes and substituting moto-taxis with e-bikes –Mobility project Tláhuac (Mexico City) Colombia: Public Policy for Bikes 2021-2039 (Bogota). Chile: National Strategy of Sustainable Mobility and National Strategy of Electro-mobility.

Sources: (Buckle et al., 2020[97]); (OECD, 2021[70]); (Gobierno de Chile, 2021[24]); (Gobierno de Panama, 2020[98]); (Greenpeace México, 2021[99]); (Gobierno de Buenos Aires, 2022[100]); (Alcaldía Mayor de Bogotá D.C., 2021[101]).

Two types of transport measures aligned with a rebound recovery pathway were the containment of energy prices, including fuels. These initiatives incentivise car use and emissions and use up resources that could bring better social and environmental value if used differently (Carlino et al., 2015[102]). The second measure found in Panama and Colombia's recovery plans is further investment in the improvement and widening of car purposed infrastructure (e.g. highways). Panama's electro mobility strategy was found to be consistent with a decoupling type of recovery, aiming to electrify 10-20% of private vehicles and only 15-50% of public buses, but doesn't transform the system's demand maintaining the prevalence of private car use vs. electric public transport.

Regarding wider well-being, Chile's National Strategy of Sustainable Mobility follows this pathway as it aims to achieve a better use of energetic resources, time, and road and urban space. It plans infrastructure projects to prioritize active transport and massive public transport over private transport. City level recovery policies stand out for being in line with the wider well-being pathway. Among these are programmes to incentivise bike-use while also improving streets for its safe and convenient use (including by increasing cycling lanes and parking), like in Bogota, Santiago, and Buenos Aires. In some cases, cycling lanes introduced during COVID-19, have been made permanent (Buckle et al., 2020[97]), as in Mexico City with Insurgentes Avenue (Greenpeace México, 2021[99]).

Table 2.2. **Rebound, Decoupling and Wider well-being pathways for the residential sector**

	Residential	Selected recovery policies implemented
Rebound	The focus is on stimulating short-term growth and employment opportunities in the construction sector. The vision of the sector is narrow and thus concentrates on the scale of the dwelling or building (focused more on delivering housing than on its quality), disregarding location or the wider living environment (e.g. surrounding areas or services and connections around the dwelling). Also focuses on short-term – and especially private – costs and benefits, disregarding full-cost accounting of wider and longer-term well-being costs and benefits.	Chile: Subsidies to limit the rise of gas prices. Panama: Support for first house purchase for lower income households (Fondo Solidario de Vivienda, with no energy efficiency or quality standards. Peru: Financing for construction of households and public spaces, but with no energy efficiency or quality standards. Colombia: Subsidies to interest rates for the financing of new urban housing, but with no energy efficiency or quality standards.
Decoupling	The main objectives are to foster growth and provide "access to shelter" while also decreasing emissions from energy use. Most efforts focus on attaining energy efficiency gains in buildings/dwellings. As in the case of Rebound, the scope of focus is narrow. Thus, Decoupling does not look beyond the dwelling and misses on options to reduce energy use by modifying the surrounding environment (e.g. use of greenspace to regulate the micro-climate, thereby reducing cooling needs). Decoupling does not integrate full-cost accounting; thus, while some deep retrofits may be pursued, these do not become the norm. Rather most efforts focus on shallow retrofits.[21]	Colombia: Programmes to replace home appliances with more efficient ones. Panama: Design of Sustainable Building Regulations and the National Cooling Plan.
Wider Well-being	This pathway considers housing as a "bundled good" that should deliver a number of functions beyond access to shelter. It prioritises measures that reduce emissions in the residential sector while also facilitating emissions reduction in other sectors. It puts emphasis on new builds and retrofits that substantially lower energy demand (e.g. passive houses)[22] and are potentially accompanied with low-carbon energy generation (e.g. rooftop solar). Full-cost accounting is embraced by public and private actors, mainstreaming considerations (short- and long-term) on health and wider well-being when evaluating projects. Wider well-being also takes into account the need to lower energy demand by considering the surrounding environment (e.g. housing location and transport connections, existence of green space to regulate the microclimate and reduce heating or cooling energy needs).	N/A

Sources: (Buckle et al., 2020[97]); (Gobierno de Chile, 2022[103]); (Departamento Nacional de Planeación, 2021[104]); (Gobierno de Panama, 2020[98]); (El Comercio, 2020[105]).

In terms of the residential sector, two types of measures were found to be aligned with a rebound type of pathway: subsidies for gas prices and housing programmes or subsidies in Panama, Peru, and Colombia that did not include any considerations of efficiency standards, nor quality of the dwelling nor the surroundings; including location and transport connections by sustainable modes, which can easily result in further fostering sprawl (Buckle et al., 2020[97]; OECD, 2021[70]). Colombia implemented a programme consistent with a Decoupling pathway, focused on identifying inefficient appliances and supporting the population in replacing them. Panama recently designed Sustainable Building Regulations that aim to save 15% of energy use in the construction of new buildings in the following two years and up to 20% in subsequent years. No recovery policies were found to be aligned with a wider well-being pathway.

Overall, revisiting recovery policies in the light of a systemic approach would be relevant for the region. In certain cases, a same country was found to implement recovery measures that align with different pathways. The risk of this is that investments will result in policy incoherence and conflicting goals not aligned with a sustainable and inclusive model. The recovery pathways here presented can serve countries as a guide to rethink the policies included in their recovery packages and redesign strategies that address social, economic, and environmental issues at the same time. If conceived systemically, investments in transport and urban systems could play a crucial role in improving well-being while also contributing to long-term collective climate goals.

Key policy messages

The green transition is a structural challenge that LAC will have to face eventually. The rising emissions coming from LAC, and the disproportionate effect that climate change has in the region and on its most vulnerable groups, are evidence of how urgent it is to address a green and just transition. At the same time, LAC is better positioned than other regions in the world to adopt a more sustainable and inclusive economic model, given the richness of biodiversity and potential for renewable energies.

The post-COVID-19 context presents itself as an opportunity for governments to align the objectives of the recovery with those of a green transition, looking beyond GDP and placing Latin American and Caribbean's well-being at the centre. A truly just transition should help close the existing social gaps that characterise LAC and avoid generating new ones. A systemic approach should guide active mitigation and adaptation policies that could help reduce inequalities across countries, socio-economic groups, territories, generations and genders. In this regard, the articulation of preventive climate change and risk management policies with universal, integral, resilient and sustainable social protection systems is key for a systemic and multi-dimensional approach.

Governments should focus on reversing unsustainable dynamics and transform territories into systems that, by their functioning, allow a sustainable and just society. With a systemic lens, governments could foster various trends by (re)designing systems able to improve people's well-being while requiring less energy, producing fewer emissions and transforming what used to be trade-offs between climate and well-being objectives into synergies.

Box 2.1. Key policy messages

- Adopt a recovery strategy based on low emissions and centred on well-being in LAC. Make use of lessons learned from recovery pathways to refocus policy decisions towards improving citizens' well-being and advancing transformative policies. The path towards net zero emissions should not be limited to follow other region's mitigation and adaptation strategies, but rather it should be the result of a comprehensive analysis of how to achieve a green and just transition taking into account LAC and its subregions' particular characteristics.
- Promote a systemic approach to reprioritise climate action towards improving systems' functioning rather than focusing on improving parts (e.g. more highways for private car use) while keeping unsustainable systems (e.g. private car-dependent territories). Some relevant actions are: designing and implementing climate, transport and land-use policies focused on reversing the unsustainable system dynamics underlying car-dependency (e.g. induced demand, urban sprawl and the erosion of share and active modes). Climate strategies that are focused on improving only specific parts fail to do so and are, as a result, ineffective in achieving emissions reduction at the pace and scale needed.
- Include the environmental dimension in education curricula to raise the environmental awareness among future generations.
- Invest in and develop data technologies and reliable information to better assess the complexity of mitigation and adaptation policies. To ensure a correct implementation and policy coherence in the short and long term, design systemic, sustainable decarbonisation strategies that address the particularities of each subregion and country in LAC. This will allow targeted programmes that include tailored solutions to reduce the GHG emissions at the sectoral level (e.g. transport, electricity, agriculture and livestock) while advancing holistic and cross-cutting adaptation measures that connect local needs with global commitments.
- Strengthen further climate risk preparedness measures and multi-hazard early warning systems, especially in the Caribbean. Gaining support from the international and the scientific and technological communities is essential to reinforce such systems.
- Identify the needs and characteristics of the communities and geographical areas most exposed to the negative impacts of climate change as a basis for strengthening climate resilience. Establishing climate risk repositories and risk maps, shared with all relevant stakeholders, is essential to inform climate change adaptation measures and prioritisation. These measures can help lower risk exposure of most vulnerable groups to climate change (e.g. women in rural areas).
- Identify the particularities and challenges of growing urbanisation in LAC to design more just, resilient and green cities. A greener economy should reduce urban pollution (particularly air pollution) and the exposure of the most vulnerable groups to their harmful health impacts. Sustainable and resilient city design should contribute to reduce coastal erosion and manage fragile coastal ecosystems to provide nature-based solutions against the impacts of climate risks (e.g. coastal floods, hurricanes, sea level rise), especially in the Caribbean.

Notes

1. As defined by the Intergovernmental Panel on Climate Change's (IPCC), "mitigation" constitutes human efforts to reduce the sources GHGs, while "adaptation" is the process of adjustment to actual or expected climate effects (IPCC, 2014[106]).
2. Whenever Historical GHG Emissions from Climate Watch (2022[14]) were used, the Climate Analysis Indicators Tool (CAIT) was chosen as the data source. The CAIT dataset is the most comprehensive on Climate Watch and includes all sectors and gases. In order to emphasise data comparability across countries, it does not use countries' official inventories reported to the United Nations Framework Convention on Climate Change. Climate Watch Historical GHG Emissions are derived from several sources. The original source for the LUCF or agriculture indicators is FAO (2022[12]), FAOSTAT Emissions. For fuel combustion data, it is OECD/IEA (2021[13]), GHG Emissions from Fuel Combustion.
3. South America includes data for Argentina, Brazil, Bolivia, Chile, Colombia, Ecuador, Paraguay, Peru, Uruguay and Venezuela. Central America includes data for Costa Rica, El Salvador, Guatemala, Honduras, Mexico, Nicaragua, and Panama. The Caribbean region includes data for Antigua and Barbuda, Bahamas, Barbados, Belize, Cuba, Dominica, Dominican Republic, Grenada, Guyana, Haiti, Jamaica, Saint Kitts and Nevis, Saint Lucia, Saint Vincent and the Grenadines, Suriname, and Trinidad and Tobago.
4. See endnote 2.
5. GHG emissions excluding LUCF, to guarantee greater accuracy in comparisons.
6. See endnote 2.
7. See endnote 2.
8. See endnote 2.
9. See endnote 2.
10. The LAC countries ranked within the top 50 most vulnerable between 2000-19 were: Bolivia (25), Colombia (38), Dominica (11), Dominican Republic (50), El Salvador (28), Grenada (24), Guatemala (16), Haiti (3), Honduras (44), Nicaragua (35), Puerto Rico (1), The Bahamas (6), St. Vincent and the Grenadines (48) (Germanwatch, 2021[26]).
11. See endnote 10.
12. As discussed in (Buckle et al., 2020[97]) while economic growth (expressed in terms of GDP) can be correlated with well-being in some respects, this relationship may also be inexistent or negative in other ways. Taking GDP growth as proxy for success has led to high energy and material demand systems that do not necessarily deliver high well-being, and increase the challenges to decarbonise at the scale and pace needed. This is why the highest GDP growth pathway is not necessarily the highest well-being pathway.
13. The term "systems innovation" was coined around 20 years ago and can be defined as the application of a systemic approach to solve real-world problems.
14. Several Latin American capitals feature among the most congested in international rankings (ITF, 2020[85]).
15. While 47% of rich households own at least one car, only 8% of low-income families do (Daude et al., 2017[108]).
16. Around 80% of public space in cities is dedicated to car use according to (Mc Arthur et al., 2022[109]).
17. More recently, Mexico City, Santiago and Lima, among others, have attempted to re-regulate public transport and expand services, especially through the introduction of bus rapid transit (BRT) systems, and initiatives to increase active modes have emerged in the region. While important, such efforts have not been able to counteract growing car ownership and use. One of the reasons for this is that they "fight" against the unsustainable dynamics illustrated in Figure 2.18 and Figure 2.19 and explained in this section.
18. Transport policy literature suggest that transport systems' contribution to human well-being lies on the provision of accessibility, i.e. on enabling ease of access to opportunities and places of interest (e.g. jobs, consumption, leisure or health services). Most transport systems today focus instead on the provision of mobility, which results in the car-dependent and sprawled territories illustrated in this section. For more on this, see Chapter 2 of the report *Transport strategies for net-zero systems by design* (OECD, 2021[70]).
19. Integrated transport subscription cards could facilitate the use of available options, and facilitate the provision of government subsidies to low-income households if needed. Government subsidies may also foster the development of shared mobility in areas where private on-demand services can bring social and environmental benefits but may not be

profitable for the private sector. Support to the development of new vehicles (e.g. innovative micro-mobility) and the expansion of services for multipurpose trips (e.g. cargo e-bikes, shared (e-)bikes with baby seats, kids' bikes) could also contribute to making shared and sustainable mobility more attractive.

20. The national recovery packages of Chile (Chile Apoya), Colombia (Nuevo Compromiso por el Futuro de Colombia), Panama (Plan para la Recuperación Económica) and Peru (Arranca Perú) were revised. Transport and residential policies were also revised individually for these countries and for cities such as Buenos Aires (Argentina), Bogota (Colombia) and Mexico City (Mexico).
21. Shallow retrofits are one-off measures instead of deep retrofits, which reduce energy usage beyond 50%
22. For passive houses, total primary energy demand should not exceed 120 kWh per m^2 annually for all services.

References

AFD (2022), *Islands and Coastal Cities Faced with Climate Change: Increasing the Resilience of Island Communities*, Agence française de développement, Paris, https://www.afd.fr/en/actualites/islands-and-coastal-cities-faced-climate-change-increasing-resilience-island-communities. [47]

AFD (2020), *Territorial and Ecological Transition – 2020-2024 Strategy*, Agence française de développement, Paris, https://www.afd.fr/en/ressources/territorial-and-ecological-transition-2020-2024-strategy?origin=/en/ressources-accueil?query=&sort=counter,desc&size=15&filter[0]=source_k=afd&filter[1]=type_k=resource&filter[2]=thematic_k=Biodiversity&filter[3]=thematic. [4]

Ajuntament de Barcelona (2014), *Metropolitan Area Urban Mobility Plans*. [89]

Alcaldía Mayor de Bogotá D.C. (2021), *Documento CONPES 15 "Política Pública de la Bicicleta 2021-2039"*, Alcaldía Mayor de Bogotá D.C., Bogotá, https://www.sdp.gov.co/sites/default/files/doc_conpes_dc_pp_bicicleta_-20210224_vconpes_0.pdf. [101]

Alejos, L. (2021), *What are the fiscal risks from extreme weather events and how can we deal with them?*, Inter-American Development Bank, Washington, DC, https://blogs.iadb.org/gestion-fiscal/en/what-are-the-fiscal-risks-from-extreme-weather-events-and-how-can-we-deal-with-them/#:~:text=It%20is%20estimated%20that%20the,income%20countries (Figure 2). [33]

Alejos, L. (2018), *Three Essays in Public Finance in Developing Countries*, University of Michigan, Ann Arbor, MI, https://deepblue.lib.umich.edu/bitstream/handle/2027.42/147524/lalejos_1.pdf?sequence=1&isAllowed=y. [30]

Bárcena, A. et al. (2020), *The climate emergency in Latin America and the Caribbean. The path ahead – resignation or action?*, Economic Commission for Latin America and the Caribbean, Santiago, https://repositorio.cepal.org/bitstream/handle/11362/45678/10/S1900710_en.pdf. [6]

Bárcena, et al. (2018), *Economics of climate change in Latin America and the Caribbean*, United Nations publication, https://repositorio.cepal.org/bitstream/handle/11362/43889/1/S1800475_en.pdf. [58]

Bleeker, A. et al. (2021), *Advancing gender equality in environmental migration and disaster displacement in the Caribbean*, Studies and Perspectives series, No. 98, Economic Commission for Latin America and the Caribbean Subregional, Santiago, https://repositorio.cepal.org/bitstream/handle/11362/46737/1/S2000992_en.pdf. [36]

Buchner, B. et al. (2021), *Global Landscape of Climate Finance 2021*, Climate Policy Initiative, San Francisco, https://www.climatepolicyinitiative.org/publication/global-landscape-of-climate-finance-2021/. [43]

Buckle, S. et al. (2020), *Addressing the COVID-19 and climate crises: Potential economic recovery pathways and their implications for climate change mitigation, NDCs and broader socio-economic goals*, OECD/IEA Climate Change Expert Group Papers, https://www.oecd-ilibrary.org/environment/addressing-the-covid-19-and-climate-crises_50abd39c-en (accessed on 16 March 2021). [97]

Buckle, S. et al. (2020), "Addressing the COVID and climate crises: Potential economic recovery pathways and their implications for climate change mitigation, NDCs and broader socio-economic goals", *OECD/IEA Climate Change Expert Group Papers*, No. 2020/04, OECD Publishing, Paris, https://doi.org/10.1787/50abd39c-en. [107]

CAF (2020), *RED 2020: Pension and health systems in Latin America. The challenges of ageing, technological change and informality*, Development Bank of Latin America, Caracas, http://scioteca.caf.com/handle/123456789/1652. [65]

CAF (2016), *Observatorio de movilidad urbana*, Development Bank of Latin America, Caracas, https://www.caf.com/es/conocimiento/datos/observatorio-de-movilidad-urbana/ (accessed on 16 May 2022). [88]

CAF (2014), *Vulnerability Index to climate change in the Latin American and Caribbean Region*, Development Bank of Latin America, Caracas, https://scioteca.caf.com/handle/123456789/509. [28]

Carlino, H. et al. (2015), "Fossil fuel subsidies in Latin America: the challenge of a perverse incentives structure", *Institut du développement durable et des relations internationales*, http://www.iddri.org (accessed on 4 July 2022). [102]

Ciudad de México (2017), *Nueva Norma de Estacionamientos*, http://www.adocac.mx/web/images/descargas/NormaEstacionamiento.pdf (accessed on 16 June 2022). [91]

Climate Watch (2022), *Historical GHG Emissions*, World Resources Institute, Washington, DC, https://www.climatewatchdata.org/ghg-emissions. [14]

Daude, C. et al. (2017), *RED 2017. Crecimiento urbano y acceso a oportunidades: un desafío para América Latina*, Development Bank of Latin America, Bogotá, http://scioteca.caf.com/handle/123456789/1090 (accessed on 16 May 2022). [108]

Departamento Nacional de Planeación (2021), *Documento CONPES 4023 Política para la reactivación, la repotenciación y el crecimiento sostenible e incluyente: Nuevo compromiso por el futuro de Colombia*, Departamento Nacional de Planeación, Bogotá, https://colaboracion.dnp.gov.co/CDT/Conpes/Econ%c3%b3micos/4023.pdf. [104]

ECLAC (2022), *How to finance sustainable development. Recovery from the effects of COVID-19 in Latin America and the Caribbean*, Economic Comission for Latin America and the Caribbean, Santiago, https://www.cepal.org/en/publications/47721-how-finance-sustainable-development-recovery-effects-covid-19-latin-america-and. [46]

ECLAC (2021), *Conceptualizing a circular economy in the Caribbean: perspectives and possibilities*, Economic Commission for Latin America and the Caribbean, Santiago, https://repositorio.cepal.org/bitstream/handle/11362/47604/LCCAR2021_07_en.pdf?sequence=1&isAllowed=y. [25]

ECLAC (2021), *Forest loss in Latina America and the Caribbean from 1990 to 2020: the statistical evidence*, https://repositorio.cepal.org/bitstream/handle/11362/47152/1/S2100265_en.pdf. [20]

ECLAC (2020), *Building a New Future: Transformative Recovery with Equality and Sustainability*, http://hdl.handle.net/11362/46226 [52]

ECLAC (2020), *The evolution of modal split in freight transport in South America, 2014–2017*, Economic Commission for Latin America and the Caribbean, Santiago, https://www.cepal.org/en/publications/45668-evolution-modal-split-freight-transport-south-america-2014-2017 (accessed on 4 July 2022). [82]

ECLAC/DFID (2010), *The Economics of Climate Change in Central America: Summary 2010*, Economic Commission for Latin America and the Caribbean/Department for International Development, Santiago/London, https://repositorio.cepal.org/bitstream/handle/11362/35229/1/lcmexl978i.pdf. [41]

ECLAC/ILO (2022), *Employment Situation in Latin America and the Caribbean. Real wages during the pandemic: Trends and challenges*, ECLAC-ILO Bulletin, No. 26, Economic Commission for Latin America and the Caribbean/International Labour Organization, Santiago/Geneva, https://hdl.handle.net/11362/47927. [50]

ECLAC/MINURVI/UN-Habitat (2017), *Regional Action Plan for the implementation of the New Urban Agenda in Latin America and the Caribbean 2016-2036*, Economic Commission for Latin America and the Caribbean, Santiago, https://unhabitat.org/regional-action-plan-for-the-implementation-of-the-new-urban-agenda-in-latin-america-and-the. [72]

Ecosystem Profile (2019), *The Caribbean Islands Biodiversity Hotspot*, Critical Ecosystem Partnership Fund, Arlington, VA, https://www.cepf.net/sites/default/files/cepf-caribbean-islands-ecosystem-profile-december-2020-english.pdf. [77]

El Comercio (2020), "Arranca Perú", *El Comercio*, https://elcomercio.pe/respuestas/en-que-consiste-arranca-peru-y-cuantos-empleos-se-generaran-arranca-peru-programa-coronavirus-covid-19-millon-de-empleos-vivienda-trabajo-agricultura-transporte-pandemia-revtli-noticia/. [105]

EM-DAT (2022), *EM-DAT Public [database]*, Emergency Events Database, Brussels, https://www.emdat.be/database (accessed on 22 May 2022). [31]

European Forest Institute (2019), "Planted forest: The big opportunity for forest recovery in Chile and Uruguay", *European Forest Institute*, https://efi.int/articles/planted-forest-big-opportunity-forest-recovery-chile-and-uruguay. [67]

FAO (2022), *FAOSTAT Emissions*, Food and Agriculture Organization, Rome, https://www.fao.org/food-agriculture-statistics/data-release/data-release-detail/en/c/1304919/. [12]

FAO (2018), *FAOSTAT Surface Area 1961-2018*, Food and Agriculture Organization, Rome, https://www.fao.org/faostat/en/#data. [34]

Florczyk, A. et al. (2019), *GHS Urban Centre Database 2015, multitemporal and multidimensional attributes*, European Commission, Brussels, https://data.jrc.ec.europa.eu/dataset/53473144-b88c-44bc-b4a3-4583ed1f547e. [74]

Germanwatch (2021), *Global Climate Risk Index 2021. Who suffers most from extreme weather events? Weather-related loss events in 2019 and 2000-2019*, Germanwatch e.V., Bonn, https://www.germanwatch.org/sites/default/files/Global%20Climate%20Risk%20Index%202021_2.pdf (accessed on 30 August 2022). [26]

Germanwatch (2020), *Global Climate Risk Index 2020: Who Suffers Most from Extreme Weather Events? Weather-Related Loss Events in 2018 and 1999 to 2018*, Briefing Paper, Germanwatch e.V., Bonn, Germany, https://www.germanwatch.org/sites/default/files/20-2-01e%20Global%20Climate%20Risk%20Index%202020_14.pdf. [35]

Gobierno de Buenos Aires (2022), *Movilidad*, https://www.buenosaires.gob.ar/jefaturadegabinete/movilidad. [100]

Gobierno de Chile (2022), *Chile Apoya Plan de Recuperación Inclusiva*, Gobierno de Chile, Santiago, https://www.gob.cl/chileapoya/. [103]

Gobierno de Chile (2021), *Chile announces that it will work to put an end to coal use by 2030 after joining the Powering Past Coal Alliance*, https://www.gob.cl/en/news/chile-announces-it-will-work-put-end-coal-use-2030-after-joining-powering-past-coal-alliance/. [24]

Gobierno de Panama (2020), *Plan para la Recuperación Económica de Pánama: Primer año de gestión. Julio 2019-Julio 2020*, Gobierno de Panama, Panama City, https://www.mef.gob.pa/wp-content/uploads/2020/07/Plan_Economico_2020.pdf. [98]

Gouveia, N. et al. (2019), "Ambient PM2.5 in Latin American cities: population exposure, trends, associated urban factors, and effects on mortality", *Environmental Epidemiology*, Vol. 3, No. 139, International Society for Environmental Epidemiology, Herndon, VA, https://doi.org/10.1097/01.EE9.0000607280.86151.6c. [75]

Greenpeace México (2021), *Los pasos hacia una movilidad sustentable en la CDMX - Greenpeace México*, https://www.greenpeace.org/mexico/blog/9959/los-pasos-hacia-una-movilidad-sustentable-en-la-cdmx/ (accessed on 4 July 2022). [99]

Guivarch, C., N. Taconet and A. Méjean (2021), *Linking Climate and Inequality*, International Monetary Fund, Washington, DC, https://www.imf.org/en/Publications/fandd/issues/2021/09/climate-change-and-inequality-guivarch-mejean-taconet. [15]

Guzmán, J. (2020), *De mínimos a máximos: cómo los estacionamientos en Ciudad de México están cambiando*, Transecto, https://transecto.com/2020/12/de-minimos-a-maximos-como-los-estacionamientos-en-ciudad-de-mexico-estan-cambiando/ (accessed on 16 June 2022). [92]

Hickey, A. and V. Wellenstein (2021), *10 key points on climate change impacts, opportunities and priorities for Latin America and the Caribbean*, World Bank Blogs, World Bank, Washington, DC, https://blogs.worldbank.org/latinamerica/10-key-points-climate-change-impacts-opportunities-and-priorities-latin-america-and. [42]

IDB (2021), *Climate Change Impacts on Agriculture in Latin America and the Caribbean: An Application of the Integrated Economic-Environmental Modeling (IEEM) Platform*, Inter-American Development Bank, Washington, DC, https://publications.iadb.org/publications/english/document/Climate-Change-Impacts-on-Agriculture-in-Latin-America-and-the-Caribbean-An-Application-of-the-Integrated-Economic-Environmental-Modeling-IEEM-Platform.pdf. [57]

IDB (2021), *Fiscal Policy and Climate Change: Recent Experiences of Finance Ministries in Latin America and the Caribbean*, Inter-American Development Bank, Washington, DC, https://publications.iadb.org/publications/english/document/Fiscal-Policy-and-Climate-Change-Recent-Experiences-of-Finance-Ministries-in-Latin-America-and-the-Caribbean.pdf. [32]

IDB (2016), *Lights On?: Energy Needs in Latin America and the Caribbean to 2040 | Publications*, https://publications.iadb.org/publications/english/document/Lights-On-Energy-Needs-in-Latin-America-and-the-Caribbean-to-2040.pdf (accessed on 4 July 2022). [95]

IDB/ITF (forthcoming), *Developing accessibility indicators for Latin American Cities: Mexico City Metropolitan Area, Bogota and Santiago de Chile*. [81]

IEA (2021), *Central & South America Total Energy Supply, 2019*, International Energy Agency, Paris, https://www.iea.org/regions/central-south-america. [19]

IEA (2021), *Global Energy Review 2021*, International Energy Agency, Paris, http://www.iea.org/reports/global-energy-review-2021. [8]

IFS (2018), *Climate change and agriculture: farmer adaptation to extreme heat*, IFS Working Paper W18/06, Institute for Fiscal Studies, London, https://ifs.org.uk/uploads/WP201806.pdf. [59]

Ined (2022), *World Population - Estimations 2022*, Institut national d'études démographiques, Paris, http://www.ined.fr/en/everything_about_population/data/all-countries/?lst_continent=900&lst_pays= (accessed on August 2022). [10]

Internal Displacement Monitoring Centre (2022), *Global Report on Internal Displacement 2022*, Internal Displacement Monitoring Centre, Geneva, Switzerland, https://www.internal-displacement.org/sites/default/files/publications/documents/IDMC_GRID_2022_LR.pdf. [49]

IPCC (2022), *Climate Change 2022: Impacts, Adaptation and Vulnerability. Working Group II Contribution to the IPCC Sixth Assessment Report*, Cambridge University Press, Cambridge/New York, https://www.ipcc.ch/report/sixth-assessment-report-working-group-ii/. [29]

IPCC (2022), *Global Warming of 1.5°C: IPCC Special Report on Impacts of Global Warming of 1.5°C above Pre-industrial Levels in Context of Strengthening Response to Climate Change, Sustainable Development, and Efforts to Eradicate Poverty*, Cambridge University Press, Cambridge/New York, https://doi.org/10.1017/9781009157940. [5]

IPCC (2021), *Climate Change 2021: The Physical Science Basis. Contribution of Working Group I to the Sixth Assessment Report of the Intergovernmental Panel on Climate Change*, Cambridge University Press, Cambridge/New York, https://www.ipcc.ch/report/ar6/wg1/. [7]

IPCC (2018), *Special Report Global Warming of 1.5°C*, Cambridge University Press, Cambridge/New York, https://www.ipcc.ch/site/assets/uploads/sites/2/2022/06/SR15_Full_Report_HR.pdf. [1]

IPCC (2014), *Annex II: Glossary in Climate Change 2014: Synthesis Report. Contribution of Working Groups I, II and III to the Fifth Assessment Report of the Intergovernmental Panel on Climate Change*, Intergovernmental Panel on Climate Change, Geneva, Switzerland, https://www.ipcc.ch/site/assets/uploads/2019/01/SYRAR5-Glossary_en.pdf. [106]

IPCC AR6 WGIII (2021), *Chapter 10. Transport*, https://report.ipcc.ch/ar6wg3/pdf/IPCC_AR6_WGIII_FinalDraft_Chapter10.pdf. [93]

ITF (2022), "Mode Choice in Freight Transport", *ITF Research Reports* [online], International Transport Forum, https://www.itf-oecd.org/sites/default/files/docs/mode-choice-freight-transport.pdf. [87]

ITF (2020), *Congestion in Latin American Cities: Innovative Approaches for a Critical Issue Discussion Paper*, OECD Publishing, Paris, https://doi.org/10.1787/938de08e-en (accessed on 16 May 2022). [85]

ITF (2019), "Benchmarking Accessibility in Cities", *International Transport Forum Policy Papers*, No. 68, OECD Publishing, Paris, http://www.itf-oecd.org/benchmarking-accessibility-cities (accessed on 10 June 2022). [80]

ITF (forthcoming), *Developing Accessibility Indicators for Latin American Cities*, International Transport Forum, Paris. [86]

IUCN/GGO (2015), *Roots for the Future: The Landscape and Way Forward on Gender and Climate Change*, International Union for Conservation of Nature/Global Gender and Climate Alliance, Washington, DC, https://portals.iucn.org/library/sites/library/files/documents/2015-039.pdf. [62]

Lamb, W. et al. (2021), "A review of trends and drivers of greenhouse gas emissions by sector from 1990 to 2018", *Environmental Research Letters*, Vol. 16/7, Institute of Physics (IOP) Publishing, Bristol, UK, https://doi.org/10.1088/1748-9326/abee4e. [69]

Landrigan, P. et al. (2017), *The Lancet Commission on Pollution and Health*, Elsevier, Amsterdam, https://doi.org/10.1016/S0140-6736(17)32345-0. [56]

Lardé, J. (2021), *Investing in sustainable, resilient and inclusive infrastructure for economic recovery*, Economic Commission for Latin America and the Caribbean, Santiago, https://www.cepal.org/en/publications/47675-investing-sustainable-resilient-and-inclusive-infrastructure-economic-recovery (accessed on 4 July 2022). [83]

López, I., J. Ortega and M. Pardo (2020), "Mobility Infrastructures in Cities and Climate Change: An Analysis Through the Superblocks in Barcelona", *Atmosphere*, Vol. 11/4, p. 410, https://doi.org/10.3390/atmos11040410. [90]

Mc Arthur, J. et al. (2022), "Better streets for better cities". [109]

McCarthy, J. (2020), "Why Climate Change and Poverty Are Inextricably Linked: Fighting one problem helps mitigate the other", *Global Citizen*, https://www.globalcitizen.org/en/content/climate-change-is-connected-to-poverty/ (accessed on 2 June 2022). [37]

OCHA (2020), *Natural Disasters in Latin America and the Caribbean 2000-2019*, United Nations Office for the Coordination of Humanitarian Affairs, New York, https://reliefweb.int/sites/reliefweb.int/files/resources/20191203-ocha-desastres_naturales.pdf. [27]

OECD (2022), *The well-being lens: An innovative process for net-zero strategies*, Brochure, OECD Publishing, Paris, https://www.oecd.org/climate-change/well-being-lens/well-being-lens-brochure.pdf. [71]

OECD (2021), *Gender and the Environment. Building Evidence and Policies to Achieve the SDGs*, OECD (2021), Gender and the Environment: Building Evidence and Policies to Achieve the SDGs, OECD Publishing, Paris, https://doi.org/10.1787/3d32ca39-en. [60]

OECD (2021), *How's Life in Latin America?: Measuring Well-being for Policy Making*, OECD Publishing, Paris, https://doi.org/10.1787/2965f4fe-en. [38]

OECD (2021), "The inequalities-environment nexus: Towards a people-centred green transition", *OECD Green Growth Papers*, No. 2021/01, OECD Publishing, Paris, https://doi.org/10.1787/ca9d8479-en. [51]

OECD (2021), *Transport Strategies for Net-Zero Systems by Design*, OECD Publishing, Paris, https://doi.org/10.1787/0a20f779-en. [70]

OECD (2020), *OECDStats: Exposure to PM2.5 in countries and regions*, https://stats.oecd.org/Index.aspx?DataSetCode=EXP_PM2_5. [61]

OECD (2019), *Accelerating Climate Action: Refocusing Policies through a Well-being Lens*, OECD Publishing, Paris, https://doi.org/10.1787/2f4c8c9a-en. [68]

OECD (2018), *Biodiversity Conservation and Sustainable Use in Latin America: Evidence from Environmental Performance Reviews*, OECD Environmental Performance Reviews, OECD Publishing, Paris, https://doi.org/10.1787/9789264309630-en. [44]

OECD (forthcoming), *Intermediary Cities and Climate Change: An Opportunity for Sustainable Development*, OECD Publishing, Paris. [96]

OECD et al. (2021), *Latin American Economic Outlook 2021: Working Together for a Better Recovery*, OECD Publishing, Paris, https://doi.org/10.1787/5fedabe5-en. [2]

OECD et al. (2019), *Latin American Economic Outlook 2019: Development in Transition*, OECD Publishing, Paris, https://doi.org/10.1787/g2g9ff18-en. [3]

OECD/FAO (2022), *OECD-FAO Agricultural Outlook 2022-2031*, OECD Publishing, Paris, https://doi.org/10.1787/f1b0b29c-en. [18]

OECD/IEA (2021), *GHG Emissions from Fuel Combustion*, OECD Publishing/International Energy Agency, Paris, https://www.oecd-ilibrary.org/energy/data/iea-co2-emissions-from-fuel-combustion-statistics_co2-data-en. [13]

OECD/UNCTAD/ECLAC (2020), *Production Transformation Policy Review of the Dominican Republic: Preserving Growth, Achieving Resilience*, OECD Development Pathways, OECD Publishing, Paris, https://doi.org/10.1787/1201cfea-en. [48]

OXFAM (2018), *The Weight of Water on Women: The Long Wake of Hurricane María in Puerto Rico*, OXFAM, Oxford, England, https://www.oxfamamerica.org/explore/research-publications/research-backgrounder-wash-gender-report-puerto-rico/. [63]

RedParques (2021), *Protected Planet Report 2020: Latin America and the Caribbean*, RedParques, https://redparques.com/modules/ecom/documentos/publicacion/INFORME-2020-final.pdf. [45]

Romanello, M. et al. (2021), "The 2021 report of the Lancet Countdown on health and climate change: code red for a healthy future", *The Lancet*, Vol. 398/10311, pp. 1619-1662, https://doi.org/10.1016/S0140-6736(21)01787-6. [64]

Sanchez, R. et al. (2017), *Inversiones en infraestructura en América Latina: tendencias, brechas y oportunidades*, Economic Commission for Latin America and the Caribbean, Santiago, https://www.cepal.org/es/publicaciones/43134-inversiones-infraestructura-america-latina-tendencias-brechas-oportunidades (accessed on 4 July 2022). [84]

Sistema de Informacion energetica de Latinoamerica y el Caribe (SieLAC) (2020), *Estadística Energética [database]*, https://sielac.olade.org/default.aspx. [22]

Tambutti, M. and J. Gómez (2020), *The outlook for the oceans, seas and marine resources in Latin America and the Caribbean: Conservation, sustainable development and climate change mitigation*, Economic Commission for Latin America and the Caribbean, Santiago, https://www.cepal.org/en/publications/46509-outlook-oceans-seas-and-marine-resources-latin-america-and-caribbean-conservation. [23]

Tubiello, F. et al. (2014), *Agriculture, Forestry and Other Land Use Emissions by Sources and Removals by Sinks*, Food and Agriculture Organization, Rome, https://www.fao.org/3/i3671e/i3671e.pdf. [17]

UNDESA (2019), *World Urbanization Prospects. The 2018 Revision*, Population Division, United Nations Department of Economic and Social Affairs, New York, https://population.un.org/wup/Publications/Files/WUP2018-Report.pdf. [76]

UNDESA (2018), *World Urbanization Prospects 2018*, United Nations Department of Economic and Social Affairs, Population Division, https://population.un.org/wup/Download/. [78]

UNDESA (2017), *Climate Change and Social Inequality*, DESA Working Paper, No. 152, United Nations Department of Economic and Social Affairs, New York, https://www.un.org/development/desa/. [53]

UNECE (2021), *UNECE Environmental Policy: Air Polution*, United Nations Economic Commission for Europe, Geneva, Switzerland, https://unece.org/air-pollution-and-health#:~:text=Air%20pollution%20is%20now%20considered,pulmonary%20illnesses%20and%20heart%20disease. [73]

UNEP, (2019), *Zero Carbon Latin America and the Caribbean 2019: The Opportunity, Costs and Benefits of the Coupled Decarbonization of the Power and Transport Sectors in Latin America and the Caribbean*, Nairobi. [94]

Vasconcelos, E. (2019), *Contribuciones a un gran impulso ambiental para América Latina y el Caribe: movilidad urbana sostenible", Documentos de Proyectos*, Economic Commission for Latin America and the Caribbean, Santiago, https://repositorio.cepal.org/bitstream/handle/11362/44668/1/S1801160_es.pdf (accessed on 4 July 2022). [79]

Vona, F. (2021), "Managing the distributional effects of environmental and climate policies: The narrow path for a triple dividend", *OECD Environment Working Papers*, No. 188, OECD Publishing, Paris, https://doi.org/10.1787/361126bd-en. [66]

Watts, N. et al. (2015), "Health and climate change: policy responses to protect public health", *The Lancet Commissions*, Elsevier, Amsterdam, pp. 1861-1914, https://doi.org/10.1016/S0140-6736(15)60854-6. [55]

WIR (2022), *World Inequeality Report 2022*, World Inequality Database, https://wir2022.wid.world/methodology/. [16]

WMO (2021), *State of the Climate in Latin America and the Caribbean 2020*, World Meteorological Organization, Geneva, Switzerland, https://library.wmo.int/doc_num.php?explnum_id=10876. [39]

World Bank (2022), *World Development Indicators - GDP (constant 2015 US$)*, World Bank, Washington, DC, https://data.worldbank.org/indicator/NY.GDP.MKTP.KD?view=chart. [11]

World Bank (2021), *Climate Change Knowledge Portal [database]*, World Bank, Washington, DC, https://climateknowledgeportal.worldbank.org/download-data. [40]

World Bank (2021), *Groundswell Part 2: Acting on Internal Climate Migration*, World Bank, Washington, DC, http://hdl.handle.net/10986/36248. [54]

World Bank (2021), *World Development Indicators - Forest area [database]*, World Bank, https://databank.worldbank.org/reports.aspx?source=2&series=AG.LND.FRST.K2,EN.MAM.THRD.NO,EN.BIR.THRD.NO,EN.FSH.THRD.NO,EN.HPT.THRD.NO,ER.LND.PTLD.ZS,ER.MRN.PTMR.ZS. [21]

World Bank (2020), *Global Action Urgently Needed to Halt Historic Threats to Poverty Reduction*, World Bank, Washington, DC, https://www.worldbank.org/en/news/feature/2020/10/07/global-action-urgently-needed-to-halt-historic-threats-to-poverty-reduction. [9]

Chapter 3

Structural change for a new development model

The recovery agenda in the LAC region calls for an integrated holistic approach that enables a green and just transition. This chapter presents three key building blocks to advance a more sustainable development model in LAC: energy, production and social protection. The chapter starts by analysing the current energy matrix and goes on to look at the possibilities industrial policies and the circular and the blue economy offer for enhancing current productive structures. Lastly, it reviews the potential impacts the green transition could have on the future of work, and the role social protection systems can play in protecting the most vulnerable during the transition.

Structural change for a new sustainable development model in LAC

More sustainable and diversified energy and production matrices will help LAC reduce GHG emissions, harness its renewable energy potential and create quality jobs

LAC is endowed with high potential for renewable energy resources

Renewables as % of total energy supply

33% LAC

13% Globally

LAC is in a strategic position to supply key minerals for the green transition

61% of **lithium** global lithium reserves

39% of global **copper**

and 32% global of **nickel and silver**

were in LAC in 2017

Universal access to electricity is a pending challenge

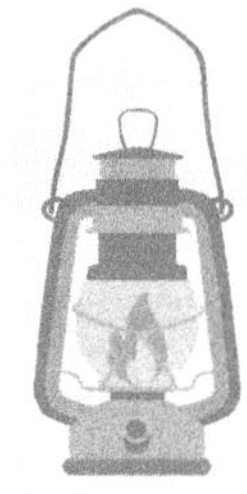

A total of **17 million people** still have **no access to electricity** especially in rural areas

Industrial, circular and blue policies can transform LAC's production structure

New industrial policies are needed to encourage investment in green innovation

Gross domestic expenditure in research and development (GERD) as % of GDP, 2018

0.3% LAC

2% OECD

The circular economy is expected to have net positive effects on GDP growth and employment by 2030, in selected countries

2.4% GDP net growth

1.9% new net jobs

The Blue economy can contribute to LAC's development

The total GDP contribution of ocean services was estimated at **USD 25 billion** for LAC in 2018

A green transition can create quality jobs for Latin Americans

In a high-impact scenario it could add **10.5% more net jobs** in the brown and green sectors

A green transition that is just should:

- Expand social protection systems to the most vulnerable
- Boost quality job creation
- Support workers in the case of job losses
- Provide necessary skills and life-long learning opportunities

Introduction

Policy makers in the LAC region – and across the globe – are confronting increasing urgency in the need to deal with the complexity of numerous and interconnected economic, social and environmental challenges. Climate change mitigation and adaptation call for transformative change[1] to solve such issues simultaneously. Transformative policies could reduce greenhouse gas emissions (GHG) by 40% to 70% globally (IPCC, 2022[1]).

Governments need to start thinking about (re)designing systems that work for people and the planet. Systems that, by design, increase people's well-being while requiring less materials, producing fewer emissions and producing better socio-economic and environmental outcomes. LAC has the opportunity to rethink what these systems should be like and what policies are needed to transition from the current situation to a more sustainable, inclusive and just development model.

A green transition goes beyond fighting climate change. It also aims to advance a more sustainable and inclusive model of production and consumption that creates new quality green jobs, generates the conditions for workers to successfully navigate the transition, and supports firms to adopt more sustainable production schemes and citizens to change their consumption habits (Chapter 2).

This chapter presents three key building blocks to advance a more sustainable development model in LAC. The first two sections focus on the need to transition to a new energy matrix and better productive structures. The third section analyses potential impacts of the green transition on the future of work and the role of social protection systems to promote a just transition. The chapter then presents a selection of policy messages to advance a more sustainable development model.

Advancing towards a more sustainable and diversified energy matrix

The transformation of the energy matrix is key to promoting greater well-being for citizens, leaving no-one behind, and avoiding environmental impacts through a green transition. LAC generated 5.8% of global total energy supply in 2018 (UN DESA, 2021[208]). The energy sector is responsible for the highest share of GHG emissions, although its contribution to global GHG is lower than in other regions. It is therefore essential to rethink the region's energy matrix to meet international climate targets and foster resiliency (Chapters 2 and 6). LAC is endowed with high potential for renewable energy resources throughout the region including: hydropower (throughout the region); wind (particularly but not limited to Patagonia, the Atlantic Coast of South America, the Isthmus of Tehuantepec and the Guajira Peninsula); solar (in different regions including Atacama and Sonora-Chihuahua); geothermal (in specific locations, such as the Andes and the Central American Cordillera) and biomass (throughout the region and, particularly, in Brazil) (UNEP, 2019[2]).

Investing in renewables technologies for LAC countries currently dependent on fossil fuels for power generation (mainly Central American and Caribbean countries), can deliver lower-cost power and reduce reliance on imported fossil fuel products, which can be subject to significant price fluctuations and risk of supply disruption and can have damaging impacts on the balance of trade.

Those LAC countries whose power generation largely comes from hydropower should also envisage other renewable energy sources. An over-dependence on generation from hydro can have negative economic impacts in times of drought (IDB, 2021[3]). Moreover, changes in rainfall patterns and intensity as a consequence of climate change will make hydropower a less reliable source of energy to meet burgeoning demand in the coming years.

Public policies to accelerate the much-needed sustainable and inclusive energy transition should advance simultaneously in five pillars: 1) increase the share of renewable sources in the energy matrix, 2) universalise access to electricity and reduce energy poverty, 3) increase energy efficiency in all buildings and economic sectors, 4) strengthen regional energy integration and interconnection, and 5) increase energy security and resilience in the face of external shocks.

Renewable electrification and integration to accelerate progress towards systemic decarbonisation

A successful transition to net zero emissions will be contingent on systemic decarbonisation through renewable electrification across all sectors, also covering the remaining more than 17 million people that lack access to electricity in mainly rural and isolated areas of the region. This will entail a massive scale-up in electricity demand – between 210% and 560% in deep decarbonisation scenarios in LAC by 2050 – as these sectors move away from reliance on fossil fuels and switch to electrification (IDB and DDPLAC, 2019[5]). Meeting such a large increase in demand for power and achieving Paris Agreement objectives will require LAC countries to undertake substantial renewable energy capacity additions while also implementing demand-side measures to increase energy efficiency and promote systems that demand less resources and materials across all sectors.

Advancing electrification will require LAC countries to implement effective power sector long-term planning through integrated resource plans. The plans should help match necessary investments in generation with forecast growth in demand, enable absorption of future variable renewables energy generation capacity, and provide incentives for timely investment in energy projects and in transmission, distribution and storage infrastructure. Long-term energy planning has proven to be a crucial instrument in the region to accelerate the transition to renewable energies, since it points out routes and generates an enabling environment for the necessary investment in each country and territory.

Digitalisation and innovation could strengthen decarbonisation, for example by digitalising payment collection from consumers or increasing the robustness of transmission infrastructure to ensure renewables generation can be added at scale. Moreover, digitalisation can help better manage demand response to eliminate short-term demand spikes, facilitate variable renewable energy integration, fostering the adoption of advanced metering infrastructure (AMI) and can support Fintech options for clean energy investment by improving electricity access or promoting new business models (e.g. energy as a service vs traditional models of providing power and other services separately).

Energy integration in LAC, particularly electricity trading, could provide benefits from economies of scale in production and a reduction of costs, thereby improving security in the supply, reducing the impact of unanticipated shocks, and achieving better service quality and environmental protection (CAF, 2021[6]). Existing international interconnections tend to be underutilised. Electricity trade among three subregions (Andean, Central and Mercosur) could boost overall trade by 13%, generating trade value of USD 1.5 billion per year. A fully integrated electricity trade system – involving 20 countries in the region – could boost electricity trade by 29%, increasing the overall trade value to USD 2 billion per year (Timilsina, Curiel and Chattopadhyay, 2021[7]). However, fully functioning and efficient interconnections and electricity trading can be challenging to establish, requiring a commitment to the free trade of electricity among countries, adequate interconnection infrastructure and harmonisation of transmission fees. Regional progress towards energy integration has been uneven. Central America has advanced with the creation of a regional electricity market and the physical interconnection of six countries. By contrast,

South America has advanced only in bilateral interconnections, with relatively more success in the Andean subregion than the Southern Cone (CAF, 2021[6]).

Different scenarios have been projected to explore the complementarity of electricity systems and the use of renewable energies in the region. Based on LAC's great renewable energy potential, the electrification of the transport and industrial sectors is key to reduce their high dependence on fossil fuels and increase energy security in the region (see section: Holistic energy policies are needed to make the green transition possible).

LAC has vast renewable energy resources but variations and risks across the region should be considered

Over the last two decades, many LAC countries have made substantial progress in building renewable energy markets and diversifying their energy mix. In 2020, 33% of total energy supply in LAC was generated by renewables compared with 13% at the global level (Chapter 2) and renewable energy accounted for 61% (952 TWh) of regional electricity generation (of which 75% came from hydroelectricity and 25% from solar, wind, biomass and geothermal) (Figure 3.1). Central America has shown the greatest increases in renewables in the last two decades, from 65% to 77%, followed by the Caribbean with a modest increase of 3 percentage points. The overall achievement in the region has been the increase in the diversification of renewable sources of power; shifting from mainly hydropower to growing shares of thermal, wind, and solar energy. However, significant variations exist across the region. For instance, Brazil generates 84% of its electric power from renewables, including 6.9% from solar, 10.9% from wind and approximately 65% from hydropower (Government of Brazil, 2022[8]). By contrast, Jamaica generates 87% of its electric power from imported oil derivatives. In Ecuador, about 60% of the installed capacity[2] is hydropower (UNEP, 2019[2]), although approximately one-third of electric power is still generated from fossil fuels (USAID, 2020[9]).

Figure 3.1. **LAC: Proportion of renewable sources in the total electricity supply matrix, 2000 and 2020**

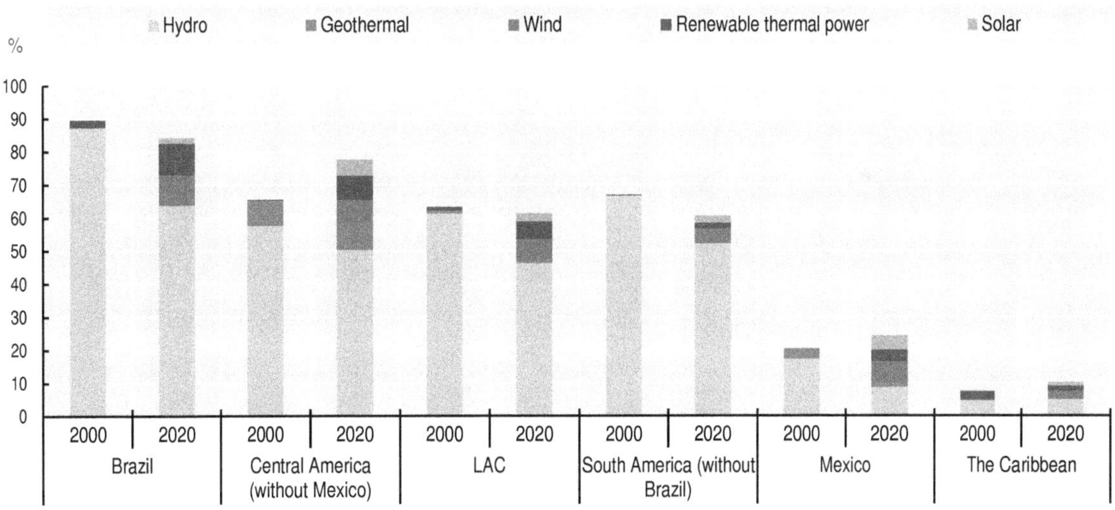

Notes: South America includes (nine countries) Argentina, Bolivia, Chile, Colombia, Ecuador, Paraguay, Peru, Uruguay and Venezuela. Central America includes (six countries) Costa Rica, El Salvador, Guatemala, Honduras, Nicaragua and Panama. The Caribbean includes (eight countries) Cuba, Grenada, Guyana, Haiti, Jamaica, Dominican Republic, Suriname, and Trinidad and Tobago. Belize and Barbados are also included in the Caribbean in 2020. Mexico generates almost six times more electricity than the other Central American countries together, which is why it is excluded from Central America and showed aside. Brazil generates more electricity than all the countries of South America together, so it was also decided to exclude it and show it aside.

Source: (sieLAC OLADE, 2022[4]).

StatLink https://stat.link/23ov56

The LAC region has seen significant investment in renewable energy in recent years, exceeding USD 35 billion over 2014-19 (excluding hydropower), with 70% of this total being directed towards Argentina, Brazil, Chile and Mexico. Moreover, Brazil (USD 13.58 billion), Mexico (USD 11.58 billion) and Chile (USD 8.16 billion) were among the top five renewables investment destinations by volume between 2009 and 2018, after India at USD 24.64 billion and the People's Republic of China (hereafter "China") at USD 18.52 billion (UNEP, 2019[2]). The composition of these investments attests to the rapid evolution of the region's energy mix towards a more diversified portfolio of renewable energy sources – particularly in Brazil, Chile and Mexico compared to other LAC countries.

The region can close its energy deficit and achieve an electricity matrix with 100% of renewable energy participation (ECLAC, 2021[11]). The RELAC (Renewables in Latin America and the Caribbean) initiative aims to reach at least 70% of renewable energy participation in the region's electricity matrix by 2030 (Box 3.1). Two conditions are needed to achieve these targets. The first is to invest 1.3% of regional gross domestic product (GDP) for ten years, equivalent to USD 114 per capita (i.e. USD 80 billion in constant 2010 prices), with some countries needing a greater or lesser proportion of GDP depending on local circumstances. The second is to increase renewables technologies (mostly solar and wind) in line with the targets of Sustainable Development Goal 7 (SDG 7) on access to affordable and clean energy. These actions could create 7 million green jobs and reduce GHG emissions by 30% by 2030 (ECLAC, 2021[11]). In line with this initiative, the Caribbean Community[3] has set a regional target of 47% renewable energy in total electricity generation by 2027. Many Caribbean countries[4] have already made significant efforts towards the adoption of renewable energy technologies, with utility-scale solar installations, wind projects and efforts to harness geothermal energy (ECLAC, 2021[12]).

Box 3.1. Renewables in LAC: The RELAC initiative

RELAC is a regional initiative created in 2019 within the framework of the United Nations Climate Action Summit. It has 15 member countries in the LAC region[5] that have voluntarily agreed to promote renewables with: 1) a concrete goal; 2) a monitoring scheme; and 3) an operating structure designed to support countries in the process.

The initiative's general goals are to accelerate the carbon neutrality of electricity systems in LAC; improve the resilience, competitiveness and sustainability of the sector; and create green jobs, improving air quality and minimising harmful health effects for Latin Americans. To do this, RELAC aims to establish a climate action platform for LAC countries and international organisations to enable sharing sustainable solutions.

Source: RELAC (2020[13]).

Governments play a key role in promoting investments towards renewables at the speed and to the depth required for the energy paradigm shift in the region. The development of renewables demands implementing long-term national policies and plans that include achievable goals with a toolbox that should include regulations and economic instruments, such as subsidies and incentives to productive enterprises, institutions and households, and the deployment of information and training on energy efficiency and renewables. LAC governments should maintain a permanent dialogue with the private sector and civil society to agree on shared roles and responsibilities to accelerate the adoption of renewables and thus generate climate resilience and energy security while also recovering the economy, employment and income in a sustainable and fair manner (Chapter 5).

Investment in alternative fuels is necessary to decarbonise hard-to-abate sectors

Investments in hydrogen and other low-carbon fuels, including sustainable biofuels, will be necessary to support decarbonisation in heavy industries and transportation – such as chemicals, steel, road freight, aviation and shipping – for which there are currently no viable alternatives to fossil fuels. Decarbonising hard-to-abate sectors is key to building the necessary demand at scale to increase the commercial viability of the hydrogen industry and can create a virtuous cycle between decarbonisation efforts and sustainable industrial development (section: Towards a new industrial policy to promote a more sustainable production model).[6]

Several LAC countries have significant possibilities for developing a competitive green hydrogen industry. Argentina, Bolivia, Brazil, Chile, Colombia, Costa Rica, Mexico and Peru have advantages in entering the hydrogen market because of their abundance of low-cost renewables and their relatively clean electricity matrices (ECLAC, 2022[14]). Mainstreaming hydrogen in the political, institutional and legal frameworks in LAC, by including it in public agendas, supporting the engagement of the private sector, an promoting a regional agenda on the topic can foster synergies to increase competitiveness and help the kick-start of the industry in the region. Several LAC countries have developed or are preparing hydrogen strategies (Box 3.2).

Box 3.2. **Hydrogen strategies in LAC**

A total of 11 LAC countries have either published or are currently preparing national hydrogen strategies and roadmaps. In turn, a pipeline of more than 25 low-carbon hydrogen projects are in the early stages of development. Several examples of national hydrogen development strategies are underway:

- **Chile** launched a green hydrogen strategy in 2020. It aims to establish 5 GW of electrolyser capacity in 2025 and 25 GW by 2030, as well as to produce the world's cheapest hydrogen by 2030, and become one of the world's top three hydrogen exporters by 2040 (IRENA/UNELCAC/GET.transform, 2022[15]). An Interministerial Committee (formed by 11 ministries and CORFO, the national economic development agency) has been created with the aim of developing the hydrogen industry, which is conceived as a national policy priority.
- **Colombia**'s National Hydrogen Strategy and Roadmap (2021) outlines plans to facilitate the development of a green hydrogen industry, taking advantage of the country's abundant renewable energy potential. It aims to deliver cost-competitive green hydrogen by 2030. The strategy also considers production of blue hydrogen, using carbon capture, utilisation and storage (CCUS) to capture emissions. Colombia's Energy Transition Law outlines fiscal incentives for the production of green and blue hydrogen (Government of Colombia, 2021[209]).
- Argentina, Bolivia, Brazil, Costa Rica, El Salvador, Panama, Paraguay, Trinidad and Tobago, and Uruguay are preparing national hydrogen plans. In particular, Argentina, Chile and Costa Rica are identifying commonalities and specificities of enabling market conditions and financing mechanisms that can foster the development of green hydrogen.

Source: IEA (2021[16]) and Cordonnier and Saygin (forthcoming[210]).

LAC countries producing electricity from renewables can position themselves as green hydrogen industrial hubs particularly in hard-to-abate sectors, such as steel and cement that will need access to abundant green hydrogen and for which global demand will continue to expand in line with the growing global population, industrialisation and urbanisation. Transforming the existing industrial and petrochemical hubs where currently grey hydrogen is consumed to kick-start green hydrogen deployment and production can help reduce countries dependence on exports, reduce energy prices volatility and provide grid stability by enabling energy storage and adding renewable energy to the grid, particularly in countries suffering intermittency issues associated with renewable energy sources. Green hydrogen in LAC also has the potential for vertical and horizontal linkages along its value chain, yielding more value added to the whole chain and avoiding it becoming only a commodity and thus fostering innovative new industries and sustainable inclusive development. Regional co-operation on hydrogen infrastructure development, cross-border regulation and free trade agreements can support demand creation to increase the commercial viability of regional hydrogen industries.

The gas sector can play a role as a "bridge" to blue and green hydrogen. LAC countries with established natural gas industries (e.g. Argentina, Bolivia, Brazil, Colombia or Peru) may be well placed to produce and export blue hydrogen (hydrogen produced from natural gas with carbon capture utilisation and storage- CCUS-), which could be an important source of foreign exchange in a scenario of declining fossil fuel exports. Countries with substantial solar and wind resources can then make the conversion to green hydrogen as renewables generation in their power matrices is gradually expanded.

LAC countries that are fossil fuel producers may have an opportunity to offset the significant capital expenditures (CAPEX) investments required to develop a hydrogen industry by repurposing existing oil and gas infrastructure. For example, pipelines for hydrogen transport or depleted oil and gas reservoirs for CCUS projects. Existing hydrogen demand in refining or petrochemical represents one of the first opportunities to start low-carbon hydrogen development. Hydrogen could help decarbonise heavy transport, for example by replacing diesel mining trucks in countries including Chile, Colombia and Peru. Although these solutions are not yet at commercial levels, costs are expected to come down in the coming years. The development of a hydrogen industry in LAC will demand government-sponsored demonstration projects, as well as collaboration with the industry at the national and regional levels to create market demand (a key factor in raising finance for hydrogen projects).

Fossil fuels decarbonisation

The transformation of the global energy mix from fossil fuels to renewable energy is necessary to meet the Paris Agreement goals (Chapter 5) and to prevent irreversible damage to the world's environment and ecosystems. However, fossil fuel use will continue in the short to medium term and will also remain an important part of the energy mix even after the world has transitioned to a low-carbon and green economy. By 2050, fossil fuels are still expected to represent 20% of the global energy supply (IEA, 2021[17]).

Several LAC countries have significant oil and gas production, and fossil fuels remain an important source of export earnings, as well as an input for domestic power generation. Key producers in the region are Argentina, Bolivia, Brazil, Colombia, Ecuador, Mexico, Trinidad and Tobago, and Venezuela. In terms of global crude oil reserves, LAC has the second-highest share (19.1%), behind the Middle East (48.3%) and ahead of North America (14.0%), the Commonwealth of Independent States (8.4%), Africa (7.2%) and the Asia-Pacific (2.6%). In addition, LAC countries currently hold a 4.3% share of the global reserves of natural gas (BP, 2021[18]).

Consequently, LAC countries should take steps to ensure that oil and gas production is as low-carbon as possible, as a transitional step towards a net-zero economy in which fossil fuels continue to form part of the energy mix. This can be achieved through a mix of regulations, policy incentives and the deployment of best-available technologies and practices to reduce flaring, venting and methane emissions across the upstream oil and gas sectors. LAC governments should implement economy-wide decarbonisation measures and structural reforms to reduce fossil fuel dependence and accelerate systemic change engaging the private sector and civil society as key stakeholders (Chapters 1, 4 and 5). In the context of Russia's invasion of Ukraine, and for a time-limited period, the European Commission's "Taxonomy Delegated Act" includes specific nuclear and gas energy activities under certain conditions, in the list of environmentally sustainable economic activities covered by the so-called "EU Taxonomy" by considering them as "transitional activities" (European Parliament, 2022[23]).

Five key actions could help LAC countries advance towards decarbonisation: 1) reducing methane emissions; 2) maximising the potential of associated gas; 3) electrification of the oil and gas industry upstream with renewables technologies; 4) advancing carbon capture utilisation and storage (CCUS); and 5) phasing out fossil fuel-fired thermal generation.

Reducing methane emissions

Reducing methane emissions is the single most important and cost-effective way to bring down GHG emissions and to improve efficiency in the oil and gas industry, which accounts for 20% of global methane emissions (GMI, 2011[19]). LAC governments can play an important role in reducing methane emissions by, first, establishing regulatory frameworks for the measurement, disclosure, and verification of methane flaring, methane venting, and carbon dioxide (CO_2) emissions. Second, they can integrate methane emissions reduction in their Nationally Determined Contributions (NDCs). Third, they can require oil and gas companies to design leak-detection and repair programmes. To foster compliance and accountability, Argentina and Mexico require public disclosure of methane emissions data by companies (IEA, 2021[20]). Colombia has integrated all methane-related measures under a unique regulatory instrument to reduce fugitive emissions from upstream oil and gas activities, thereby becoming the first South American nation to regulate methane emissions from oil and gas (Banks and Miranda-González, 2022[21]).

LAC governments should consider methane emissions in the liquefied natural gas (LNG) value chain. The global LNG industry is rapidly expanding with LNG projects projected to account for around 80% of the increase in global gas trade up to 2040 (Stern, 2019[22]). LAC countries that currently export LNG (Peru and Trinidad and Tobago) and those exploring the development of an LNG export market (Argentina, Colombia and Mexico) should consider how the introduction by importer countries of new GHG reduction regulations may affect LNG projects over their operating lives. In fact, several of the largest LNG importing countries (including France, Japan, Korea, Spain and the United Kingdom) have pledged to become carbon-neutral by 2050. The European Union is creating country profiles for methane emissions from oil and gas to inform EU's purchasing choices (Banks and Miranda-González, 2022[21]) (Chapter 6).

Making the most of associated gas

Flaring of associated gas[7] not only contributes to climate change but also wastes a valuable energy resource that could be used to advance sustainable development and the low-carbon transition in producing countries. LAC governments can implement a number of policies and incentives to ensure that associated gas is monetised rather than flared. For example, Brazil's National Agency of Petroleum, Natural Gas and Biofuels encourages the capture and use of associated gas by charging royalties for all gas that is flared (IEA, 2021[20]). The government is working to develop a midstream market and to drive domestic demand for natural gas, notably to electrify rural communities and support industrial growth (IEA, 2021[20]).

Upstreaming electrification of the oil and gas industry with renewables technologies

Electricity generation at oil and gas facilities needs to be fully decarbonised using electricity supplied by renewable energy sources in order to meet climate objectives.

Grid-based electricity could be an option, when possible, but for remote oil and gas operations, off-grid electricity options will be necessary. LAC governments could provide incentives or requirements for oil and gas operators to take advantage of the steady cost decline over the past years in renewable energy technology and to integrate off-grid electricity generation into their upstream operations, for example through a mix of off-grid solar photovoltaics (PV), wind, hydro, small modular reactors and battery storage systems. A similar logic applies to the mining sector (section on sustainable mining in Annex 3.A1).

Advancing carbon capture, utilisation and storage

The Intergovernmental Panel on Climate Change (IPCC) and the International Energy Agency (IEA) have recognised the critical role for CCUS[8] in reducing CO_2 emissions and achieving net-zero emissions by 2050. Of 135 CCUS facilities in operation worldwide, only one is located in the LAC region (IOGP, 2022[24]). The Petrobras's Santos Basin Pre-Salt Oil Field CCUS facility, opened in 2011, is located 300 kilometres off the coast of Brazil. The Santos Basin programme is the third-largest CCUS project in the world, accounting for around 12% of global capacity (IOGP, 2022[24]). Potential policy measures to scale up CCUS in LAC are: 1) undertake geological mapping to establish a national register of potential CO_2 storage sites; 2) determine whether highly concentrated, large point-source emitters of CO_2 (e.g. petroleum, cement and fertiliser industries) are relatively close and well connected to potential storage sites; and 3) implement robust CCUS regulatory frameworks that include an independent third-party verification role to provide the private sector with the necessary confidence to invest (Global CCS Institute, 2020[25]). Risks and trade-offs with broader environmental objectives associated with CCUS should be addressed accordingly.

Phasing out fossil fuel-fired thermal generation

LAC countries need to consider phasing out fossil fuel-fired thermal generation capacity earlier than intended. Committed emissions from existing and planned infrastructure in the LAC power sector will be 6.9 Gt CO_2 by 2050, more than is consistent with limiting global warming to either 1.5°C or 2°C (IDB and DDPLAC, 2019[5]). Retiring power plants early can be politically contentious because the nature of project finance relies on capital being recouped over the full life of a project (normally a 20- to 40-year period) and early retirement is likely to entail significant losses for a utility and investor. Early planning for retirement, as well as defining clear criteria based on which plants will be selected and ongoing dialogue with operators, can help mitigate negative market perceptions and open up avenues for financing early retirement. In specific contexts, repurposing older, high-polluting plants with renewable energy and storage can significantly reduce overall CAPEX requirements for the transition and provide a more efficient avenue for the renewables capacity addition because projects can rely on existing infrastructure, including substations and transmission and evacuation lines, rather than building renewable energy infrastructure from scratch. However, given the varying local conditions, types, sizes and ages of thermal power plants – and the varying roles they play in the local economy – and the need to meet national electricity demand and to stabilise the grid, the feasibility of repurposing will be contingent on local circumstances, and a one-size-fits-all approach is unlikely to bear fruit.

In Central America, a low-carbon energy pathway could be more cost-effective than current polluting alternatives. The Renewable Energy Roadmap for Central America finds that the decarbonised pathway would cost the subregion around USD 20 billion less than the planned scenario 2018-50. Central America has a unique opportunity to ensure sustainable development with renewable energy resources. They can bolster energy security, mitigate fossil fuel dependence while reducing costs, stimulate the subregion's post-coronavirus (COVID-19) recovery and help address climate change (IRENA, 2022[26]).

Rising demand for minerals and LAC's strategic position

The green energy transition, boosted by the unstable global and geopolitical context, is driving a surge in demand for minerals that are abundant in LAC. Raw materials for green energy technologies are set to become the fastest-growing segment of mineral demand. Based on the Sustainable Development Scenario projection, by 2040 their share of total demand edges up to over 40-50% for copper and rare earth elements; 60-70% for nickel and cobalt; and almost 90% for lithium (Figure 3.2).

Several LAC countries will be strategically positioned as suppliers of these minerals. The region includes the world's biggest copper producer (Chile), the world's biggest silver producer (Mexico), the third-biggest steel producer (Brazil), the world's second and third largest Lithium producers (Chile and Argentina, respectively), and the seventh-biggest bauxite producer (Jamaica). In 2017, a total of 61% of global lithium reserves, 39% of copper reserves and 32% of nickel and silver reserves, respectively, were in LAC (ECLAC, 2018[28]) (section on "Other key selected sectors for the green transition"). Over the period 2020 to 2021, investments in lithium in LAC increased by 117%, with increases of 559% in Chile and 77% in Argentina (S&P Global Market Intelligence, 2022[29]).

Figure 3.2. **The energy transition is driving a global demand surge for minerals**

Clean energy technologies share of global energy demand for selected minerals, actual and projected

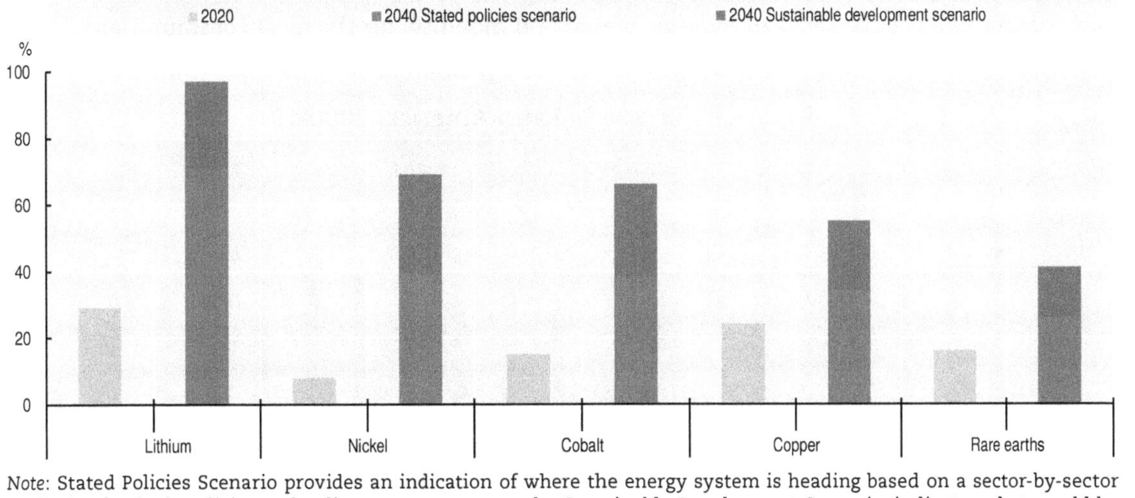

Note: Stated Policies Scenario provides an indication of where the energy system is heading based on a sector-by-sector analysis of today's policies and policy announcements; the Sustainable Development Scenario, indicates what would be required in a trajectory consistent with meeting the Paris Agreement goals. Neodymium demand is used as indicative for rare earth elements.

Source: (IEA, 2022[27]).

StatLink https://stat.link/tk3qdi

LAC countries should apply a sustainable mining approach as they transform exploration spending into mineral production and processing growth to take this opportunity as a lever in the green transition. The challenge will be to not repeat the mistakes of previous transitions. This time, the region should aim to integrate into global value chains in a more sophisticated way, putting sustainability, citizens' well-being and the potential for productive integration at the centre. Mining activity has faced high social opposition in the region, highlighting the need for good governance, consultation and ensuring that local communities benefit from projects as a part of the low-carbon transition (Chapter 5).

Closing the energy access gap: Universal access to electricity as a pending challenge

Addressing energy poverty is a crucial element of a sustainable, inclusive and just transition as it can help overcome historical social inequalities and provide a path for local economic growth. Universal electricity access is a key enabler for improved livelihoods, inclusive local economic growth and increased well-being (in terms of access to health and education, increasing the time available for new activities, and contributing to the creation of productive processes, among other positive outcomes).

Access to electricity in LAC has increased 15.7% in the last two decades, reaching 95.5% in 2019 (Figure 3.3). Still, a total of 17 million people have no access to electricity, especially in rural areas, where the electrification rate is around 76%. Substantial amounts of additional grid capacity (from gas or renewables) may not alone solve the problem, given the prohibitive costs of infrastructure development. Off-grid solutions, such as off-grid solar PV, have contributed to improving rural electrification in Peru (IRENA, 2018[30]), and could be a good alternative for the region. In Ecuador's Amazon basin, off-grid electrification efforts through solar PV in local communities enabled the provision of an electromobility solution to improve the transport of children to school along the Tupungayo River. The replacement of the gasoline outboard engine for electric ones improved the regularity of the boat service, reduced emissions, facilitated the decoupling between the growth of mobility demand and the demand for fossil fuels, mitigated the noise and the risk of water pollution. The project meant a paradigm shift in the provision of basic services (energy, water, and mobility) for the local communities (Wilmsmeier and Jaimurzina, 2017[31]).

Figure 3.3. **Electricity access in Latin America, 2000-19**

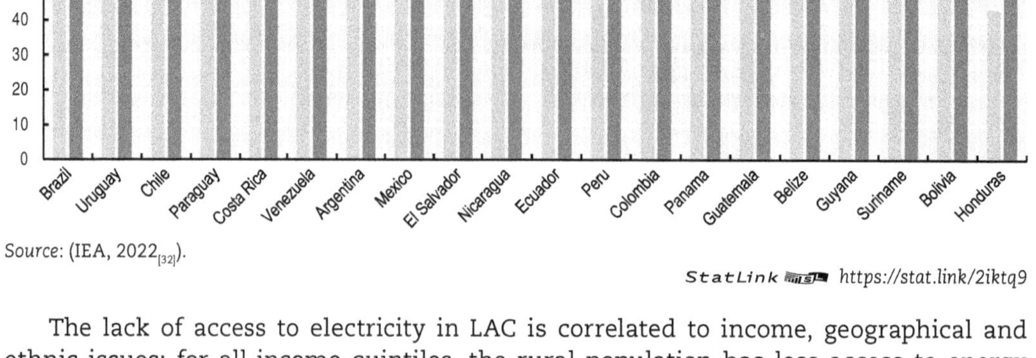

Source: (IEA, 2022[32]).

StatLink https://stat.link/2iktq9

The lack of access to electricity in LAC is correlated to income, geographical and ethnic issues: for all income quintiles, the rural population has less access to energy resources (Figure 3.4). On average, 15% of the population living in precarious housing has no access to electricity. In Bolivia, Chile, El Salvador, Honduras, Guatemala and Nicaragua, this figure reaches between 30% and 40%. Moreover, the percentage of total household budget that the poorest quintiles of the region's population spend on electricity and gas is twice – and in some cases three times – that of the richest quintiles. The proportion of the indigenous and Afro-descendant population without access to electricity is, on average, double and, in some cases, triple the respective proportion of the non-indigenous and Afro-descendant population (sieLAC OLADE, 2022[4]).

Figure 3.4. **Lack of access to electricity in LAC is higher for the first quintiles and in rural areas**

Proportion of the population without access to electricity by income quintile (rural, urban and total), latest available year

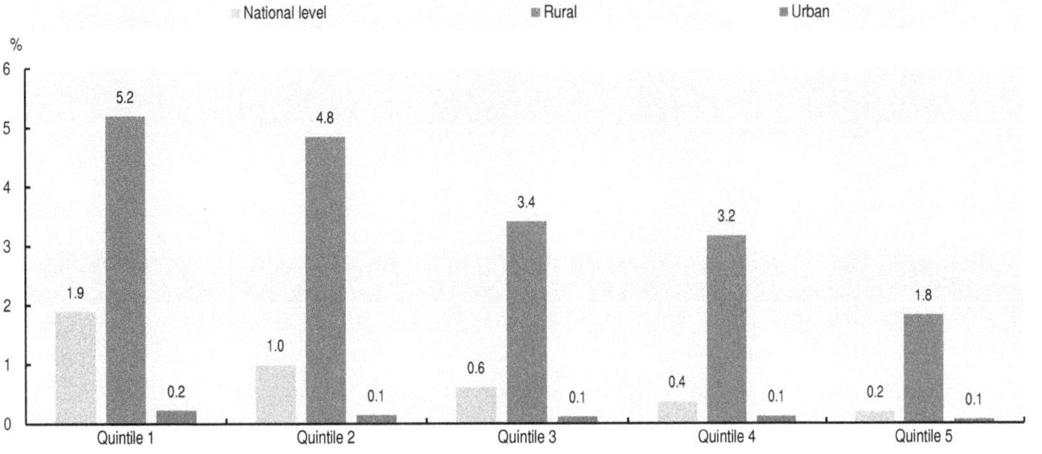

Note: Latest year available: 2017 for Bolivia, Brazil, Chile, Colombia, Costa Rica, Ecuador, El Salvador, Paraguay, Peru, and Uruguay; 2016 for Honduras, México and Dominican Republic; 2014 for Guatemala, Nicaragua and Venezuela.
Source: Authors' elaboration based on latest household surveys, Banco de Datos de Encuestas de Hogares (BADEOHG).

StatLink *https://stat.link/hkutp3*

Achieving universal access to electricity based on renewables in LAC by 2030 will require an investment of approximately USD 852 billion (ECLAC, 2020[34]). To make it possible, LAC governments may consider establishing an energy access fund, capitalised by development finance institutions and fossil fuel export revenues, to roll out energy access programmes via mini-grids. Financing for off-grid entrepreneurs and affordability for poorer households are among the biggest challenges that must be addressed to achieve energy equality throughout the region (Chapter 4).

The combined use of renewable technologies has the capacity to serve on-site and decentralised electricity to rural, isolated and remote communities where interconnected systems do not currently reach, thus achieving the universalisation sought in the energy transition. It is therefore a matter of generating electricity based on renewables in the territories, achieving "the last mile" to universalise access in a sustainable way, leaving no one behind. This is possible as long as public-private partnerships are established to attract investment in these local energy markets.

Holistic energy policies are needed to make the green transition possible

LAC needs bold energy plans to effectively transform its energy mix. Based on the current objectives of energy policies applied in the region, the changes achieved by 2040 in the regional energy mix total energy supply will only be marginal, since the investments provided for national energy plans are not sufficiently transformative (Figure 3.5). According to a review of policies to be applied, the transition will not occur fast enough to comply with NDCs unless negative externalities of fossil fuels are properly priced, incentives for renewables are applied, and governments provide clear guidance on the way forward.

Figure 3.5. **LAC: Total energy supply and demand mix in current policies, actual and projected**

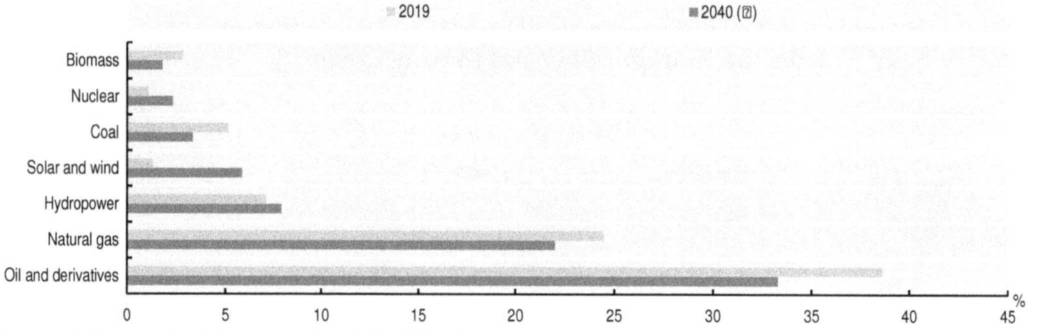

Source: Authors' calculations based on (IEA, 2021[10]).

StatLink https://stat.link/6t4bq8

Policies focused on energy efficiency, upstreaming methane reductions, re-evaluating fossil-fuels subsidies and investing in renewable energy could reduce more than 90% of LAC regional emissions (IEA, 2015[33]). In particular, energy efficiency policies applied to industry, buildings and transport could reduce the region's emissions by 40%. Moreover, energy efficiency could reduce GHG emissions if applied to industrial motors, road transport, heating and cooling appliances, and lighting (IEA, 2015[33]). Going beyond efficiency, applying a systemic approach to combine energy policies with productive, social and environmental ones will be needed to improve environmental and social results.

A regional sustainable, inclusive and just energy transition requires a framework of innovation, co-operation and integration across LAC. The progressive construction of innovative ecosystems, shaped by a new culture of renewability, energy efficiency, security and resilience, in which education policies, investment instruments and regulations converge will be key. These innovation ecosystems need to be articulated regionally, promoting co-operation and integration to increase economies of scale and overcome barriers to the development of renewable energies.

Governance, civil society participation and public-private co-operation are fundamental to accelerating the energy transition. Better integrated governance of energy resources – including stability, clear regulations and guarantees for investors and financiers – are needed to underpin the energy transition. Strengthening the role of national regulatory and planning bodies to implement adequate mechanisms and instruments that resolve information asymmetries between regulators and private agents, provide guidance and clear signals to attract infrastructure investments will be key. These investments, both private and public, should be oriented to meet three key characteristics: sustainability, quality and resilience.

Structural change in the energy transition also requires greater citizen participation from the outset of projects, in turn requiring decentralisation and democratic forms of governance aimed at improving the distribution of power and decisions (Chapter 5). Energy citizenship has a key role to play in individual capacities and willingness to participate, especially in what some define as prosumers, highlighting the importance of access to smart and small-scale technology.

LAC countries are using long-term scenarios and energy planning tools to inform national planning and advance clean energy transitions (IRENA/UNELCAC/GET.transform, 2022[15]). Good practices in the region show that long-term scenarios are:

- Being developed with a broad scope incorporating social and environmental factors, for example, in Argentina (Towards a Shared Vision of Argentina's Energy Transition to 2050), Ecuador (National Energy Plan 2050) and the Dominican Republic (prioritising geographic security of its power system).
- Linked to climate goals. For example, Chile has institutionalised its long-term energy planning with ambitious targets, such as achieving carbon neutrality by 2050.
- Incorporating participation processes in energy planning. Brazil, Costa Rica and Panama involved key stakeholders (e.g. regional community leaders, academia and businesses) to take part in the development of their National Energy Plans 2050 (NEPs).
- Promoting more renewable energy and more efficient energy consumption. Chile plans to develop its renewable energy potential in solar, wind, hydropower and green hydrogen, while Mexico has developed clean energy generation and demand simulations and modelling.
- Incorporating transparent energy data and statistics. For example, Colombia shares all data used in the National Energy Plan 2020-2050 on a publicly accessible government website.
- Being supported by international co-operation. El Salvador, for example, received support from the Latin American Energy Organization and the International Renewable Energy Agency (IRENA) to develop its National Energy Plan 2020-2050. With financial support from Canada, management support from the IDB and technical support from Brazilian consultants, Peru developed a software tool for optimising long-term integrated energy planning.

Going forward, integrated and holistic long-term scenarios can help advance the systemic changes that the green transition demands in the energy sector, the transformation of LAC's production structure, its labour market (next sections) and sustainable territorial development (Chapter 2).

The energy transition can be accelerated in LAC countries by enabling proper ecosystems for investment. The implementation of renewable technologies, the achievement of a 100% renewable electricity matrix, and greater regional electricity integration will be key to reduce the region's high dependence on fossil fuels, a situation that implies great energy insecurity for the region, and which can be addressed through the electrification of different sectors, particularly the transport and industrial sectors, by taking advantage of the region's great renewable potential (ECLAC, 2020[34]).

Three scenarios of renewable energy adoption have been defined for the LAC region.[9] First, the Base Scenario (BS)[10] in which the adoption of renewables is calculated based on the 2020 national long-term renewable energy expansion plans of LAC countries (solar and wind increase their share of total electricity generation from 12% to 24.6%). Second, the High Share of Renewable Energy (RE) scenario incorporates a high proportion of renewable energy generation by 2032 (89% renewables, including large-scale hydro), but energy interconnections are maintained as in the Base Scenario (low regional transmission integration). The solar and wind (non-hydro) would increase their share of electricity generation from 12% to 41.1%.The third scenario is the High Renewable Energy Adoption and High Regional Transmission Integration (RE+INT) Scenario. This scenario incorporates in a cost-effective way a high proportion of renewable energy generation by 2032 (achieving 100% renewables, including large-scale hydro), and a high degree of regional interconnection that allows for a high degree of renewable energy integration and a more efficient electricity system (ECLAC, 2020[34]).

The analysis of the different scenarios[11] shows that achieving decarbonisation in the electricity sector in LAC is possible. It requires investing 1.3% of the region's annual GDP over the next ten years to incorporate renewable energies, universalise access to electricity for all, and increase regional electricity integration (ECLAC, 2020[34]). Greater regional electricity integration and the development of a regional electricity market are vital to achieving greater energy security and independence from fossil fuels. Investment to promote the adoption of renewables (solar and wind) would provide a more flexible and efficient electricity grid in LAC. The complementarity between these sources, together with hydropower and the potential use of storage in the medium term, is vital for the correct functionality of a new sustainable and inclusive electricity system (ECLAC, 2020[34]). Moreover, in the ER+INT Scenario: 1) generation of GHG emissions from the regional electricity system would decrease by 31.5% (compared to -30.1% in the RE Scenario, and -4.8% in the BS); 2) approximately 7 million new jobs will be created by 2032; and 3) if the renewable energy industry were located in LAC, manufacturing the solar panels and wind turbines needed to achieve this scenario would represent almost 1 million new jobs by 2032 (ECLAC, 2020[34]). Given the geographic characteristics of the Caribbean, it is impossible to carry out this type of study in the subregion. Nonetheless, it would be important to explore the opportunities and costs presented by the potential for electricity integration via the use of subsea cables with generation based on geothermal energy (e.g. in St. Lucia, St. Vincent and the Grenadines, St. Kitts and Nevis, and Dominica) and maximise the benefits of distributed generation based on solar and wind technology (ECLAC, 2020[34]).

Industrial, circular and blue policies to transform LAC's production structure

The green transition offers an opportunity to address one of the region's greatest challenges: transforming the production structure. Productivity has remained stagnant in recent decades, with the productive structure biased towards activities that have high intensity for materials and natural resources, plunging the region into a productivity trap that reinforces the environmental trap (OECD et al., 2019[35]). Currently, 75% of the region's total exports are primary products and natural resource-based manufactures (OECD et al., 2021[36]).

Productive policies for a green transition must advance a more sustainable production model that promotes regional competitiveness and formal employment. Therefore, LAC countries should enhance innovation, embrace green technologies and diversify the energy and productive matrix towards less resource-intensive sectors. It will be key to attract greener investments and take advantage of new trade opportunities to foster regional integration and to join global chains in higher-value segments, while gradually complying with environmental criteria in exports and in sustainable and responsible sourcing of materials (Chapter 6).

The production and export of environmental goods could facilitate structural transformation and improve international competitiveness by increasing the intensity of green technologies and innovation. Green policies have the potential to increase the competitiveness of LAC economies by setting green standards and certifications that enable firms to differentiate their products vertically. These standards demand a period of transition towards this model and renewed international co-operation (Chapter 6). Firms can increase revenues by selling and adopting green technologies; in turn, that can lead to productivity improvements and knowledge spillovers in the innovation processes as a way to enhance competitiveness (Altenburg and Assmann, 2017[37]). All these processes are gradual and should be accompanied by due diligence, respect for environmental standards and transparency in public procurement (ECLAC, 2020[35]).

Towards a new industrial policy to promote a more sustainable production model

Industrial policies in LAC should shift towards a green transition that prioritises technical change, generates new quality employment and reduces the region's environmental footprint. The transformation of production requires a combination of policies on investment, foreign trade, science, technology and innovation, and training and skills development, with a special focus on micro-, small- and medium-sized enterprises (MSMEs). Policies to close the gaps in the infrastructure, transport and energy sectors, in both urban and rural areas, are also key (ECLAC, 2020[35]). Strategic sectors for developing green industrial policies include renewable energy, sustainable transportation systems, digital transformation, bioeconomy, the circular economy, sustainable tourism, and sustainable agriculture and livestock, water and waste management, plastics and sustainable mining (see also section: "Other key selected sectors for the green transition").

The transition towards more inclusive and sustainable economies, therefore, depends largely on the opportunities associated with the future disruptive changes in the new technological cycle where a new advanced manufacturing (Industry 4.0) is needed. The current and accelerating "fourth industrial revolution" coupled with post-globalisation, demands sound industrial policies to avoid increasing existing production and technological gaps and their associated negative consequences. In the aftermath of the COVID-19 crisis and the complex current global context, a special need exists to implement mission-oriented policies aimed at industrial transformation, innovation and productive resilience, with job formalisation at the centre.

Developing renewable energy value chains can be vectors for economic development and energy security. The LAC region is endowed with sufficient skilled human capital and critical raw materials for renewables including production and storage (i.e. lithium batteries) to create more value-added across all value chains to ensure renewables and clean energy can be deployed at their fullest potential. However, these efforts require planning and co-ordination to achieve industrial and energy scales and policies for the expansion and integration of renewable technology value chains. Investment and funding and technical support to firms are also key.

LAC should promote regional production of inputs and equipment for the manufacturing, storage and distribution of renewable energies to develop true regional energy security, instead of just importing equipment and technology from global suppliers. The more that renewable inputs, technologies and know-how are sourced from within the region, the greater the energy security and resilience to global events (e.g. pandemics) and geopolitical conflicts will be. Renewable energies show the greatest dynamism and potential as vectors of development for the region, particularly solar PV, wind and, increasingly, green hydrogen. To bring about this paradigm shift, it is necessary to foster and maintain a continuous debate with policy makers, the private sector and stakeholders in each country and across the region. Promoting investment in R&D and industrial promotion programmes through higher and technical education programmes is also key.

Investment in innovation is a first key step for industrial policies and a green innovation ecosystem in LAC. Nonetheless, the region's gross domestic expenditure in research and development (GERD) has remained flat in the past decade, amounting to 0.3% of GDP in 2018. By contrast, GERD has consistently grown in the countries belonging to the Organisation for Economic Co-operation and Development (OECD), from 1.5% of GDP in 2000 to 2% of GDP in 2018. Brazil is the only LAC country that spends more than

1% of GDP in research and development (R&D); in the rest of the region, GERD ranges from 0.5% in Cuba to as low as 0.03% in Guatemala (UNESCO, 2021[38]).

R&D in LAC remains highly government driven (56.5% of total GERD), highlighting the need to increase private investment in R&D to promote innovation in the region. In 2019, business enterprises in LAC represented only 22.7% of total expenditure in R&D on average vs. 49.1% in the OECD area, although heterogeneity remains high in LAC (Figure 3.6). A new industrial policy with the environmental dimension at the centre should engage the private sector by increasing the co-ordination and dialogue across various actors, including national and sub-national authorities, the private sector and academia, on a new development strategy for each country of the region (OECD/UNCTAD/ECLAC, 2020[39]; OECD et al., 2019[40]). Entrepreneurs and start ups can be a source of innovation by the creation of new more sustainable businesses models. There are some examples of public-private collaboration in LAC that connect firms with entrepreneurs working on cross-cutting solutions to issues such as social inclusion, education, support for MSMEs and environmental protection. A total of 60% of the technological solutions promoted by this initiative will benefit vulnerable communities and the remaining 40% will promote environmental protection (IDB, 2021[42]).

Figure 3.6. **Share of gross domestic expenditure in research and development (GERD) by source of funds, selected LAC countries**

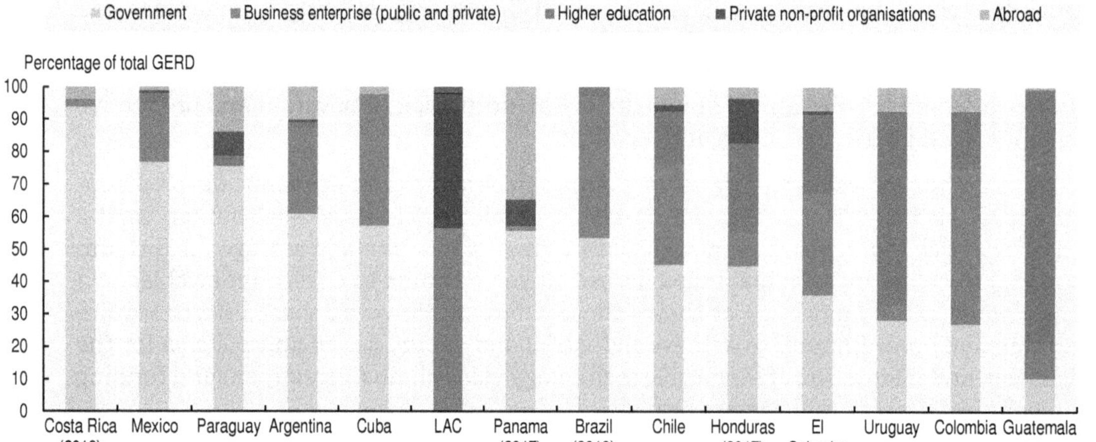

Notes: Data for 2019 or latest year available. Costa Rica: company data correspond to amounts financed by private companies but not reported as R&D amounts executed by companies (data collected through the business innovation questionnaire). El Salvador: the information provided corresponds to the expenditure made by the higher education and government sectors. Guatemala: the information provided corresponds to the expenditure of the government and higher education sector. Paraguay: as a source of financing, the government sector includes public universities.

Source: (RICYT , 2021[41]).

StatLink https://stat.link/zeh35t

In particular, the proper design and implementation of environmental policy can improve the economic performance of firms through innovation. Environmental regulation may help managers overcome behavioural biases and draw their attention to inefficiencies or new opportunities in production processes (Porter and van der Linde, 1995[43]). Environmental policy can help firms reduce input costs of energy or raw materials through process innovation or facilitate access to new markets through the development of new products if the right policies are in place (Box 3.3) (Dechezleprêtre et al., 2019[44]; Lanoie et al., 2011[45]).

Box 3.3. Innovation for sustainable agriculture and livestock: The case of Brazil

In the last decades, Brazil has implemented different types of innovations to promote more sustainable agricultural and livestock sectors through different stages, from technological innovation to the production of data, the creation of monitoring capacity, and social innovation work with local communities or public-private collaboration experiences. Some selected examples are:

- **The ABC and ABC+ plans.** The Sectoral Plan for Adaptation to Climate Change and Low Carbon Emissions in Agriculture (the ABC Plan) was designed in 2010 to reinforce the innovation system in tropical soils responsible for the so-called "Brazilian agricultural revolution", which transformed the country from an importer to an exporter of food. From 2010-20, the ABC focused on sustainable agricultural production technologies: (i) pastures recovery, (ii) crop-livestock-forestry and agroforestry systems, (iii) no-tillage system, (iv) biological nitrogen fixation, (v) planted forests and (vi) animal waste treatment. These technologies helped mitigate 170 million tons of CO_2e in two decades. For 2020-30, the plan was updated (ABC+) to include new technologies: (vii) bio-inputs, (viii) sustainable irrigation systems, and (ix) intensive fattening on pastures aiming to expand them across 72 million extra hectares to curb more than 1 billion tons of CO_2e. The ABC+ includes an Integrated Landscape Approach which considers different elements of rural landscapes to ensure that natural elements such as carbon, water, soil and biodiversity, can work sideways with agricultural production in a sustainable way.
- **Observatory of Brazilian Agriculture.** This public access portal includes data and statistics from more than 200 databases on Brazilian agriculture, such as: 1) a statistical platform containing data of various agricultural indices; and 2) a geospatial platform, which presents territorial and cartographic data on national agribusiness. The observatory aims to transform these data into ways to increase productivity and transparency throughout all agribusiness production and value chains, promote sustainable practices and improve monitoring actions. Key areas of consultation available are: sustainable agriculture and the environment; fishery and aquaculture; rural credit; agricultural products; agricultural zoning of climate risk; and Brazilian soils.
- **Public-private collaboration.** Brazil has been investing in the conservation of biodiversity and natural resources for more than 50 years. The Brazilian Ministry of Agriculture, Livestock and Food Supply (MAPA), the Agricultural Research Corporation (Embrapa), and the Agriculture and Livestock Confederation have been collaborating on the implementation of integrated systems and low-carbon emission agriculture. One concrete experience is the Integrated Crop-Livestock-Forestry System, which currently covers 45% of farms, contributes to minimising GHG emissions by up to 40% and helped develop the production of verified carbon-neutral beef.
- **Support of local communities to promote forest protection and the bioeconomy.** The Federal Programme for Community and Family Forest Management and the Bioeconomy and Sociobiodiversity, run by the Secretariat of Family Agriculture and Cooperativism (part of MAPA), have implemented a strategy to support local communities, thereby strengthening value chains and consolidating sustainable markets for non-timber forest products while conserving agrodiversity and providing renewable energy (mainly solar) to family agriculture.

Sources: Ministry of Agriculture, Livestock and Food Supply (2022[45]; 2021[46]; 2019[47] and Michail, 2019[48]).

MSMEs: A fundamental player for production transformation

Micro-, small- and medium-sized enterprises (MSMEs) are major job creators in LAC. They represent 60% of jobs, 99.5% of businesses and 25% of total production in LAC (Herrera, 2020[49]) but are characterised by low productivity and low competitiveness (Dini and Stumpo, 2019[50]). A new industrial policy should support MSMEs to increase their participation in regional and global value chains by promoting innovation, disseminating new knowledge and production linkages, creating and strengthening clusters, and partnering networks (ECLAC, 2020[34]).

Policies focused on MSMEs should be part of the general industrial sustainable development strategies implemented through horizontal approaches that focus on building capacities in priority production chains and territories, taking into account the needs of (private and/or social) MSMEs. The governance of these policies needs to have a local dimension while also guaranteeing a space for regional action (Chapters 5 and 6) (ECLAC, 2020[34]). For instance, in 2009, Uruguay launched an auctioning of small wind farms that required at least 20% locally produced content, 80% local jobs and a control centre based in Uruguay (IRENA, 2015[51]). These local content requirements make the green transition more inclusive and can contribute to promoting local quality jobs in greener sectors as the economies disengage from natural resource-intensive production and polluting activities (section: "Social policies for a just transition: the role of the labour market").

Institution building is essential for a sustainable industrial policy and its implementation. Stronger intergovernmental co-ordination at the regional and subregional levels would improve governance results. Areas including data and information generation, climate change adaptation, water resources management, environment and health, sustainable production and consumption, and biodiversity management would be strengthened through regional co-ordination and a renewed multilateralism (Chapter 6).

Shifting towards circular productive processes to promote a structural sustainable change

The circular economy approach can contribute to the implementation of the mentioned green industrial policies and of key sectoral policies that are needed to advance towards a more sustainable development model and rethink the way goods and services are produced and consumed.

The circular economy seeks to preserve the value of materials and products for as long as possible and minimise waste generation, as opposed to the dominant paradigm of the linear production-consumption-disposal economy (OECD, 2022[52]). Promoting circularity means designing and manufacturing products that have a longer useful life and can be upgraded, repaired, reused, reconditioned or remanufactured. It also means promoting eco-design policies that minimise the use of resources, take advantage of secondary resources, and promote re-use and recycle high-quality materials. It also involves combating planned obsolescence and standardising design elements (e.g. building universal chargers for electrical and electronic equipment or designing "circular" buildings) (Bárcena et al., 2018[53]).

There are at least 100 definitions of the circular economy (Kirchherr, Reike and Hekkert, 2017[54]), of which four have been selected for this publication. A first definition conceives the circular economy as one that "helps to keep resources flowing within rather than through the economy by modifying the flow of products and materials through three main mechanisms: Closing resource loops through the substitution of secondary materials and second-hand, repaired or remanufactured products in place of their virgin equivalents; slowing resource loops through the emergence of products which remain in the economy for longer, usually due to more durable product design; and narrowing resource flows through more efficient use of natural resources, materials and products, including the development and diffusion of new production technologies, an increased utilisation of existing assets and shifts in consumption behaviour" (McCarthy, A., R. Dellink and R. Bibas, 2018[55]; Yamaguchi, 2018[56]; OECD, 2022[52]). The second definition understands the term as: "an industrial system that is restorative or regenerative by intention and design. It replaces the 'end-of-life' concept with restoration, shifts towards the use of renewable energy, eliminates the use of toxic chemicals, which impair reuse, and aims for the elimination of waste through the superior design of materials, products, systems, and, within this, business models" (Ellen MacArthur Foundation, 2013[57]). A third approach, defines the circular economy as an economic system based on business models that replace the 'end-of-life' concept with reducing, alternatively reusing, recycling and recovering materials in production/distribution and consumption processes, thus operating at the micro level (products, companies, consumers), meso level (eco-industrial parks) and macro level (city, region, nation and beyond), to accomplish sustainable development, which implies creating environmental quality, economic prosperity and social equity, to the benefit of current and future generations (Kirchherr, Reike and Hekkert, 2017[54]). Finally, the circular economy is also seen as one that promotes systemic change through a new economic model that works for and with the planet (UNEP, 2021[58]).

The circular economy approach reinforces climate change mitigation actions. While the transition to renewable energy and energy efficiency would help reduce 55% of total GHG emissions, the circular economy can help eliminate the remaining 45% that are generated by the way goods are manufactured and used (Ellen MacArthur Foundation, 2019[59]). Other research estimates that materials management activities account for up to two-thirds (67%) of global GHG emissions (UNDP, 2017[60]) and projects that by 2060 materials management activities will be responsible for two-thirds of GHG emissions, mainly coming from the combustion of fossil fuels for energy from agriculture, manufacturing and construction (OECD, 2019[61]).

The innovation process behind the circular economy could translate into more sustainable economic growth through new activities due to more productive and efficient use of natural resources. These processes require skilled labour for new material recovery processes, employment generation and investments in innovation and technology incorporation.

The circular economy can be a driver of sustainable development. Its transformative, systemic and functional characteristics can foster several SDGs, including SDG 12 on sustainable and responsible production and consumption patterns; SDG 6 on water; SDG 7 on energy; SDG 9 on infrastructure, industrialisation and innovation; SDG 11 on sustainable cities and communities; SDG 13 on climate action; and SDG 15 on life on land (OECD, 2020[62]).

The transition to a circular economy is expected to have net positive effects on GDP growth and employment while reducing GHG emissions (Chateau and Mavroeidi, 2020[63]). Although the transition to the circular economy and cleaner production imposes economic costs for certain sectors, the net effects expected for Chile, Colombia, Mexico and Peru are increased GDP (from 0.82% in Chile to 2.4% in Peru) and job creation (from 1.1% in Chile and Colombia to 1.9% in Peru). These figures also grow over time, in line with those in Europe, although at a slightly lower pace. The effects on emissions reduction depend on each country's energy matrix characteristics, emission factors, fuel use reduction goals and size of the effect on GDP (Figure 3.7). GHG emissions are expected to fall in Chile (6,8%), Colombia (1.2%) and Mexico (1.4%). In the case of Peru, economic growth is still very dependent on fossil fuels and the fossil fuel reduction goal set by the country was only 5% (compared to 30% in Chile, 18% in Colombia or 15% in Mexico). That is why a 2.4% increase in GDP would have a slightly positive effect on the overall emissions levels (0.5%) (Econometría Consultores, 2022[64]).

Figure 3.7. **Potential effects of the circular economy transition by 2030 on GDP, employment and GHG emissions, selected LAC countries**

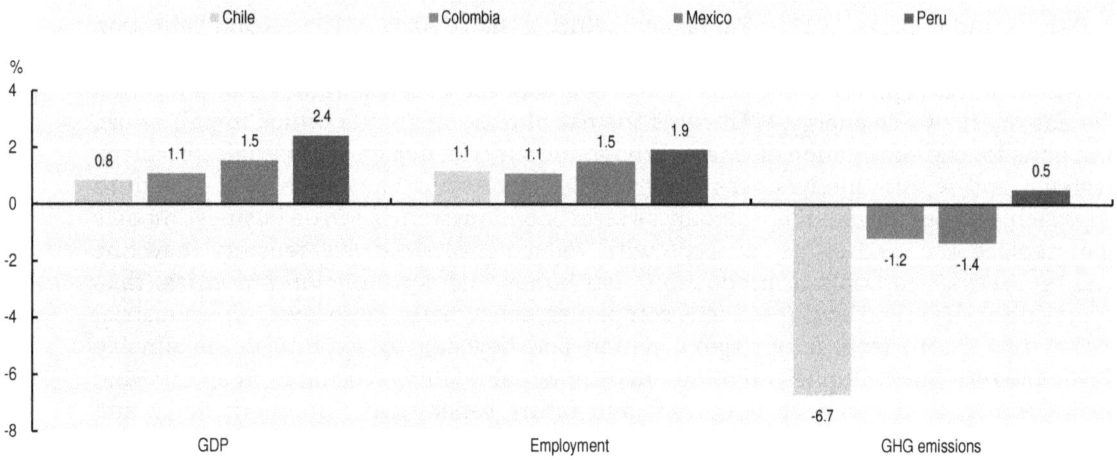

Source: (Econometría Consultores, 2022[64]).

StatLink https://stat.link/c0q64g

The circular economy goes beyond recycling and could help reduce informality in the waste management sector in LAC. Strategic solid waste management has the potential to have positive economic and social effects. Indeed, if LAC's waste and recycling sector were to develop into a key sector with a municipal waste recycling rate equivalent to that of Germany, it could contribute to a green economic revival: almost 450 000 stable jobs would be created and the region's GDP would increase by 0.35% (ECLAC, 2020[34]).

The circular economy is gaining momentum in LAC. Although still in their initial phase, more than 80 circular economy public policy initiatives are being implemented in the region, and an increasing number of national circular economy roadmaps and strategies are under development. Examples of national circular economy policy strategies include the Roadmap for a Circular Chile by 2040 (2021), the Circular Economy National Strategy of Colombia (2019), Ecuador's Law for an Inclusive Circular Economy (2021), Mexico's General Circular Economy Law (2021), Peru's Circular Economy Roadmap for Industry (2020), and the Circular Economy Action Plan of Uruguay (2019) (UNEP, 2021[58]). Moreover, the Circular Economy Coalition for Latin American and the Caribbean was officially launched in February 2021 to accelerate the circular transition in the region and has published keys ideas to develop a shared circular economy vision (UNEP, 2021[58]).National circular economy initiatives can become a pathway for the creation of regional or state initiatives such as the city of Querétaro Circular Economy System. This framework has served as a benchmark for sectoral circularity initiatives such as the Querétaro's Circular Economy System (SECQ), led by the Querétaro Automotive Cluster. The SECQ is expected to implement 100 circular projects by the end of 2022 (800 projects by the end of 2025) and reach 1 000 companies by 2027, fundamentally reducing environmental impacts in terms of materials, carbon footprints, energy use, water consumption, and waste generation (Estado de Querétaro, 2022[65]).

Governance is key to enable the circular transition in LAC. Governments can promote the circular economy through several economic, regulatory, voluntary, information, education and research and co-operation measures. Economic incentives (e.g. favourable tax or extended producer responsibility [EPR] schemes, incentive subsidies, tradeable permits or deposit-refund systems) and financing initiatives (e.g. public circular contests to finance circular business model initiatives or dedicated public budgets as in Peñalolén, Chile) can help boost local entrepreneurship and circular innovation by providing market signals to influence the behaviour of producers and consumers (OECD, 2020[62]; OECD, 2022[52]). Regulatory instruments foster legislation and regulation to remove potential barriers (e.g. adapting waste regulations). They can limit or ban polluting activities (e.g. introducing plastic bans or setting waste reduction targets), or promote circular production and consumption (e.g. introducing standards and certification for reused, remanufactured or recycled products). Green public procurement can be an essential tool for implementing national, regional and local circular economy initiatives. It can encourage the use of circular business models, promote circular construction developments, incorporate secondary materials, and encourage repair and reuse actions through public purchases (OECD, 2020[62]). The adoption of circular strategies and programmes, long-term visions for the circular economy (e.g. the Roadmap for a Circular Chile by 2040 or the National Strategy for the Circular Economy in Colombia) can help define clear responsibilities, objectives and actions, within the public sector and help guide the needed co-operation with the private sector, academia and civil society (triple and quadruple helix innovation processes).

Other governance measures that can help promote the circular economy in LAC are: voluntary measures to help sectors find more affordable means to achieve objectives (e.g. voluntary certification standards in the construction industry, or in agro-food chains); information instruments to enable consumers, companies and public authorities to make responsible purchases (e.g. via eco-labelling); education and research promotion (e.g. capacity-building through education and training programmes, creating circular economy observatories and knowledge networks); and measures to facilitate co-operation (e.g. private-public, coalition building across the private sector via circular hotspots and across value chains (OECD, 2022[52]).

Extended Producer responsibility (EPR)[12] is a useful instrument implemented by several LAC countries aiming to advance towards circularity: 14 LAC countries[13] currently have an EPR law (de Miguel et al., 2021[66]; Van Hoof et al., 2022[67]). EPR legislations are fundamentally waste management rules designed to mitigate the volume of waste sent to landfill, incentivising its re-use and improving its commercial value. However, waste collection, sorting and management require the development of cost-intensive infrastructure, as well as the use of natural resources. Specific legislation and waste collection infrastructure are needed to mitigate the costs generated by waste collection, attributing more responsibility to the producers in waste generation (Forti et al., 2020[68]; Wagner et al., 2022[69]). Electronic waste is the fastest-growing waste stream in LAC. Between 2000 and 2019, per-capita waste generation from electrical and electronic equipment almost tripled, from 3.4 kg to 8.8 kg per inhabitant, surpassing the global average of 7.3 kg per inhabitant indicated by the United Nations Institute for Training and Research. Only 1.5% of e-waste was recycled in LAC in 2017, falling to 1.3% in 2019 (ECLAC, 2021[70]). In 2018, a regional initiative for the management of electrical and electronic waste and the dismantling of dangerous chemical compounds was launched across 13 LAC countries (UNIDO, 2018[71]).

In the case of the energy sector, circular economy strategies are needed to help manage the waste associated with the production of renewable energies in line with the waste hierarchy. The solar PV panels and wind turbines installed to date were designed with a life-cycle of roughly 25 to 30 years, without much forethought about their eventual decommissioning. At the global level, waste from the solar PV sector alone is expected to reach between 1.7 Mt and 8 Mt in 2030 and between 60 Mt and 78 Mt in 2050 (IRENA, 2016[72]). Along with reducing waste and toxins, circular economy principles applied to renewable technologies could open new market opportunities, using effective technology and economically viable methods to separate the materials embedded in the renewable energy technologies (e.g. some companies have started trying to recycle crystalline PV panels). Despite current efforts, renewable energy technologies recovering and recycling in LAC are still in their early stages (Contreras-Lisperguer et al., 2017[73]) Although renewable energy technology waste offers a rich source of materials that can be re-used and converted into inputs for the production of new renewable energy devices and/or other products, e-waste legislation in the LAC region has given little consideration to treatment and re-use of e-waste, amid generally low awareness on the issue among policy makers and the public. Public policies should create the conditions for manufacturers to include designs for non-energy intensive disassembly and total reusability of materials embedded in energy technologies, replacing toxic components with non-toxic ones to assess the potential to implement up-cycling at a commercial scale (Contreras-Lisperguer et al., 2017[73]).

Other necessary conditions for the circular economy are capacity-building initiatives (e.g. training, research and networking events), digitalisation efforts (e.g. exchange and awareness platforms, waste tracking apps or open-access online tools), and the production and sharing of data (e.g. on energy consumption, air quality or waste production). The promotion of circular business models (e.g. circular supply, collaborative consumption, service systems or hiring and leasing instead of buying) is crucial (OECD, 2020[62]; OECD, 2019[74]). Stakeholder engagement is key for the circular transition. The public and private sectors, citizens, and academia need to be involved through communication, consultation, participation, representation, partnership or co-decision/-production mechanisms to promote innovative circular business models, advance towards a more sustainable production matrix, and change unsustainable consumption patterns.

The blue economy as a means for development

The blue economy is a driver of economic activity, employment and other social benefits. In 2018, the total GDP contribution of ocean services was estimated at USD 25 billion for LAC and USD 7 billion for Caribbean countries (IDB, 2021[75]). More than 2.5 million people in the region worked directly in fisheries or aquaculture in 2018 (IDB, 2021[75]). Blue carbon ecosystems protect coasts from flooding, buffer ocean acidification, enhance water filtration, promote biodiversity conservation, help carbon sequestration, and boost sustainability in fishery productivity, which in turn creates jobs and increases economic returns (IPCC, 2019[76]).

Given increasing awareness of the importance of the ocean for human well-being and economic activities, a blue approach must be adopted in the LAC region to spur sustainable growth opportunities. The blue economy approach has two complementary goals: it protects marine and coastal ecosystem assets and services while simultaneously addressing the economic challenges of coastal countries (ECLAC, 2020[77]).

LAC's marine area is considered one of the most important and productive in the world. With unique marine biodiversity, it is home to 47 of the 258 global marine ecoregions and to the second-largest barrier reef in the world (ECLAC, 2020[77]). Marine and coastal activities are particularly important in LAC given that more than 27% of the population lives in coastal areas and the sea accounts for a large share of the region's territory (IDB, 2021[75]).

Ocean warming and acidification are damaging marine ecosystems and compromising the ability of the ocean to provide food, livelihoods and safe coastal living (IPCC, 2019[76]). Increased coastal erosion, greater bleaching of corals and increasing flooding of ecosystems are some of the expected effects of climate change on the ocean. Despite this alarming situation, policies related to the management of ocean resources are still underdeveloped and many countries have not incorporated blue carbon into their climate change mitigation strategies (UNESCO, 2020[78]). SDG 14 on the protection of life below water remains the most underfunded of all 17 SDGs, despite its immense potential to be a game changer in addressing the triple planetary crisis: climate change, air pollution and biodiversity loss (UNDP, 2022[79]).

The blue economy for climate change mitigation and adaptation

Protecting, conserving and restoring coastal, river and marine ecosystems directly contributes to climate change mitigation. Blue carbon coastal ecosystems (seagrass meadows, tidal marshes and mangroves) are habitats that uptake and store carbon from the atmosphere in the ocean (carbon sequestration). They are among the most intensive carbon sinks in the biosphere. The degradation of these ecosystems can cause the release of billions of tonnes of GHG; they should be conserved and restored to enhance their sequestration potential (UNESCO, 2020[78]). Mangroves represent a promising mitigation opportunity, particularly in LAC, because they have the capacity to store between three and four times more carbon than most of the forests on the planet (ECLAC, 2020[77]).

LAC is one of the regions with the greatest water availability: up to 33% of the world's total available water (Vargas, 2021[80]). Nevertheless, a great part of the region suffers from water scarcity, given the varied spatial distribution of water resources. Integrated management of river basins and lakes is key to the blue economy approach in LAC, to protect communities' well-being. Freshwater ecosystems provide the foundation to fulfil basic necessities, such as drinkable water, food, health, sanitation, and water for irrigation systems and agriculture. The effects of climate change on water resources compromise water and food security in LAC. Rivers and lakes in the region are highly exposed to polluting sources and the management of water for agriculture is inefficient. The sector uses around 70% of available water (Vargas, 2021[80]). A sustainable blue approach that guarantees water security requires an integrated regional response to protect rivers, decrease deforestation and pollution, and promote sustainable economic activities that minimise harm to ecosystems, as well as a management strategy that includes measures to mitigate and adapt to the effects of climate change. A comprehensive social approach to water management is necessary, especially in rural areas where people are highly exposed to climate events and rely heavily on freshwater resources (Vargas, 2021[80]).

The adoption of a blue carbon approach is a way to help countries meet their NDCs (IDB, 2021[75]). Yet further international action is needed to internalise the positive externalities of those activities, for instance by developing the new blue carbon markets. The ratification of Article 6 of the Paris Agreement at the 2021 UN Climate Change Conference in Glasgow (COP26), which recognises voluntary co-operation in the implementation of NDCs to allow for higher ambition, could mark a milestone for establishing these markets. Participating governments have the chance to mainstream blue carbon solutions by including them in their NDCs (Claes et al., 2022[81]).

Blue economy opportunities in LAC: Fishery and aquaculture, sustainable tourism, and renewable energy

Blue policies have great potential aside from climate-related benefits. Human activities, including overfishing, destructive fishery practices, coastal development, and domestic and industrial contamination, have damaged marine and river ecosystems. Transitioning to blue sustainable economic opportunities in LAC presents particular prospects in three key sectors: fishery and aquaculture, sustainable tourism, and renewable energy generation.

Advancing towards a more sustainable fishery sector, with due regard for the protection of marine and river ecosystems, offers an opportunity to increase employment, improve food security and promote exports. To achieve this, management plans that help restore fish stocks are needed, alongside supportive regional and international policy and regulatory frameworks. It is crucial to end unsustainable overfishing by tackling illegal, unreported or unregulated fishing, banning destructive fishing practices, and eliminating harmful subsidies that contribute to overfishing (Rustomjee, 2016[82]). An example of blue innovation for fishing is the Mexican smartphone app PescaData, which enables local fishers to monitor populations of birds, sea mammals, fish and crustaceans to manage overfishing, fight non-selected fishing and define some fishing areas as sanctuaries (AFD, 2022[83]).

Despite the COVID-19 pandemic roughly halving tourism's contribution to LAC GDP from 2019 to 2020, economies are recovering and tourism demand is increasing (Annex Table 3.A.6 Sustainable tourism). Ecotourism, as part of the blue economy, offers great potential to lift the region out of its current economic downturn (UNWTO/CAF, 2021[84]). Costa Rica has earned an international reputation for its unique marine natural assets and has managed to boost coastal and biodiversity-based tourism (UNCTAD, 2019[85]) (Annex Table 3.A.1).

As a promising component of the blue economy, the ocean also offers various renewable energy options, such as offshore wind, wave and tidal and the use of temperature and salinity gradients to produce energy. Among these, offshore wind is a highly promising but still largely underfunded option, given that LAC[14] is the region with the highest technical potential[15] globally (6 830 GW), and many of its countries show particularly suitable conditions (Figure 3.8). Argentina could benefit from the golden combination of windswept waters and relatively shallow sea territory, however, its potential still has not been exploited (BNamericas, 2021[86]). Brazil is currently the leading country with six offshore wind power projects under review for licences. The country has the second highest technical potential, followed by Chile and Mexico (ECLAC, 2020[77]). Colombia has an offshore technical potential of 110 GW and recently approved the roadmap for offshore wind projects in the Pacific Ocean. The explored areas could produce 1 GW of energy by 2030 and up to 9 GW by 2050 (currently, the country generates 0.725 GW of solar energy), and are expected to attract an investment of USD 27 billion and create 50 000 jobs (Hidalgo, Fontecha Mejía and Escobar, 2022[87]). Offshore wind energy offers a great opportunity for the Caribbean as well; it has the potential to generate around 560 GW (World Bank/ESMAP, 2020[88]).

Figure 3.8. **Technical potential for offshore wind generation in LAC**

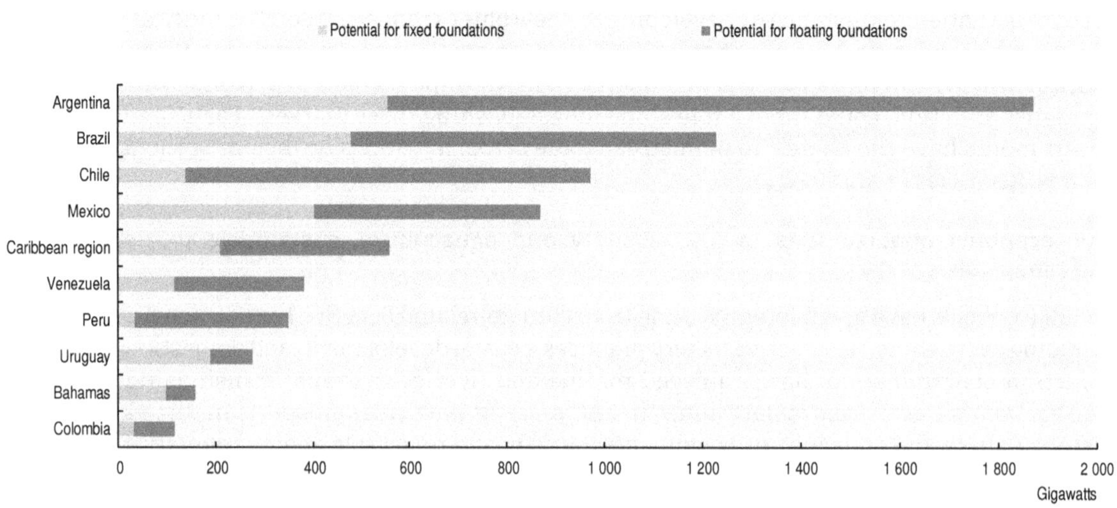

Notes: The offshore wind technical potential is an estimate of the amount of generation capacity that could be technically feasible with current technology, considering only wind speed and water depth, and it is expressed in terms of installed power capacity in gigawatts (GW) within 200 kilometres of the shoreline. Results report the total fixed foundations potential and the total floating foundations potential. Above 50 m depth is considered optimal for fixed foundations and, 50-1 000 m depth is where floating wind can be considered. The Caribbean region includes Antigua and Barbuda, Barbados, Bahamas, Belize Cuba, Dominica, Dominican Republic, Grenada, Jamaica, Haiti, Saint Kitts and Nevis, Saint Vincent and the Grenadines, Saint Lucia, Suriname, and Trinidad and Tobago.

Source: Authors' calculations based on (World Bank/ESMAP, 2020[88]).

StatLink https://stat.link/z91ng5

Policy options against marine plastic waste

Plastic threatens marine ecosystem services that underpin the blue economy, as it accelerates climate change and is a massive source of waste (Diez et al., 2019[89]). Lifecycle emissions from plastic, over 99% of which is derived from fossil fuels, could reach 10-13% of the entire remaining global carbon budget by 2050 (Hamilton and Feit, 2019[90]). In the form of waste, plastic items and microplastics harm marine organisms through entanglement, ingestion, smothering and leakage of associated chemicals (Tekman et al., 2022[91]). The OECD estimated a total inflow of macroplastics into the ocean of 1.7 Mt in 2019 and projects this number to reach 4 Mt by 2060 (OECD, 2022[92]). This poses a major risk to the LAC region, where beaches show the second-highest levels of pollution in the world in terms of macro litter density (after Asia) (Haarr, Falk-Andersson and Fabres, 2022[93]).

The region lags in terms of waste management, including plastics. Almost 90% of plastics are mismanaged or sent to landfill in LAC, although this could be prevented at moderate costs. Only around 10% of plastic waste generated in LAC countries[16] was recycled in 2019, while 47% was landfilled and almost 42% mismanaged. The region's yearly plastic waste generation is expected to more than double by 2060. With current policies in place, these figures will improve insufficiently by that year, reaching only 17.5% recycled and still 25% mismanaged. This could be prevented at moderate costs: it is estimated that LAC could almost entirely prevent plastic leakages into its aquatic environments by 2060 at an annual cost of only 0.74% of its GDP (OECD, 2022[92]).

Marine plastic pollution is becoming a policy priority for the LAC region (UNEP, 2021[94]). Countries have set specific goals to reduce plastic waste (Box 3.4). Plastic-specific policies should target different lifecycle stages: 1) restrain demand (fiscal instruments that disincentivise the production and use of plastics, and other policies that enhance product design to increase their durability and favour reuse and repair). For example, taxes on plastics, packaging, promotion of circular design or repair services); 2) enhance recycling (includes instruments such as recycled content targets, extended producer responsibility (EPR) schemes or region-specific recycling rate targets); and 3) close leakage pathways (aims to decrease and eliminate mismanaged plastic waste by investing in waste management infrastructure and increasing litter collection rates) (OECD, 2022[92]). For plastic that cannot be avoided, the focus should be on circular design, recycling and closing leakage pathways (OECD, 2022[92]; Geyer, 2020[95]). Four critical levers for "bending the plastic curve" are: recycled (secondary) plastics markets; technological innovation for more circular plastics value chains; more coherent and ambitious domestic policy measures; and greater international co-operation.

Box 3.4. Selected LAC commitments on plastics

Chile's Plastic Pact commits (by 2025) to eliminating unnecessary and problematic single-use plastic packaging, ensuring that 100% of plastic packaging is reusable, recyclable, or compostable and incorporates 25% recycled content (Pacto Chileno de los Plásticos, 2020[96]). The National Strategy for Marine Waste and Microplastics Management commits to preventing 40% of waste from entering aquatic ecosystems by 2030 and to taking recovery actions (Ministerio del Medio Ambiente, 2021[97]). The EPR legislation requires 45% of plastic packaging to be recovered and recollected in households and 55% in other establishments by 2034 (Ministerio del Medio Ambiente, 2021[98]). The Roadmap for a Circular Chile by 2040 includes a 65% recyclability goal (Ministerio del Medio Ambiente, 2022[99]).

Box 3.4. **Selected LAC commitments on plastics** *(cont.)*

Colombia has launched a national plan for single-use plastics, all of which should be reusable, recyclable or compostable and have a minimum average recycled content of at least 30% by 2030 (Mesa Nacional para la Gestión Sostenible del Plástico, 2021[100]). The National Circular Economy Strategy (adopted in 2019) reinforces this goal (Gobierno de la Republica de Colombia, 2019[101]).

Mexico is collaborating with Canada and the United States in a project to transform recycling and solid waste management, to reduce waste (particularly plastics), close material loops and help minimise environmental impacts throughout value chains. North America has the highest per capita plastic consumption in the world. The region represents 21% of total global plastics consumption and four times the global average per capita paper consumption (Commission for Environmental Cooperation, 2021[102]).

Chile, Colombia and Mexico support the process towards a binding global treaty, which was launched in March 2022 with the adoption of the "End plastic pollution" resolution at United Nations Environment Assembly 5.2 (UNEP, 2022[103]).

Integrated collaborative policy responses are needed

Due to the variety of its sectors and the relationship between ecosystem services and human welfare, the blue economy requires comprehensive and integrated management of human activities to ensure that ecosystem goods and services are used sustainably and that balances environmental, economic and social concerns (Le Tissier, 2020[104]).

As a response to problems in coastal LAC – such as chaotic urbanisation, threats to indigenous cultures, degradation of ecosystems, loss of biodiversity, and effects of climate change on the socio-ecological system – integrated coastal zone management (ICZM) has been emerging as an effective strategy for ecosystem-based adaptation (Barragán Muñoz, 2020[105]). ICZM also offers important policy instruments for plastic waste abatement such as co-ordinated cleanups and the creation of marine protected areas (UNEP, 2021[106]). ICZM performance in LAC countries is still unsatisfactory, although results are heterogeneous. While most countries lack an adequate set of ICZM policies, strategies, regulation or institutions, Belize, Brazil and Puerto Rico are in developed stages of its application. This heterogeneity and the fact that coastal ecosystems are shared among countries call for greater international co-operation to achieve a joint response (Barragán Muñoz, 2020[105]).

An integrated solution also needs to highlight the role of the marine space. To prevent conflict among policy priorities and to reconcile nature conservation with economic development, human activities can be regulated and organised through maritime spatial planning (MSP). The status of MSP adoption differs greatly across LAC countries. Ecuador has applied MSP to the management and zoning planning of the Galapagos Islands while many countries in Central America and the Caribbean have initiated MSP, often with international funding support (e.g. the Caribbean Regional Oceanscape Project funded by the World Bank and the Global Environmental Facility) (World Bank, 2021[107]). Mexico is at a more advanced stage: it controls one of the largest exclusive economic zones in the world, which it has divided into planning regions based on ecosystem considerations for which it is developing regional marine spatial plans (Ehler, 2021[108]).

Due to their dependence and impact on ecosystems, ICZM and MSP should be applied as ecosystem-based approaches (EBA) to manage the interactions and trade-offs among the goods and services provided by ecosystems and various economic and social objectives. This implies the development of long-term ecosystem objectives and indicators for adequate monitoring (Altvater and Passarello, 2018[109]; European Commission, 2021[110]). Monitoring tools, such as the Ocean+ Habitats Platform (UNEP-WCMC, 2022[111]) are crucial to enable data-based decision making. The Sustainable Ocean Economy Database provides policy analysis and guidance for the blue economy, sustainable use and conservation of the oceans, and management of climate-related risks (OECD, 2022[112]).

Key selected sectors for the green transition

Annex 3.A1 presents a list of selected sectors[17] identified as strategic for the LAC region to advance the green transition towards a more sustainable development model. These sectors are: 1) nature-based solutions, use of land, and biodiversity and forestry preservation (Annex Table 3.A.1); 2) sustainable agriculture and livestock (Annex Table 3.A.2); 3) the bioeconomy and regenerative food systems (Annex Table 3.A.3); 4) water management (Annex Table 3.A.4); 5) waste management and plastics (Annex Table 3.A.5); 6) sustainable tourism (Annex Table 3.A.6); and 7) sustainable mining (Annex Table 3.A.7). For each sector, some key facts are presented to show their importance for the region and their relevance to advance the green transition. Policy instruments and relevant experiences adopted are also detailed, to contribute to the policy-making process of the green transition.

Social policies for a just transition: The role of the labour market

The green transition will have a great impact on LAC societies. The job market, for example, will experience deep transformations. If properly addressed, broad green policies can create quality job opportunities for Latin Americans. If well-designed social and labour policies are implemented, better-paid formal jobs could be created by innovative firms, which will emerge in the green industries. Workers can benefit from this green dividend by transitioning to jobs with better working conditions. In LAC, where more than half of workers are informal, the green agenda can be an opportunity to create formal jobs (OECD, 2021[112]). However, the green transition will also entail economic costs. Fossil fuel extractive sectors will face considerable job losses, especially for the affected regions. Displaced workers may face skills obsolescence, which results in fewer opportunities to find new jobs (Dutz, Almeida and Packard, 2018[114]).

In this context, labour market and social policies play a crucial role both in stimulating the creation of high-quality new jobs and in cushioning the downside consequences of the transition to cleaner economies. Unemployment benefits and unemployment assistance should help workers affected by the transition's costs. In parallel, well-designed active labour market policies (ALMPs), such as training programmes, hiring incentives or placement services, are crucial both to promote green jobs and to boost the skills of workers who will lose their jobs. Non-contributory social protection policies, such as income support measures or conditional cash transfer programmes, may play a positive role in minimising the income losses of families with workers negatively affected by the green policies. The green transition can be an opportunity to move towards comprehensive and universal social protection systems, which include universal health insurance programmes, thanks to the shift towards formal jobs (ECLAC, 2022[14]; Grundke and Arnold, 2022[115]).

Quality green jobs for greater inclusion

The green agenda will be an unprecedented opportunity to improve well-being in LAC (Chapter 2). The effects of mitigation and adaptation policies on the labour markets of LAC countries will involve both the creation of new job opportunities and the loss of jobs due to the obsolescence of current technologies (Vona et al., 2018[116]). This section presents a sectoral forecast exercise to estimate the impact of green policies (Box 3.5).

Two types of economic sectors, the green and the brown sectors, have been identified. The rationale behind the empirical strategy is that adaptation to climate change policies, technological transformation, investment, and new green skills would boost job creation in green sectors. By contrast, some mitigation policies and regulations aimed at cutting GHG emissions and other pollutants would lead to job destruction in brown sectors, which have nonetheless recently been showing very low net employment growth in the region. The net effect on jobs will depend on both the industrial structure of each country and the effectiveness of the reform packages in boosting green job creation while softening the negative impact of mitigation policies on firms and the labour market.

Box 3.5. **Estimating the impact of the green agenda on net job creation**

The methodology implemented to analyse the impact of broad green agendas on net job creation in LAC countries is described in detail in OECD (forthcoming[117]).

Green jobs and green sectors

Green jobs are defined according to the methodology of Vona, Marin, Consoli, & Popp (2018[116]). The green transition will transform technologies. Jobs created will increasingly involve new tasks associated with less polluting and greener production processes. The Bureau of Labor Statistics O*NET database contains standardised, occupation-specific descriptors for nearly 1 000 occupations, covering the US economy. Detailed information is presented on the tasks that workers carry out in the labour market. The proportion of tasks that are green as a percentage of the total tasks carried out in each occupation is an indicator of the intensity of green tasks – i.e. of how green each occupation will be, based on the technological frontier. Under the assumption that green tasks performed by workers in LAC do not differ extensively from the ones at the frontier, for each LAC country, tasks identified in O*NET are matched to the labour force survey data on occupation, using the crosswalk provided in Hardy, Keister and Lewandowski (2018[118]). The level of detail of each match depends on the original occupation coding of the national labour force survey, resulting in

The definition of green sectors is based on the definition of green jobs. The distribution of green occupations across sectors is calculated. The ten industries that employ the most of green jobs are defined as green sectors. In LAC countries, green sectors represent between 6% and 90% of green jobs. These sectors should be considered "potentially green", as not all jobs in these industries are green, given the present distribution of tasks.

Brown sectors

Brown sectors will experience the destruction of jobs (Figure 3.12 and Figure 3.13). They are defined following the Climate Analysis Indicators Tool (CAIT) database on total GHG emissions (Climate Watch, 2020[119]). Brown sectors account for the most emissions. They are: 1) agriculture; 2) energy and heat production; 3) extraction and production of fossil fuels plus construction; 4) industrial processes; 5) transportation; and 6) waste management. The data on emissions are matched with national accounts data on value added by activity and employment time series from labour force surveys using ISIC Rev 3.1 or ISIC Rev 4 classifications of industries, depending on the availability in each LAC country.

Methodology of the forecasts

For both brown and green sectors, the forecasts present three scenarios compared to the business-as-usual (BAU) scenario. BAU assumes that value added and employment in green sectors for each LAC country will grow as in the past ten years (from 2010 and 2020). The effect of change in value added on employment is estimated at the sectoral level using a dynamic panel regression model (Arellano-Bond estimator) (Arellano and Bond, 1991[120]).

Box 3.5. **Estimating the impact of the green agenda on net job creation** *(cont.)*

Based on realistic assumptions found in the literature (OECD, forthcoming[117]), for green sectors, three policy scenarios assume the following impacts of green policies on investment in fixed and human capital: 1) in the high-impact scenario, value added will grow in each sector by 3 percentage points more than what would have been the case in the absence of any policy; 2) the medium-impact scenario assumes an additional sectoral value-added growth of 2 percentage points; and 3) in the low-impact scenario, the additional growth will be just 1 percentage point. In all three scenarios, total factor productivity (TFP) will grow by 1% due to the technological shifts induced by the green transition.

For brown sectors, it is assumed that green policies will reduce total emissions by 5% per year in each of the top emissions-intensive industries defined above. This would imply a reduction of approximately 40% in total CO_2 emissions in 2030 with respect to 2020 levels. Three future scenarios are assumed for each Latin American economy: 1) in the high-impact scenario, value added will decrease by 5 percentage points each year; 2) in the medium-impact scenario, it will decrease by 4 percentage points each year; and 3) in the low-impact scenario, it will decrease by 3 percentage points each year. In all three scenarios, total factor productivity (TFP) will grow by 1 percentage point, due to the technological shifts induced by the green transition.

The estimated effects on employment are presented in proportional change compared to the BAU employment levels in the green and brown sectors in 2020. They represent the additional change compared to what would be in the absence of any policy change.

The overall effect of effective green policies on the Latin American labour market can be substantial. In the case of high-impact green policies, employment in green sectors could grow by 15% in LAC by 2030 compared to the baseline scenario (Figure 3.9). In absence of any policy intervention, the yearly average growth of employment would be 0.9% in green sectors. Green policies with high impact in stimulating private and public investment in new technologies and human capital would increase the yearly growth rate to 2.3%. Employment in potential green sectors represents 55% of total employment across LAC. The forecasts of employment growth in green sectors by 2030 are 11.0% in the case of medium impacts and 7.2% for low impacts compared to BAU. In the case of high impacts, estimates range from 18.9% for Bolivia to 12.6% for Brazil (Figure 3.10), signalling the potential of job creation in green sectors in all the countries of the region. The green transition can also be beneficial as a tool to boost overall economic growth and enhance productivity. Among the countries that would benefit the most are Ecuador, Guatemala and Paraguay, all with GDP per capita lower than the regional average.

The identification of green sectors depends on the distribution of green tasks across occupations and on the industrial structure of each LAC country, under the assumption that the tasks' content of jobs is similar to the one observed in the United States. However, some insights emerge at the regional level. Five out of ten sectors are present in at least six out of the nine countries covered. They are food production, construction, retail and wholesale trade, transport, and public administration. In total, they account for 67% of employment in potential green sectors in LAC, in 2020; as such, they would contribute the most to job creation over the next decade. Compared to the 15% of the total of green sectors, food production should add more jobs, with a forecasted 18.8% deviation from BAU in 2030, in the high-impact scenario. The other sectors would show a deviation from BAU as follows: public administration (14.6%), construction (14.3%), trade (14.1%) and transport (14.1%).

Figure 3.9. **Job creation in green sectors in LAC, 2020-30**

Change in employment in green sectors, under various green policies scenario, in LAC, compared to the BAU, as percentage of 2020 baseline employment in green industries

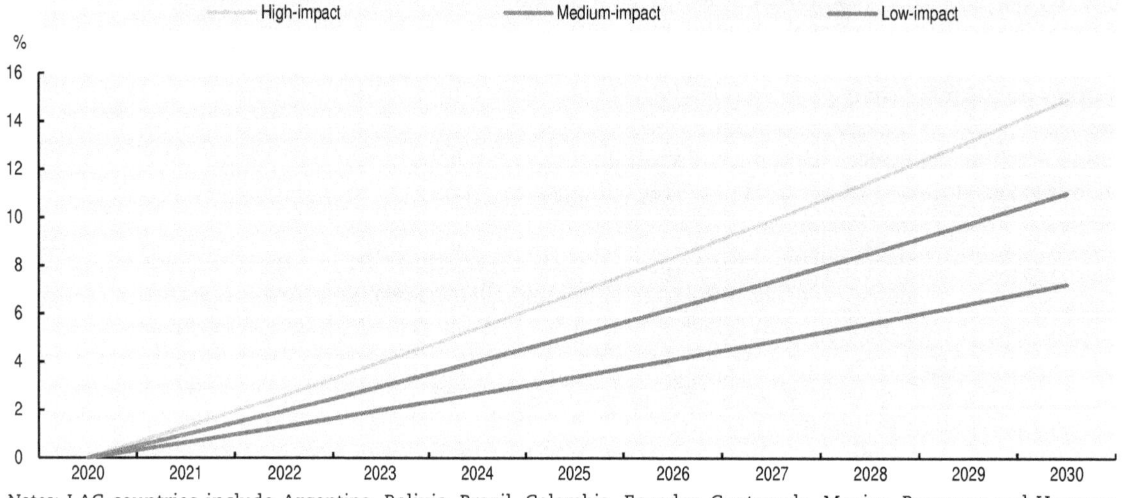

Notes: LAC countries include Argentina, Bolivia, Brazil, Colombia, Ecuador, Guatemala, Mexico, Paraguay and Uruguay. The data refer to an unweighted average over the countries' forecasts. Green sectors are defined in each country by first identifying the number of green tasks that workers perform in their occupations and then by looking at the top ten industries in which those jobs are distributed. The baseline scenario assumes that, in each green sector, value added and employment will follow the same dynamic as in the past ten years. The counterfactual scenarios are defined according to the impact of a green policy that aims to boost investment in fixed and human capital, with a positive impact on value added growth in each green sector. The high-impact scenario assumes that the value added in each sector will increase by 3 percentage points each year, adjusting to the new equilibrium. The medium-impact scenario assumes that the value added will increase by 2 percentage points, while the low-impact scenario assumes that it will increase by 1 percentage point each year. In all forecasts, Total factor productivity will increase by 1 percentage point due to lower climate damages and new technology-induced change. Employment change is forecasted using the estimated short-term elasticity to the value added, using a panel dynamic model, defined by each sector and country, in the last ten years.

Sources: Authors' estimates based on Labour Force Surveys, National Accounts data by industries, (Vona et al., 2018[115]) and (Hardy, Keister and Lewandowski, 2018[117]).

StatLink https://stat.link/gpamzt

Using a different methodology, the International Labour Organization (ILO) reports that agriculture will have more job creation (Saget, Vogt-Schilb and Luu, 2020[120]). Green policies would add 19 million full-time equivalent jobs by 2030, in this sector, with respect to the high-emissions scenario, a 54% increase. Employment in the renewable energy sector would add 22% more jobs in the decarbonisation scenario compared to the high-emissions scenario, representing 100 000 additional jobs. The construction sector would contribute 540 000 additional jobs linked to energy efficiency investments, representing a growth of 2% in employment in the sector compared to the baseline scenario. Some 120 000 jobs would be created in manufacturing to support low-carbon technologies, a net addition of 0.4% of employment in the sector. The forestry sector would create 60 000 jobs (6% of the sector's jobs). Green policies could increase the energy sector's productivity and create direct formal employment if they promote a green fiscal stimulus aimed at the energy industry. This is particularly the case in developing and middle-income countries with segmented and informal labour markets. After suffering adverse aggregate demand shocks it will be important to seek a bigger and longer-lasting boost to green labour-intensive projects. In these circumstances, the green transition and job creation may well go hand in hand (Bowen, 2012[123]). If green policies aiming at increasing the cost of carbon emissions are coupled with a reduction in labour tax wedge, the green transition may grant a double dividend in terms of both environmental goals and net job creation in LAC (OECD, 2018[124]; Willis et al., 2022[125]).

Only 0.8% of the LAC workforce was employed in the energy production sector in 2020, similar to the OECD average of 0.9% (Figure 3.11). Already today, more than half of all energy-related jobs in Central and South America are related to clean energy (IEA, 2022[122]) and this share is bound to increase further. Some 11.7% of workers were employed in manufacturing, which accounts for the majority of total CO_2 production in the economy, less than the OECD average of 15.2%. In LAC, the share of employment in the transport sector, another growing contributor to overall GHG emissions and pollution, stood at 6.7%, higher than the OECD average of 5.5%. Agriculture (the second-largest emitter of the region) accounts for 18% of the workforce, much higher than the OECD average of close to 6% in 2020.

Figure 3.10. **Job creation in green sectors in LAC countries**

Change in employment in green sectors under various impacts of green policy scenarios in selected LAC countries, compared to the BAU, as percentage of 2020 baseline employment in green industries

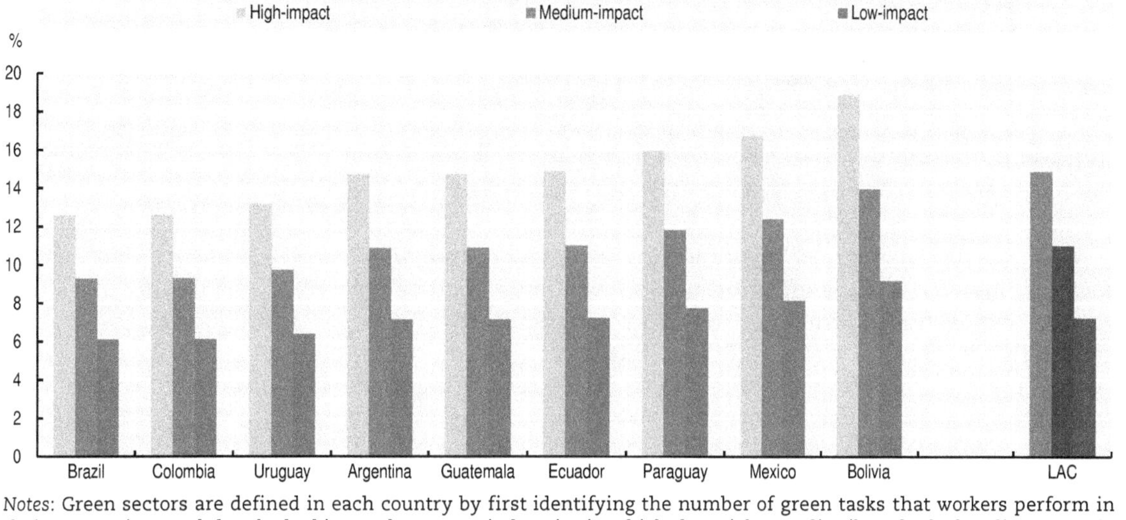

Notes: Green sectors are defined in each country by first identifying the number of green tasks that workers perform in their occupations and then by looking at the top ten industries in which those jobs are distributed. The baseline scenario assumes that, in each green sector, value added and employment will follow the same dynamic as in the past ten years. The counterfactual scenarios are defined according to the impact of a green policy that aims to boost investment in fixed and human capital, with a positive impact on value added growth in each green sector. The high-impact scenario assumes that the value added in each sector will increase by 3 percentage points each year, adjusting to the new equilibrium. The medium-impact scenario assumes that the value added will increase by 2 percentage points, while the low-impact scenario assumes that it will increase by 1 percentage point each year. In all forecasts, Total factor productivity will increase by 1 percentage point due to lower climate damages and new technology-induced change. Employment change is forecasted using the estimated short-term elasticity to the value added, using a panel dynamic model, defined by each sector and country, in the last ten years.

Sources: Authors' estimates based on Labour Force Surveys, National Accounts data by industries, (Vona et al., 2018[115]) and (Hardy, Keister and Lewandowski, 2018[117]).

StatLink https://stat.link/w4v0up

Figure 3.11. **Employment shares in industries with high intensity of GHG emissions**

As a percentage of total employment, 2010 and 2020

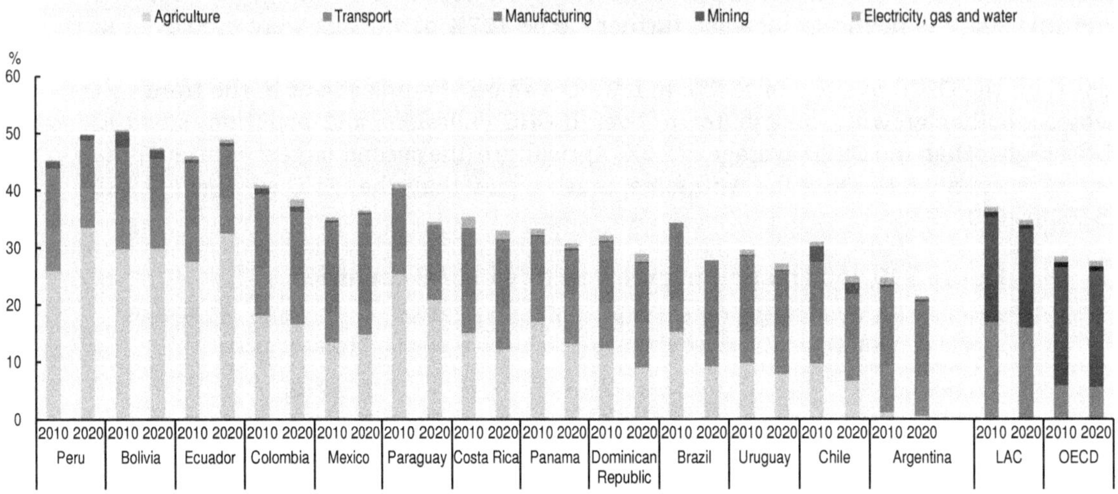

Notes: Data for Argentina refer to the urban areas. LAC and OECD refer to an unweighted average of the LAC and OECD countries. The industry categorisation follows the ISIC Rev. 4 codification of economic sectors.

Sources: Authors' estimates based on (ECLAC, 2022[122]) and (OECD, 2022[123]).

StatLink https://stat.link/3y8own

The green transition will inevitably have heterogeneous impacts in the agricultural sector in LAC, even if the effect on net job creation should be positive. Transitioning to cleaner and capital-intensive technologies in agriculture will translate into jobs losses, especially for informal workers (ECLAC/ILO, 2018[125]), even if the total net effect should be positive. The local dimension should be at the centre of green policies, as agriculture accounted for 53.5% of total employment (most being informal workers) in rural areas in 2020.

LAC countries have experienced a 20% growth in employment in industries linked to energy production, water supply and mining activities over the last decade. Manufacturing firms, which consume more energy, have added only a few jobs thanks in part to the process of digitalisation and robotisation, which normally implies more capital-intensive and labour-saving technologies. By contrast, employment shrunk in agriculture (-4.5%).

In the transition to a net-zero carbon economy, many jobs will be destroyed, and workers dismissed due to the technological shifts needed to achieve lower emissions, especially in brown industries. On average across LAC, brown industries could experience a higher decline in jobs – as much as 13.3% compared to the BAU scenario, in which no green policy is effectively put in place (Figure 3.12). All forecasted scenarios imply a bold target of -5% GHG emissions each year. If these targets were met, GHG emissions would decline by 40% in 2030 compared to 2020 levels. If firms invest in fixed capital more proactively and workers acquire greener skills and human capital, job losses will be significantly lower. For instance, in the medium- and low-impact scenarios, deviation from BAU in 2030 would be 10.1% and 6.9%, respectively. Public policies may help alleviate the transition. Public investment supporting technological transformation and adoption may help firms. ALMPs, such as retraining and education programmes, may help workers retain their jobs in the new technological environment or switch to new ones, with lower aggregate losses.

Figure 3.12. **Job losses in brown sectors in LAC, 2020-30**

Change in employment in brown sectors, under various mitigation policies scenarios in LAC, compared to the BAU, as percentage of 2020 baseline employment in brown industries

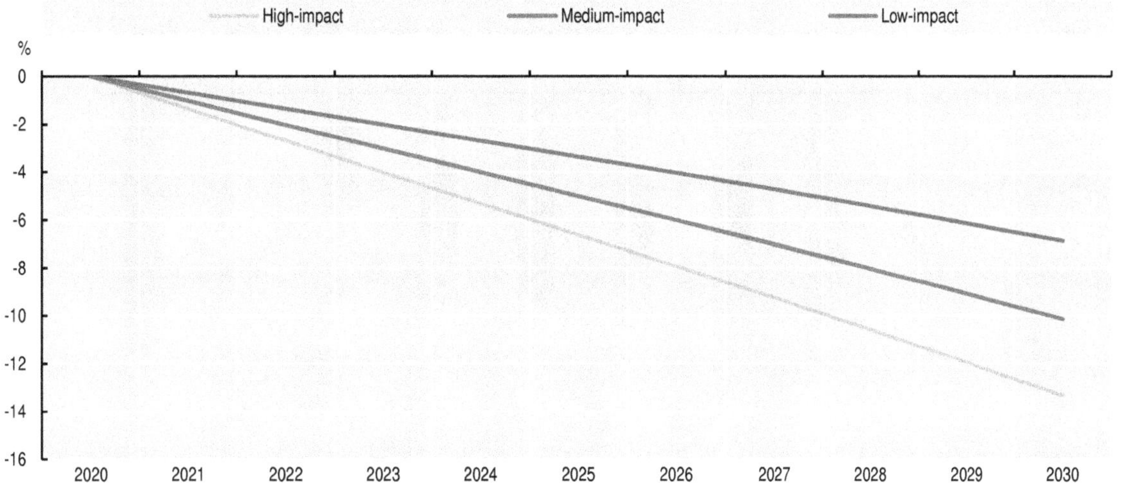

Notes: LAC countries include Argentina, Bolivia, Brazil, Colombia, Ecuador, Guatemala, Mexico, Paraguay and Uruguay. The data refer to an unweighted average over the country forecasts. Brown sectors are defined according to the CAIT definition (https://datasets.wri.org/dataset/cait-country). The baseline scenario assumes that, in each brown sector, GHG emissions, value added and employment will follow the same dynamic as in the past ten years. The counterfactual scenarios are defined according to the impact of a green policy that aims to reduce total GHG net emissions by 50% in 2030 compared to the 2020 levels in each brown sector. The high-impact scenario assumes that the value added in each brown sector will decrease by 5 percentage points each year, adjusting to the new equilibrium. The medium-impact scenario assumes that the value added will decrease by 4 percentage points, while the low-impact scenario assumes that it will decrease by 3 percentage points each year. In all forecasts, TFP will increase by 1 percentage point due to lower climate damages and new technology-induced change. Employment change is forecast using the estimated short-term elasticity to the value added, using a panel dynamic model, defined by each sector and country, in the last ten years.

Source: Authors' estimates based on CAIT data, labour force surveys and national accounts data by industries.

StatLink https://stat.link/fpgdl7

The net effect of the green transition on employment could be positive and will depend on the adaption mechanisms to create formal jobs resulting from the implementation of green policies. Even in the worst-case scenario, the effects would be positive compared to BAU. This is because brown sectors in LAC represent 35% of total employment, compared to 55% for green sectors. For instance, if green sectors created jobs according to the low-impact scenario and brown sectors destroyed jobs according to the high-impact scenario, this would still translate into an additional 1.8% of total employment in those sectors in 2030. In the case of medium-impact and high-impact policies for green sectors, in 2030, additional job creation would be 6.0% and 10.5% of total employment in the brown and green sectors, respectively. There is a clear incentive for governments to promote an active transition for green sectors, as it will increase job creation and formalisation.

Job losses would be felt among the brown industries of LAC, especially in agriculture and manufacturing, with 37% and 30% of the total losses, respectively. The transport sector would account for 12% of the total.

Across countries, jobs losses would range from 19% in Bolivia to 7% in Argentina (Figure 3.13) in the high-impact scenario, as it depends heavily on each industrial structure. In Ecuador, Guatemala and Paraguay, agriculture would account for the biggest part of the contraction (71%, 39% and 47% of the total, respectively). In Argentina, Brazil, Mexico and Uruguay, manufacturing would likely be the industry most affected, as it would account for 40%, 41%, 43% and 36% of total jobs by 2030, respectively.

Figure 3.13. **Job losses in brown sectors in a high-impact scenario in LAC**

Change in employment by 2030, compared to the BAU, as percentage of 2020 baseline employment in brown industries

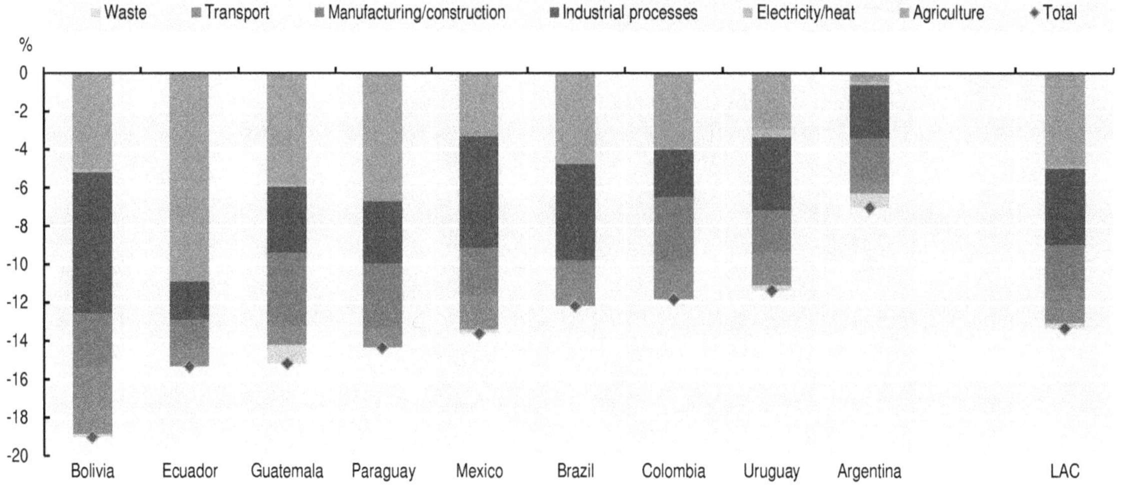

Notes: LAC countries include Argentina, Bolivia, Brazil, Colombia, Ecuador, Guatemala, Mexico, Paraguay and Uruguay. The data refer to an unweighted average over the country forecasts. Brown sectors are defined according to the CAIT definition (https://datasets.wri.org/dataset/cait-country). The baseline scenario assumes that, in each brown sector, GHG emissions, value added and employment will follow the same dynamic as in the past ten years. The counterfactual scenarios are defined according to the impact of a green policy that aims to reduce total GHG net emissions by 50% in 2030 compared to the 2020 levels in each brown sector. The high-impact scenario assumes that the value added in each brown sector will decrease by 5 percentage points each year, adjusting to the new equilibrium. TFP will increase by 1 percentage point due to lower climate damages and new technology-induced change. Employment change is forecast using the estimated short-term elasticity to the value added, using a panel dynamic model, defined by each sector and country, in the last ten years. For Brazil, Colombia, Ecuador, and Paraguay, the waste sector could not be identified due to lack of disaggregated data either for value added or employment by industry.

Source: Authors' estimates based on CAIT data, labour force surveys and national accounts data by industries.

StatLink https://stat.link/kmnz6c

The IDB and the ILO show similar forecasts in terms of job losses. By 2030, 7.5 million jobs would be destroyed in fossil fuel electricity, fossil fuel extraction and animal-based food production (Saget, Vogt-Schilb and Luu, 2020[120]). More specifically, 4.3 million jobs would be lost in the livestock, poultry, dairy, fishing and animal-based food processing sectors compared to the high-emissions scenario, representing 29% of jobs in the sector. Fossil fuel extraction would lose more than 520 000 jobs (46%), while electricity from fossil fuel generation would also suffer a relatively important downsizing, with 60 000 fewer jobs (51%) compared to the baseline scenario.

Wages and job quality of jobs in brown sectors

Globally, the energy sector demands more high-skilled workers than other industries, with 45% of the workforce requiring some degree of tertiary education, from university degrees to vocational certifications. Less than 10% of energy employment is in low-skilled labour (IEA, 2022[122]). For the latter reason, energy sector wages typically see a premium over economy-wide average wages, though this premium ranges substantially from 10% to 50% across advanced economies alone. These premiums remain true across all regions, but the differences between wages in advanced economies and EMDEs remain pronounced, with the range of wages between geographies being larger than the range of jobs within the energy sector within the same region (IEA, 2022[122]).

Across the LAC region, jobs in the energy production industries are generally well paid, with wages standing close to 2 000 international dollars per month in 2020. Over the last decade, those wages experienced a 51% growth in real terms, one of the highest among all industries. In industries that are normally highly dependent on energy and are the greatest net GHG emissions contributors, wages are generally higher than the mean across all sectors. They stand at 600 international dollars per month in agriculture, 1 500 in mining, 1 200 in manufacturing and 1 300 in transport. Great variation exists across countries. For instance, in the energy production sector, monthly earnings vary from USD 3 200 in Argentina to USD 650 in Guatemala. Jobs created in the sectors more exposed to the green policies are normally of high quality, but any job loss resulting from the technological transformation will also entail high costs in terms of income. Plant workers employed by large electricity generators in countries with good collective bargaining coverage, such as Chile, may benefit from agreements with their employers that allow them to keep their jobs and transition to other power plants in the country. This can be the case more generally for upstream fossil fuel workers employed by firms that diversify into renewable energy production (Saget, Vogt-Schilb and Luu, 2020[120]).

How to ensure a green and just transition for all

The green transition will inevitably affect groups of people in dissimilar ways. Jobs losses in high-carbon energy production firms may well be concentrated among older workers or workers lacking up-to-date skills to transition to cleaner technology firms. Moreover, workers in fossil base sectors are relatively few compared to the whole economy but concentrated in certain regions. It is essential to identify the socio-demographic groups and the regions that will benefit and those most at risk (Chapter 2), to design the best labour market and social policies to build an inclusive green agenda, where the most vulnerable people are not left behind and share the overall economic benefit of the transition. When assessing the just transition, the quality of clean energy jobs is as important as their quantity. Key criteria determining job quality include wages, medical insurance, retirement and other benefits, job security, terms of employment, occupational safety standards, union membership, and overall scope of labour rights. Globally, energy jobs pay more than the median national wage, however there are disparities between segments. Workers in less established clean energy industries typically earn less than jobs in the fossil fuel or nuclear industries (IEA, 2022[122]). Moreover, they high degree of labour informality in LAC poses further social issues (OECD et al., 2021[36]).Green policies must be designed to tackle successfully the redistributive effects of the transition.

Informal work and the green transition

Informality is a widespread concern in LAC (Chapter 1). The share of informal employment excluding the agricultural sector is still close to 50% on average and close to 80% in several lower middle-income countries, such as Bolivia, Guatemala, Honduras and Nicaragua (OECD et al., 2021[37]). Informality rates among women are even higher (ILO, 2019c). Moreover, informal employment is higher in the agricultural sector (69% across LAC). In agriculture, child labour is also widespread; 71% of child labour globally occurs in agriculture (ILO, UNICEF, 2020[126]).

Across LAC, almost four out of ten workers in mining, manufacturing and transport services are informal. In agriculture, almost seven out of ten workers are informal (Figure 3.14). The heterogeneity is large across LAC countries. Mining industries are affected by large labour informality in Bolivia (36.0% of workers are informal), Colombia (50.3%), Costa Rica (60.9%) and El Salvador (86.7%). In manufacturing, the incidence of informal work is lower than or close to 30% in Brazil, Chile, Mexico and Uruguay. In transport services, informal workers represent more than half of the workforce in Bolivia, Colombia, Costa Rica and El Salvador. In these industries, the green transition will entail challenges, as technological transformation will be needed, and many informal workers may find themselves without jobs or forced to find new ones in other informal industries.

On the positive side, across LAC, less than 10% of workers in the electricity, gas, steam and air conditioning supply industries are informal. The incidence of labour informality is particularly low in Argentina (9.2%), Colombia (5.3%), and Costa Rica and Uruguay (less than 1%). Therefore, any direct transformation in the energy production matrix induced by policy changes will probably entail the creation of new formal jobs and opportunities for skilled workers, in the case of dismissal (OECD et al., 2021[37]).

Figure 3.14. **Informal work in sectors with a high intensity of GHG emissions**

As a percentage of total employment in each sector, 2019 or latest available year

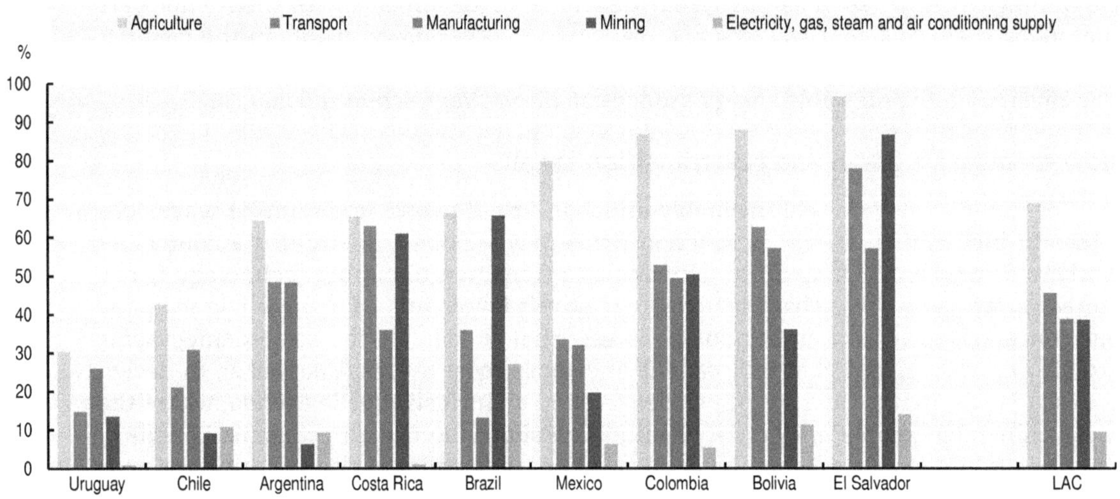

Note: The LAC average is an unweighted average of the countries shown.
Source: Authors' estimates based on national household income surveys.

StatLink https://stat.link/7eskd2

A crucial step in extending contributory social insurance to the informal economy has been the extension of social and labour rights to domestic workers through a mix of: 1) enforcement and simplification measures (for instance in Argentina, Brazil, Ecuador and Uruguay); 2) including self-employed workers in social insurance schemes through adapted mechanisms and simplified registration, tax and contribution payment mechanisms (Argentina, Peru and Uruguay); 3) adapting contribution calculation and payment modalities to the characteristics of workers and employers affected by particular business conditions, such as seasonality (Brazil); 4) harnessing digital and mobile technology to facilitate access to social protection (Brazil and Uruguay); and 5) extending pension coverage through existing or new schemes adapted to the needs of self-employed workers (Brazil and Costa Rica) (OECD/ILO, 2019[127]).

Skills and life-long learning

New skills will have to be developed to attract new green investments (Cedefop, 2021[128]; ILO, 2019[129]). "Skills bottlenecks" and the supply of competences that do not meet firms' demand are among the barriers to investment in green industries. Severe skills shortages are already evident in fast-growing sectors, such as renewable energies and energy efficiency. Increased investments in a green industry may result in increased demand for a certain occupation, with no change in the skills needed for the job. In this case, skills are quantitatively scarce, and policies should focus on providing more specific training for potential new workers (ILO, 2018[130]). On the other hand, the green transition may change the bundles of skills needed to perform the same job, or it may cause the loss of certain occupations. In these cases, skills policies should focus on upskilling or retraining the workers involved in the technological transformation. Sustainability and nature preservation should be considered new skills in the green transformation (OECD/Cedefop, 2014[131]). Skills anticipation and forecasting have been used extensively to define future developments of the tasks and the skills needed to perform greener jobs (Consoli et al., 2016[132]; Vona et al., 2018[116]).

Green jobs need an adaptation of existing on-the-job training courses. Effective examples already exist, in some OECD countries, like the Green Jobs Programme funded by the EU (Cedefop, 2022[133]). Nonetheless formal education and work experience are still important for green skills formation. Formal course offerings and degree programmes for these jobs are not yet well developed. Policies that promote learning by doing can fill the short-term gaps in current education policy (OECD/Cedefop, 2014[131]). Workforce development in green sectors is an area in which the private sector can play a key role. The private sector has put in place initiatives across the region, to allow companies to discover and connect with local talent to respond to emerging challenges with an open and collaborative mind-set. These strategic innovation hubs work with stakeholders, including business incubators, universities, and government institutions, to transform traditional businesses, effectively upskilling and developing these new green business models in the LAC region (OECD, forthcoming[134]).

Skills requirements of the new green jobs

In the green transition, workers need new skills to perform tasks in the new jobs created (ILO, 2019[129]; Cedefop, 2018[135]). Moreover, compliance with green energy regulations requires specialised skills and knowledge (Vona et al., 2018[116]). Many countries have adopted regulations on renewable energy or energy efficiency, including rules on skills certification and/or professional training. These rules often target specific occupations in certain sectors more linked to the green transformation, such as energy auditors, inspectors, assessors, energy managers, installers and operators of equipment and buildings (ILO, 2018[130]). Establishing regional policies on skills certification and training provision may increase investor confidence in a country's skills capital and help boost investment.

Although the sectoral approach has many benefits, including the relative ease of stakeholder co-ordination and the identification of specific skills needs, it is not sufficient to ensure comprehensive skills development for the green transition (Cedefop/OECD, 2015[136]; OECD/Cedefop, 2014[131]). From an economy-wide perspective, all sectors have potential for greening. It is crucial to identify skills needs arising from both direct and indirect job creation along supply chains and, in turn, design and implement training programmes. Because of the difficulties in co-ordinating the major players, there are few examples of good practices across countries (ILO, 2018[130]).

The assessment of skills needs can be both quantitative and qualitative. The green transition will cause changes in the number of workers in various occupations (quantitative assessment), as well as changes in the skills required for existing occupations (qualitative assessment) (Gregg, Strietska-Ilina and Büdke, 2015[137]). All major stakeholders should be involved in the green transition. For instance, the Chamber of Industry in Costa Rica carried out a study covering 100 of its 800 members to identify their skills needs for the green transition (ILO, 2018[130]). On the managerial side, new entrepreneurial skills will be essential to foster the adoption of innovative, environmentally friendly technology, human resource designs and higher productivity. Updating curricula is also key to ensure that future workers receive an education that allows them to take part in the transition (Saget, Vogt-Schilb and Luu, 2020[120]).

The role of lifelong learning in the acquisition of skills needed in the green transition

Skills gaps are a persistent issue over the working life cycle in LAC, and life-long training systems must be adapted to face the challenges of the green transition. The percentage of workers who receive some form of training is around 15% in LAC, much lower than the 56% across OECD countries (Alaimo et al., 2016[138]). In addition, life-long training is often offered to workers with higher education levels and those in formal and full-time employment, who are also the ones with more incentives and interests in skills' development along their careers. This perpetuates and amplifies the inequities acquired within the education system, feeding a vicious cycle of low investment in human capital, inadequate or obsolete skills, and low productivity levels (González-Velosa, Rosas and Flores, 2016[139]).

Expansion of social protection systems for a just green transition

This section explores the labour market policies and social protection mechanisms needed to boost job opportunities and other social outcomes of the green agendas, as well as to overcome and minimise the transition costs. In LAC, the unemployment rate is relatively low but with high labour turnover. Many workers are unemployed at some point, often resulting in losses in income and wages, which leads to significant welfare costs (Alaimo et al., 2016[138]; OECD et al., 2021[36]). The lack of employment and other social safety nets is a costly feature of LAC labour markets. The green transition will likely result in substantial job reallocations. It can be an opportunity to rethink social protection systems, boost job creation and protect workers in the case of job losses (OECD et al., 2021[36]; Saget, Vogt-Schilb and Luu, 2020[121]). OECD countries have put in place a broad range of policy actions towards green growth (OECD, 2015[140]; OECD, 2011[141]). LAC countries can benefit from the experience and design effective labour market mechanisms, along with social safety nets, aiming at an inclusive green growth strategy.

Unemployment benefits for workers dismissed due to the green transition

Unemployment benefits are inadequate in the LAC region. In 2020, 700 000 individuals received unemployment benefits (ECLAC, 2022[14]). Even where there is official coverage (Argentina, Brazil, Chile, Colombia, Ecuador, Mexico and Uruguay), unemployment insurance schemes often cover only a small proportion of workers and exclude underemployed and informal workers. Public expenditure on social security is limited, in part because of fiscal constraints. Structural fiscal reforms are therefore needed (Chapter 1) (ILO, 2018[130]). The green transition can be an opportunity to foster unemployment benefit schemes across the region. Some countries, such as Brazil, have invested considerably in creating a national registry of people and families to target and follow up on all social protection schemes. The main obstacles to adequate unemployment benefit schemes come not only from the high degree of labour informality but also from the financing, which increases labour costs for firms, with obvious impacts on employment levels (OECD et al., 2021[37]).

The traditional way in which most LAC countries protect workers from unemployment risks is through a combination of high severance payments and low unemployment insurance or unemployment assistance, with very low coverage overall. Through these systems, workers obtain job security at the cost of reduced employment creation, less capacity by firms to adapt to change and engage in technical innovation and increased atypical contracts and informal jobs (Cortázar, 2001[142]). Designing and implementing new unemployment insurance schemes may prove challenging, in the context of the green transition, where many sectors of the economy will be involved in profound technological transformations.

The competitiveness of many LAC countries is based partly on wages being lower than those of developed economies, which is not favourable for sustainable green growth (OECD et al., 2021[36]; OECD et al., 2020[143]). Governments must foster labour institutions, such as unemployment protection mechanisms, that transform employment relations through progressive structural change. In Latin America, unemployment insurance or similar schemes only operate in some countries, including Argentina, Brazil, Chile, Colombia, Ecuador, Mexico, Uruguay and Venezuela, and have undergone deep structural changes in the last two decades. In Argentina, Uruguay and Venezuela, they operate as pay-as-you-go systems, financed mainly from the affiliates' monthly contributions. Brazil's system relies on non-contributory schemes financed by general government revenues. In Ecuador, the two systems co-exist (Isgut and Weller, 2016[144]). In the rest of the region, the increase in coverage of social insurance has faced steep financing constraints. Individual saving accounts to cope with unemployment risks in Latin America are one of the possible policy options (Ferrer and Riddell, 2009[145]). Such accounts operate in a small but growing number of middle and high-income countries, where firms are legally obliged to make periodic contributions. The corresponding deposits earn interest and are paid as a lump sum or in monthly instalments subject to some eligibility conditions. When a worker loses his or her job, a certain amount can be withdrawn each month. However, individuals precariously attached to the labour market (e.g. those with part-time jobs, piecework contracts or fixed-term jobs, or employed in the informal sector, many of whom are women or young people) have very limited possibilities for accumulating the savings needed to cope with unemployment episodes. Moreover, these people tend to be unemployed most frequently. Individual accounts are rarely useful to them.

Active Labour Market Policies for greener economies

ALMPs refer to a broad range of policies that aim to activate workers with low employability or those who have lost their jobs due to dismissal. Policies include placement services, training programmes, hiring incentives, job rotation schemes or direct employment creation by public authorities (OECD/EU, 2020[146]).

ALMPs in LAC generally have a broader range of objectives (including poverty reduction, community development and equity promotion) compared to OECD countries, where ALMPs have mostly been seen as tools to address inefficiencies in labour markets, such as suboptimal investment in training and other labour market frictions (Escudero et al., 2018[147]). Despite the interest in ALMPs in developing countries, evidence of their effectiveness is still scarce (McKenzie, 2017[148]). In contrast to developed countries, ALMPs in developing countries, particularly in LAC, generally show positive, although small, effects on vulnerable groups (Card, Kluve and Weber, 2018[149]). Recent evidence from LAC shows that ALMPs are statistically more effective for women and youth than for the rest of the population (Escudero et al., 2018[147]). In the context of the green transition, in segmented labour markets with pervasive informality, induced structural change, green or otherwise, should be accompanied by active labour market policies, in order to grant an employment dividend (Bowen, 2012[123]).

If well designed, ALMPs can prove effective in the context of high informality. Providing job seekers with information about job vacancies and wage subsidies to work in a formal firm may improve their chances of employment, formality and earnings. These schemes are particularly important in the context of green agendas, as they may offset the negative effects of dismissal, especially for the most vulnerable groups, such as youth, informal workers and women (Novella and Valencia, 2019[150]). Such programmes could benefit from a component of skills training, which would help those who lose jobs gain the skills needed to re-enter the labour market. Active labour market policies targeted to dismissed workers, particularly those implemented during the COVID-19 pandemic, need to include conditional clauses on skills training and, more broadly, education outcomes (Chapter 1). Certainly, these skills training mechanisms should emphasise factors affecting the future of jobs, such as green and digital transformations.

Given the high prevalence of labour informality, self-employment and entrepreneurship programmes are another tool to support the start-up and development of independent work activities or microenterprises. Usually, self-employment and microenterprise creation programmes include technical services, such as counselling, training and assistance with business planning, and direct financial support for the newly created business. All reviewed studies that have evaluated the employment impacts of self-employment and microenterprise creation programmes find positive effects. By contrast, findings are mixed concerning raising earnings or profits (Escudero et al., 2018[147]). Support to informal workers or micro-enterprises in the context of the green transition may be a viable solution, in case marginal firms face financial difficulties in coping with more stringent environmental regulations.

Expanding social protection systems to the most vulnerable

Social security coverage remains insufficient in LAC. More than half of workers in the region do not participate in any contributory social security system against risks related to illness, unemployment and old age (ILO, 2018[130]). In 2020, on average, only 40% of vulnerable people received some form of social assistance and around 60% of the population was covered by at least one social protection benefit. However, over the last 15 years, LAC countries have expanded the coverage of both contributory (financed by wages) and non-contributory (financed by taxes) social protection schemes (OECD et al., 2021[37]).

While significant progress has been made in building LAC social protection systems, many informal workers often remain excluded from them (OECD et al., 2021[36]; OECD/ILO, 2019[127]; ECLAC, 2022[14]). In many LAC countries, large groups in the population are not covered. Despite lower incomes and greater need for protection, informal workers often fall through the cracks of social protection systems, making many incomes insecure or vulnerable to income poverty, affecting families.

Recent trends show that extension of social protection coverage often occurs through the development of both contributory and non-contributory schemes (OECD/ILO, 2019[127]; OECD et al., 2021[36]). Many countries rely largely on public resources, including for subsidising contributions, which puts growing pressure on government budgets. In most LAC countries, the funding gap to extend social protection to informal workers remains particularly pronounced (Chapter 1).

Some Latin American countries have extended the coverage of contributory social protection schemes to informal workers. Success owes to several measures, such as combining support for the formalisation of enterprises with access to social protection schemes, extending statutory coverage to previously uncovered workers, adapting benefits, contributions and administrative procedures to reflect the needs of informal workers, and subsidising contributions for those with very low incomes. In addition, several countries expanded the fiscal space needed to scale up social protection programmes financed through general government revenues. These efforts have significantly contributed to building safety nets that guarantee universal health coverage and at least basic income security throughout the life cycle, for instance through tax-financed pensions, disability benefits, child benefits, maternity benefits or employment guarantee schemes (OECD et al., 2021[36]).

Individuals and households in LAC have a long tradition of informal networks of mutual support to cope with risks and uncertainty, especially in contexts where public options are absent or limited, as in rural areas. Informal support is often organised around life cycle or livelihood risk and vulnerability. Private transfers received from friends, relatives and other households are another element of this form of inter-household informal protection. Around the mid-2010s, the share of private transfers in household income ranged from 4% in Bolivia and Honduras to around 15% in Costa Rica (OECD/ILO, 2019[127]). However, informal social protection has limitations. Studies suggest that informal risk-sharing mechanisms are close to efficiency when they protect from idiosyncratic shocks linked to individuals, households or life-cycle events, such as illness or death. They may fall short when it comes to broader shocks that affect a wider geographical area, such as a neighbourhood or community, which is likely the case for environmental health risks and the broad changes brought by green agendas. Income shocks may hurt poorer households, which are already financially constrained (Watson, 2016[151]). It is therefore crucial that public policies complement the informal mechanisms in place, to ensure a just, green transition for all people (OECD/World Bank, 2020[152]; ITF, forthcoming[153]; OECD, 2021[113])

Key policy messages

Box 3.6 presents preliminary policy messages to advance towards a more sustainable development model in LAC, based on the analysis presented in this chapter.

Box 3.6. Key policy messages

Towards a more sustainable energy matrix for LAC

- Transition towards diversified and adapted energy systems to include higher shares of renewable energy.
- Unlock non-hydro renewable energy potential, creating the necessary conditions in terms of regulation, economic incentives and promoting investment.
- Foster electrification to accelerate progress towards systemic decarbonisation by implementing integrated and effective power sector planning.
- Advance towards a systemic approach to the energy sector within national and regional economies, promoting energy systems that are better integrated across sectors, are more energy efficient and reduce total energy demand. The industrial and transport sectors have considerable savings potential.
- Increase energy integration between countries to generate economies of scale. Regional electricity integration could help incorporate variable renewables into energy systems and address potential vulnerabilities related to climate change (e.g. if a drought affects hydropower capacity in producing countries, alternative energy sources could be provided by neighbouring countries).
- Promote investments in electricity grids (transmission and distribution) to close territorial gaps between energy generation and energy demand areas.
- Advance towards universal access to electricity in LAC by, for example, establishing an energy access fund to roll out energy access programmes via mini-grids and to finance off-grid entrepreneurs and improve affordability for low-income households.
- Harness the opportunities arising from the surge in global demand for critical minerals that are abundant in LAC. The region should aim to integrate into global value chains in a more strategic way that in past transitions, putting sustainability, citizens' well-being and the potential for productive integration at the centre.

Box 3.6. **Key policy messages** *(cont.)*

- Build regional energy security and resilience in the face of external shocks. Increase efforts, particularly in the Caribbean subregion, to transition towards a more renewable energy matrix, profiting from solar, wind, ocean, geothermal and biomass potential, as a strategy to ensure energy security and mitigate GHG emissions.

Transformation of the production structure

- Develop industrial policies to advance a more sustainable production structure including policies on investment, foreign trade, science, technology and innovation, and training and skills development, with a special focus on MSMEs
- Promote innovation, adoption of green and energy-efficient technologies, and diversification towards less resource-intensive sectors, by connecting representatives from governments, industry, academia and civil society.
- Encourage and attract investments in green innovation, while taking advantage of new trade opportunities, both to foster regional integration and to join global chains in higher-value segments, ensuring environmental criteria in exports, and sustainable and responsible sourcing.
- Develop or update national sustainable and circular economy strategies that are open to all stakeholders and to all levels of government, advancing towards place-integrated and interconnected policies.
- Promote R&D investment to foster innovation to increase the competitiveness of industrial sectors, enabling solutions in products, services, business models and behaviour (consumption/use) with lower emissions and resource intensity.
- Unlock the blue economy potential for the green transition by adopting an ecosystem-based approach that manages trade-offs and pays particular attention to fishery and aquaculture, sustainable tourism, renewable energy generation, integrated management of river basins and lakes, and marine ecosystems protection.
- For the service-based economies of the Caribbean, adopt a circular economy strategy that minimises material and energy use and promotes sustainable tourism models that generate quality formal jobs, and reduce emissions and negative externalities.

Quality green jobs for greater inclusion

- Foster job creation in the new green technologies with a tailored mix of innovation and employment incentives, training schemes and job placement services.
- Promote additional public and private investments that contribute to increase the necessary value added of green sectors to boost the creation of formal jobs.
- Protect workers from job losses due to the green transformation, through well-designed and co-ordinated social assistance schemes, individual unemployment accounts and ALMPs to activate the most vulnerable workers negatively affected by the transformation.
- Create a social protection floor to protect living standards for those who have no access to unemployment benefits or assistance; guarantee at least universal access to essential health care and targeted basic income, while ensuring sustainable and equitable financing of these measures.
- Incentivise the transition of informal workers to new, productive firms involved in green technologies. Reinforce self-employment and entrepreneurial programmes through ALMPs to help the formalisation of micro-firms negatively impacted by the transition.

Box 3.6. **Key policy messages** *(cont.)*

- Ensure continued social protection coverage during labour market transitions, including by ensuring coverage for workers in all types of employment and by facilitating portability of entitlements between schemes.
- Tackle environmental health risks by expanding general health coverage and/or targeting measures to the most exposed people, such as those who lack sanitation services or adequate access to good quality air or water.

Notes

1. Transformative change is defined as "a system-wide change that requires more than technological change through consideration of social and economic factors that, with technology, can bring about rapid change at scale" (IPCC, 2022[1]), thus associating transformative change to the notion of systemic change or change in the system structure and its interactions.
2. Installed capacity is the maximum amount of electricity that a generating station can produce under specific conditions. While electricity generation is the amount of electricity that is produced over a specific period of time. Given that some renewable sources of energy depend on the sun or the wind, one refers to their installed capacity as a measure of their potential contribution to electricity generation.
3. The Caribbean Community is an intergovernmental organisation with 15 member states: Antigua and Barbuda, Bahamas, Barbados, Belize, Dominica, Grenada, Guyana, Haiti, Jamaica, Montserrat, Saint Kitts and Nevis, Saint Lucia, Saint Vincent and the Grenadines, Suriname, and Trinidad and Tobago.
4. Antigua and Barbuda, Aruba, Barbados, Curacao, Guadeloupe, Guyana, Jamaica, Saint Lucia, Saint Kitts and Nevis, and Saint Vincent and the Grenadines are some of the countries that already have or are actively pursuing RE installations (ECLAC, 2021[12]).
5. Bolivia, Chile, Colombia, Costa Rica, the Dominican Republic, Ecuador, El Salvador, Guatemala, Haiti, Honduras, Nicaragua, Panama, Paraguay, Peru and Uruguay. The technical secretariat of RELAC is led by the Inter-American Development Bank (IDB).
6. The OECD "Equitable Framework and Finance for Extractive-based Countries in Transition (EFFECT)" assists policy makers in designing comprehensive strategies to advance the low-carbon transition, avoid high-carbon lock-in and leave no-one behind in a global low-carbon economy.
7. Associated gas is natural gas produced along with crude oil, which is often seen as an inconvenient by-product of oil production.
8. CCUS refers to the process of capturing, utilising, transporting and storing underground CO_2 to avoid its release into the atmosphere.
9. The study has been carried out using the PLEXOS Integrated Energy Model, a simulation software designed for energy market analysis by Energy Exemplar. It was first developed as an electricity market simulator. Later, its functionality was extended so that the latest versions of PLEXOS integrate electricity, gas, heat and water (https://energyexemplar.com/solutions/plexos/).
10. This scenario considers only existing binational interconnections and low transmission integration between countries in the region. The supply seeks to meet the projected demand of the region by 2032.
11. The main results in the scenario of greater integration (RE+INT) show lower levels of solar PV and wind technology adoption than in the non-integrated scenario (RE and BE). This is due to higher system efficiencies and the ability to reduce the number of new generation plants.
12. Extended Producer Responsibility (EPR) is a policy approach under which producers are given a significant responsibility – financial and/or physical – for the treatment or disposal of post-consumer products. Assigning such responsibility could in principle provide incentives to prevent waste at the source, promote product design for the environment and support the achievement of public recycling and materials management goals (OECD, 2016[207]).
13. Argentina, Bolivia, Brazil, Chile, Colombia, Costa Rica, Ecuador, Guatemala, Honduras, Mexico, Panama, Peru, Dominican Republic and Uruguay.

14. LAC includes territories in Antigua and Barbuda, Argentina, Bahamas, Barbados, Belize, Brazil, Chile, Colombia, Costa Rica, Cuba, Dominica, Dominican Republic, Ecuador, El Salvador, Grenada, Haiti, Honduras, Jamaica, Peru, Mexico, Nicaragua, Panama, Saint Kitts and Nevis, Saint Lucia, Saint Vincent and the Grenadines, Suriname, Trinidad and Tobago, Uruguay, and Venezuela.
15. Technical potential is a term used to describe the energy that is extractable with current technology. It specifically refers to the total Installed Capacity [GW] for fixed and floating foundations within 200 kilometres of the shoreline (World Bank/ESMAP, 2020[88]).
16. The average for LAC countries does not include Chile, Colombia, Costa Rica and Mexico, since these countries were included in the OECD group for analysis (OECD, 2022[92]).
17. The selection of sectors was based on discussions held with LAC delegates to the Governing Board of the OECD Development Centre, public officials, experts, academics and private-sector representatives at the following events: Experts Meetings (28-29 April 2022 and 3 August 2022); LAC Forum 2022 (8 July 2022); OECD Emerging Markets Network LAC Roundtable Consultation on the Latin American Economic Outlook 2022 (17 June 2022); and desk research.

References

AFD (2022), *A smart App for sustainable fishing in Mexico*, Agence Française de Développement, Paris, http://www.afd.fr/en/actualites/smart-app-sustainable-fishing-mexico (accessed on 25 July 2022). [83]

Alaimo, V. et al. (2016), *Jobs for Growth*, Inter-American Development Bank, Washington, DC, https://doi.org/10.18235/0000139. [138]

Alova, Galina (2018), "Integrating renewables in mining: Review of business models and policy implications", *OECD Development Policy Papers*, No. 14, OECD Publishing, Paris, https://doi.org/10.1787/5bbcdeac-en. [206]

Altenburg, T. and C. Assmann (eds.) (2017), *Green Industrial Policy: Concept, Policies, Country Experience*, UN Environment; German Development Institute / Deutsches Institut für Entwicklungspolitk (DIE), Geneva/Bonn, https://wedocs.unep.org/20.500.11822/22277. [37]

Altvater, S. and C. Passarello (2018), *Implementing the Ecosystem-Based Approach in Maritime Spatial Planning*, European MSP Platform, Brussels, https://maritime-spatial-planning.ec.europa.eu/sites/default/files/20181025_ebainmsp_policybrief_mspplatform.pdf. [109]

Aramendis, R., A. Rodríguez and L. Krieger Merico (2018), *Contribuciones a un gran impulso ambiental en América Latina y el Caribe: bioeconomía*, Economic Commission for Latin America and the Caribbean, Santiago, https://repositorio.cepal.org/handle/11362/43825. [174]

Arellano, M. and S. Bond (1991), *Some Tests of Specification for Panel Data: Monte Carlo Evidence and an Application to Employment Equations*, Oxford University Press, Oxford, http://www.jstor.org/stable/2297968. [120]

Banks, J. and A. Miranda-González (2022), *A Methane Champion: Colombia becomes first South American country to regulate methane from oil and gas*, Clean Air Task Force, Boston, http://www.catf.us/2022/02/methane-champion-south-america-colombia-becomes-first-south-american-country-regulate-methane/ (accessed on 1 August 2022). [21]

Bárcena, A. (2020), *Mobilizing international solidarity, accelerating action and embarking on new pathways to realize the 2030 Agenda and the Samoa Pathway: Small Island Developing States*, Economic Commission for Latin America and the Caribbean, Santiago, http://www.cepal.org/en/presentations/mobilizing-international-solidarity-accelerating-action-and-embarking-new-pathways. [168]

Bárcena, A. et al. (2018), *Economics of climate change in Latin America and the Caribbean: a graphic view*, Economic Commission for Latin America and the Caribbean, Santiago, https://repositorio.cepal.org/bitstream/handle/11362/43889/1/S1800475_en.pdf. [53]

Barragán Muñoz, J. (2020), "Progress of coastal management in Latin America and the Caribbean", *Ocean and Coastal Management*, Vol. 184, https://doi.org/10.1016/j.ocecoaman.2019.105009. [105]

BIOFIN Costa Rica (2021), *Finance for Biodiversity*, Habitat Restoration Specialist for Biodiversity Finance Initiative Costa Rica, San José, https://biofin.cr/en/ (accessed on 15 July 2022). [197]

BNamericas (2021), *Unlocking Argentina's leviathan offshore wind potential*, BNamericas, Santiago, https://www.bnamericas.com/en/interviews/unlocking-argentinas-leviathan-offshore-wind-potential (accessed on 10 July 2022). [86]

Borges, M. et al. (2021), *Contribución de la bioeconomía a la recuperación pospandemia de COVID-19 en el Uruguay*, Recursos Naturales Ydesarrollo 208, Economic Commission for Latin America and the Caribbean, Santiago, https://www.cepal.org/sites/default/files/publication/files/47255/S2100314_es.pdf. [177]

Bowen, A. (2012), *'Green' Growth, 'Green' Jobs and Labor Markets*, World Bank, https://doi.org/10.1596/1813-9450-5990. [123]

BP (2021), *bp Statistical Review of World Energy 2021*, BP, London, https://www.bp.com/en/global/corporate/energy-economics/statistical-review-of-world-energy.html. [18]

CAF (2021), *RED 2021: Pathways to integration: trade facilitation, infrastructure, and global value chains*, Development Bank of Latin America, Caracas, http://scioteca.caf.com/handle/123456789/1907. [6]

CAF (2017), *Agua y saneamiento en la nueva ruralidad de América Latina*, Development Bank of Latin America, Caracas, https://www.caf.com/es/actualidad/noticias/2017/02/agua-y-saneamiento-en-la-nueva-ruralidad-de-america-latina/ (accessed on 3 August 2022). [178]

Card, D., J. Kluve and A. Weber (2018), "What Works? A Meta-Analysis of Recent Active Labor Market Program Evaluations", *Journal of the European Economic Association*, Vol. 16/3, pp. 894-931, https://doi.org/10.3386/w21431. [149]

Cedefop (2022), https://www.cedefop.europa.eu/en/tools/matching-skills/all-instruments/green-jobs-programme. [133]

Cedefop (2021), *Get your skills together for Europe's green deal*, https://www.cedefop.europa.eu/en/news/get-your-skills-together-europes-green-deal#group-related. [128]

Cedefop (2018), *Skills for green jobs: 2018 update*, https://www.cedefop.europa.eu/files/3078_en.pdf. [135]

Cedefop/OECD (2015), *Green skills and innovation for inclusive growth*, Publications Office of the European Union, Luxembourg, https://doi.org/10.2801/49143. [136]

Chateau and Mavroeidi (2020), "The jobs potential of a transition towards a resource efficient and circular economy", *OECD Environment Working Papers*, No. 167, OECD Publishing, Paris, https://doi.org/10.1787/28e768df-en. [63]

Circular Economy Coalition of Latin America and the Caribbean (2022), *Circular Economy in Latin America and the Caribbean: A Shared Vision*, United Nations Evironment Programme Circular Economy Coalition of Latin America and the Caribbean, https://emf.thirdlight.com/link/5fhm4nyvnopb-e44rhq/@/#id=0. [189]

Claes, J. et al. (2022), *Blue carbon: The potential of coastal and oceanic climate action*, McKinsey & Company, https://www.mckinsey.com/capabilities/sustainability/our-insights/blue-carbon-the-potential-of-coastal-and-oceanic-climate-action. [81]

Climate Watch (2020), *Country Greenhouse Gas Emissions Data [database]*, Climate Watch, Washington, DC, https://datasets.wri.org/dataset/cait-country (accessed on 20 June 2022). [119]

Climate-ADAPT (2021), *Water management*, Climate-ADAPT, https://climate-adapt.eea.europa.eu/eu-adaptation-policy/sector-policies/water-management (accessed on 6 July 2022). [180]

Commission for Environmental Cooperation (2021), *Transforming Recycling and Solid Waste Management in North America*, Commission for Environmental Cooperation, Montreal, http://www.cec.org/transforming-recycling-and-solid-waste-management-in-north-america/ (accessed on 29 June 2022). [102]

Consoli, D. et al. (2016), "Do green jobs differ from non-green jobs in terms of skills and human capital?", *SPRU Working Paper*, University of Sussex Business School, Brighton, https://www.sussex.ac.uk/webteam/gateway/file.php?name=2015-16-swps-consoli-etal.pdf&site=25. [132]

Contreras-Lisperguer et al. (2017), *Cradle-to- Cradle approach in the life cycle of silicon solar photovoltaic panels*. [73]

Cordonnier, J. and D. Saygin (forthcoming), "Green hydrogen opportunities for emerging and developing economies: Identifying success factors for market development and building enabling conditions", *OECD Environment Working paper*, OECD Publishing, Paris. [210]

Cortázar, R. (2001), "Unemployment Insurance Systems for Latin America", in *Labor Market Policies in Canada and Latin America: Challenges of the New Millennium*, Springer, Boston, https://doi.org/10.1007/978-1-4757-3347-1_4. [142]

CTAGUA (2022), *Centro Tecnológico del Agua*, Centro Tecnológico del Agua, Montevideo, https://ctagua.uy/ (accessed on 12 May 2022). [183]

de Miguel, C. et al. (2021), "Economía circular en América Latina y el Caribe: oportunidad para una recuperación transformadora", *Documentos de Proyectos*, (LC/TS.2021/120), Economic Commission for Latin America and the Caribbean, Santiago, https://hdl.handle.net/11362/47309. [66]

Dechezleprêtre, A. et al. (2019), "Do environmental and economic performance go together? A review of micro-level empirical evidence from the past decade or so", *International Review of Environmental and Resource Economics*, Vol. 13/1-2, pp. 1-118, https://doi.org/10.1561/101.00000106. [43]

Diez, S. et al. (2019), *Marine Pollution in the Caribbean: Not a Minute to Waste*, World Bank, Washington, DC, https://documents1.worldbank.org/curated/en/482391554225185720/pdf/Marine-Pollution-in-the-Caribbean-Not-a-Minute-to-Waste.pdf. [89]

Dini and Stumpo (2019), "Mipymes en América Latina: un frágil desempeño y nuevos desafíos para las políticas de fomento", *Project Documents*, (LC/TS.2019/20), Economic Commission for Latin America and the Caribbean, Santiago, https://www.cepal.org/es/publicaciones/44603-mipymes-america-latina-un-fragil-desempeno-nuevos-desafios-politicas-fomento. [50]

Dominish, E., N. Florin and S. Teske (2019), *Responsible Minerals Sourcing for Renewable Energy*, Report prepared for Earthworks, Institute for Sustainable Futures, University of Technology Sydney, https://earthworks.org/wp-content/uploads/2021/09/MCEC_UTS_Report_lowres-1.pdf. [203]

Dutz, M., R. Almeida and T. Packard (2018), *The Jobs of Tomorrow: Technology, Productivity, and Prosperity in Latin America and the Caribbean*, Directions in Development--Information and Communication Technology, World Bank, Washington, DC, https://openknowledge.worldbank.org/handle/10986/29617. [114]

ECLAC (2022), *CEPALSTAT [database]*, Economic Commission for Latin America and the Caribbean, Santiago, https://statistics.cepal.org/portal/cepalstat/dashboard.html?lang=en (accessed on 30 July 2022). [123]

ECLAC (2022), *Social Panorama of Latin America 2021*, (LC/PUB.2021/17-P), Economic Commission for Latin America and the Caribbean, Santiago, https://repositorio.cepal.org/bitstream/handle/11362/47719/1/S2100654_en.pdf. [14]

ECLAC (2022), *Water resources*, Economic Commission for Latin America and the Caribbean, Santiago, https://www.cepal.org/en/topics/water-resources (accessed on 28 May 2022). [179]

ECLAC (2021), "Conceptualizing a circular economy in the Caribbean: perspectives and possibilities", *Policy Brief*, No. LC/CAR/2021/7, Economic Commission for Latin America and the Caribbean, Santiago, https://repositorio.cepal.org/bitstream/handle/11362/47604/LCCAR2021_07_en.pdf?sequence=1&isAllowed=y. [12]

ECLAC (2021), *Latin America and the Caribbean Has All the Right Conditions to Become a Renewable Energy Hub with Great Potential in Green Hydrogen*, Economic Commission for Latin America and the Caribbean, Santiago, https://www.cepal.org/en/news/latin-america-and-caribbean-has-all-right-conditions-become-renewable-energy-hub-great (accessed on 10 August 2022). [11]

ECLAC (2021), *SDGs in Latin America and the Caribbean: Statistical knowledge management hub: Regional SDG statistical profiles by target [database]*, Economic Commission for Latin America and the Caribbean, Santiago, https://agenda2030lac.org/estadisticas/regional-sdg-statistical-profiles-target-1.html?lang=en (accessed on 9 May 2022). [70]

ECLAC (2020), *Building a New Future: Transformative Recovery with Equality and Sustainability. Summary*, Economic Commission for Latin America and the Caribbean, Santiago, https://www.un-page.org/files/public/eclac_recovery_modelling.pdf. [34]

ECLAC (2020), *Economics of climate change in Latin America and the Caribbean: A Graphic View*, Economic Commission for Latin America and the Caribbean, Santiago, https://repositorio.cepal.org/bitstream/handle/11362/43889/1/S1800475_en.pdf. [186]

ECLAC (2018), *Gobernanza de los Recursos Naturales en América Latina y el Caribe para el desarrollo sostenible*, Economic Commission for Latin America and the Caribbean, Santiago, https://www.cepal.org/sites/default/files/presentation/files/181119-final_final_corta-giz_revisada_alicia_barcena_ministros_mineria_limarev.pdf. [28]

ECLAC/ILO (2018), *Employment Situation in Latin America and the Caribbean: Environmental sustainability and employment in Latin America and the Caribbean*, No. 19 (LC/TS.2018/85), Economic Commission for Latin America and the Caribbean, Santiago, https://repositorio.cepal.org/handle/11362/44186. [125]

Ecoins (2022), *Preguntas frecuentes*, Ecoins, San José, https://ecoins.eco/preguntas-frecuentes/ (accessed on 2 August 2022). [193]

Econometría Consultores (2022), *Modeling of the macroeconomic effects of the transition to the circular economy for Latin America: cases of Chile, Colombia, Mexico and Peru*, Economic Commission for Latin America and the Caribbean, Santiago, https://www.euroclima.org/images/2022/03/02/Modelo_Economia_Circular.pdf (accessed on 27 May 2022). [64]

Ehler, C. (2021), "Two decades of progress in Marine Spatial Planning", *Marine Policy*, Vol. 132, https://doi.org/10.1016/j.marpol.2020.104134. [108]

Ellen MacArthur Foundation (2019), *Completing the Picture: How the Circular Economy Tackles Climate Change*, Ellen MacArthur Foundation, Cowes, https://emf.thirdlight.com/link/w750u7vysuy1-5a5i6n/@/preview/1?o. [59]

Ellen MacArthur Foundation (2013), *Towards The Circular Economy*, Ellen MacArthur Foundation, Cowes, https://www.werktrends.nl/app/uploads/2015/06/Rapport_McKinsey-Towards_A_Circular_Economy.pdf. [57]

Escudero, V. et al. (2018), "Active Labour Market Programmes in Latin America and the Caribbean: Evidence from a Meta-Analysis", *Research Department Working Paper*, No. 20, International

Labour Office, Geneva, https://www.ilo.org/wcmsp5/groups/public/---dgreports/---inst/documents/publication/wcms_577292.pdf. [147]

Estado de Querétaro (2022), *Sistema Economía Circular Querétaro*, https://autoqro.mx/economia-circular (accessed on 20 July 2022). [65]

European Commission (2021), *Communication from the Commission on a new approach for a sustainable blue economy in the EU: Transforming the EU's Blue Economy for a Sustainable Future*, European Commission, Brussels, https://eur-lex.europa.eu/legal-content/EN/TXT/?uri=COM%3A2021%3A240%3AFIN. [110]

European Parliament (2022), *Taxonomy: MEPs do not object to inclusion of gas and nuclear activities*, European Parliament, Luxembourg City, https://www.europarl.europa.eu/news/en/press-room/20220701IPR34365/taxonomy-meps-do-not-object-to-inclusion-of-gas-and-nuclear-activities (accessed on 10 August 2022). [23]

FAO (2021), *Climate Smart Agriculture Sourcebook*, Food and Agriculture Organization, Rome, https://www.fao.org/climate-smart-agriculture-sourcebook/production-resources/module-b6-water/chapter-b6-4/en/. [181]

FAO (2021), *Hacia una agricultura sostenible y resiliente en América Latina y el Caribe. Análisis de siete trayectorias de transformación exitosas*, Food and Agriculture Organization, Regional Office for Latin America and the Caribbean, Santiago de Chile, https://doi.org/10.4060/cb4415es. [154]

FAO (2014), *Agriculture, Forestry and Other Land Use Emissions by Sources and Removals by Sinks*, FAO Statistics Series, ESS/14-02, Food and Agriculture Organization, Rome, http://www.fao.org/3/i3671e/i3671e.pdf. [159]

FAO/UNEP (2020), *The State of the World's Forests 2020: Forests, biodiversity and people*, Food and Agriculture Organization/United Nations Environment Programme, Rome, https://doi.org/10.4060/ca8642en. [155]

Ferrer, A. and W. Riddell (2009), "Unemployment Insurance Savings Accounts in Latin America: Overview and Assessment", *Social Protection Discussion Paper*, No. SP 0910, World Bank, Washington, DC, http://documents.worldbank.org/curated/en/438061468044956180/Unemployment-insurance-savings-accounts-in-Latin-America-overview-and-assessment. [145]

Food Navigator (2019), *'A major step forward': Brazil founds trade group for carbon-neutral meat*, Food Navigator, Montpellier, France, http://www.foodnavigator-latam.com/Article/2019/02/25/A-major-step-forward-Brazil-founds-trade-group-for-carbon-neutral-meat?utm_source=copyright&utm_medium=OnSite&utm_campaign=copyright. [172]

Forti, V. et al. (2020), *The Global E-waste Monitor 2020: Quantities, flows and the circular economy potential*, United Nations University (UNU)/United Nations Institute for Training and Research (UNITAR) – co-hosted SCYCLE Programme, International Telecommunication Union (ITU) & International Solid Waste Association (ISWA), Bonn/Geneva/Rotterdam, https://ewastemonitor.info/wp-content/uploads/2020/11/GEM_2020_def_july1_low.pdf (accessed on 5 May 2021). [68]

Global CCS Institute (2020), *Global Status Report 2019: Targeting Climate Change*, Global CCS Institute, Melbourne, https://www.globalccsinstitute.com/wp-content/uploads/2019/12/GCC_GLOBAL_STATUS_REPORT_2019.pdf. [25]

GMI (2011), *Oil and Gas Systems Methane: Reducing Emissions, Advancing Recovery and Use*, Global Methane Initiative, Washington, DC, http://www.globalmethane.org/documents/oil-gas_fs_eng.pdf. [19]

Gobierno de la Republica de Colombia (2019), *Estrategia nacional de economía circular. Cierre de ciclos de materiales, innovación tecnológica, colaboración y nuevos modelos de negocio*, Presidencia de la República/Ministerio de Ambiente y Desarrollo Sostenible/Ministerio de Comercio, Industria y Turismo, Bogotá, http://andi.com.co/Uploads/Estrategia%20Nacional%20de%20EconA%CC%83%C2%B3mia%20Circular-2019%20Final.pdf_637176135049017259.pdf. [101]

Gobierno Federal Sectur (2020), *Programa de Turismo sustentable 2030*, Gobierno Federal Sectur, Mexico City, https://adaptur.mx/pdf/Mexico-Estrategia-de-Turismo-Sostenible-2030.pdf. [200]

González-Velosa, C., D. Rosas and D. Flores (2016), "On-the-Job Training in Latin America and the Caribbean: Recent Evidence", in *Firm Innovation and Productivity in Latin America and the Caribbean: The Engine of Economic Development*, Palgrave Macmillan, New York, https://doi.org/10.1057/978-1-349-58151-1_5. [139]

Government of Brazil (2022), *Renewable energy in the Brazilian energy matrix: the share of solar energy reached 6.9% and wind energy, 10.9%*, Government of Brazil, Brasília, https://www.gov.br/en/government-of-brazil/latest-news/2022/renewable-energy. [8]

Government of Colombia (2021), *National Hydrogen Strategy and Roadmap*, https://www.minenergia.gov.co/es/micrositios/enlace-ruta-hidrogeno/ (accessed on 25 April 2022) [209]

Gregg, C., O. Strietska-Ilina and C. Büdke (2015), *Anticipating skill needs for green jobs: A practical guide*, International Labour Organization, Geneva, http://www.ilo.org/wcmsp5/groups/public/---ed_emp/---ifp_skills/documents/publication/wcms_564692.pdf. [137]

Grundke, R. and J. Arnold (2022), *Mastering the transition: A synthetic literature review of trade adaptation policies*, OECD Publishing, Paris, https://doi.org/10.1787/5fad3487-en. [115]

Gütschow, J. et al. (2016), "The PRIMAP-hist national historical emissions time series", *Earth System Science Data*, pp. 571-603, https://essd.copernicus.org/preprints/essd-2016-12/essd-2016-12.pdf. [170]

Haarr, M., J. Falk-Andersson and J. Fabres (2022), "Global marine litter research 2015–2020: Geographical and methodological trends", *Science of The Total Environment*, Vol. 820, pp. 153-162, https://doi.org/10.1016/j.scitotenv.2022.153162. [93]

Hamilton, L. and S. Feit (2019), *Plastic & Climate: The Hidden Costs of a Plastic Planet*, Center for International Environmental Law, Geneva, https://www.ciel.org/plasticandclimate/. [90]

Hardy, W., R. Keister and P. Lewandowski (2018), "Educational upgrading, structural change and the task composition of jobs in Europe", *Economics Of Transition*, Vol. 26, https://doi.org/10.1111/ecot.12145. [118]

Heidbreder, L. et al. (2019), "Tackling the plastic problem: A review on perceptions, behaviors, and interventions", *Science of the Total Environment*, Vol. 668, Elsevier, Amsterdam, pp. 1077-1093, https://www.sciencedirect.com/science/article/abs/pii/S0048969719309519. [191]

Herrera, D. (2020), *MSME Financing Instruments in Latin America and the Caribbean During COVID-19*, Inter-American Development Bank, Washington, DC, https://doi.org/10.18235/0002361. [49]

Hidalgo, C., J. Fontecha Mejía and S. Escobar (2022), *Principales aspectos del despliegue de energía eólica costa afuera en Colombia*, Holland & Knight, Tampa, https://www.hklaw.com/en/insights/publications/2022/05/principales-aspectos-del-despliegue-de-energia-eolica-costa-afuera (accessed on 23 May 2022). [87]

IDB (2021), "BID y Telefónica lanzan iniciativa para conectar emprendedores y corporaciones", https://www.iadb.org/es/noticias/bid-y-telefonica-lanzan-iniciativa-para-conectar-emprendedores-y-corporaciones (accessed on 15 July 2022). [42]

IDB (2021), *Climate policies in Latin America and the Caribbean: success stories and challenges in the fight against climate change*, Inter-American Development Bank, Washington, DC, https://doi.org/10.18235/0003239. [3]

IDB (2021), *Una visión azul para América Latina y El Caribe*, Inter-American Development Bank, Washington, DC, https://blogs.iadb.org/sostenibilidad/es/una-vision-azul-para-america-latina-y-el-caribe/. [75]

IDB (2020), *Plastic Waste Management and Leakage in Latin America and the Caribbean*, Inter-American Development Bank, Washington, DC, https://publications.iadb.org/en/plastic-waste-management-and-leakage-latin-america-and-caribbean. [185]

IDB and DDPLAC (2019), *Getting to Net-Zero Emissions: Lessons from Latin America and the Caribbean*, Inter-American Development Bank, Washington, DC, https://doi.org/10.18235/0002024. [5]

IEA (2022), *SDG7: Data and Projections*, International Energy Agency, Paris, http://www.iea.org/reports/sdg7-data-and-projections. [32]

IEA (2022), *The Role of Critical Minerals in Clean Energy Transitions*, International Energy Agency, Paris, http://www.iea.org/reports/the-role-of-critical-minerals-in-clean-energy-transitions/executive-summary. [27]

IEA (2022), *World Energy Employment*, IEA, Paris, https://www.iea.org/reports/world-energy-employment. [122]

IEA (2021), *Driving Down Methane Leaks from the Oil and Gas Industry: A Regulatory Roadmap and Toolkit*, International Energy Agency, Paris, http://www.iea.org/reports/driving-down-methane-leaks-from-the-oil-and-gas-industry. [20]

IEA (2021), *Hydrogen in Latin America: From near-term opportunities to large-scale deployment*, International Energy Agency, Paris, https://iea.blob.core.windows.net/assets/65d4d887-c04d-4a1b-8d4c-2bec908a1737/IEA_HydrogeninLatinAmerica_Fullreport.pdf. [16]

IEA (2021), *Net Zero by 2050*, International Energy Agency, Paris, http://www.iea.org/reports/net-zero-by-2050. [17]

IEA (2021), *World Energy Balances [database]*, International Energy Agency, Paris, http://www.iea.org/data-and-statistics/data-product/world-energy-statistics-and-balances. [10]

IEA (2015), *World Energy Outlook 2015*, International Energy Agency, Paris, http://www.iea.org/reports/world-energy-outlook-2015. [33]

IFPRI (2021), "Achieving sustainable agricultural practices: From incentives to adoption and outcomes", *IFPRI Policy Brief*, International Food Policy Research Institute, Washington, DC, https://ebrary.ifpri.org/utils/getfile/collection/p15738coll2/id/134262/filename/134472.pdf. [161]

IICA (2021), *Transformar el conocimiento en evidencia para la transformación sostenible de los sistemas alimentarios de América Latina y El Caribe*, IICA Blog, Inter-American Institute for Cooperation on Agriculture, San José, https://bit.ly/389vHB4. [166]

ILO (2021), *Hacia una recuperación sostenible del empleo en el sector del turismo en América Latina y el Caribe*, International Labour Organization, Geneva, https://www.ilo.org/wcmsp5/groups/public/---americas/---ro-lima/documents/publication/wcms_809290.pdf. [195]

ILO (2019), *Skills for a greener future: A global view based on 32 country studies*, https://www.ilo.org/wcmsp5/groups/public/---ed_emp/documents/publication/wcms_732214.pdf. [129]

ILO (2018), *World Employment and Social Outlook 2018: Greening with jobs*, International Labour Office, Geneva, https://www.ilo.org/wcmsp5/groups/public/---dgreports/---dcomm/---publ/documents/publication/wcms_628654.pdf. [130]

ILO, UNICEF (2020), *Child Labour: Global estimates 2020, trends and the road forward*, https://www.ilo.org/ipec/Informationresources/WCMS_797515/lang--en/index.htm. [126]

IOGP (2022), *Map of Global CCUS Projects*, International Association of Oil & Gas Producers, London, https://www.iogp.org/bookstore/product/map-of-global-ccs-projects/ (accessed on 18 May). [24]

IPCC (2022), *Global Warming of 1.5°C: IPCC Special Report on Impacts of Global Warming of 1.5°C above Pre-industrial Levels in Context of Strengthening Response to Climate Change, Sustainable Development, and Efforts to Eradicate Poverty*, Cambridge University Press, Cambridge/New York, https://doi.org/10.1017/9781009157940. [1]

IPCC (2019), *IPCC Special Report on the Ocean and Cryosphere in a Changing Climate*, Cambridge University Press, Cambridge/New York, https://doi.org/10.1017/9781009157964. [76]

IRENA (2022), *Renewable Energy Roadmap for Central America: Towards a Regional Energy Transition*, International Renewable Energy Agency, Abu Dhabi, https://www.irena.org/publications/2022/Mar/Renewable-Energy-Roadmap-for-Central-America. [26]

IRENA (2018), *Policies and Regulations for Renewable Energy Mini-Grids*, International Renewable Energy Agency, Abu Dhabi, https://www.irena.org/publications/2018/Oct/Policies-and-regulations-for-renewable-energy-mini-grids. [30]

IRENA (2016), *End-of-life Management: Solar Photovoltaic Panels*, International Renewable Energy Agency, Abu Dhabi, https://www.irena.org/publications/2016/Jun/End-of-life-management-Solar-Photovoltaic-Panels. [72]

IRENA (2015), *Renewable Energy Policy Brief: Uruguay*, International Renewable Energy Agency, Abu Dhabi, https://www.irena.org/-/media/Files/IRENA/Agency/Publication/2015/IRENA_RE_Latin_America_Policies/IRENA_RE_Latin_America_Policies_2015_Country_Uruguay.pdf. [51]

IRENA/UNELCAC/GET.transform (2022), *Scenarios for the Energy Transition: Experience and Good Practices in Latin America and the Caribbean*, International Renewable Energy Agency, Abu Dhabi, https://www.irena.org/publications/2022/Jul/Scenarios-for-the-Energy-Transition-LAC. [15]

IRP (2020), *Mineral Resource Governance in the 21st Century: Gearing extractive industries towards sustainable development*, United Nations Environment Programme, Nairobi, https://www.resourcepanel.org/reports/mineral-resource-governance-21st-century. [205]

Isgut, A. and J. Weller (2016), *Institutions for improving workforce integration in Latin America and Asia*, Economic Commission for Latin America and the Caribbean, Santiago, https://repositorio.cepal.org/bitstream/handle/11362/40661/6/S1600550_en.pdf. [144]

Isla Urbana (2021), *Isla Urbana*, Isla Urbana, Mexico City, https://islaurbana.org/isla-urbana/ (accessed on 17 June 2022). [182]

ITF (forthcoming), *Developing Accessibility Indicators for Latin American Cities*, International Transport Forum, Paris. [153]

Karasik, R. et al. (2020), *20 Years of Government Responses to the Global Plastic Pollution Problem: The Plastics Policy Inventory*, Nicholas Institute for Environmental Policy Solutions, Duke University, Durham, NC, https://nicholasinstitute.duke.edu/publications/20-years-government-responses-global-plastic-pollution-problem. [190]

Kaza, S. et al. (2018), *What a Waste 2.0: A Global Snapshot of Solid Waste Management in 2050*, Urban Development Series, World Bank, Washington, DC, https://doi.org/10.1596/978-1-4648-1329-0. [184]

Kirchherr, J., D. Reike and M. Hekkert (2017), "Conceptualizing the Circular Economy: An Analysis of 114 Definitions", *Resources, Conservation and Recycling*, Vol. 127, pp. 221-232, https://doi.org/10.1016/j.resconrec.2017.09.005. [54]

Lanoie, P. et al. (2011), "Environmental policy, innovation and performance: New insights on the Porter Hypothesis", *Journal of Economics & Management Strategy*, Vol. 20/3, pp. 803-842, https://doi.org/10.1111/j.1530-9134.2011.00301.x. [44]

Le Tissier, M. (2020), "Unravelling the Relationship between Ecosystem-Based Management, Integrated Coastal Zone Management and Marine Spatial Planning", in *Ecosystem-Based Management, Ecosystem Services and Aquatic Biodiversity*, Springer, Cham, https://doi.org/10.1007/978-3-030-45843-0_20. [104]

Letcher, T. (ed.) (2020), *Production, use and fate of synthetic polymers*, Academic Press, Cambridge, MA, https://doi.org/10.1016/B978-0-12-817880-5.00002-5. [95]

McCarthy, A., R. Dellink and R. Bibas (2018), *The Macroeconomics of the Circular Economy Transition: A Critical Review of Modelling Approaches*. [55]

McKenzie, D. (2017), "How effective are active labor market policies in developing countries? A critical review of recent evidence", *Policy Research Working Paper*, No. 8011, World Bank, Washington, DC, http://hdl.handle.net/10986/26352. [148]

MCTI (2021), *Programa Brasil-Biotec vai fortalecer pesquisa e desenvolvimento da biotecnologia nacional*, Ministério da Ciência, Tecnologia e Inovações, Government of Brazil, Brasília, https://www.gov.br/mcti/pt-br/acompanhe-o-mcti/noticias/2021/02/programa-brasil-biotec-vai-fortalecer-pesquisa-e-desenvolvimento-da-biotecnologia-nacional. [176]

Mesa Nacional para la Gestión Sostenible del Plástico (2021), *Plan Nacional para la gestión sostenible de plásticos de un solo uso*, Ministerio de Ambiente y Desarrollo Sostenible, Bogotá, http://vip.acoplasticos.com.co/_lib/file/doc/PLAN_PLASTICOS.pdf. [100]

Michail, N. (2019), "A major step forward: Brazil founds trade group for carbon-neutral meat", *FoodNavigator-LATAM*, William Reed, Crawley, https://www.foodnavigator-latam.com/Article/2019/02/25/A-major-step-forward-Brazil-founds-trade-group-for-carbon-neutral-meat?utmsource=copyright&utm_medium=OnSite&utm_campaign=copyright (accessed on 20 July 2022). [48]

Ministério da Agricultura, Pecuária e Abastecimento (2019), *Serviço Florestal Brasileiro (2019) Bioeconomia da Floresta A Conjuntura da Produção Florestal Não Madeireira no Brasil*, https://ava.icmbio.gov.br/mod/data/view.php?d=17&rid=3228. [47]

Ministerio de Ambiente (2020), *Plan Maestro de Turismo Sostenible 2020-2025*, Ministerio de Ambiente, Panama City, https://www.atp.gob.pa/Plan_Maestro_de_Turismo_Sostenible_2020-2025.pdf. [201]

Ministerio de Industria y Comercio (2020), *Política de Turismo Sostenible: "Unidos por la Naturaleza"*, Government of Colombia, Bogotá, https://www.mincit.gov.co/minturismo/calidad-y-desarrollo-sostenible/politicas-del-sector-turismo/politica-de-turismo-sostenible. [199]

Ministerio del Medio Ambiente (2022), *Roadmap for a Circular Chile by 2040*, Ministerio del Medio Ambiente, Santiago, https://economiacircular.mma.gob.cl/wp-content/uploads/2022/01/HOJA-DE-RUTA-PARA-UN-CHILE-CIRCULAR-AL-2040-EN.pdf. [99]

Ministerio del Medio Ambiente (2021), *Decreto 12: Establece Metas de Recolección y Valorización y Otras Obligaciones Asociadas de Envases y Embalajes*, Ministerio del Medio Ambiente, Santiago, https://www.bcn.cl/leychile/navegar?idNorma=1157019. [98]

Ministerio del Medio Ambiente (2021), *Estrategia Nacional para la Gestión de Residuos Marinos y Microplásticos*, Ministerio del Medio Ambiente, Santiago, https://mma.gob.cl/wp-content/uploads/2021/08/Estrategia-Nacional-para-la-gestion-de-residuos-marinos-y-microplasticos.pdf. [97]

Ministry of Agriculture, Livestock and Food Supply (2022), *Observatory of Brazilian Agriculture*. [45]

Ministry of Agriculture, Livestock and Food Supply (2021), *ABC+: Plan for adaptation and low carbon emission in agriculture strategic vision for a new cycle*, https://www.gov.br/agricultura/pt-br/assuntos/sustentabilidade/plano-abc/arquivo-publicacoes-plan. [46]

Novella, R. and H. Valencia (2019), "Active Labor Market Policies in a Context of High Informality: The effect of PAE in Bolivia", *Inter-American Development Bank Working Paper Series*, Inter-American Development Bank, Washington DC, https://publications.iadb.org/en/active-labor-market-policies-in-a-context-of-high-informality-the-effect-of-pae-in-bolivia. [150]

OECD (2022), "Closing the loop in the Slovak Republic: A roadmap towards circularity for competitiveness, eco-innovation and sustainability", *OECD Environment Policy Papers*, No. 30, OECD Publishing, Paris, https://doi.org/10.1787/acadd43a-en. [52]

OECD (2022), *Global Plastics Outlook: Policy Scenarios to 2060*, OECD Publishing, Paris, https://doi.org/10.1787/aa1edf33-en. [92]

OECD (2022), *OECD work for a sustainable ocean*, OECD, Paris, https://www.oecd.org/ocean/ (accessed on 23 September 2022). [112]

OECD (2022), *OECD.Stat [database]*, Labour Indicators, OECD, Paris, https://stats.oecd.org/Index.aspx. [124]

OECD (2021), *How's Life in Latin America?: Measuring Well-being for Policy Making*, OECD Publishing, Paris, https://doi.org/10.1787/2965f4fe-en. [113]

OECD (2020), *The Circular Economy in Cities and Regions: Synthesis Report*, OECD Urban Studies, OECD Publishing, Paris, https://doi.org/10.1787/10ac6ae4-en. [62]

OECD (2020), *Towards Sustainable Land Use: Aligning Biodiversity, Climate and Food Policies*, OECD Publishing, Paris, https://doi.org/10.1787/3809b6a1-en. [162]

OECD (2019), *Business Models for the Circular Economy: Opportunities and Challenges for Policy*, https://doi.org/10.1787/g2g9dd62-en. [74]

OECD (2019), *Global Material Resources Outlook to 2060: Economic Drivers and Environmental Consequences*, OECD Publishing, Paris, https://doi.org/10.1787/9789264307452-en. [61]

OECD (2018), *Impacts of green growth policies on labour markets and wage income distribution: a general equilibrium application to climate and energy policies*, https://www.oecd.org/officialdocuments/publicdisplaydocumentpdf/?cote=ENV/EPOC/WPIEEP(2016)18/FINAL&docLanguage=En. [124]

OECD (2016), *Extended Producer Responsibility: Updated Guidance for Efficient Waste Management*, OECD Publishing, Paris, https://doi.org/10.1787/9789264256385-en. [207]

OECD (2015), *Towards Green Growth?: Tracking Progress*, OECD Green Growth Studies, OECD Publishing, Paris, https://doi.org/10.1787/9789264234437-en. [140]

OECD (2011), *Towards Green Growth*, OECD Green Growth Studies, OECD Publishing, Paris, https://doi.org/10.1787/9789264111318-en. [141]

OECD (forthcoming), *Business Insights on Emerging Markets 2023*, OECD Emerging Markets Network, OECD Development Centre, Paris. [134]

OECD (forthcoming), *Transición verde y formalización laboral en Colombia*, Making Development Happen series, OECD Development Centre, Paris https://www.oecd.org/dev/mdh.htm. [117]

OECD et al. (2020), *Latin American Economic Outlook 2020: Digital Transformation for Building Back Better*, OECD Publishing, Paris, https://doi.org/10.1787/e6e864fb-en. [143]

OECD/Cedefop (2014), *Greener Skills and Jobs*, OECD Green Growth Studies, OECD Publishing, Paris, https://doi.org/10.1787/9789264208704-en. [131]

OECD et al. (2021), *Latin American Economic Outlook 2021: Working Together for a Better Recovery*, OECD Publishing, Paris, https://doi.org/10.1787/5fedabe5-en. [36]

OECD et al. (2019), *Latin American Economic Outlook 2019: Development in Transition*, OECD Publishing, Paris, https://doi.org/10.1787/g2g9ff18-en. [35]

OECD et al. (2019), *Production Transformation Policy Review of Colombia: Unleashing Productivity*, OECD Development Pathways, OECD Publishing, Paris, https://doi.org/10.1787/9789264312289-en. [40]

OECD/EU (2020), *Impact evaluation of labour market policies through the use of linked administrative data*, OECD/European Union, Paris/Brussels, https://www.oecd.org/els/emp/Impact_evaluation_of_LMP.pdf. [146]

OECD/FAO (2022), *OECD-FAO Agricultural Outlook 2022-2031*, OECD Publishing, Paris, https://doi.org/10.1787/f1b0b29c-en. [171]

OECD/ILO (2019), *Tackling Vulnerability in the Informal Economy*, Development Centre Studies, OECD Publishing, Paris, https://doi.org/10.1787/939b7bcd-en. [127]

OECD/UNCTAD/ECLAC (2020), *Production Transformation Policy Review of the Dominican Republic: Preserving Growth, Achieving Resilience*, OECD Development Pathways, OECD Publishing, Paris, https://doi.org/10.1787/1201cfea-en. [39]

OECD/World Bank (2020), *Health at a Glance: Latin America and the Caribbean 2020*, OECD Publishing, Paris, https://doi.org/10.1787/6089164f-en. [152]

Pacheco, P. et al. (2021), *Deforestation Fronts: Drivers and Responses in a Changing World*, World Wildlife Fund, Gland, Switzerland, https://www.worldwildlife.org/publications/deforestation-fronts-drivers-and-responses-in-a-changing-world-full-report. [158]

Pacto Chileno de los Plásticos (2020), *El Pacto Chileno de los Plásticos*, Fundación Chile/Ministerio del Medio Ambiente, Santiago, https://fch.cl/wp-content/uploads/2020/01/roadmap-pacto-chileno-de-los-plasticos.pdf. [96]

Perrine, T. et al. (2020), *Don't Throw Caution to the Wind: In the Green Energy Transition, Not All Critical Minerals Will Be Goldmines*, Columbia Center on Sustainable Investment, New York, https://ccsi.columbia.edu/sites/default/files/content/docs/Dont%20Throw%20Caution%20to%20the%20Wind_0.pdf. [204]

Pietrobelli, C. and B. Calzada (2018), *En busca de una minería innovadora y sustentable en Latinoamérica*, Inter-American Development Bank, Washington, DC, https://blogs.iadb.org/innovacion/es/en-busca-de-una-mineria-innovadora-y-sustentable-en-latinoamerica/. [202]

Porter, M. and C. van der Linde (1995), "Toward a new conception of the environment-competitiveness relationship", *The Journal of Economic Perspectives*, Vol. 9/4, pp. 97-118, https://www.jstor.org/stable/2138392. [42]

Rateau, M. and L. Tovar (2019), "Formalization of wastepickers in Bogota and Lima: Recognize, regulate, and then integrate?", *EchoGéo*, No 47, Pôle de recherche pour l'organisation et la diffusion de l'information géographique, Aubervilliers, France, https://journals.openedition.org/echogeo/16614. [194]

RELAC (2020), *What is RELAC?*, REnovables in Latin America and the Caribbean, Energy HUB, Inter-American Development Bank, Washington, DC, https://hubenergia.org/en/relac#home_section--one (accessed on 15 March 2022). [13]

RICYT (2021), *Network for Science and Technology Indicators –Ibero-American and Inter-American [database]*, Ibero-American and Inter-American Network of Science and Technology Indicators, Buenos Aires, http://app.ricyt.org/ui/v3/comparative.html?indicator=GASIDSFPER&start_year=2010&end_year=2019. [41]

Rodriguez, A., A. Mondaini and M. Hitschfeld (2017), *Bioeconomía en América Latina y el Caribe: contexto global y regional y perspectivas*, Economic Commission for Latin America and the Caribbean, Santiago, https://www.cepal.org/es/publicaciones/42427-bioeconomia-america-latina-caribe-contexto-global-regional-perspectivas. [156]

Rodríguez, A., M. Rodrigues and O. Sotomayor (2019), "Towards a sustainable bioeconomy in Latin America and the Caribeean: Elements for regional vision", *Natural Resources and Development series*, N°193 (LC/TS.2019/25), Economic Commission for Latin America and the Caribbean, Santiago, https://www.cepal.org/en/publications/44994-towards-sustainable-bioeconomy-latin-america-and-caribbean-elements-regional. [175]

Rustomjee, C. (2016), "Developing the Blue Economy in Caribbean and other Small States", *CIGI Policy Brief*, No. 75, Centre for International Governance Innovation, Waterloo, https://www.cigionline.org/sites/default/files/pb_no.75web_1.pdf. [82]

S&P Global Market Intelligence (2022), *Lithium and Cobalt Exploration Trends*, S&P Global Market Intelligence, New York, https://pages.marketintelligence.spglobal.com/Lithium-and-cobalt-infographic-interactive-MS.html (accessed on 6 July 2022). [29]

Saget, C., A. Vogt-Schilb and T. Luu (2020), *Jobs in a net-zero emissions future in Latin America and the Caribbean*, International Labour Organization/Inter-American Development Bank, Geneva/Washington, DC, https://publications.iadb.org/publications/english/document/Jobs-in-a-Net-Zero-Emissions-Future-in-Latin-America-and-the-Caribbean.pdf. [121]

sieLAC OLADE (2022), *Sistema de Informacion energetica de Latinoamerica y el Caribe (SieLAC)*, http://sielac.olade.org. [4]

Steiner, A., I. Andersen and Q. Dongyu (2020), *Paraguay demonstrates benefits of forests as a nature-based solution to climate change*, UN-REDD Programme, Geneva, https://www.un-redd.org/news/paraguay-demonstrates-benefits-forests-nature-based-solution-climate-change#:~:text=In%20particular%2C%20Paraguay%20engaged%20in,from%20deforestation%20and%20forest%20degradation. [165]

Stern, J. (2019), *Challenges to the Future of LNG: decarbonisation, affordability and profitability*, Oxford Institute for Energy Studies, Oxford, https://doi.org/10.26889/9781784671464. [22]

Sturzenegger, G. (2021), *The Plastics Pandemic*, Inter-American Development Bank, Washington, DC, https://blogs.iadb.org/agua/en/plastics-pandemic/#_ftnref7. [188]

Tambutti, M. and J. Gómez (eds.) (2020), "The outlook for oceans, seas and marine resources in Latin America and the Caribbean: conservation, sustainable development and climate change mitigation", *Project Documents*, (LC/TS.2020/167), Economic Commission for Latin America and the Caribbean, Santiago, https://repositorio.cepal.org/bitstream/handle/11362/46509/4/S2000911_en.pdf. [77]

Tekman, M. et al. (2022), *Impacts of plastic pollution in the oceans on marine species, biodiversity and ecosystems*, World Wide Fund for Nature Germany, Berlin, https://doi.org/10.5281/zenodo.5898684. [91]

Timilsina, G., I. Curiel and D. Chattopadhyay (2021), "How Much Does Latin America Gain from Enhanced Cross-Border Electricity Trade in the Short Run?", *Policy Research Working Paper*, No. 9692, World Bank, Washington, DC, https://openknowledge.worldbank.org/bitstream/handle/10986/35729/How-Much-Does-Latin-America-Gain-from-Enhanced-Cross-Border-Electricity-Trade-in-the-Short-Run.pdf?sequence=1&isAllowed=y. [7]

UNCTAD (2019), *Costa Rica on course for a vibrant and inclusive blue economy*, United Nations Conference on Trade and Development, Geneva, https://unctad.org/news/costa-rica-course-vibrant-and-inclusive-blue-economy (accessed on 24 August 2022). [85]

UN DESA (2021), 2021 *Energy Statistics Pocketbook*, Statistics Paper, United Nations Publications, New York, https://unstats.un.org/unsd/energystats/pubs/documents/2021pb-web.pdf [208]

UNDP (2022), *The ocean and the blue economy are fundamental to addressing the triple planetary crisis - says UNDP*, United Nations Development Programme, New York, https://www.undp.org/press-releases/ocean-and-blue-economy-are-fundamental-addressing-triple-planetary-crisis%E2%80%94says-undp (accessed on 5 July 2022). [79]

UNDP (2021), *Considerations for integrating Nature-based Solutions into Nationally Determined Contributions: Illustrating the potential through REDD+*, United Nations Development Programme, New York, https://www.undp.org/publications/consideration-integrating-nature-based-solutions-nationally-determined-contributions-illustrating-potential-through-redd. [163]

UNDP (2017), *Circular economy strategies for Lao PDR A metabolic approach to redefine resource efficient and low-carbon development*, https://www.undp.org/sites/g/files/zskgke326/files/publications/undp-lecb-circular-economystrategies-for-laos-pdr-20. [60]

UNEP (2022), *End plastic pollution: Towards an international legally binding instrument*, United Nations Environment Programme, Nairobi, https://wedocs.unep.org/bitstream/handle/20.500.11822/38522/k2200647 - unep-ea-5-l-23-rev-1 - advance.pdf?sequence=1&isAllowed=y. [103]

UNEP (2021), *Addressing Single-use Plastic Products Pollution Using a Life Cycle Approach*, United Nations Environment Programme, Nairobi, http://www.unep.org/fr/node/29018. [192]

UNEP (2021), *Circular Economy in Latin America and the Caribbean: A Shared Vision*, United Nations Environment Programme, Nairobi, https://emf.thirdlight.com/link/5fhm4nyvnopb-e44rhq/@/#id=0. [58]

UNEP (2021), *From Pollution to Solution: a global assessment of marine litter and plastic pollution*, United Nations Environment Programme, Nairobi, https://www.unep.org/resources/pollution-solution-global-assessment-marine-litter-and-plastic-pollution. [106]

UNEP (2021), *Las ciudades de América Latina y el Caribe pueden reducir a la mitad su consumo de recursos mientras combaten la pobreza*, United Nations Environment Programme, Nairobi, https://www.unep.org/es/noticias-y-reportajes/comunicado-de-prensa/las-ciudades-de-america-latina-y-el-caribe-pueden#:~:text=%2D%20Si%20impulsan%20una%20transformaci%C3%B3n%20sostenible,de%20las%20Naciones%20Unidas%20para. [187]

UNEP (2021), *Policies, Regulations and Strategies in Latin America and the Caribbean to Prevent Marine Litter and Plastic Waste*, United Nations Environment Programme, Nairobi, https://wedocs.unep.org/bitstream/handle/20.500.11822/34931/Marine EN.pdf?sequence=1&isAllowed=y. [94]

UNEP (2019), *Zero Carbon Latin America and the Caribbean: The opportunity, cost and benefits of the coupled decarbonization of the power and transport sectors in Latin America and the Caribbean*, United Nations Environment Programme, Office for Latin America and the Carribbean, Panama City, https://wedocs.unep.org/20.500.11822/34532. [2]

UNEP (2018), *Waste Management Outlook for Latin America and the Caribbean*, United Nations Environment Programme, Latin America and the Caribbean Office, Panama City, https://www.unep.org/ietc/resources/publication/waste-management-outlook-latin-america-and-caribbean. [173]

UNEP-WCMC (2022), *Ocean+ Habitats*, [On-line], [insert month/year of the version downloaded]. Available at: habitats.oceanplus.org. [111]

UNESCO (2021), *Science, technology and innovation (database)*, United Nations Educational, Scientific and Cultural Organization, Paris, http://data.uis.unesco.org/Index.aspx?DataSetCode=SCN DS&lang=en#. [38]

UNESCO (2020), *Custodians of the globe's blue carbon assets*, UNESCO Marine World Heritage, Paris, https://unesdoc.unesco.org/ark:/48223/pf0000375565. [78]

UNIDO (2018), *Cooperación regional en gestión de residuos electrónicos en países de América Latina*, United Nations Industrial Development Organization, Vienna, https://www.unido.org/news/cooperacion-regional-en-gestion-de-residuos-electronicos-en-paises-de-america-latina (accessed on 6 June 2022). [71]

UNWTO/CAF (2021), *UNWTO and CAF Boost Sustainable and Innovative Tourism in Latin America*, World Tourism Organization, Madrid, https://www.unwto.org/news/unwto-and-caf-boost-sustainable-and-innovative-tourism-in-latin-america (accessed on 18 July 2022). [84]

UNWTO/Organization of American States (2018), *El turismo y los Objetivos de Desarrollo Sostenible – Buenas prácticas en las Américas*, United Nations World Tourism Organization/Organization of American States, Madrid/Washington, DC, https://www.e-unwto.org/doi/pdf/10.18111/9789284419937. [198]

USAID (2020), *Ecuador Energy Sector Assessment*, United States Agency for International Development, Washington, DC, https://pdf.usaid.gov/pdf docs/PA00WQNF.pdf. [9]

Van Hoof et al. (2022), *Metodología para la evaluación de avances en la economía circular en los sectores productivos de América Latina y el Caribe*, Economic Commission for Latin America and the Caribbean, Santiago, https://repositorio.cepal.org/bitstream/handle/11362/47975/S2200477_es.pdf?sequence=1&isAllowed=y. [67]

Vargas, J. (2021), "Recursos hídricos: América Latina y El Caribe al 2050", *iAgua*, Madrid, https://www.iagua.es/blogs/juan-carlos-castro-vargas/recursos-hidricos-america-latina-y-caribe-al-2050-0 (accessed on 17 August 2022). [80]

Vona, F. et al. (2018), "Environmental Regulation and Green Skills: An Empirical Exploration", *Journal of the Association of Environmental and Resource Economists*, Vol. 5/4, pp. 713-753, https://doi.org/10.1086/698859. [116]

Wagner, M. et al. (2022), *Monitoreo regional de los residuos electrónicos para América Latina: resultados de los trece países participantes en el proyecto UNIDO-GEF 5554*, United Nations Industrial Development Organization/Global Environment Facility, Bonn, https://www.unido.org/sites/default/files/files/2022-01/REM_LATAM_2021_ESP_Final_dec_10.pdf. [69]

Watson, C. (2016), *Shock-Responsive Social Protection Systems Research - Working Paper 3: Shock-Responsive Social Protection in the Sahel: Community Perspectives*, Oxford Policy Management, Oxford, https://www.opml.co.uk/files/Publications/a0408-shock-responsive-social-protection-systems/wp3-community-perspectives-sahel-en.pdf?noredirect=1. [151]

WEF (2020), *Latin America and Caribbean Travel & Tourism Competitiviness Landscape Report: Assessing Regional Opportunities and Challenges in the Context of Covid-19*, World Economic Forum, Cologny, Switzerland, https://www.weforum.org/reports/latin-america-and-caribbean-travel-tourism-competitiveness-landscape-report-in-the-context-of-covid-19/. 196]

Willis, W. et al. (2022), *Economic and social effectiveness of carbon pricing schemes to meet Brazilian NDC targets*, Taylor & Francis, https://dx.doi.org/10.1080/14693062.2021.1981212. [125]

Wilmsmeier and Jaimurzina (2017), *Energy Efficiency and Electric-Powered Mobility by River: Sustainable Solutions for Amazonia*, FAL Bulletin, No. 353, Santiago, Economic Commission for Latin America and the Caribbean (ECLAC), https://www.cepal.org/en/publications/41858-en (accessed on 25 July [online] 2022). [31]

World Bank (2021), *Caribbean Regional Oceanscape Project*, World Bank, Washington, DC, https://projects.worldbank.org/en/projects-operations/project-detail/P159653. [107]

World Bank (2020), *Future Foodscapes: Re-imagining Agriculture in Latin America and the Caribbean*, World Bank Group, Washington, DC, http://documents.worldbank.org/curated/en/942381591906970569/Future-Foodscapes-Re-imagining-Agriculture-in-Latin-America-and-the-Caribbean. [169]

World Bank (2012), *Expanding Financing for Biodiversity Conservation: Experiences from Latin America and the Caribbean*, International Bank for Reconstruction and Development/The World Bank, Washington, DC/New York, https://www.worldbank.org/content/dam/Worldbank/document/LAC-Biodiversity-Finance.pdf. [164]

World Bank/ESMAP (2020), *Offshore Wind Technical Potential [database]*, World Bank/Energy Sector Management Assistance Program, Washington, DC, https://energydata.info/dataset/offshore-wind-technical-potential (accessed on 23 July 2022). [88]

World Travel and Tourism Council (2022), *Economic Impact Reports. Latin America and The Caribbean 2022 Annual Research: Key Highlights*, World Travel and Tourism Council, London, https://wttc.org/Research/Economic-Impact. [167]

WRI/IDB (2021), *Nature-Based Solutions in Latin America and The Caribbean: Regional Status and Priorities for Growth*, World Resources Institute/Inter-American Development Bank, Washington, DC, https://files.wri.org/d8/s3fs-public/2021-10/nature-based-solutions-in-latin-america-and-the-caribbean-regional-status-and-priorities-for-growth_1.pdf?VersionId=.3xcu8RuodnxfSmw9wCUAYgdEK6evOMa. [160]

WWF (2020), *Living Planet Report 2020: Bending the curve of biodiversity loss*, World Wildlife Fund, Gland, Switzerland, https://www.wwf.fr/sites/default/files/doc-2020-09/20200910_Rapport_Living-Planet-Report-2020_ENGLISH_WWF-min.pdf. [157]

Yamaguchi, S. (2018), "International Trade and the Transition to a More Resource Efficient and Circular Economy: A Concept Paper", *Trade and Environment Working Papers – 2018/03*, OECD Publishing, Paris, https://www.oecd.org/environment/waste/policy-highlights-international-trade-and-the-transition-to-a-circular-economy.pdf. [56]

Annex 3.A. Key selected sectors for the green transition

Annex Table 3.A.1. Nature-based solutions, use of land, and biodiversity and forestry preservation

Key facts	Relevance for the green transition in LAC
• LAC contains 50% of the world's biodiversity and 23% of the total forest cover (FAO, 2021[154]). • LAC includes six of the ten countries with most tree species in the world (FAO/UNEP, 2020[155]). • The region is host to 8 of the 17 most megadiverse countries on the planet (Rodriguez, Mondaini and Hitschfeld, 2017[156]): ○ Andean-Amazon: Bolivia, Brazil, Colombia, Ecuador, Peru and Venezuela. ○ Central America: Costa Rica and Mexico. • Biodiversity has declined by 94% in LAC since 1975, more than in any other region in the world (WWF, 2020[157]). • Almost 40% of global deforestation fronts are in LAC (Pacheco et al., 2021[158]). • More than 43 million hectares were deforested in LAC between 2004 and 2017 due to fires, livestock production, agriculture, mining and transport (Pacheco et al., 2021[158]).	• Current productive practices contribute to climate change through: ○ Deforestation, desertification and biodiversity loss. ○ GHG emissions from land use change and forestry sectors, which are more than three times larger in LAC than in the rest of the world (19.3% vs. 5.8%) (FAO, 2014[159]). • If properly addressed, nature-based solutions can help protect and restore ecosystems and increase human well-being: ○ Sustainable land and forest management could increase the resilience of ecosystems and societies (WRI/IDB, 2021[160]). ○ Green infrastructure can help preserve ecosystems and promote sustainable urban development and buildings (WRI/IDB, 2021[160]). • Environmental justice principles could contribute to avoiding indigenous communities' relocation and loss of livelihoods.
Policy instruments	**Relevant experiences**
• Implement regulations, such as green certifications, environmental laws and standards, to assign specific budgets to preserve forests and prevent unsustainable land uses (e.g. "Native forest law" in Argentina) (IFPRI, 2021[161]), or encourage public-private collaborations (e.g. concessions for sustainable forest and land management) (OECD, 2020[162]). • Promote participation processes with local communities and civil society organisations (CSOs) to identify needs and increase the legitimacy of policy. • Develop national strategies and promote activities within the United Nations Framework Convention on Climate Change to reduce emissions from deforestation and forest degradation in developing countries (REDD+) to obtain results-based payments (UNDP, 2021[163]).	• **Costa Rica and Mexico:** payments for the Ecosystem Services Programme have allowed small and medium-sized landowners to help conserve land and biodiversity through cash transfers since 1997 (World Bank, 2012[164]). • **Colombia:** the Mosaic Conservation project allowed local communities to work on the restoration and preservation of damaged territories surrounding national parks and protected areas. • **Paraguay:** the National Strategy of Forests for Sustainable Growth (2019) and the national forest monitoring system enabled the quantification of emissions reduction and generated reliable data on changes on forest areas (Steiner, Andersen and Dongyu, 2020[165]). • **Peru:** Public administration contracts targeted to non-governmental organisations and local entities devoted investments of USD 20 million in ten protected areas.

Annex Table 3.A.2. **Sustainable agriculture and livestock**

Key facts	Relevance for the green transition in LAC
• LAC is the major net food-exporting region of the world (17% of the net export value of global agricultural and fish commodities in 2020) (IICA, 2021[166]; OECD, 2021[113]). • The region contains 12% of the world's land currently under cultivation. • The agricultural sector is the second-most polluting sector in LAC (22.9% of all regional GHG emissions), and emissions are rising (Bárcena, 2020[168]). • Agriculture represented 14% of total employment in LAC in 2019.	• Current practices contribute to climate change through: ○ Land degradation: there is accelerated soil erosion, salinisation and loss of soil organic matter, mainly due to widespread monoculture production. ○ Unsustainable demand for resources: agriculture in LAC consumes large and unsustainable amounts of fresh water (World Bank, 2020[169]). ○ Rising GHG emissions: agriculture emissions in LAC increased by 32% between 1990 and 2019, reaching a total of 1.04 Gt CO_2e (Climate Watch, 2020[119]). • Proper policies could help: ○ Preserve ecosystem services and reduce pressure on the environment. ○ Reduce poverty and hunger and create quality formal jobs, particularly in rural areas.
Policy instruments	**Relevant experiences in LAC**
• Implement consensual regulations to: ○ promote sustainable spatial planning (OECD, 2020[162]) ○ establish certifications, environmental laws and standards (IFPRI, 2021[161]) ○ foster public-private agreements to promote the large-scale adoption of climate-smart production processes and technologies, especially in the livestock sector (e.g. *Acuerdos de Producción Limpia* in Chile) (OECD/FAO, 2022[171]) • Allocate resources by: ○ Applying taxes to discourage unsustainable practices (OECD, 2020[162]). ○ Defining charges and fees (OECD, 2020[162]) ○ Redesigning subsidy schemes to discourage the use of pesticides or fossil fuels and promote biodiversity (OECD, 2020[162]). • Invest in technology R&D for better use of resources and green farming practices (e.g. Observatory of Brazilian Agriculture and Livestock).	• **Brazil:** has integrated low-carbon sustainable agriculture and livestock experiences. In 2020, it launched the National Bioinputs Program, a policy to foster the use of bio-based products in agriculture. The Brazilian Association of Carbon-Neutral Meat Producers unites farmers, who are implementing integrated crop-livestock-forestry systems to link sustainable farming, cattle and forestry activities to continue producing high-quality products that have a low carbon footprint. For example, a carbon-neutral beef certification was developed and is already available on a commercial scale (Food Navigator, 2019[172]). • **Ecuador:** the climate-smart livestock farming project allowed an increase in milk production, improved soil quality on 40 000 ha and paid higher wages in the sector, while avoiding 24 000 tonnes of GHG emissions thanks to rotational grazing and pasture compost production. • **Uruguay:** good practices and alternatives to the use of pesticides allowed reduced use of herbicides (by up to 70%) in the soybean production cycle without affecting yields and saving up to USD 40 per cultivated hectare. • **Mexico:** efficient and low-emission technologies in agriculture and agro-industry allowed a total of 1 842 agribusinesses to reduce their net GHG emissions by 6 Mt of CO_2 equivalent and to produce energy from biomass (FAO, 2021[154]).

Annex Table 3.A.3. Bioeconomy and regenerative food systems

Key facts	Relevance for the green transition in LAC
• LAC has high potential for biomass production due to the availability of land, adequate soils and water (Rodriguez, Mondaini and Hitschfeld, 2017[156]). • The bioeconomy raises the need for new relationships between agriculture and food, due to the fact that 50% of municipal waste is organic and food consumption waste represents 34% in LAC (UNEP, 2018[173]). • LAC presents advantages in six thematic areas for the development of the bioeconomy: 1) exploitation of biodiversity resources; 2) eco-intensification of agriculture; 3) biotechnology applications; 4) biorefineries and bioproducts; 5) improved efficiency of agrifood chains; and 6) ecosystem services (Aramendis, Rodríguez and Krieger Merico, 2018[174]).	• The bioeconomy can contribute to meeting the challenges of productive diversification and structural change associated with changing the economic dependence on primary commodity producing in agriculture, mining and fossil resources (Rodríguez, Rodrigues and Sotomayor, 2019[175]). • The bioeconomy promotes new production, enabling the development of new products that can be used as inputs in other sectors, such as biomaterials for construction, bio-inputs for agriculture, enzymes for industry and substitutes for petrochemical products. It also helps change consumer behaviour or satisfy new consumer demands (e.g. functional foods, biocosmetics) (Rodríguez, Rodrigues and Sotomayor, 2019[175]). • The bioeconomy is a feasible alternative towards decarbonisation. It provides an adequate framework for the harmonisation of policies required for the implementation of the 2030 Agenda for Sustainable Development (Rodríguez, Rodrigues and Sotomayor, 2019[175]).
Policy instruments	**Relevant experiences in LAC**

• Promote the development of national bioeconomy strategies by:
  ○ Recognising territorial and landscape specificities, to increase a just distribution of benefits.
  ○ Creating or adjusting education programmes, technical training and the use of information and communications technology to develop necessary capacities.
  ○ Focusing on nature-based solutions as an option to harmonise development and ecosystem protection.
	• **Brazil**: the Biotec Initiative (2021) prioritises four areas of biotechnology: human health, agriculture, industrial and environment (MCTI, 2021[176]).
	• **Colombia:** the international bioeconomy mission (2020) aims to provide the grounds to take advantage of biodiversity, and to promote development of an environmentally sustainable agricultural and livestock sector, by adopting a zero food waste approach and promoting advanced technologies in the health sector.
	• **Uruguay:** the bioeconomy contributed to the COVID-19 recovery, to the advancement of biotechnology and the valorisation of agricultural and agro-industrial products development in the meat and dairy chains (Borges et al., 2021[177]).

Annex Table 3.A.4. Water management

Key facts	Relevance for the green transition in LAC
• LAC is host to one-third of the world's freshwater resources. • Cities and productive areas in LAC have high levels of water stress: ○ 166 million people do not have access to a safely managed potable water service (ECLAC, 2022[179]). ○ 46 million do not have basic sanitation facilities (CAF, 2017[178]). ○ Less than 50% of wastewater is adequately treated. ○ A total of 43% of countries in the region report low levels of Integrated Water Resources Management (ECLAC, 2022[179]).	• Access to water is a cornerstone of development and a strong engine for reducing inequalities. It is a key determinant of economic growth, environmental health and social well-being. • Climate change affects water management in multiple ways, including changes in precipitation and therefore in seasonal and annual patterns of floods and droughts, water availability or dilution capacity, while affecting health, economic activities and water-dependent ecosystems (Climate-ADAPT, 2021[180]). • With large parts of the world dependent on rain-fed agriculture, reduction in water availability, coupled with unpredictable changes in rainfall, could affect millions of farm-based livelihoods and jeopardise food security (FAO, 2021[181]).
Policy instruments	**Relevant experiences in LAC**
• Modern water legislation should focus on basin planning, climate change adaptation measures, water conservation and prioritisation of human consumption and water supply over other uses. Energy regulation is required to encourage the establishment and use of renewable energy and efficiency without overexploiting water resources. • Promote nature-based solutions, such as mangroves. Protecting shorelines from storms, lakes storing large water supplies, and floodplains absorbing excess water runoff is a key part of these initiatives. • Address the need for a water transition. Water should be considered a natural asset. The human right to water should be guaranteed and negative externalities reversed, while moving towards circular water management.	• **Peru:** the Lacomes Project provides potable water and sanitation systems, powered by solar energy and built with local materials, in accordance with the dry forest climate (MINAGRI, 2022). • **Mexico:** Isla Urbana is dedicated to contributing to water sustainability in Mexico through rainwater harvesting (Isla Urbana, 2021[182]). • **Uruguay:** CTAGUA is the technological centre for water development in Uruguay, using the internet of things and big data to help companies, technical institutes and universities address the country's main water challenges (CTAGUA, 2022[183]).

Annex Table 3.A.5. **Waste management and plastics**

Key facts	Relevance for the green transition in LAC
• The LAC region produces 541 000 t of municipal waste per day. This figure is projected to grow by 25% by 2050 (UNEP, 2018[173]). • Almost 95% of waste goes to sanitary landfills (52.0%), open dumps (26.8%) or other landfills (16.5%). Only 4.5% is recycled and less than 1% is composted, incinerated or digested anaerobically (Kaza et al., 2018[184]). • In LAC, plastics represent 12.4% of municipal solid waste – the fourth-largest waste stream in the region (Kaza et al., 2018[184]). • A total of 40% of the plastic used in LAC is disposed of after single use (IDB, 2020[185]).	• High GHG emissions: waste and industrial processes are the fourth-largest GHG emissions producer in the region, accounting for 10% of total emissions (ECLAC, 2020[186]). • Unsustainable consumption patterns: urban household material consumption is projected to increase to 25 t per capita in LAC by 2050 (well above the range of 6-8 t per capita of the sustainable scenario established by the UN International Resource Panel) (UNEP, 2021[187]). • High potential to create formal jobs: there are over 1.8 million informal waste pickers in LAC, recovering up to 50% of the material that is recycled (Sturzenegger, 2021[188]). • Plastic waste in the ocean: it is estimated that the ocean contains about 75-199 Mt of plastic waste and the yearly additional inflow could nearly triple by 2040. This threatens ecosystem services that are fundamental to many livelihoods and the blue economy (UNEP, 2021[106]).
Policy instruments	**Relevant experiences in LAC**
• Set ambitious goals as part of co-ordinated strategies to reduce solid waste generation and disposal in landfills while promoting the formalisation of informal recycling workers. • Co-ordinate actions across levels of government and key stakeholders. Cities and local governments tend to be responsible for waste management services. • Support and promote informal businesses and workers operating in circular business models (e.g. repair, refurbishment, remanufacture and recycling) (Circular Economy Coalition of Latin America and the Caribbean, 2022[189]). • Expand plastic bans, which are often narrowly focused on certain types of single-use products (Karasik et al., 2020[190]). • Increase the low price of virgin plastic (Geyer, 2020[95]), and reduce the costs of alternatives (Karasik et al., 2020[190]). • Internalise waste externalities and require producers to take responsibility for waste treatment or disposal (EPR). • Engage in stakeholder consultations (UNEP, 2021[106]). • Combine information instruments with clean-ups and measures that provide convenient alternatives to plastic products (Heidbreder et al., 2019[191]). • Provide incentive schemes to reduce waste production (e.g. pay-as-you-throw system or differentiated tariffs). • Use public procurement to promote sustainable practices in the waste sector. • Generate information and data to monitor total waste streams, their recovery and the use of secondary raw materials in the economy (Kaza et al., 2018[184]).	• Prohibitive regulatory instruments: there are bans on the sale and/or use of single-use plastic products in Chile, Colombia, Panama, Peru and Uruguay, in addition to import bans in highly affected small island states, such as Antigua and Barbuda and Saint Lucia (UNEP, 2021[192]; UNEP, 2021[106]). • Affirmative regulatory instruments: there are obligations for plastic producers in Colombia and Peru to use recycled material, and EPR schemes in Argentina, Brazil, Chile, Colombia and Mexico that provide positive incentives (UNEP, 2021[192]; IDB, 2020[185]). • Economic instruments: examples include exemption of plastic alternatives from import duties in Saint Lucia (UNEP, 2021[192]); a tax on the consumption of plastic bags and single-use plastic products in Colombia; a green tax on polyethylene terephthalate bottles in Ecuador (IDB, 2020[185]); and an environmental protection levy on packaging, as an advanced disposal fee, in Jamaica (UNEP, 2021[106]). "Ecoins" purchasing power parity in Costa Rica rewards recycling with a digital coin that can be exchanged for discount coupons from sponsoring companies (Ecoins, 2022[193]). • Information instruments: examples include the #MenosPlásticoMásVida (#lessplasticmorelife) initiative on responsible plastic consumption in Peru (UNEP, 2021[192]) and Ecuador's permanent campaign Plásticos en el mar… NO MÁS! (No more plastics in the sea) to educate fishermen and associated parties (IDB, 2020[185]). • There are efforts in Chile, Colombia and Peru to recognise informal waste pickers and recyclers and integrate them into formal waste management (Rateau and Tovar, 2019[194]; IDB, 2020[185]).

Annex Table 3.A.6. Sustainable tourism

Key facts	Relevance for the green transition in LAC
• In 2021, tourism contributed 6.1% to Latin America's GDP (+26.5% change from 2020) and 13.4% of total jobs (14.25 million), representing an 8% recovery. In the Caribbean, tourism contributed 9.1% of GDP (+36.6% change from 2020) and 13.4% of total jobs (2.35 million), representing a 15.2% increase (World Travel and Tourism Council, 2022[167]). • In 2019, informal employment in tourism reached 63.3% in the region. Individuals under age 24 account for 20.9% of total employment in the sector (ILO, 2021[195]).	• By 2030, CO_2 emissions from tourism corresponding to transport will increase by 25% over 2016 levels (from 1 597 Mt CO_2 to 1 998 Mt CO_2) (World Travel and Tourism Council, 2022[167]). • Potential environmental impacts caused by activities related to tourism (World Economic Forum, 2020[1]) include: ○ Unco-ordinated land use due to rapid urban growth. ○ Destruction of fragile ecosystems. ○ Contamination of bodies of water. ○ Aesthetic deterioration of the landscape and the urban environment. • The tourism sector plays an essential role in realising circularity in the use of plastics. Tourism contributes to plastic pollution through the single use of water bottles, disposable toiletries, plastic bags, bin liners, food packaging and cups. • The tourism sector can contribute directly to the achievement of SDG 8 on decent work and economic growth, SDG 11 on sustainable cities and communities, SDG 12 on responsible consumption and production, SDG 14 on life below water, and SDG 15 on life on land.
Policy instruments	**Relevant experiences in LAC**
• The environmental consequences of tourism are largely associated with a lack of information and unsustainable behavioural patterns. Some key policy instruments used to reduce GHG emissions in the sector in LAC are: ○ Regulatory approaches, e.g. emission standards, bans on toxic substances, and land planning instruments. ○ Communication campaigns to provide information to citizens and enterprises for adequate use and disposal of resources ○ Market-based instruments, e.g. environmental taxes, financing, payment for environmental services, GHG emission trading, green bonuses, public-private partnerships, concessions, and seed capital initiatives (BIOFIN Costa Rica, 2021[197]).	• Sustainable tourism projects: ○ **Guatemala:** in 2015, IMPULSA was created to protect 334 areas (32% of the territory). ○ **Mexico:** tourism development Mayakoba aims to increase biological diversity by preserving and strengthening terrestrial ecosystems, as well as creating aquatic habitats. ○ **Peru:** with the Project of the Tingana Association, inhabitants changed their behaviour from indiscriminate logging, fishing and hunting to ecotourism activities and safeguarding the jungle (UNWTO/Organization of American States, 2018[198]). • Policy guidelines: to promote ecosystem-friendly tourism with minimal impact on the environment and local culture, several LAC countries have developed regulatory framework guidelines. Examples include: ○ **Columbia:** United for the Environment sustainable tourism policy (Ministerio de Industria y Comercio, 2020[199]) ○ **Mexico:** sustainable tourism policy 2030 (Gobierno Federal Sectur, 2020[200]) ○ **Panama:** Sustainable Tourism Master Plan (Ministerio de Ambiente, 2020[201]).

Annex Table 3.A.7. **Sustainable mining**

Key facts	Relevance for the green transition in LAC
• LAC countries are well placed to take advantage of demand for critical minerals during the green transition. Argentina, Bolivia, Brazil, Chile, Mexico and Peru have significant reserves of copper, iron ore, lithium, nickel, magnesium, molybdenum, silver and zinc. • LAC includes the world's biggest copper producer (Chile), the world's biggest silver producer (Mexico), the third-biggest steel producer (Brazil) and the seventh-biggest bauxite producer (Jamaica). • In 2017, a total of 61% of global lithium reserves were in LAC, along with 39% of global copper reserves and 32% of nickel and silver reserves (ECLAC, 2018[28]). • Investments in lithium in 2021 in LAC increased by an average of 117% compared to 2020, reflecting increases of 559% in Chile and 77% in Argentina (S&P Global Market Intelligence, 2022[29]). • The mining sector accounts for 21% of total exports from Peru, 60% from Chile and 46% from Brazil (Pietrobelli and Calzada, 2018[202]).	• The demand for materials associated with a low-carbon transition is projected to grow by 110% by 2060 (from 2015 levels), requiring resource extraction to more than double, reaching 190 Mt per year. • Critical minerals present in LAC are essential inputs for the scaling up of the renewable energy technologies needed for the green transition. Lithium-ion batteries require cobalt, lithium, nickel, manganese; electric vehicles require rare earth elements; solar PV requires cadmium, indium, gallium, selenium, silver and tellurium; wind turbines require rare earth elements; and aluminium and copper are needed across all renewable energy technologies (Dominish, Florin and Teske, 2019[203]). • The development of a sustainable mining sector can be an opportunity for LAC countries to generate substantial revenue from tax and royalties, lead to improved infrastructure, and facilitate uptake of renewable energy generation, creating jobs and generating revenues that could support local investment in education, health care and other community benefits.
Policy instruments	**Relevant experiences in LAC**
• In the development of a sustainable mining sector, address environmental, social and governance risks throughout the value chain – from extraction to end use and recycling – to avoid the environmental damage, fragility, conflict and human rights abuses that have often characterised the mining sector in LAC. • Develop regulatory frameworks that encourage private investment (including related power and transport infrastructure), eliminate corruption, and engage and deliver concrete benefits for local communities throughout the mining life cycle. This could be a catalyst for inclusive development, economic growth and structural transformation of economies. • Use triple and quadruple helix co-ordination to promote innovation in the mining and energy sectors through interaction among mining companies, suppliers, governments, academia and CSOs. • Use geological mapping to help governments understand the extent of endowments of critical minerals and metals, and present geoscience data in an accessible way to attract private-sector investment. • Use the mining sector to support the roll-out of energy access to surrounding communities. The sector can provide anchor electricity offtake in rural areas not covered by the national electricity grid.	• **Chile:** has implemented several policy and investment measures to develop its local lithium value chain. For example, the Ministry of Mining has set out provisions to incentivise downstream industries and clarify existing policies, to encourage public and private investment, and to double lithium carbonate production to 230 000 mt per year by 2023 (Perrine et al., 2020[204]). • **Peru:** developed the Towards a Vision for Mining in Peru in 2030 strategy in 2016. It sets out measures to enhance the economic contribution of mining and emphasises the need to ensure alignment with the SDGs and territorial development priorities (IRP, 2020[205]). • **Chile** and **Mexico:** the two countries are leading the way in the integration of utility-scale renewable energy projects for mines. Notable examples include Antofagasta's 115 MW wind power and Compañía de Acero del Pacífico's 100 MW solar PV projects in Chile, and Industrias Penoles' 180 MW wind power plant in Mexico (Alova, Galina, 2018[206]).

Chapter 4

How to make it possible? Financing a green and just transition

To transition to net-zero emissions economies, the LAC region must mobilise substantial resources in a tight fiscal space. To do this, the region needs an effective financing strategy that involves the public and private sectors. This chapter analyses a sustainable fiscal policy that aims to invest more and better in the green transition, with a focus on the energy sector. It proposes ways in which the region can mobilise further resources, specifically through environmental taxes, innovative debt tools, and phasing out fossil fuel subsidies (without neglecting the most vulnerable). It stresses the need for sustainable financial strategies that channel public and private investment towards projects with greater environmental benefits. It focuses on the role of finance ministries and that of sub-national, national and international development finance institutions (DFIs) in helping mobilise resources for the green transition. It also suggests strategies to help the public sector mobilise private-sector investments towards sustainable projects. Finally, it looks at the importance of sustainable finance frameworks in developing and improving regulatory guidelines that facilitate private- and public-sector investments.

Both the public and the private sectors are needed to finance the green transition in LAC

To fund a green transition the region must levy further resources through environmental taxes, emissions trading systems and a step-by-step phase-out of fossil-fuel subsidies

Environmentally-related tax revenue in LAC countries, by main tax base, 2020

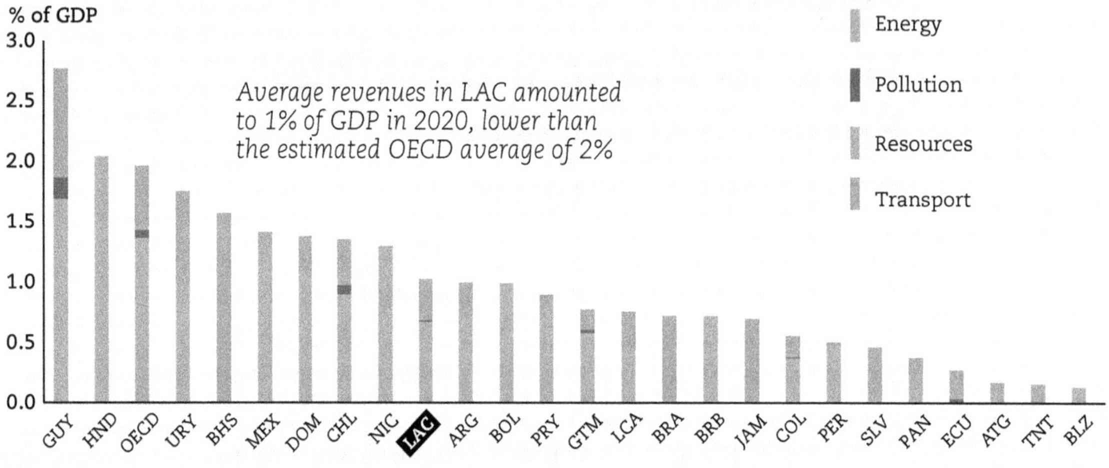

To mobilise the vast amount of funds needed for the green transition, the pool of stakeholders and tools will also have to be enlarged

In LAC, the GSSS market has been growing since 2015 reaching an accumulated USD 73 billion in September 2021

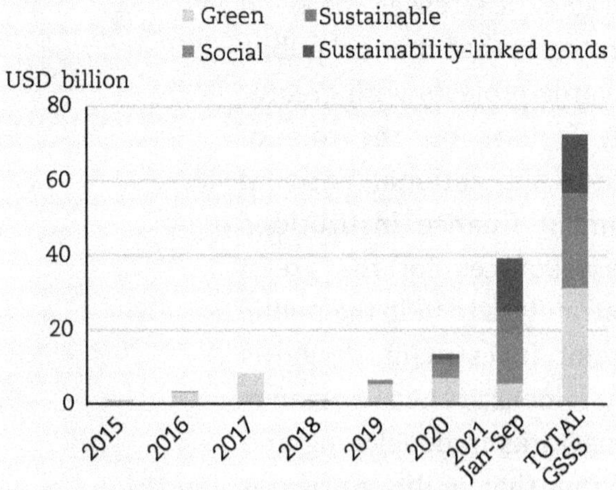

Total LAC GSSS bond issuance in international markets, by type of issuer, December 2014 to September 2021

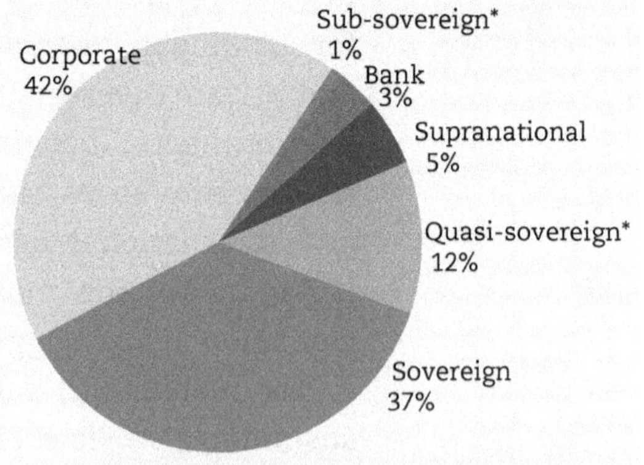

*Sub-sovereign refers to states, cities and provinces. Quasi-sovereign issuers are defined as companies with full or partial government ownership.

The role of national and sub-national development banks is key in leveraging private-sector resources

Sustainable finance frameworks should maintain transparency and avoid greenwashing by improving regulatory tools, such as sustainability and green bond standards and taxonomies

LAC faces the challenge of financing the green transition under a tight fiscal space in a context of persistent social inequality and historic development challenges. In answer to the coronavirus (COVID-19) crisis, the region expanded social spending while experiencing a drop in revenues, resulting in a strong increase in public debt. With reduced fiscal leeway, most LAC countries must align recovery stimulus with achieving sustainability targets, while protecting the most vulnerable (Chapter 1).

To finance the green transition, the region needs to invest more, while mobilising further resources from public and private sources. The cost of inaction is high, for instance a 2.5°C rise in temperature could cost the region between 1.5% and 5.0% of gross domestic product (GDP) by 2050 (Bárcena et al., 2015[1]). Thus, it is important to seek and scale up innovative financial instruments and develop strategies that enable LAC countries to achieve both their Nationally Determined Contributions (NDCs) and their commitment to the United Nations Sustainable Development Goals (SDGs). Specifically, countries need to increase and improve public and private spending on clean energy and energy efficiency, since these are the most cost-effective ways to reduce emissions on a global scale (IEA, 2021[2]). This should be coupled with the development of well-defined renewable energy infrastructure investment plans, also known as "robust low-carbon infrastructure project pipelines", that specify what and where project investments are needed, when they should be built, how to finance them, or if they are sufficient to meet long-term objectives (OECD, 2018[3]). The region also needs a "big push" towards developing innovative ways to mobilise public and private resources. Environmentally related taxes present themselves as an important opportunity, as they can raise additional revenues, create the right behavioural incentives, and accelerate the green transition. Similarly, rationalising and phasing out fossil fuel subsidies, particularly the ones that benefit the affluent population, is also a way to liberate further resources. Likewise, scaling up debt tools, such as green, social, sustainable, and sustainability-linked (GSSS) bonds, debt-for-nature swaps, catastrophe (CAT) bonds, and natural disaster clauses, will be critical.

Developing the right compensation schemes will be necessary as the green transition advances. The effects of climate change and the policies necessary to address them will leave certain sectors of the population in a more vulnerable state, for instance as a result of job loss in certain sectors. Compensation schemes need to include cash transfers, in-kind transfers, unemployment benefits for workers dismissed active labour market policies (ALMPs) (Chapter 3) and expanded social protection systems for the most vulnerable (Chapter 1). Compensation policies can also include facilitating relocation and retraining of workers, promoting decent work in rural areas, offering new business models, and supporting displaced workers.

Sustainable finance strategies that involve a diverse pool of stakeholders are essential to guide the green transition. Finance ministries have the important role of developing fiscal frameworks that protect green investments and thus ensuring that the ecological transition becomes a long-term priority. An important step is to align national expenditure and revenue processes with climate and other environmental goals. Similarly, the use of innovative tools such as a "green golden rule" could also prove to be useful. National and sub-national development banks can also help by providing technical and financial support for the design of climate and financial strategies (Finance in Common, 2021[4]; Galindo, Hoffman and Vogt-Schilb, 2022[5]).

Climate-related development finance also plays a key role in increasing investment in projects with environmental benefits. Here, the expansion of green financing by bilateral, multilateral, and private donor sources is critical to help countries meet their ambitious climate-related goals. This includes tapping into the growing resources of multilateral climate funds, which include a variety of donors such as multilateral and bilateral actors, the private sector and donations. Moreover, the improvement of sustainable finance frameworks through public-private co-operation is crucial. These frameworks should enhance regulatory tools, such as sustainable standards and taxonomies that increase the flow of private resources to sustainable projects. Last, since most of the investment for the green transition will come from the private sector, the public sector must put in place the necessary incentives to redirect private-sector investments towards green projects.

While the costs of inaction to fiscal stability have been addressed in Chapter 1, this chapter highlights the importance of developing an environmentally sustainable fiscal policy in the LAC region to drive a green and just transition. This involves a focus on more and better spending on clean energy and energy efficiency, as well as seeking new ways to mobilise additional revenues, such as environmental taxes, emissions trading systems (ETS), phasing out subsidies and scaling up debt tools (e.g. GSSS bonds, debt-for-nature swaps, CAT bonds and natural disaster clauses). The chapter then addresses compensation mechanisms to protect the most vulnerable and those who stand to lose, at least temporarily, from the impacts of green policies. It then focuses on sustainable finance strategies including enhancing green fiscal frameworks (e.g. through green golden rules) and mobilising green investments by key actors (e.g. finance ministries and sub-national, national and international DFIs). The chapter also focuses on financial strategies to increase the mobilisation of private-sector resources and stresses the importance of developing and expanding sustainable finance frameworks that ensure that public and private investments effectively reach environmentally sustainable projects. Lastly, it ends with messages and conclusions.

Developing environmentally sustainable fiscal policies that favour the green transition

To achieve a sustainable and climate resilient net-zero economy, LAC countries have set their own climate mitigation and adaptation targets through their NDCs for 2030, prioritising the energy sector, among others. The energy sector represents the largest source of greenhouse gas (GHG) emissions, accounting for 43.5% of LAC's emissions (Climate Watch, 2022[6]; FAO, 2022[7]; OECD/IEA, 2021[98]). As a result, countries have been shifting their focus towards the energy sector (Chapters 2 and 3).

A successful green fiscal policy in the region aimed at achieving the NDCs, including the more ambitious energy-related emissions targets, will necessarily have to increase spending and incentives on clean energy and energy efficiency while mobilising vast resources through specific tax strategies and novel financial instruments. This needs to be coupled with compensation mechanisms for the most vulnerable to guarantee a just transition.

Increasing spending on clean energy and energy efficiency would be a cost-effective way for the region to meet its net-zero emissions targets

Responding to the climate crisis in LAC requires immediate and better investment in clean energy and energy efficiency (IRENA, 2022[8]; OECD, 2018[3]). Globally, these two types of investments could provide more than 90% of the required reductions in energy-related carbon emissions, driven by substantial electrification (IRENA, 2022[9]). For LAC, this presents a unique opportunity due to the rapid nature of the region's growth and the amount of new equipment and infrastructure being built and purchased, including buildings, factories, vehicles, and networks (IEA, 2021[2]). The capital intensity of clean energy investment also requires keeping costs from public and private sources low, which will be critical to the speed and affordability of this transformation (IEA, 2021[2]).

In emerging markets and developing economies, including LAC, clean energy investment needs to increase substantially, especially in electricity generation. If emerging economies are to become net-zero by 2050,[1] investment needs to increase from an average per year of USD 150 billion in 2020 to over USD 1 trillion by 2030. Over the next decade, the largest increase lies in electricity generation, with annual investment going from around USD 0.5 trillion in 2020 to USD 1.6 trillion by 2030 (IEA, 2021). Likewise, if these net- zero economy targets are to be achieved, a transition of the manufacturing industry to low-carbon technologies will be necessary given that this sector is the largest energy consumer and is one of the main sources of CO_2 emissions worldwide, accounting for 40% of total CO_2 emissions (OECD, 2022[10]).

Additionally, investment in electricity grids needs to increase substantially to accommodate higher electricity demands and the rise in renewables deployment. Since electricity grids are the backbone of power systems, it is necessary to invest in their expansion and modernisation to integrate renewable energies (IEA, 2021[2]). Similarly, countries, along with the private sector, need to increase spending on electrification via greener buildings, appliances and electric vehicles (EVs). For the latter, governments need to promote investment in public EV chargers to accommodate growing electric mobility needs, develop tax incentives and purchase subsidies, and expand green auto loans and leasing models (IEA, 2021[2]). A further, more efficient effort, which also deals with the increasing urban traffic congestion in LAC cities, is to promote the vast deployment of structured public electromobility systems (ECLAC, 2020[11]).

Improving and investing in energy efficiency is the cheapest and most immediate way for LAC countries to start reducing their use of fossil fuels (EESI, 2022[12]). Increasing spending on eliminating energy waste entails, for instance, investing in energy-efficient, digitally connected buildings driven by climate-based scenarios. These scenarios consider projections of future GHG emissions to assess infrastructures' future vulnerability to climate change. This will have to be accompanied by a substantial shift towards clean solutions to manage the significant rise in demand for cooling. To achieve the goal, governments also have to develop appropriate regulations to increase affordable consumer finance, improve building codes and reduce fossil fuel energy subsidies (IEA, 2021[2]).

A 'twin transition' approach recognising the importance of progressing both the digital transformation and the green transition in order to reduce emissions will be critical to boost efficiency and serve to build a more inclusive and just energy system in the long term (OECD, 2022[13]; OECD, forthcoming[14]). Accelerating the use of digitalised technologies can optimise the deployment of decentralised renewable energy (DRE) by being able to identify who needs energy and deliver it at the right time, in the right place, and at the lowest cost (IEA, 2022[15]; OECD, 2022[13]). DRE is renewable energy that, instead of being generated at an energy plant and distributed through a national grid, is generated close to where it will be used and distributed through the grid, mini grids and off-grid installations (UN, 2018[16]). Governments in the region have the important role of setting up conducive frameworks for the digital transformation of the energy system (IEA, 2022[15]).

Investing in smart grids with smart meters for better energy distribution in electricity systems is another way to boost energy efficiency. Smart meters enhance efficiency by improving customer demand forecasts and consumer awareness. At least a quadrupling of investment in these is needed by 2030 if LAC countries are to meet their accelerated decarbonisation and electrification aims from their SDG commitments (2030) and their NDCs (IEA, 2021[2]).

Countries in the LAC region should advance development of their pipelines of robust, low-carbon energy infrastructure projects. By providing more security to investors, well-defined renewable energy infrastructure plans can guarantee that investments reach sustainable projects (OECD, 2018[3]). Pipelines and plans help investors identify opportunities among low-carbon infrastructure options that match their needs. Governments can facilitate these processes, for example, by fast-tracking valuable projects or supporting certain projects to overcome barriers to their development (OECD, 2018[3]).

Phasing out fossil fuel subsidies and introducing environmental taxes can mobilise more revenues while supporting environmental objectives

New investments in the green transition require the mobilisation of additional public resources and fiscal reforms. The LAC region is currently under tight fiscal constraints characterised in most countries by low tax revenues. Thus, to finance the green transition, fiscal policy has to support the recovery. For the green transition, the power of appropriate fiscal reform is that it not only generates more resources but also drives the needed productive transformation, which generates quality formal employment, pushes the green agenda, and protects the most vulnerable (OECD et al., 2021[17]). Besides alternative options to raise additional resources, further exploiting other policies can achieve multiple targets. These include environmental taxes, ETS or the phase-out of fossil fuel subsidies, particularly to the most affluent population.

To date, the rationalisation and sequential phase-out of subsidies is an incomplete fiscal policy in the LAC region, even though both could free up resources for projects that have a positive green impact. While some countries have made efforts to eliminate fossil fuel subsidies, such as Mexico through the reduction of fiscal stimulus for the lowest octane fuel in 2022 (Ministry of Finance and Public Credit, 2022[18]), much more remains to be done. Even if subsidies are intended to protect vulnerable households and firms, evidence shows that, in addition to having a high fiscal cost, they often carry negative distributional impacts (high regressivity); if not adequately targeted, they tend to favour wealthier households that use more fuel and energy.

Many LAC countries provide subsidies for the use of energy products that have negative effects on environment and social dimensions. Reasons for implementing fossil fuel subsidies may include mitigating the impacts of high and volatile petroleum prices, controlling inflation, boosting competitiveness and protecting the poorest segments of the population. However, fossil fuel subsidies can put a strain on national budgets while also benefiting high-income households (Puig and Salinardi, 2015[19]), increasing air pollution (with associated high health costs), and sending the wrong signals to the markets (negatively affecting social-environmental goals) (Rentschler and Bazilian, 2017[20]). Generalised fossil fuel subsidies can also contribute directly to urban sprawl, which renders mass transport less effective and results in higher emissions due to greater use of personal vehicles. A green and just transition in LAC requires phased elimination of fossil fuel subsidies, the establishment or adjustment of environmental taxes, and the promotion of productive diversification that expands the tax base. All of these measures should be within the framework of a policy that provides support to firms and households most vulnerable to energy price volatility (ECLAC, 2022[21]).

Public funds coming from fossil fuels could be redirected to green projects. In 2022, however, the macroeconomic context is hindering this opportunity. Despite an overall downward trend of fossil fuel support measures over recent years, their use is rebounding, and some countries still offer generous support measures that could be redirected towards more efficient sustainable projects. Of the five LAC countries analysed, Argentina and Mexico provided the highest support packages to both consumers and producers in 2020, such that fossil fuel support measures accounted for 1.1% of GDP (Figure 4.1). The LAC average ranged from 1.0% of GDP in 2010 to 0.7% in 2020, peaking in 2016 at 1.4%. So far, energy tax revenues are greater than the cost of subsidies in the five countries analysed and thus represent, on average, a net positive impact on public finances (OECD, 2021[23]).

Figure 4.1. **Fossil Fuel support measures in LAC**

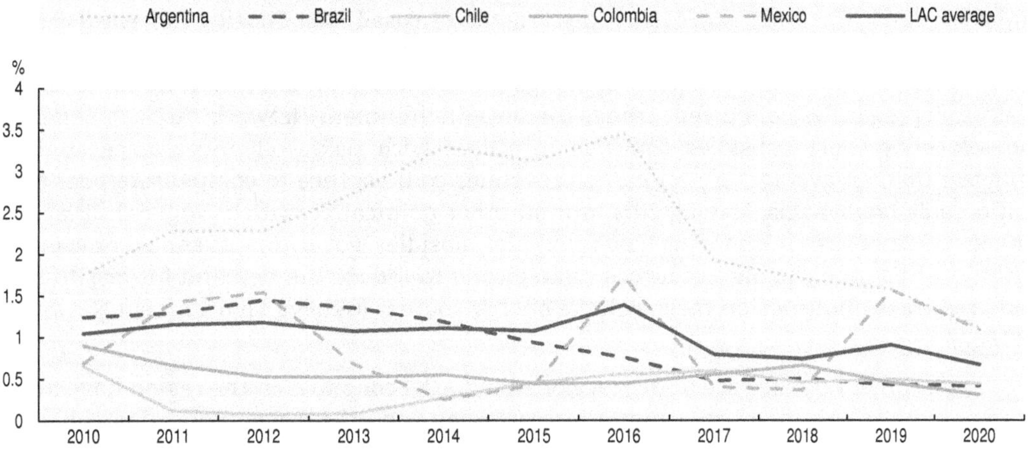

Note: Data include measures benefitting producers or consumers collectively, as both are measures that do not increase current fossil fuel production or consumption but may do so in the future. Examples of General Services Support Estimate (GSSE) measures would include public support for industry-specific infrastructure development (e.g. public support for construction of coal or natural-gas terminals) and government funding for sector-wide R&D related to fossil fuel exploration and transformation. LAC average reflects five countries: Argentina, Brazil, Chile, Colombia and Mexico.

Source: Authors' calculations based on (OECD.Stat, 2020[22]).

StatLink *https://stat.link/mk0tdc*

Understanding the real impact of fossil fuel subsidies and support measures on the most vulnerable should be a priority. Instruments such as direct conditional and unconditional cash transfers offer governments more targeted and cost-efficient approaches to help lower-income households. Even when oil prices remain elevated, governments should shift to such targeted measures, noting that to ensure effective targeting, such a shift will require improvements to existing transfer and social welfare systems (OECD, 2022[24]).

Taking into account the impacts that phasing out these subsidies will have on the most vulnerable populations, as well as the need for gradual reform, is essential to contain potential political backlash. It is important that countries start limiting these types of interventions that dampen incentives to reduce fossil-based energy use. They should instead focus more on building capacity to better address household vulnerabilities to price shocks and accelerate development of alternative sources of energy (OECD, 2022[24]). Countries should also carry out *ex ante* assessments of the impacts of phasing out subsidies on different segments of the population and provide compensatory measures to mitigate any negative effects. Rather than quickly eradicating subsidies, governments should streamline the reform through a systematic transition that includes: ministerial co-ordination; building trust gradually through socialisation; promoting active government and stakeholder participation; gaining clarity on potential winners and losers through macroeconomic modelling exercises; better targeting support; and communicating changes to all stakeholders (Chapter 5). A step-by-step change in fiscal policy, developed through joint formulation with stakeholders, legitimises the new policy and prioritises medium- and long-term benefits over and above the negative side effects it may have (Coady et al., 2010[25]).

Any changes to the subsidies and support measures should take into account the current context. Disclosing information on the subsidies through a transparent and clear roadmap could also be key in assessing the benefits and costs for all stakeholders (Coady et al., 2015[26]). The LAC region is experiencing high energy prices, rising inflation (Chapter 1) and social protests. While the common practice has been to implement such reforms only when the international price of fossil fuels does not show an imminent upward trend (Coady et al., 2015[26]), the Friends of Fossil Fuel Subsidy Reform (FFFSR), a group of 45 World Trade Organization (WTO) member countries, expressed the need to superimpose long-term environmental goals on short-term ones by working towards reforming regardless of current trends (Geneva Environment Network, 2022[27]). Benefits to reforming these subsidies continue to accrue with rising oil prices since studies suggest that, on average, 30% of carbon revenues could suffice to compensate poor and vulnerable households, leaving 70% to fund other political priorities (Vogt-Schilb et al., 2019[28]). Rather than introducing new fossil fuel subsidies, governments should now more than ever use their public resources strategically to reduce the demand for fossil fuels and reduce dependency on these volatile energy sources (Geneva Environment Network, 2022[27]).

Environmental taxes are underdeveloped in LAC. Economies in the region have been slow to implement them, but environmental taxes and price-based policy instruments are playing an increasing role. These types of policies include price signals to guide consumer decisions, for example by encouraging businesses and households to consider the environmental costs of their behaviour. On average in LAC, environmentally related tax revenues amounted to 1.0% of GDP in 2020, lower than the estimated OECD average of 2.0% of GDP (Figure 4.2) (OECD et al., 2022[29]). In 2020, revenues from energy taxes (most commonly excises from diesel and petrol) generated the highest share of total environmentally related tax revenues (65.5%), followed by revenues from motor vehicle and transport services (32.5%). There is strong heterogeneity in the region, as environmentally related tax revenues ranged from 0.1% of GDP in Belize to 2.8% in Guyana (OECD et al., 2022[29]). Most LAC economies do not levy an explicit carbon tax; fuel excise taxes are the most common form of energy tax, while electricity is sometimes also taxed (OECD et al., 2021[17]). These types of taxes should help modify consumer behaviour towards cleaner energies and transportation modes, thus playing a role in the green transition. As the transition advances, the amount levied by environmental taxes should be reduced.

Enhancing carbon pricing could generate multiple benefits in the LAC region. It provides an incentive for private actors to take production or consumption decisions consistent with global goals to limit climate change and limit health damage from local pollution. Depending on its design features, carbon pricing can also increase revenues that can be used to finance green public-sector investment and guarantee a green and just transition (OECD, 2021[23]). In a net-zero emission scenario, for hydrocarbon producers such as Brazil, Colombia and Mexico, implementing carbon taxes could generate additional revenues and may alleviate some of the projected fall in hydrocarbon revenues (Chapter 1). However, these revenues could be small for countries that are relatively low consumers of energy, such as Bolivia, Ecuador, and Trinidad and Tobago (Titelman et al., 2022[30]).

Figure 4.2. **Environmentally related tax revenue in LAC countries, by main tax base, 2020**

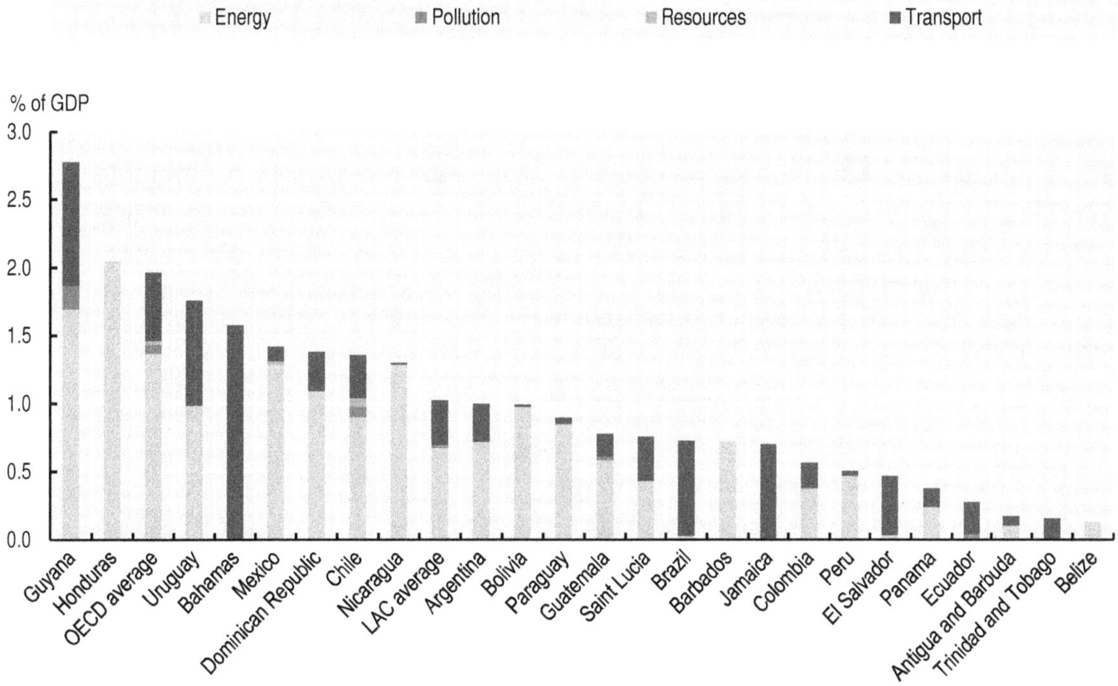

Note: The LAC average represents the unweighted average of 24 LAC countries included in this publication and excludes Cuba, Costa Rica and Venezuela due to data availability issues. The figure does not include Jamaica's revenues from the special consumption tax on petroleum products (estimated to be more than 2.0% of GDP in 2018) (OECD, 2021) as the data were not available. Chile, Colombia and Mexico are also part of the OECD.

Source: (OECD et al., 2022[29]).

StatLink *https://stat.link/ftj342*

Carbon pricing can also have negative effects, if certain conditions are not met. If clean and inexpensive alternatives are not readily available, implementing carbon pricing is likely to increase costs for households (IEA, 2021[2]). Moreover, even if some countries can generate higher revenues from carbon pricing schemes, these can be offset by the need for higher expenditure to cushion the impacts on consumers (section below on protection schemes) (Titelman et al., 2022[30]). The Helsinki Principles of the Coalition of Finance Ministers for Climate Action is a step in the right direction to enhance carbon pricing. These principles call, above all, for progress towards implementing carbon pricing mechanisms, reducing subsidies that are detrimental to the fight against climate change, and better monitoring of climate finance by governments and financial systems (Bárcena et al., 2020[31]).

As part of their decarbonisation strategies, some LAC countries have started introducing carbon pricing instruments, including either some kind of carbon tax or an ETS. In the region, efficient fossil fuel pricing mechanisms can raise substantial revenues up to around 2% of GDP (Parry, Black and Vernon, 2021[36]) (Chapter 6). Among the carbon pricing instruments, ETS are the most cost-effective market-based instruments in generating incentives to reduce emissions. They consist of an emission cap set by the government in specific industries, through which entities covered are allowed to trade emission permits (IEA, 2020[34]). They also facilitate emissions reduction as polluters for whom it is hard to reduce emissions can buy allowances from polluters who can minimise them at lower costs (OECD, 2022[35]). Argentina, Chile, Colombia and Mexico have already introduced carbon taxes. In 2020, Mexico became the first country in the region to introduce an ETS pilot programme (World Bank, 2022[32]). While covering around 40% of national emissions, the new mitigation policy instrument will give participants time to become familiar with the carbon market, letting regulators test its overall design and sending an initial carbon price signal through the economy (Castro, Vogt-Schilb and Santikarn, 2020[33]). Colombia, Brazil and Chile have set a timeline to develop an ETS in the next few years (World Bank, 2022[32]). More recently, hybrid systems with elements of both carbon taxes and ETS have emerged as the most effective way to meet decarbonisation goals (IEA, 2020[34]).

Green policies need to be accompanied by protection schemes for vulnerable households to limit negative impacts of the green transition and climate change

The effects of climate change, together with some of the undesired consequences of the green transition and its policies, will further expose the most vulnerable, highlighting the need for compensation and mitigation schemes. These can include cash and in-kind transfers, complemented by active labour market policies (ALMPs) and self-employment and entrepreneurship programmes. These are all essential tools to cushion the social costs of decarbonisation, limit the negative impacts on vulnerable households and communities, and allow for a progressive strengthening of countries' social protection systems.

Phasing out fossil fuel subsidies should be done gradually under a wider transition policies package, with good communication mechanisms and accompanied by other compensation and mitigation measures to contain backlash and backsliding. For example, in the case of Indonesia's reform of electricity subsidies and fuel pricing, complementary measures of increased funding for social assistance programmes and infrastructure projects have been key in containing backlash and guaranteeing equity (D'Arcangelo, F. et al., 2022[37]). To ameliorate the negative impacts on the poor and to increase public acceptability of reform, a robust mix of both mitigation and adaptation measures, together with public information campaigns that highlight rationales and benefits should be considered. Above all, to increase transparency, it is important to highlight how the consequent substantial budget savings could be reallocated to increase welfare. Reallocation could include increased spending and benefits in sectors such as infrastructure, rural development, health, education and agriculture (D'Arcangelo, F. et al., 2022[37]). Similarly, existing social protection mechanisms should be bolstered with the additional resources (Chapter 1).

Complementary transition support policies should also include helping households cope with energy price hikes through the development of capabilities to adapt to a world of higher energy costs. It is key that governments enable support mechanisms, such as technological change and the development of public transport, that help households develop the capabilities to reduce their energy consumption levels and their energy bills. Income support policies stand out as a good transitional option that protects households to a certain degree without blurring energy price signals, given tax pricing. However, more progressive tax systems alone cannot solve the issue of the negative distributional consequences of carbon taxes and the phasing out of fossil fuels subsidies. Thus, more targeted mechanisms, including social transfers on the basis of constrained energy consumption requirements, need to be developed using carbon tax revenues (Chancel and Ilse, 2014[38]).

Compensation policies should be complemented by ALMPs and self-employment and entrepreneurship programmes. This would facilitate relocation and retraining of workers, promote decent work in rural areas, offer new business models and support displaced workers (Chapter 3) (IDB/ILO, 2020[39]). During the transition to a net-zero economy, on average across LAC, brown industries may experience job losses of up to 13.3% of employment compared to the business-as-usual scenario (Chapter 3). However, new employment opportunities will more than offset these losses by creating up to 15% more jobs in green industries, which normally employ more workers (Chapter 3).

Most countries in the LAC region have cash transfer programmes that are much more cost-effective than energy subsidies and do not provide incentives for increased fossil fuel use (IDB, 2021[40]). During the COVID-19 pandemic, countries were forced to develop demand policies swiftly, mainly through non-conditional cash transfers and other innovative measures to provide rapid support to public health systems, households and firms. Leveraging this infrastructure and the newly developed targeted transfer mechanisms becomes essential at a time when it is important that compensatory aid reaches the most vulnerable populations (Chapter 1) (OECD et al., 2021[17]).

Although the cost is lower than energy subsidies, badly designed cash transfer programmes can have some drawbacks. A major fiscal problem is that they can foster informality if they are, in effect, conditional on lacking a formal job (IDB, 2021[40]). Moreover, once implemented, they can become permanent and represent a fiscal cost year after year. Additionally, their targeting is imperfect, and they may reach some non-poor households, which increases their cost. Last, even though they have multiple social benefits (IDB, 2021[40]), they do not always cover marginalised populations and households that do not meet the requirements (e.g. when the benefits of low energy prices reach many households).

Tapping into additional resources by scaling up novel debt tools to finance the transition

Green, social, sustainability, and sustainability-linked (GSSS) bonds

As the LAC region faces a tight fiscal space, there is a need to promote and scale up innovative financial instruments to mitigate climate change and achieve sustainable development. On the one hand, the development and strengthening of GSSS bonds increasingly consolidates a capital markets-based approach to sustainable finance that contributes to increased private sector financing. On the other hand, scaling up other mechanisms (e.g. debt-for-nature swaps, catastrophe bonds and catastrophe clauses) is also key to addressing important challenges such as fiscal volatility, stabilising budgets and the need to build up large budget reserves.

Novel financial instruments, such as GSSS bonds, are already playing an important role in the international debt market. They are incentivising greater participation of governments, corporates and supranational institutions in the debt market. They are also effectively fostering the mobilisation of resources towards sustainable projects in the region. In particular, there are two types of structures in the sustainability debt market: use of proceeds and target-linked. As defined by the International Capital Market Association (ICMA), green, social and sustainability bonds belong to the first type of structure. These are fixed-income instruments whose proceeds are exclusively applied to finance or refinance, entirely or partially, environmental and social projects or a combination of both. In the case of sustainability-linked bonds (SLBs) they are target linked, and proceeds are used for general purposes. They differ from green, social and sustainability bonds in that they are target-linked and allow financing outside of specific projects or use of proceeds categories. SLBs are also more easily tracked through the assessment of key performance indicators (KPI). For SLBs, the issuers choose the associated targets they want to achieve through the issuance of the bond, accruing additional payments to bondholders should they not meet the set target (Núñez, Velloso and Da Silva, 2022[42]; ICMA, 2022[41]).

Even though sovereign GSSS bond issuance remains a small part of total external debt, it can play a key role in driving the corporate sustainability debt market. Sovereign issuers that have the capacity can serve as role models for other types of issuers to generate sustainability standards that are harmonised and have monitoring mechanisms. Sovereign bonds can also broaden and diversify the investor base of issuances. Different stakeholders such as development banks, central banks, bond market associations, structuring consultants, and NGOs are already coming together to support the various stages of issuance, including the design of the specific frameworks (Núñez, Velloso and Da Silva, 2022[42]).

The region's GSSS bond issuances in international markets have grown substantially since 2015. The GSSS market reached an accumulated USD 73 billion from 2014 to September 2021 of which green bond issuance accounted for USD 31 billion alone followed by social with USD 17 billion. In particular, about 36 distinct LAC issuers accessed the sustainable finance market in 2021, with LAC GSSS bond issuances in international markets reaching a record high of USD 39 billion in the first nine months of the year (Figure 4.3) (Núñez, Velloso and Da Silva, 2022[42]).

Figure 4.3. **Total LAC GSSS bond issuance in international markets, by type of instrument**

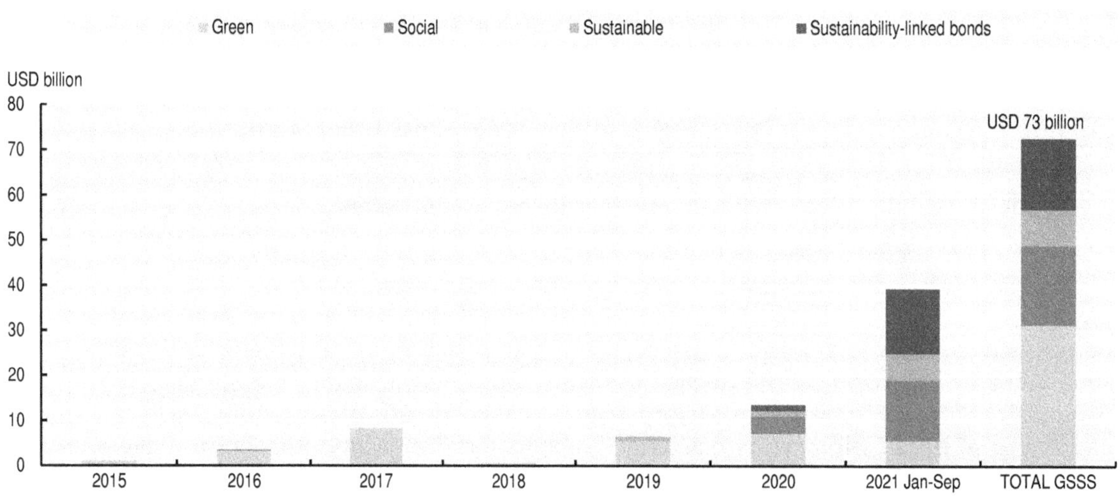

Source: (Núñez, Velloso and Da Silva, 2022[42]).

StatLink *https://stat.link/8bw9ph*

In the early stages, green bonds were the predominant tool, but social, sustainable and sustainability-linked bonds have recently been more common. Since the issuance of the first ten-year green bond of USD 204 million by the Peruvian company *Energía Eólica* in 2014, green bonds became the most used instruments in the region. In 2020, the energy and transport sectors together accounted for 79% of green bond allocations, which went to renewable energy and sustainable mobility projects (ECLAC, 2022[21]). Although green bonds accounted for the largest share (43%) of the region's total GSSS bond issuance in international markets in the December 2014 to September 2021 period, there was a significant growth of GSSS bonds in the first nine months of 2021. As a result of the market's expansion and the onset of the COVID-19 pandemic, attention has moved from an environment-only focus to a broader perspective that includes addressing social and sustainability concerns. Within GSSS bond in 2021, whereas green bonds accounted for only 14%, SLBs accounted for 37%, the largest share, followed by social bonds (34%) (Núñez, Velloso and Da Silva, 2022[42]). This follows the incremental growth of GSSS bonds in emerging markets and developing economies (EMDEs) in 2021, largely driven by issuance in LAC, which accounted for 66% of the total (IFC/Amundi, 2022[43]).

The region's GSSS bond issuances come mainly from ten countries[2] and two supranational entities, the Development Bank of Latin America (CAF) and the Central American Bank for Economic Integration (CABEI). Chile was the region's largest GSSS bond market at USD 20.9 billion in cumulative issuance in 2021, followed by Brazil (USD 11.1 billion) and Mexico (USD 7.8 billion) (Núñez, Velloso and Da Silva, 2022[42]; ECLAC, 2022[44]). While most of Chile's GSSS bond issuance comes from the sovereign sector, Brazilian and Mexican GSSS bond issuances are mostly from the corporate sector (Núñez, Velloso and Da Silva, 2022[42]). The role of the latter has become increasingly significant in the region, leading the GSSS bond volumes between December 2014 to September 2021 with a share of 42% of the total LAC GSSS bond issuance, while sovereign represented 37%, quasi-sovereign 12% and supranational issuers 5% (Figure 4.4) (Núñez, Velloso and Da Silva, 2022[42]).

Figure 4.4. **Total LAC GSSS bond issuance in international markets, by type of issuer, December 2014 to September 2021**

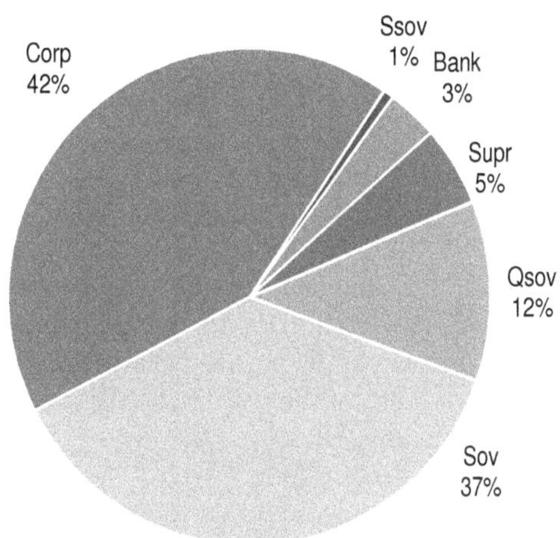

Note: Sov = sovereign. Corp = corporate. Ssov = sub-sovereign (states, cities and provinces). Supr = supranational. Qsov = quasi-sovereign. Quasi-sovereign issuers are defined as companies with full or partial government ownership or control, and supranational issuers as entities formed by two or more central governments to promote economic development for the member countries. The "bank" category refers to commercial banks. Other non-bank financial institutions are included in corporates.

Source: (Núñez, Velloso and Da Silva, 2022[42]).

StatLink *https://stat.link/anshz6*

LAC SLBs issued in international bond markets reached USD 17 billion in 2021, all from the corporate sector (Núñez, Velloso and Da Silva, 2022[42]; ECLAC, 2022[44]). South America and the Caribbean currently represent approximately 10% of the global market (Environmental Finance Data, 2022[45]). LAC market leaders when it comes to SLB issuance in international bond markets are Brazil (54%), Mexico (32%) and Chile (8%) (Núñez, Velloso and Da Silva, 2022[42]). Overall, SLB issuances have been dominated by corporates in LAC; however, sovereigns are starting to use these instruments. Chile issued the world's first sovereign SLB in March 2022, a USD 2 billion issuance, with the two key performance indicators being GHG emissions reduction and scale-up of power generation (S&P Global, 2022[46]; BNP Paribas, 2022[47]).

In a COVID-19 context, SLBs appear as an alternative that can help countries tackle economic, social and environmental challenges in a holistic way. SLBs can also be a promising innovation to sharpen investors' focus on supporting the transition strategies of entire companies. SLBs can further develop the key role that debt markets can play in funding and encouraging companies that contribute to sustainability (Núñez, Velloso and Da Silva, 2022[42]; ICMA, 2022[41]). Supporting the expansion and fine-tuning of this kind of instrument offers the opportunity to redirect capital flows towards projects that advance climate change mitigation and adaptation while also addressing social and sustainable dimensions to ensure a green and just transition.

To continue scaling up debt market instruments, governments in the region must work on innovative approaches, for instance issuing green bonds in local currency or fostering digital and technological advances (Box 4.1). The latter can increase debt markets' transparency and make capital much more traceable. While sovereign green bonds can foster investment for the energy transition, leveraging private finance, they also need financial backup from an enhanced fiscal space. Therefore, a fiscal reform, together with deeper changes in the global financial architecture, needs to be orchestrated. Blockchain technology can play a part in solving some of the problems associated with conventional bonds. This involves the ability of various stakeholders to monitor the flow of money, obtain or provide updates on the development status in real time, or demonstrate the impacts of GSSS bonds. A blockchain-supported bond issuance platform can be used to digitalise the whole bond issuing process, which can allow for the establishment of transparent nodes for its effective supervision (Chen and Volz, 2021[48]). Other key supporting instruments to develop the sustainable bond market include strengthening the role of local and external reviewers and second opinion providers.

Overall, scaling up GSSS and promoting a capital markets-based approach to sustainable finance – coupled with comprehensive frameworks that improve its effectiveness, transparency, comparability and credibility – can contribute to raising vast resources in the region (section below on sustainable finance frameworks) (Núñez, Velloso and Da Silva, 2022[42]). In the case of LAC countries, meeting the growing demand for quality public services and infrastructure in a tight fiscal context while making them sustainable and climate friendly will require catalysing other sources of financing, especially from the private sector (section below on mobilising private investment flows) (Núñez, Velloso and Da Silva, 2022[42]).

Box 4.1. Innovative financing tools to enhance local markets and advance in the digital transformation of the region

Upscaling the issuance of sovereign green bonds in the local market

In 2021, Colombia became the first emerging economy to issue a sovereign green bond in local currency in its domestic market (TES Verdes). Due to its innovative nature, it received the "green bond of the year" award from *Environmental Finance* (Environmental Finance, 2022[49]). To achieve it, the government carried out co-ordinated work between the Ministry of Finance and Public Credit and the National Planning Department, together with other public-sector entities. This process also received technical support from the World Bank and the Inter-American Development Bank (IDB). The first portfolio of eligible green expenditures amounted to COP 2.3 billion (Colombian peso), distributed across 27 projects and 6 categories. Of these, 40% are focused on water management, 27% on the transition of transport towards a cleaner and more sustainable system, 16% on the protection of diversity, and 14% on the transition to non-conventional and renewable energies. The remainder was distributed over waste and circular economy, and sustainable agricultural production.

In addition, the project portfolio was found to be robust in terms of environmental impact and ESG risk management. The total amount issued in 2021 was COP 1.49 trillion, with an initial issuance of COP 750 billion in September. One-third of the issuance (i.e. COP 250 billion) was added thanks to the activation of 50% over-allotment clauses, as there was significant demand for the bond equivalent to 4.61 times the amount initially offered. These clauses worked as a provision that granted the right to sell investors more shares than initially planned. The second

Box 4.1. **Innovative financing tools to enhance local markets and advance in the digital transformation of the region** *(cont.)*

auction (in October for COP 650 billion) saw demand from local and foreign investors for 1.5 times the amount initially offered. This type of investment allows the country to provide resources for initiatives with a high socio-environmental impact, thereby strengthening its capacity to respond to unexpected climate and environment events. This type of initiative also facilitates the arrival of new investors that see in these issuances' benefits related to transparency and, in turn, compliance with the guidelines on investments with ESG standards (Ministerio de Hacienda y Crédito Público, 2022[50]).

Developing digital crowdfunding platforms

In line with the advances in digital transformation in the region, public and private stakeholders can use digital technology as an instrument to mobilise small amounts of domestic savings for sustainable infrastructure investment. Municipalities, along with private finance-sector investors, can come together to develop a digital crowdfunding platform that entails responsible blockchain-based project bonds. The platform can be used to raise finance, while the blockchain is able to record transparently and to certify the use of proceeds, the sustainability impact and the revenue streams of the project.

Blockchain technology continues to attract great attention from financial institutions, energy companies, technical developers, national governments and academia. The Pathfinder Initiative think-tank, along with the Government of Bangladesh, developed an initiative that envisages transforming micro savers into micro investors and reducing the need for international borrowing, using blockchain as a technical backbone to improve the accountability of the funds and returning the dividends from infrastructure investment to the citizens. Scaling this up could be key to mobilising much-needed investment in sustainable and green infrastructure (Chen and Volz, 2021[48]). Nevertheless, as blockchain is an emerging technology still under development, it is important to consider its downfalls regarding security issues and its potential for high energy consumption. Consideration should be given to addressing these downfalls as these technologies are developed and implemented.

Debt-for-nature swaps

Debt-for-nature swaps can also play an important role in providing additional resources to face the fiscal challenges of the green transition. These swaps can step up as voluntary transactions, where an amount of debt owed by a developing country in the region is cancelled or reduced by a creditor in exchange for the debtor making financial commitments to conservation. For richly biodiverse and highly indebted countries, climate or nature swaps can serve as important sustainable financing instruments. Argentina, Colombia and Ecuador, three heavily indebted countries, have put forward requests for considering these types of initiatives (Arauz, Larrea and Ramos, 2022[51]). These swaps can result in win-win situations, where countries are allowed to both protect the environment and contribute to face their fiscal challenges. The benefits of redirecting resources towards sustainable infrastructure and green investments are far greater than the costs, as it can avoid the worst impacts of climate change and generate economic, social, fiscal and environmental benefits.

Debt-for-nature swaps have already been used in the region, and new proposals are being drawn up. In the case of Peru, debt swap operations resulted in USD 881.5 million of foreign debt being cancelled between 1992 and 2015, half of which was linked to debt-for-nature swaps, mobilising approximately USD 115 million for conservation.

More recently, a nature swap has been designed between the People's Republic of China (hereafter, 'China') and Ecuador. This proposal considers an Ecuadorian debt exchange with China for conservation through a project with the potential to reduce deforestation by 47% in ten years. This could save 200 000 ha of Amazon rainforest and avoid the emission of 117 million tonnes of carbon dioxide (Mt CO_2) in exchange for a reduction of USD 440 million of debt (Arauz, Larrea and Ramos, 2022[51]). Debt renegotiations between the two countries are ongoing, taking lessons from the Amazon Fund, an initiative between Brazil and Norway that prompted a decline in deforestation in the Amazon in 2005-12 (Birdsall, Savedoff and Seymour, 2014[52]).

To scale up this instrument in the present, countries and financial institutions can work on reducing transaction costs. Legal fees and environmental expertise to structure the debt deal significantly increase the costs. A memorandum of understanding (MoU) detailing how the broad parameters of the debt swap will be translated into revised lending agreements can reduce the amount of bilateral back-and-forth needed to agree on the general aspects of debt suspension before addressing the country-specific technical terms and conditions. This type of MoU can help address issues of scale and coverage, and significantly reduce transaction costs (Steele and Pate, 2020[53]).

CAT bonds

Catastrophe (CAT) bonds are another tool that could help finance the transition. This type of financial instrument provides insurance against losses caused by natural disasters. Additionally, CAT bonds transfer natural disaster risks to global capital markets and thus help governments manage fiscal volatility, stabilise budgets and ease the need to build up large budget reserves (World Bank, 2021[54]). This tradable financial instrument promises to be one of the most innovative disaster insurance mechanisms (Cavallo, 2017[55]) by providing key advantages, such as pay-outs based on the severity of the events rather than on the estimates of damages. The benefits encompass pay-outs that can be made quickly and without much contention as soon as catastrophes strike, allowing governments to provide emergency relief before official assistance arrives. Governments and multilateral institutions in the region could subsidise the necessary research associated with calculating the likelihood of natural disasters and the related costs of these events to help face this challenge and grow the market (IDB, 2017[56]).

CAT bonds can be a useful tool for countries with a higher risk of being struck by natural disasters and are, therefore, more vulnerable to default. Economies vulnerable to default due to natural disasters have rising default risk because of potentially higher fiscal needs in the case of extreme natural events. Consequently, they have higher costs of capital in financial markets and therefore must sell their debt at lower prices with higher yields. CAT bonds can help countries in the region reverse this equation, especially in the Caribbean, the most indebted region in the world, where 50% of the increase in debt is attributable to natural disasters (Persaud, 2022[57]). CAT bonds can enable these governments to increase their external borrowing from around 30% to more than 60% of GDP, providing welfare gains equivalent to several percentage points of consumption (IDB, 2017[56]). In 2021, the World Bank issued a USD 185 million cat bond for Jamaica that addresses its financing gap by securing coverage for three hurricane seasons. By entering into an insurance-like risk transfer agreement, Jamaica will be able to receive the funds it needs if future storms exceed pre-defined intensity thresholds (World Bank, 2021[58]). Work is underway to develop a CAT bond for the Caribbean region with the support of the World Bank, with four countries currently engaged and with the possibility of four others to join (Evans, 2022[59]).

Natural disaster clauses

Natural disaster clauses can also be an innovative financing instrument to link countries' repayment capacity to their risk exposure. For instance, hurricane clauses were included as part of a comprehensive restructuring of Grenada's public debt (2013-15) and significantly helped reduce public debt levels from 94% of GDP in 2013 to 56% of GDP in 2019. Hurricane clauses were also part of the debt restructuring (postponement of interest payments) in Barbados (2018-19). These types of clauses are designed to provide cash flow relief at crucial moments when countries experience high needs in terms of financing, and new financial sources might be limited. Embedding hurricane-linked clauses in debt contracts can help LAC countries tap into extended maturity periods in the event of a natural disaster by allowing them to defer either interest or principal payments (or both) for a defined period (ECLAC, 2021[60]).

Disaster clauses may carry a cost, however, as they incentivise governments to borrow more and pay higher yields (Malucci, 2020[61]). Introducing disaster clauses entails investors being subjected to the risk of delayed repayments and the need to be compensated accordingly through higher spreads. However, recent analyses suggest that borrowing terms generally improve with the introduction of these clauses (Malucci, 2020[61]). Study cases evidence how governments take advantage of the better borrowing terms and expand their borrowing up until the point that default risk reaches levels like those observed in the economy without disaster clauses. Since governments expect to postpone repayments, even when spreads increase due to delay risk, the expected cost of servicing debt declines. Thus, with the introduction of disaster clauses, governments tolerate higher spreads, as they have little impact on the overall borrowing costs (Malucci, 2020[61]).

Designing sustainable financial strategies to support and guide the green transition

Finance ministries have the important role of developing fiscal frameworks that promote and protect green investments

As a fundamental tool for the transition, LAC economies need to develop fiscal frameworks that promote and protect investments, especially green investments, from unsustainable economic and political inertia. In fiscal consolidations, governments find it easier to reduce capital expenditure more than current expenditure (public consumption), despite its potential impact on long-term growth. As composition matters in a fiscal consolidation, when the ratio of public investment declines relative to public consumption, it can produce negative long-term effects on growth. For instance, a consolidation of 1% of GDP reduces output by about 0.5% during fiscal consolidation. This figure increases cumulatively to 0.7% in the three years following the start of consolidation. Protecting investment during the fiscal consolidation can mitigate the economic contraction and, in some cases, lead to an expansion. As a result, the region must protect investment through fiscal frameworks where fiscal rules can be an important tool (Ardanaz et al., 2022[62]; Ardanaz et al., 2021[63]).

Even if there is regional heterogeneity in their implementation, fiscal rules are widely used in LAC. Most rules are concentrated in expenditure or budget balance rules, with some economies using a combination of these. For instance, Argentina, the Bahamas, Brazil, Colombia and Costa Rica are currently implementing expenditure rules that aim to set a limit on total, primary or current government expenditures. Similarly, the Bahamas, Brazil, Chile, Colombia, Costa Rica and Peru are using budget balance rules. Last, the Bahamas and Peru are using the debt rule. Some economies base their budget balance rules on fiscal balances that consider the (structural) cycle. Some economies also have clauses that aim to protect investment. For instance, Costa Rica included a golden rule in its framework according to which borrowing can be used only to finance investment spending (Hamid et al., 2022[64]).

• Going forward, fiscal rules must have the flexibility to protect green investment and green golden rules could be considered, especially during fiscal consolidation events. LAC will need to mobilise further resources to finance the transition, while at the same time, some economies of the region will be undergoing a fiscal consolidation process. In this context, a green golden rule (which allows for green investment to be funded by deficits that would not count in the fiscal rules) can be a useful tool to ensure that the adjustment composition does not affect the green transition and that, just like fiscal rules, it sets predictable and consistent capital spending over time (Pekanov and Schratzenstaller, 2020[65]; Ardanaz et al., 2022[62]). To be flexible enough in accommodating exogenous shocks to protect investment, fiscal rules can include cyclically adjusted fiscal targets, well-defined escape clauses and differential treatment of investment expenditures. The importance of flexible fiscal rules is evidenced by the fact that economies with no fiscal rules or rigid ones can reduce by 10% their public investment in a 2% of GDP fiscal consolidation. By contrast, in countries with flexible fiscal rules, fiscal consolidation does not affect investment (Ardanaz et al., 2021[66]). Although the positive effects of protecting investment are well documented, the impact of green investment requires further analysis as, in many cases, the investment would be to replace existing brown infrastructure rather than adding new and clean needed infrastructure (Guntram and Zsolt, 2022[67]).

Other action areas for finance ministries include developing green budgeting and channelling public investment into environmentally beneficial projects

Evaluating the environmental impacts of budgetary and fiscal policies is essential for governments to achieve their national and international environmental goals. Green budgeting involves the use of budgetary policy-making tools to assess the environmental impact of budgetary and fiscal policies. This can be an important tool for finance ministries to improve the alignment of national expenditure and revenue with climate and other environmental goals (OECD, 2017[68]). Green budgeting requires that LAC countries effectively incorporate environmental dimensions into their fiscal frameworks (including the annual and multiannual budget documents), evaluate tax and expenditure policies, and consolidate long-term sustainability analyses (OECD, 2017[68]). Additionally, green budgeting can substantially strengthen green accountability mechanisms, allowing for greater transparency, and thus foster the citizen trust needed to support any future green fiscal reform or climate policy.

Another important role of finance ministries is to channel public investment towards projects with greater environmental benefits. One of the tools to make this possible are lowering social discount rates (SDRs) used to evaluate these projects. SDRs are a type of interest rate applied to specific projects that bring benefits in the future and costs in the present. In the context of climate change policy making, SDRs estimate how much today's society should invest in trying to limit the future impacts of climate change (Grantham Research Institute, 2018[69]). For instance, the Peruvian Ministry of Economy and Finance uses a general SDR of 8% to evaluate public investment projects and a 4% SDR for projects that provide environmental services that reduce or mitigate GHG emissions (Bárcena et al., 2020[31]). Applying differentiated rates to low-carbon projects from the outset, when project comparisons are made, is a promising way to shift relative returns in favour of these types of projects. Thus, these rates could be complemented by measures such as efficiency standards and the tax or non-tax price of carbon (Bárcena et al., 2020[31]). In cost-benefit analyses, studies indicate that keeping SDRs low in the long term supports future generations' well-being, since environmentally important investments, particularly those related to climate change, have inter-temporal and inter-generational equity effects (Bárcena et al., 2020[31]). The importance of environmental sustainability has led several LAC countries to consider climate change criteria, such as the social cost of carbon, when evaluating public investment projects (Bárcena et al., 2020[31]). The Economic Commission for Latin America and the Caribbean (ECLAC), the EUROCLIMA+ programme and the Network of the National Public Investment Systems in Latin America and the Caribbean are currently implementing a regional initiative to promote the use of the social cost of carbon in the evaluation of public investment.

The role of national and sub-national development banks is key in mobilising resources to drive the transition

National development banks (NDBs) can have a significant impact in mobilising resources to drive and support a sustainable recovery. They can provide technical and financial support for the design of climate and financial strategies. In addition, they can promote regulatory and institutional reform plans to help realign both public and private flows, whether domestic or international, with climate change and sustainability goals (Galindo, Hoffman and Vogt-Schilb, 2022[5]). Through this realignment, they could help address both short- and long-term needs, setting market signals and, when required and feasible, leveraging private-sector financing for sustainable projects. As of February 2021, NDBs in LAC committed to the equivalent of USD 90 billion in financial support to confront the effects of COVID-19 in the region, which includes laying the groundwork for a climate-smart and resilient recovery. This groundwork includes NDBs helping to foster counter-cyclical finance and financial inclusion and promoting environmental sustainability, particularly by combating climate change (Griffith Jones et al., 2021[70]).

NDBs can also help mobilise and leverage private flows towards low-carbon and climate-resilient pathways, specifically through de-risking and credit enhancement. Credit financing for small and medium-sized enterprises can also foster private-sector participation in new green projects. NDBs can serve as the principal interface between the public and private sectors and are in a unique position to promote economic or social development by financing activities with social returns. They have already been instrumental over the last decade in helping governments green their economies and have developed innovative financial solutions for micro, small and medium enterprises and for infrastructure projects with terms tailored to the financial profile for low-carbon investments. These investments typically require longer tenors, lower (or at least not higher) interest rates, flexible amortisation profiles and alternative collateral approaches (including non-/limited recourse financing). However, most LAC economies are market based, and private capital therefore dominates the financial landscape. Thus, shifting the trillions in private-sector financing into sustainable infrastructure is key to meeting the investment challenge. Many NDBs in the region have taken a leadership role in this domain based on their public development mandates. However, NDBs remain to date an underused conduit for the mobilisation of commercial capital and the intermediation of international climate finance (OECD, 2020[71]). Finance ministries can be key actors in helping governments align policies and regulations with such investments (Griffith Jones et al., 2021[70]).

Sub-national development banks (SDBs) can also provide value added within the financing chain. Since less than 10% of international climate finance is currently allocated to local investments, SDBs can help provide an efficient response to close the subnational financial gap (Finance in Common, 2021[4]). They can also offer a wide range of benefits at the local level such as providing a large panel of financial instruments to unlock and broaden the sources of funding directly or through commercial banks, offer local currency financing and pave the way for the development of stronger subnational financial markets in the long term, especially for intermediary cities (Finance in Common, 2021[4]). Following the Finance in Common Summit in 2021, the Alliance of Subnational Development Banks in Latin America and the Caribbean was launched, a major step forward to strengthen financial flows for local and regional governments. One of the main objectives is to align their strategies, norms, standards, investments and portfolios with the 2030 Agenda and its SDGs and the Paris Agreement. This represents a major opportunity to increase the engagement of SDBs in sustainable investments to boost urban and municipal financial markets. It also presents a window of opportunity to help build future low-carbon and climate-resilient cities and territories, and to provide equal access to high-quality services for all (Finance in Common, 2021[4]).

Climate-related development finance can play an important role in the green transition

Climate-related development finance should increase its role in expanding investments that guide the green transition in the LAC region. Total climate-related development finance from bilateral (OECD Development Assistance Committee [DAC] member countries),[3] multilateral (MDBs and other multilateral funds), and private donor sources in LAC reached USD 17 billion in 2020 (OECD, 2020[78]). MDBs committed most of these resources reaching a total of USD 11.8 billion (69.4%) while DAC members committed USD 4.4 billion (26%), other multilateral funds USD 694.6 million (4.1%), and private donors USD 85.6 million (0.5%) (Figure 4.5). The main financial instruments used by MDBs were debt instruments (77.4%) followed by grants (22.4%), and equity and shares in collective investment vehicles (0.2%). DAC members used grants the most (97.8%) followed by debt instruments (1.8%) and equity and shares (0.3%). Other multilateral funds used grants the most (91.1%), followed by debt instruments (8.0%) and mezzanine finance instruments (0.9%). Private donors' most-used instrument was grants (57%) followed by debt instruments (43%) (OECD, 2020[78]). Prior to 2017, the share of DAC countries and MDBs in the resource's commitment was almost equal; MDBs have since taken the lead and become key players.

Figure 4.5. **Climate-related development finance to LAC by provider type, 2010-20**

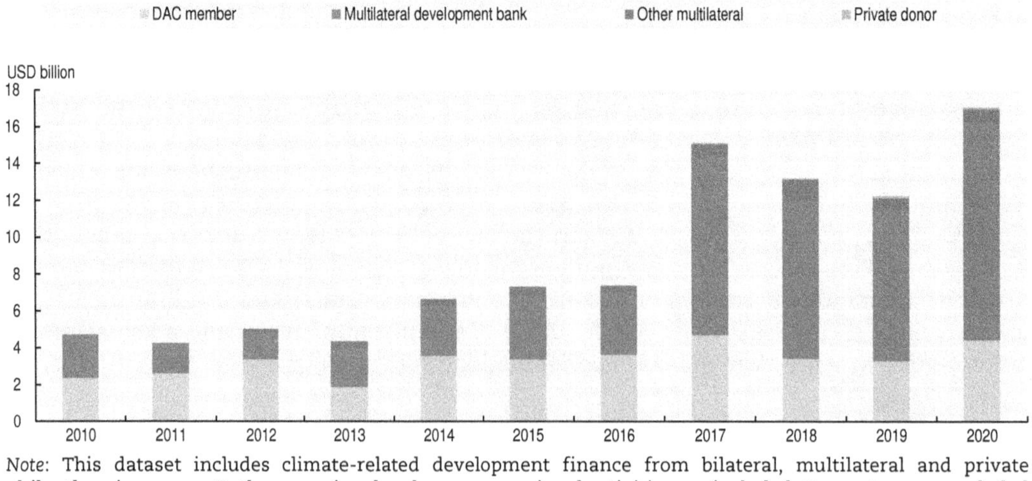

Note: This dataset includes climate-related development finance from bilateral, multilateral and private philanthropic sources. Both concessional and non-concessional activities are included. Guarantees are excluded because they are categorised as non-flow operations. The total amount of climate-related development finance for each activity is indicated in the field "Climate-related development finance - Commitment". The total amount of climate-related development finance corresponds to the sum of the values of mitigation and of adaptation, minus the Overlap value. For World Bank, Climate Adaptation and Mitigation percentages (%) reflect the share of financial resources committed by the World Bank at Board Approval in support of activities eligible for climate mitigation and/or adaptation finance as per the Joint MDB methodologies for tracking climate change finance. Providers who committed less than USD 120 million total during 2010-20 were not included.

Source: (OECD, 2020[78]).

StatLink *https://stat.link/l1f3us*

MDBs can structure projects that attract more private lenders while also offering the right protection mechanisms to borrowers. These multilateral funds are instrumental in unlocking additional resources for various projects (see section on blended finance). For example, evidence suggests that, for every USD 1.00 invested in climate action projects in LAC by the IDB, an additional USD 2.60 were leveraged from external climate finance, additional donors, and public and private sources (Viguri et al., 2020[73]).

Fostering partnerships and co-ordination between MDBs and development agencies is crucial to ramping up green financing (CAF, 2021[74]). Effective co-ordination with national development agencies can also increase the flow of resources aimed at combatting climate change in the region. For instance, since 2009, the French Development Agency (*Agence française de développement* [AFD]) has committed EUR 11 billion and, in 2020, allocated EUR 2 billion, along with its private-sector arm, Proparco. Not only does its budget support green projects in the region to access loans and credit lines, but 70% of its co-operation projects also focus on the environment, including topics of energy transition, transition towards green taxation and water recycling (AfD, 2022[75]).

Tapping into the resources of other multilateral institutions, such as climate funds will also be key if countries are to achieve their climate mitigation and adaptation objectives. Multilateral climate funds enable support via funds provided mainly by developed countries targeting a variety of activities (e.g. adaptation, mitigation, REDD, capacity-building) (OECD, 2022[76]). These funds are also one way in which developed countries are distributing the climate finance they committed to at the 2009 UN conference in Copenhagen, "where by 2020 they would be jointly mobilising USD 100 billion a year to help developing countries tackle climate change" (Carbon Brief, 2017[72]).

The Green Climate Fund committed the highest amount of resources for the LAC region with USD 538.5 million in 2020, followed by other funds such as the Global Environment Facility General Trust Fund (GEF), the Global Green Growth Institute, and the Climate Investment Funds - Clean Technology Fund (OECD, 2020[78]). More recently, the Forest Carbon Partnership Facility (FCPF) also took a leading role by paying Costa Rica USD 16.4 million for reducing 3.28 million tons of carbon emissions ($MtCO_2$) during 2018 and 2019 – making it the first LAC economy to receive payments from a World Bank trust fund. Costa Rica is on track to unlock up to USD 60 million for reducing up to 12 $MtCO_2$ by the end of 2025 (World Bank, 2022[77]). Countries must continue ramping up efforts to tap into these funds by developing greater capacity-building and project preparation skills as well as enhancing their use of the National Adaption Plan (NAP) process.

Further expansion of DAC members' development assistance is also crucial at a moment when this aid lags in some key priority areas, such as climate adaptation (OECD, 2020[78]). Over the past decade, this assistance has almost doubled from 2.3 billion in 2010 to 4.4 billion in 2020 (Figure 4.5). In 2020, total climate-related development assistance for mitigation dominated climate-related financing efforts with 56% of the share compared to 34% to climate change adaptation (the remaining 10% corresponds to the overlap of both) (OECD, 2020[78]). On average from 2010 to 2020, adaptation support continued to be less than half of mitigation development assistance. However, in 2021 the OECD DAC Declaration evidenced efforts to increase development assistance for adaptation from developed countries to developing countries, offering a new approach to align development co-operation with the goals of the Paris Agreement. DAC countries committed to use their resources as well as mobilise others to help developing countries access more technical opportunities to enable and accelerate a clean, sustainable and just energy transition (OECD, 2021[79]).

A shift from the traditional development assistance model to blended finance can multiply the impact of development finance

Blended finance is an innovative tool that can mobilise private resources to finance the green transition. The mechanism's main objective is to attract commercial capital to projects that benefit society while also providing financial returns to investors. Blended finance is defined as *the strategic use of development finance for the mobilisation of additional finance towards sustainable development in developing countries* (OECD, 2018[80]). In other words, development resources are used as to improve the risk-return profile of individual investments to attract commercial private financing, demonstrate project viability and build markets that ultimately can attract further commercial capital for development (OECD, 2018[80]). In LAC, since there are insufficient regulatory mechanisms and misperceptions of risk hinder access to finance, especially for SMEs, blended finance could offer a way out. It can also address issues such as financing for renewable mega-projects as well as problems of access to transmission infrastructure since it often involves a high cost of financing.

One way to determine the amount of blended finance that is being deployed in LAC, is by observing the mobilisation of private climate finance by DFIs and MDBs. Despite the benefits of this mobilisation, it has been decreasing in LAC, contrary to other regions. While Africa has experienced a steady increase in the mobilisation of private climate finance by DFIs and MDBs since 2014, this has been decreasing since 2017 for the LAC region (Figure 4.6) (OECD, 2020[71]). In order to revert this trend continued collaboration between MDBs and NDBs is required, as well as with policy makers and the private sector.

Figure 4.6. **Amounts mobilised from the private sector by official development finance interventions, 2014-20**

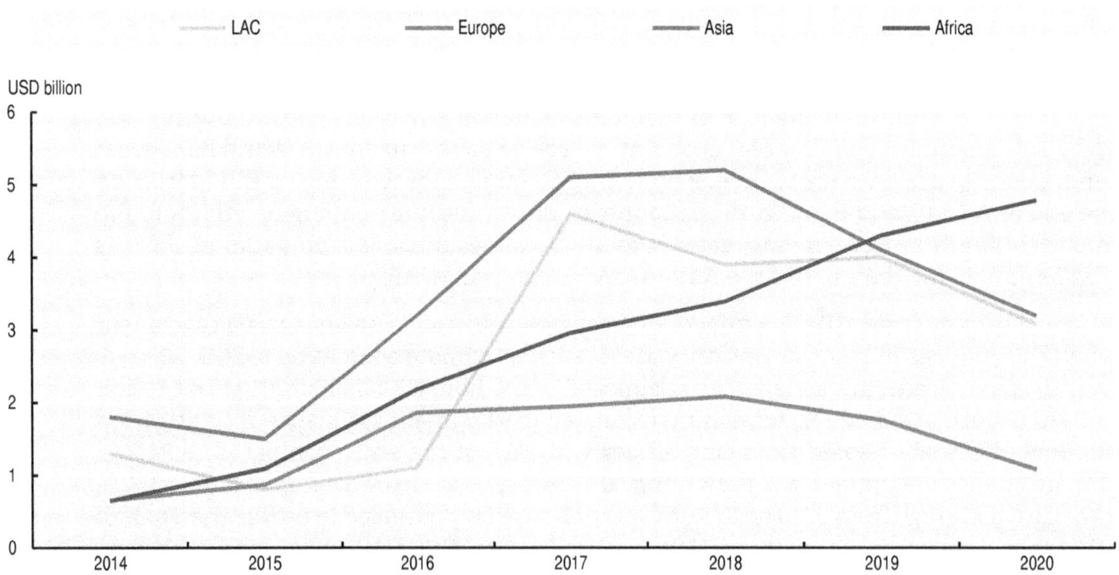

Note: The term "mobilisation" describes the causal link between private finance made available for a specific project and an official intervention. Data are collected following instrument-specific methodologies, covering all leveraging mechanisms used by Development Finance Institutions (DFIs) and Multilateral Development Banks (MDBs): guarantees, syndicated loans, project finance schemes, shares in collective investment vehicles, direct investment in companies, credit lines and simple co-financing.

Source: (OECD.Stat, 2020[82]).

StatLink *https://stat.link/ihcyfv*

The collaborations of private investors with regional development financial institutions can be instrumental to enhance the reach of sustainable investment to markets where it is yet underdeveloped. Since there is a need to spread the risk burden, these types of partnerships are essential to partially de-risk investments in frontier economies with very shallow capital markets (OECD, 2022[13]). While MDBs enable the largest share of private-sector investment mobilisation through dedicated private-sector operations, a wide array of actors is also engaging in blended finance, including foundations, philanthropic investors, commercial actors, institutional investors, commercial banks, private equity and venture capital funds, hedge funds, and companies (OECD, 2021[81]). Development banks and development finance institutions (DFIs) play a critical role in the deployment of the needed instruments and structuring mechanisms (OECD, 2021[81]).

The public sector is uniquely positioned to mobilise the private investment flows essential for the transition

The public sector and key stakeholders are fundamental to mobilise further private funds for the green transition. LAC has attracted the highest share of private-sector financing for clean energies compared to, for example, sub-Saharan Africa and Southeast Asia. It is crucial for the region to build upon this experience and generate certainty for current and future investments by continuing to strengthen the development of policies, green finance, and regulations that create the adequate conditions (IEA, 2021[2]).

Governments should support overcoming common risks and barriers to private investment. In energy projects, this includes developing stable, clear and non-retroactive regulations and policies. There is also a need to improve the transparency of processes and timelines for issuing licences and permits to develop projects. Governments should improve the quality and inclusiveness of projects' previous informed consent processes, when required. Similarly, it is essential to enhance local administrative capacity in terms of land acquisition and local content requirements for project approvals. In some markets with enabling reforms, developing new business models is key to attracting private financing that can help bridge existing investment gaps (IEA, 2021[2]). Public authorities can also help redirect private investment towards climate solutions through concrete regulatory reforms. These include auctions to back the deployment of solar and wind power generation, for example, or effective tax schemes that incentivise renewable energy transport or reforms in bidding processes for bus services (Beltrán et al., 2021[83]). Governments and finance ministries can enable market structures that improve the participation of private actors and increase the role of competition and transparent pricing in the energy sector. Over time, these actions tend to support investment in clean energies (IEA, 2021[84]).

Governments should foster and prepare the groundwork for the adoption of responsible banking mechanisms throughout the private banking sector. The role of private banks is vital, as they have the main responsibility for redirecting private flows towards a carbon-neutral scenario. Under the newly installed responsible banking mechanisms, private banks have many ways to reduce climate impacts through their activities. Among the most important channels are: increasing lending to renewable energy projects and public electromobility; eliminating financing for fossil fuel-based projects; issuing SLBs or green bonds where corporate borrowers receive better credit and loans (with lower interest rates) if, for example, they meet climate-related targets; advising and helping investors find and finance green projects; helping green businesses raise funds and pursue deals; and helping corporations find and adopt new technologies that help reduce CO_2 emissions (Santander, 2021[85]). However, a more systemic way to redirect private finance towards the energy transition entails the incorporation of climate-related financial risks in their daily operations, as it is for their own interest to mitigate and manage these risks (see sub-section on corporate sustainability standards: approaches such as that of the Task Force on Climate-Related Financial Disclosure and the International Accounting Standards Board).

Improving sustainable finance frameworks is crucial to enable and scale up green investments

Sustainable finance frameworks are essential to maintain and increase public and private green investments. These are regulatory guidelines developed by both public and private finance-sector stakeholders that serve to reduce and manage environmental, social and governance (ESG) risks from financial activities, and encourage the flow of capital to assets, projects, sectors and businesses that have environmental, climate and social benefits (SBFN/IFC, 2021[86]). These guidelines are essential to achieve transparency, reduce transaction costs, and facilitate private and public sector flows. They are also conducive to an increase and more efficient allocation of resources to sustainable projects. Given that the transition will mainly be financed by the private sector, regulatory tools (e.g. sustainability standards or taxonomies) will also be essential for investors to meet regulatory requirements, reduce risks exposure and avoid greenwashing.

There has been a progressive increase in implementing sustainable finance frameworks in LAC. Since 2010, the number of frameworks has almost doubled with governments playing a leading role in the formulation of green and sustainable finance protocols. Overall, the non-banking sector has had the lowest participation, which highlights the need for LAC countries to deepen and broaden implementation of sustainable finance frameworks in the rest of the financial ecosystem. For instance, Colombia's framework extended its coverage from the banking sector to the pensions, capital markets and asset management sectors. Broadening the understanding of national sustainable finance roadmaps is essential to recognising the diversity of approaches and starting points, including the potential leadership role of all parts of the financial sector (SBFN/IFC, 2021[86]).

The Sustainable Banking and Finance Framework (SBFN) and the International Finance Corporation (IFC) of the World Bank Group have developed a methodology to assess developing countries' progress on sustainable finance frameworks. The SBFN, the first global network of its kind focused on sustainable finance at the market level, represents 43 developing countries and USD 43 trillion (86%) of the total banking assets in emerging markets (SBFN/IFC, 2021[86]). Its methodology consists of assessing three main pillars: 1) ESG integration, which refers to the management of ESG risks in the governance, operations, lending and investment activities of financial institutions; 2) climate risk management, which refers to new governance, risk management and disclosure practices that financial institutions can use to mitigate and adapt to climate change; and 3) financing sustainability, which refers to initiatives by regulators and financial institutions to unlock capital flows for activities that support climate, green economy and social goals. This pillar includes the development of new products and initiatives, such as green bonds, sustainability-linked loans, definitions, guidance, taxonomies, monitoring and incentives (SBFN/IFC, 2021[86]).

Most LAC countries are at a formulation or early implementation stage of financial frameworks, with a setback in climate risk management. Brazil, Colombia and Mexico are leading efforts in these areas. Through the formulation of agreements among the banking, non-banking and government sectors, these three countries are either at an advanced stage of implementation or a maturing stage of consolidation regarding at least two of the three pillars of financial sustainability. The main challenge remains in climate risk management as the region is one of the most prone to increasingly frequent natural disasters. It is therefore critical that countries work to enhance this pillar by requiring financial institutions to identify, measure and report the exposure of the sectors that are most vulnerable to transition and physical risks (SBFN/IFC, 2021[86]).

Given that the transition in the LAC region will mostly be financed by the private sector, it is key to develop a variety of mitigation and adaptation policies to ensure investments flow into assets aligned with transition objectives. To do this, it is essential to develop regulatory tools (e.g. sustainability, green bond standards, or taxonomies) within national sustainable finance frameworks. These tools promote transparency, comparability and credibility for investors while helping to avoid greenwashing. As the number of investors wanting to participate in the debt markets is increasing, these tools have become extremely necessary so that investors effectively meet disclosure requirements for financial and non-financial undertakings.

Countries in the region should continue to develop clear green bond and corporate sustainability standards. These are voluntary, usually third party-assessed, norms and standards relating to environmental, social, and governance issues adopted by concerned stakeholders (e.g. producers, traders, manufacturers, retailers or service providers) and used to qualify their performance in achieving sustainability-related targets. These can foster investor protection and avoid greenwashing of bonds, products or services marketed as sustainable. Governments must also ensure that both green and corporate sustainability standards are aligned with international practice:

- First, regarding green bond standards, the European Union Green Bond Standard can serve as a good example for LAC to continue consolidating a harmonised standard that all the countries in the region can follow. This could guarantee the development of quantifiable green targets with which issuers can ultimately reach climate neutrality by 2050 with reduced transaction costs (EU Green Bond Standard Working Group, 2019[87]).
- Second, regarding corporate sustainability standards, approaches such as that of the Task Force on Climate-Related Financial Disclosure and the International Accounting Standards Board are key to expanding efforts in standardising climate-related financial disclosures. These have been widely used as a reference by companies in their ESG standards (Núñez, Velloso and Da Silva, 2022[42]). In particular, the creation of the International Sustainability Standards Board, launched at COP26 in November 2021 and signed by three Latin American countries (Brazil, Mexico and Uruguay), was a first step towards harmonising sustainable methodologies with a benchmark that all countries can follow. The initiative focuses on developing a global sustainability baseline to meet investor needs for information on company ESG strategies (including physical and transition risks), which have an impact on the value of their business. This would address investor demand for globally comparable sustainability information that is consistent with financial statements.
- While the implementation of voluntary standards has been the common practice, in the last few years, mandatory standards have become increasingly important amongst large companies and bond issuers. However, it is important that governments consider that mandatory standards as a condition for issuance could discourage some issuers (Núñez, Velloso and Da Silva, 2022[42]). Mandatory standards should be imposed gradually and on pace with the development and strengthening of capacities in the local sustainable finance ecosystem.

Regulatory tools and classification systems such as green, transition or sustainable taxonomies, can also increase transparency and comparability for the private sector's financial activities. These can help identify sectors associated to specific portfolios that are directly exposed to physical and transition climate risks (Box 4.2). For instance, private banks' internal climate change risk taxonomies enable them to track their activities, support product development, avoid greenwashing and reinforce their transparency and commitment to promote and increase their green, social and sustainability-linked activity (Santander, 2021[85]). These mechanisms also help banks decarbonise their portfolios and reduce climate-related risk, including via new criteria that prohibit financing and advising on new upstream oil clients (except for transactions for financing renewable energy) and direct financing of upstream oil greenfield projects (Santander, 2021[85]).

Working to harmonise such taxonomies across the region is also key to fostering certainty, credibility integrity and transparency in the market. This could enable further mobilisation of capital aligned to countries' environmental goals. To achieve this, the Working Group on Sustainable Finance Taxonomies in Latin America and the Caribbean[4] is developing a common regional framework of sustainable finance taxonomies. As there are at least six taxonomies under development or implemented in LAC, the initiative seeks to support policy makers, financial regulators and supervisors, banks, investors, and international organisation specialists to understand and build a common language and a science-based definition of what sustainable finance is – and what it is not (UNDP, 2022[88]). This would improve interoperability with global taxonomies and reduce transaction costs to interested investors while fostering participatory processes and collaborative work across financial stakeholders.

Box 4.2. **Green, transition and sustainable taxonomies in LAC**

The increasing ambition of national and international climate targets, including in response to the already severe impacts of climate change in many regions, helped drive the creation of sustainable and green finance frameworks and standards. While the most focus to date has been on green finance and defining demonstrably low-emission activities as part of green taxonomies, increasing attention is now being paid to transition finance and the need to provide more capital to enable high-emitting activities to progressively shift or transition to lower emissions while avoiding emission lock-in. Three core eligibility criteria for transition finance can be distilled from currently existing approaches: (i) substitutability (absence of a zero or near-zero alternative), (ii) a commitment by the borrower/issuer to a low-emissions trajectory; and (iii) avoiding lock-in of emissions (Tandon, 2021[89]). However, differences in existing transition finance approaches developed to date can fragment markets, reduce investor confidence and create greenwashing risks. To address these gaps and ensure environmental integrity, the OECD is currently developing Guidance on Transition Finance, with a focus on corporate climate transition plans (OECD, 2022[90]).

While there is currently no specific transition taxonomy in LAC, a few countries have developed or are in the process of developing sustainable or green finance taxonomies, including Brazil, Colombia, Chile, the Dominican Republic and Mexico. The Colombian government is leading the way with the publication of its green taxonomy in April 2022 (Responsible Investor, 2022[91]). The taxonomy seeks to facilitate the identification of projects with environmental objectives, develop capital markets, and promote the effective mobilisation of private and public resources (Government of Colombia, 2022[92]).

In other countries, various initiatives to implement a national taxonomy are under development (Table 4.2). In Chile, the Climate Bonds Initiative (CBI), along with Chile's Ministry of Finance, the Green Finance Public-Private Roundtable and the IDB, has created a roadmap for a national taxonomy, focusing on key economic sectors, including the high-emitting sectors of construction, energy, transport and mining (Climate Bonds Initiative, 2021[93]). In Mexico, a comprehensive effort involving many bilateral and multilateral actors is supporting the Mexican Committee on

Box 4.2. **Green, transition and sustainable taxonomies in LAC** *(cont.)*

Sustainable Finance (composed of the Ministry of Finance, the Central Bank [Banxico] and the financial regulatory commissions) in the development of a national taxonomy. Many international organisations, including ECLAC, the World Bank and bilateral aid agencies are supporting the Mexican Ministry of Finance in the development of a national taxonomy. *The Deutsche Gesellschaft für Internationale Zusammenarbeit* (GIZ) worked both with the Mexican banks association on a green taxonomy for the financial system (Muller et al., 2020[94]) and with the Ministry of Economy in Brazil on the development of a green taxonomy (GIZ, 2018[95]). In Peru, the Climate & Company think tank, in partnership with the Peruvian Ministry of Environment and the GIZ, is exploring the benefits of a national sustainable taxonomy based on the World Bank's recommendations (Climate & Company, 2022[96]). The IFC signed an MOU with the Securities Superintendence of the Dominican Republic and the Ministry of the Environment, which is expected to lead to the development of a consolidated taxonomy (Climate Bonds Initiative, 2021[93]).

Table 4.1. **Overview: Selected LAC countries with taxonomy initiatives**

Country	Chile	Colombia	Dominican Republic	Mexico	Peru	Brazil
Stage of taxonomy development	Roadmap proposed	Initial documents published	Under development	Under development	Research underway	Under development
Participants	CBI, IDB, Ministry of Finance	Government ministries	IFC, government ministries	GIZ, AfD, ECLAC, IFC, GGGI, UK PACT, Bank of Mexico, World Bank, Ministry of Finance	Ministry of Environment, GIZ	Ministry of Finance, GIZ

Sources: (Government of Colombia, 2022[92]); (Climate Bonds Initiative, 2021[93]); (Climate & Company, 2022[96]); (IFC, 2021[97]); (GIZ, 2018[95]).

It is important that LAC countries push for further implementation and integration of these frameworks and tools to enable more effective and transparent flow of sustainable finance, especially from the private sector. Advancing towards more mature, sustainable finance ecosystems in the region can contribute to generating more consistent and comparable data about sustainable finance implementation by financial institutions on both the risk management and opportunity sides. It also contributes to better reporting requirements for the purposes of regulation and supervision and allow investors and stakeholders to better understand the sustainability performance of financial institutions and companies (SBFN/IFC, 2021[86]). With a view to promoting sustainable finance frameworks that are anchored in the region's digital transformation, they can also help increase digitisation in the region by recognising the role of telecommunication networks and favouring their financing. Countries that act late to strengthen their frameworks and tools risk putting their financial institutions at a disadvantage. Lack of alignment with other markets may create costs and inefficiencies for cross-border sustainable finance activities. The expansion and deepening of these frameworks throughout the finance sector can serve as a foundation for increasing competitiveness, investment opportunities and impact (SBFN/IFC, 2021[86]).

Key policy messages

Under a tight fiscal space, LAC countries must develop and scale up sustainable fiscal frameworks and financial strategies that can effectively guide a green and just transition.

To finance the green transition, LAC economies need to ramp up investment and mobilise further resources from both public and private sources. To achieve a net-zero emissions economy, increased spending on clean energy and energy efficiency will be essential. To mobilise the necessary funds, the correct incentives must be put in place, fossil fuel subsidies phased-out, and further resources levied through environmental taxes and ETS. During this process, and because of, climate change policies must also aim to ensure a just transition by compensating the most vulnerable. To mobilise further resources, the upscaling of debt instruments such as GSSS bonds, debt-for-nature swaps, CAT bonds, and natural disaster clauses will be crucial. Investments in the green transition must be channelled and protected by fiscal frameworks and green budgeting. In addition, sub-national, national and international development institutions must be supported since they play a crucial role in further mobilising public and private resources. Finally, to maintain transparency and avoid green washing on green investments, finance frameworks will be essential through several regulatory tools (e.g. sustainability and green bonds standards and taxonomies).

Climate-related development finance from bilateral, multilateral and donor sources also play a key role in increasing investment in projects with environmental benefits. Here, tapping into the growing resources of multilateral climate funds – which include funds provided mainly by developed countries – is also key to achieving environmental goals.

Box 4.3. **Key policy messages**

Encourage environmentally sustainable fiscal policies

- Promote more and better spending on clean energy and energy efficiency:
  - Increase investment in technologies for electricity generation and storage, including electricity grids and battery storage, to accommodate higher electricity demand and the surge in renewables deployment.
  - Invest in and improve energy efficiency since it is the cheapest and most immediate way to reduce the use of fossil fuels. Investing in digitalised energy systems will also boost efficiency and serve to build a more inclusive and just energy transition in the long term.
  - Develop pipelines/plans for low-carbon infrastructure projects to ensure available private and public investments for renewable energy reach sustainable projects. Such pipelines entail developing investment-ready and bankable projects that investors and project developers can trust and back.
- Better align environmentally related taxes and subsidies with the cost of pollution:
  - Rationalise and phase out unjustified fossil fuel subsidies, particularly to the most affluent population, to free additional revenues to finance the transition and avoid perverse incentives for fossil fuel use.
  - Increase revenues from environmental taxes. These are crucial as they include price signals that aim to factor into consumer decisions and encourage businesses and households to modify behaviour. The revenues of these taxes will start declining when consumers modify their behaviour towards cleaner energies and transportation modes.
  - Work further in the consolidation of carbon pricing instruments that generate an ambitious climate policy and include hybrid systems with elements of both carbon taxes and ETS.

Box 4.3. **Key policy messages** *(cont.)*

- Develop compensation mechanisms for vulnerable households affected negatively by climate reform policies:
  - Further help vulnerable households cope with energy price hikes through complementary transition support policies that enable them to adapt to a world of higher energy costs.
  - Leverage already developed targeted transfer infrastructure to make sure compensatory aid reaches the most vulnerable populations.
- Promote and scale-up debt tools that can help raise additional revenues to ensure flows of resources target climate action, comply with climate spending targets, and build sustainable and economically viable portfolios:
  - Scale up GSSS bonds by enhancing the domestic debt market, working towards the issuance of bonds in local currency and fostering digital and technological advances.
  - Support the expansion and improvement of SLBs, as they offer an opportunity to redirect capital flows towards projects that address climate change mitigation and adaptation while accounting for social and sustainable dimensions to ensure an inclusive and just transition.
  - Participate in debt-for-nature swaps that can further finance the green transition by releasing debt burdens and increasing spending on environment-oriented projects. This mechanism can be scaled up by reducing transaction costs through MoUs, which can help address issues of scale and coverage.
  - Develop tradable financial instruments, such as CAT bonds, that can help spread risk across global capital markets while swiftly providing necessary natural disaster relief resources. Increase subsidies for necessary research associated with calculating the likelihood of natural disasters and the related costs of these events.
  - Use natural disaster clauses in debt contracts to link the country's repayment capacity to its risk exposure. These are key instruments in debt contracts, allowing countries to defer either interest or principal payments (or both) for a defined period.

Advance sustainable financial strategies to guide the transition

- Implement fiscal frameworks that protect green investment:
  - Develop tools, such as fiscal rules, that include green golden rules to protect green investment from economic and political cycles.
  - Apply green budgeting as an important tool for finance ministries and financial regulators to improve the alignment of national expenditure and revenue processes with climate and other environmental goals. These tools help evaluate the environmental impacts of budgetary and fiscal policies to achieve national and international environmental goals.
  - Channel public investment towards projects with greater environmental benefits by using lower Social Discount Rates (SDRs) to evaluate projects.
- Deliver the necessary support for NDBs and SDBs to enhance the mobilisation of public and private resources toward sustainable projects:
  - Optimise NDBs to help mobilise and leverage private flows towards low-carbon and climate-resilient pathways through de-risking and credit enhancement.
  - Back SDBs to help unlock and broaden local sources of funding and develop stronger subnational financial markets in the long-term, especially for intermediary cities.

Box 4.3. **Key policy messages** *(cont.)*

- Help further expand climate-related development finance to foster higher investments that guide green transition:
  - Support MDBs in structuring projects that attract more private lenders and offer the right protection mechanisms to borrowers.
  - Promote partnerships and co-ordination between MDBs and development agencies.
  - Tap into the growing resources of multilateral climate funds by developing greater capacity-building and project preparation skills as well as enhancing the use of the National Adaptation Plan (NAP) process.
- Advance the blended finance agenda. This requires continued collaboration between international and domestic development banks, as well as with policy makers and the private sector.
- Support the public sector in mobilising private sector resources:
  - Develop concrete regulatory reforms and enable market structures that improve participation of private actors while also increasing the role of competition and transparent pricing in the energy sector.
  - Further promote responsible banking mechanisms so that private banks reduce climate impacts through their business activities and redirect private flows towards a carbon-neutral scenario.
- Support the improvement and expansion of sustainable finance frameworks:
  - Expand regulatory tools (e.g. sustainability or green bond standards and green, sustainable or transition taxonomies) to increase the flow of private and public sustainable investments and avoid greenwashing.
  - Promote public-private co-operation to harmonise standards and taxonomies across the region to avoid inefficiencies for cross-border sustainable finance activities.
  - Support the broadening of sustainable finance frameworks in the non-banking sector (e.g. pensions, capital markets and asset management sectors) to boost competitiveness and investment opportunities.
  - Advance transition finance taxonomies alongside corporations in order to provide more capital to enable high-emitting activities to progressively shift or transition to lower emissions.

Notes

1. Following the Paris Agreement, to keep global warming to no more than 1.5°C, emissions need to be reduced by 45% by 2030 and reach net zero by 2050.
2. Argentina, Brazil, Colombia, Costa Rica, Ecuador, Guatemala, Mexico, Panama, Paraguay and Peru.
3. As of September 2022, there are 30 members of the DAC, including the European Union which acts as a full member of the committee. In addition, there are "Participants" and "Observers". The listed Participants at this time are: Azerbaijan, Bulgaria, Kuwait, Qatar, Romania, Saudi Arabia and the United Arab Emirates. The Observers are: World Bank, the IMF, UNDP, the African Development Bank, the Asian Development Bank, and the Inter-American Development Bank.
4. The LAC Taxonomy Working Group is an initiative of the Interagency Technical Committee of the Forum of Ministers of Environment of Latin America and the Caribbean and is constituted by the United Nations Environment Programme, the ECLAC, the United Nations Development Programme, the World Bank Group, the IDB, CAF – Development Bank of Latin America and the Food and Agriculture Organization of the United Nations. The LAC Taxonomy Working Group is financially supported by the European Commission through the EUROCLIMA+ programme.

References

AfD (2022), *AFD and Latin America: Combating Climate Change and Inequalities*, French Development Agency, Paris, https://www.afd.fr/en/page-region-pays/latin-america. [75]

Arauz, A., C. Larrea and J. Ramos (2022), *Propuesta para un canje de deuda por naturaleza con China*, Centro de Estudios China-México, Facultad de Economía, Universidad Nacional Autónoma de México, Mexico City, https://dusselpeters.com/CECHIMEX/20220209_1_CECHIMEX_Canje_de_deuda_por_naturaleza_con_China_Arauz_Larrea_Ramos.pdf. [51]

Ardanaz, M. et al. (2022), *Los beneficios de proteger la inversión pública durante la consolidación fiscal*, Inter-American Development Bank, Washington, DC, https://blogs.iadb.org/gestion-fiscal/es/los-beneficios-de-proteger-la-inversion-publica-durante-la-consolidacion-fiscal/. [62]

Ardanaz, M. et al. (2021), "Growth-friendly fiscal rules? Safeguarding public investment from budget cuts through fiscal rule design", *Journal of International Money and Finance*, Elsevier, Amsterdam, https://www.sciencedirect.com/science/article/pii/S0261560620302758. [66]

Ardanaz, M. et al. (2021), "The Output Effects of Fiscal Consolidations: Does Spending Composition Matter?", *IDB Working Paper Series*, Inter-American Development Bank, Washington, DC, https://publications.iadb.org/en/output-effects-fiscal-consolidations-does-spending-composition-matter. [63]

Bárcena et al. (2020), *The climate emergency in Latin America and the Caribbean: the path ahead – resignation or action?*, Economic Commission for Latin America and the Caribbean (ECLAC), Santiago, https://repositorio.cepal.org/bitstream/handle/11362/45678/10/S1900710_en.pdf. [31]

Bárcena, A. et al. (2015), *The economics of climate change in Latin America and the Caribbean: Paradoxes and challenges of sustainable development*, Economic Commission for Latin America and the Caribbean, Santiago, https://repositorio.cepal.org/bitstream/handle/11362/37311/4/S1420655_en.pdf. [1]

Beltrán et al. (2021), *Lecciones aprendidas en la implementación de modelos de negocio para la masificación de buses eléctricos en Latinoamérica y el Caribe*, Inter-American Development Bank, Washington, DC, https://publications.iadb.org/publications/spanish/document/Lecciones-aprendidas-en-la-implementacion-de-modelos-de-negocio-para-la-masificacion-de-buses-electricos-en-Latinoamerica-y-el-Caribe.pdf. [83]

Birdsall, N., W. Savedoff and F. Seymour (2014), *The Brazil-Norway Agreement with Performance-Based Payments for Forest Conservation: Successes, Challenges, and Lessons*, Center for Global Development, Washington, DC, https://www.cgdev.org/sites/default/files/brazil-norway-agreement-performance-based-payments-forest-conservation-brief. [52]

BNP Paribas (2022), *Chile sets a trend with first sovereign sustainability-linked bond*, BNP Paribas, Paris, https://cib.bnpparibas/chile-sets-a-trend-with-first-sovereign-sustainability-linked-bond/#:~:text=Supported%20by%20the%20country%27s%20SLB,increasing%20to%2060%25%20by%202032. [47]

CAF (2021), *CAF and the World Bank to increase support for climate action and biodiversity to Latin America and the Caribbean*, Development Bank of Latin America, Caracas, https://www.caf.com/en/currently/news/2021/11/caf-and-the-world-bank-to-increase-support-for-climate-action-and-biodiversity-to-latin-america-and-the-caribbean/. [74]

Carbon Brief (2017), *Mapped: Where multilateral climate funds spend their money*, https://www.carbonbrief.org/mapped-where-multilateral-climate-funds-spend-their-money/. [72]

Castro, M., A. Vogt-Schilb and M. Santikarn (2020), *Can carbon pricing help Latin America and the Caribbean engage in a net-zero carbon future?*, https://blogs.worldbank.org/climatechange/can-carbon-pricing-help-latin-america-and-caribbean-engage-net-zero-carbon-future. [33]

Cavallo, E. (2017), *Natural Disaster Insurance: Why Don't More Countries Have It?*, Inter-American Development Bank, Washington, DC, https://blogs.iadb.org/ideas-matter/en/natural-disaster-insurance-why-dont-more-countries-have-it/. [55]

Chancel, L. and S. Ilse (2014), *Environmental taxes and equity concerns: A European perspective*, Background paper prepared for the Spring Alliance, https://www.academia.edu/5651522/Environmental_taxes_and_equity_concerns_A_European_perspective. [38]

Chen, Y. and U. Volz (2021), "Scaling up sustainable investment through blockchain-based project bonds", *Development Policy Review*, Wiley-Blackwell, Hoboken, NJ, https://onlinelibrary.wiley.com/doi/10.1111/dpr.12582. [48]

Climate & Company (2022), *Towards a Green Finance Taxonomy in Peru*, Climate & Company, Berlin, https://climateandcompany.org/consulting-services-towards-a-green-finance-taxonomy-in-peru?msclkid=1ab881a8b96811ec82dba979599cd2e0. [96]

Climate Bonds Initiative (2021), *Latin America & Caribbean: State of the market*, Climate Bonds Initiative, https://www.climatebonds.net/files/reports/cbi_lac_2020_04e.pdf. [93]

Climate Watch (2022), *Historical GHG Emissions*, World Resources Institute, Washington, DC, https://www.climatewatchdata.org/ghg-emissions. [6]

Coady, D. et al. (2010), *Petroleum Product Subsidies: Costly, Inequitable, and Rising*, IMF Staff Position Report, International Monetary Fund, Washington, DC, https://www.imf.org/external/pubs/ft/spn/2010/spn1005.pdf. [25]

Coady, D. et al. (2015), "How large are Global Energy Subsidies?", *IMF Working Paper*, No. 5814, International Monetary Fund, Washington, DC, https://www.imf.org/external/pubs/ft/wp/2015/wp15105.pdf. [26]

D'Arcangelo, F. et al. (2022), "A framework to decarbonise the economy", *OECD Economic Policy Papers*, No. 31, OECD Publishing, Paris, https://doi.org/10.1787/4e4d973d-en. [37]

ECLAC (2022), *Capital flows to Latin America and the Caribbean: 2021 year-in-review and first four months of 2022*, https://www.cepal.org/sites/default/files/news/files/22-00485_was_-_capital_flows_2021_year_in_review_web.pdf. [44]

ECLAC (2022), *How to finance sustainable development: Recovery from the effects of COVID-19 in Latin America and the Caribbean*, COVID-19 Special Report, No. 13, Economic Commission for Latin America and the Caribbean, Santiago, https://www.cepal.org/en/publications/47721-how-finance-sustainable-development-recovery-effects-covid-19-latin-america-and. [21]

ECLAC (2021), *An innovative financing for development agenda for the recovery in Latin America and the Caribbean*, COVID-19 Special Report, Economic Commission for Latin America and the Caribbean, Santiago, https://repositorio.cepal.org/bitstream/handle/11362/47490/1/S2100627_en.pdf. [60]

ECLAC (2020), *Hacia una electromovilidad pública en México*, Economic Commission for Latin America and the Caribbean, Santiago, https://www.cepal.org/es/publicaciones/46060-electromovilidad-publica-mexico. [11]

EESI (2022), *Energy Efficiency*, Environmental and Energy Study Institute, Washington, DC, https://www.eesi.org/topics/energy-efficiency/description. [12]

Environmental Finance (2022), *Environmental Finance's Bond Awards 2022*, Environmental Finance, London, https://www.environmental-finance.com/content/awards/environmental-finances-bond-awards-2022/winners/green-bond-of-the-year-sovereign-republic-of-colombia.html. [49]

Environmental Finance Data (2022), *Sustainability-linked Bonds and Loans – Key Performance Indicators (KPIs)*, Environmental Finance, London, https://www.environmental-finance.com/assets/files/research/sustainability-linked-bonds-and-loans-kpis.pdf. [45]

EU Green Bond Standard Working Group (2019), *Report on EU Green Bond Standard*, European Commission, Brussels, https://ec.europa.eu/info/sites/default/files/business_economy_euro/banking_and_finance/documents/190618-sustainable-finance-teg-report-green-bond-standard_en.pdf. [87]

Evans, S. (2022), *Caribbean regional cat bond planned with World Bank support*, https://www.artemis.bm/news/caribbean-regional-catastrophe-bond-world-bank/. [59]

FAO (2022), FAOSTAT *Emissions*, Food and Agriculture Organization, Rome, https://www.fao.org/food-agriculture-statistics/data-release/data-release-detail/en/c/1304919/. [7]

Finance in Common (2021), *Joint Declaration of the Alliance of Subnational Development Banks*, https://financeincommon.org/sites/default/files/2021-08/Declaration%20of%20the%20Alliance%20of%20Subnational%20Development%20Banks%20in%20Latin%20America%20and%20the%20Caribbean_0.pdf. [4]

Galindo, L., B. Hoffman and A. Vogt-Schilb (2022), "How Much Will It Cost to Achieve the Climate Goals in Latin America and the Caribbean?", *IDB Working Paper Series*, No. IDB-WP-01310, Inter-American Development Bank, Washington, DC, https://publications.iadb.org/publications/english/document/How-Much-Will-It-Cost-to-Achieve-the-Climate-Goals-in-Latin-America-and-the-Caribbean.pdf. [5]

Geneva Environment Network (2022), *Subsidizing Fossil Fuels in Times of High Energy Prices | Fossil Fuel Subsidy Reform in Trade and Climate Discussions*, https://www.genevaenvironmentnetwork.org/fr/evenements/subsidizing-fossil-fuels-in-times-of-high-energy-prices-fossil-fuel-subsidy-reform-in-trade-and-climate-discussions/#scroll-nav__4. [27]

GIZ (2018), *Financiamento verde e sustentável (FiBraS)*, Deutsche Gesellschaft für Internationale Zusammenarbeit, Bonn, Germany, https://www.giz.de/en/worldwide/76093.html. [95]

Government of Colombia (2022), *Taxonomía Verde de Colombia – Fase I*, Government of Colombia, Bogotá, https://www.minhacienda.gov.co/webcenter/portal/TaxonomiaVerdeColombia/pages_taxonomiavercolombia. [92]

Grantham Research Institute (2018), *What are social discount rates?*, London School of Economics, London, https://www.lse.ac.uk/granthaminstitute/explainers/what-are-social-discount-rates/#:~:text=Social%20discount%20rates%20(SDRs)%20are,climate%20change%20in%20the%20future. [69]

Griffith Jones, S. et al. (2021), *The future of national development banks and the green recovery*, Inter-American Development Bank, Washington, DC, https://blogs.iadb.org/sostenibilidad/en/the-future-of-ndbs-and-the-green-recovery/. [70]

Guntram, W. and D. Zsolt (2022), *How to reconcile increased green public investment needs with fiscal consolidation*, VoxEU, London, https://voxeu.org/article/how-reconcile-increased-green-public-investment-needs-fiscal-consolidation. [67]

Hamid et al. (2022), *Fiscal Rules Dataset: 1985-2021*, International Monetary Fund, Washington, DC, https://www.imf.org/external/datamapper/fiscalrules/map/map.htm. [64]

ICMA (2022), *Guidance Handbook January 2022*, International Capital Market Association, Zürich, Switzerland, https://www.icmagroup.org/assets/GreenSocialSustainabilityDb/The-GBP-Guidance-Handbook-January-2022.pdf. [41]

IDB (2021), *Fiscal Policy and Climate Change: Recent Experiences of Finance Ministries in Latin America and the Caribbean*, Inter-American Development Bank, Washington, DC, https://publications.iadb.org/publications/english/document/Fiscal-Policy-and-Climate-Change-Recent-Experiences-of-Finance-Ministries-in-Latin-America-and-the-Caribbean.pdf. [40]

IDB (2017), *Fact sheet: the economic impacts of natural disasters in Latin America and the Caribbean*, Inter-American Development Bank, Washington, DC, https://www.iadb.org/en/news/fact-sheet-economic-impacts-natural-disasters-latin-america-and-caribbean. [56]

IDB/ILO (2020), *Jobs in a net-zero emissions future in Latin America and the Caribbean*, Inter-American Development Bank/International Labour Organization, Washington, DC/Geneva, Switzerland, https://www.ilo.org/wcmsp5/groups/public/---americas/---ro-lima/documents/publication/wcms_752069.pdf. [39]

IEA (2022), *Distributed renewable energy and the digital transformation of energy systems - Challenges and opportunities for Latin America*, International Energy Agency, Paris, https://www.iea.org/events/distributed-renewable-energy-and-the-digital-transformation-of-energy-systems-challenges-and-opportunities-for-latin-america. [15]

IEA (2021), *Driving Down Methane Leaks from the Oil and Gas Industry*, International Energy Agency, Paris, http://www.iea.org/reports/driving-down-methane-leaks-from-the-oil-and-gas-industry. [84]

IEA (2021), *Financing Clean Energy Transitions in Emerging and Developing Economies*, International Energy Agency, Paris, https://iea.blob.core.windows.net/assets/6756ccd2-0772-4ffd-85e4-b73428ff9c72/FinancingCleanEnergyTransitionsinEMDEs_WorldEnergyInvestment2021SpecialReport.pdf. [2]

IEA (2020), *Implementing Effective Emissions Trading Systems*, https://www.iea.org/reports/implementing-effective-emissions-trading-systems. [34]

IFC (2021), *Caribbean Green Financing Frameworks*, International Finance Corporation, Washington, DC, https://disclosures.ifc.org/project-detail/AS/605795/caribbean-green-financing-frameworks. [97]

IFC/Amundi (2022), *Emerging Market Green Bonds Report 2021: Riding the Green Wave*, International Finance Corporation/Amundi Asset Management, Washington, DC/Paris, https://www.ifc.org/wps/wcm/connect/f68a35be-6b49-4a86-9d65-c02e411de48b/2022.06+-+Emerging+Market+Green+Bonds+Report+2021_VF.pdf?MOD=AJPERES&CVID=o54kW0D. [43]

IRENA (2022), *Progress has been made but more needs to be done*, International Renewable Energy Agency, Abu Dhabi, https://www.irena.org/DigitalArticles/2019/Apr/How-To-Transform-Energy-System-And-Reduce-Carbon-Emissions. [8]

IRENA (2022), *World Energy Transitions Outlook 2022: 1.5°C Pathway*, International Renewable Energy Agency, Abu Dhabi, https://www.irena.org/-/media/Files/IRENA/Agency/Publication/2022/Mar/IRENA_World_Energy_Transitions_Outlook_2022.pdf. [9]

Malucci, E. (2020), *FEDS Notes: Natural Disasters, Climate Change, and Sovereign Risk*, Board of Governors of the Federal Reserve System, Washington, DC, https://www.federalreserve.gov/econres/notes/feds-notes/natural-disasters-climate-change-and-sovereign-risk-20201218.htm. [61]

Ministerio de Hacienda y Crédito Público (2022), *Marco Fiscal del Mediano Plazo 2022*, Ministerio de Hacienda y Crédito Público, Bogotá, https://www.minhacienda.gov.co/webcenter/ShowProperty?nodeId=%2FConexionContent%2FWCC_CLUSTER-197963%2F%2FidcPrimaryFile&revision=latestreleased. [50]

Ministry of Finance and Public Credit (2022), *Acuerdo 143/2022*, https://dof.gob.mx/nota_detalle.php?codigo=5668433&fecha=14/10/2022#gsc.tab=0. [18]

Muller, A. et al. (2020), *Presentación de Taxonomía Verde para el Sistema Financiero Mexicano*, IKI Alliance Mexico Blog, International Climate Initiative (IKI) in Mexico, Berlin, https://iki-alliance.mx/presentacion-de-taxonomia-verde-para-el-sistema-financiero-mexicano/). [94]

Núñez, G., H. Velloso and F. Da Silva (2022), *Corporate governance in Latin America and the Caribbean: Using ESG debt instruments to finance sustainable investment projects*, Economic Commission for Latin America and the Caribbean, Santiago, https://repositorio.cepal.org/handle/11362/47778. [42]

OECD (2022), *Business Insights on Emerging Markets 2022*, OECD Emerging Markets Network, OECD Development Centre, Paris, http://www.oecd.org/dev/oecdemnet.htm. [13]

OECD (2022), *Climate Fund Inventory*, OECD Publishing, Paris, https://www.oecd.org/g20/topics/energy-environment-green-growth/database-climate-fund-inventory.htm#:~:text=The%20CFI%20database%20currently%20covers%2099%20climate%20funds. [76]

OECD (2022), *Emission trading systems*, OECD Publishing, Paris, https://www.oecd.org/env/tools-evaluation/emissiontradingsystems.htm. [35]

OECD (2022), *Framework for industry's net-zero transition: Developing financing solutions in emerging and developing countries*, OECD Publishing, Paris, https://www.oecd-ilibrary.org/docserver/0c5e2bac-en.pdf?expires=1663921114&id=id&accname=ocid84004878&checksum=3D35729149DF9E0B86CB35E822EBBB4D. [10]

OECD (2022), *Why governments should target support amidst high energy prices*, OECD Publishing, Paris, https://www.oecd.org/ukraine-hub/policy-responses/why-governments-should-target-support-as-energy-prices-continue-to-soar-40f44f78/. [24]

OECD (2021), *OECD DAC Declaration on a new approach to align development co-operation with the goals of the Paris Agreement on Climate Change*, OECD Publishing, Paris, https://www.oecd.org/dac/development-assistance-committee/dac-declaration-climate-change-cop26.pdf. [79]

OECD (2021), *Taxing Energy Use for Sustainable Development: Opportunities for energy tax and subsidy reform in selected developing and emerging economies*, OECD Publishing, Paris, https://www.oecd.org/tax/tax-policy/taxing-energy-use-for-sustainable-development.htm. [23]

OECD (2021), *The OECD DAC Blended Finance Guidance*, OECD Publishing, Paris, https://doi.org/10.1787/ded656b4-en. [81]

OECD (2020), *Climate-related development finance at the activity level: Recipient perspective, 2010-2020*, OECD Publishing, Paris, https://www.oecd.org/dac/financing-sustainable-development/development-finance-topics/climate-change.htm. [78]

OECD (2020), *The role of domestic DFIs in using blended finance for sustainable development and climate action – The case of Brazil*, OECD Publishing, Paris, https://www.oecd.org/dac/financing-sustainable-development/blended-finance-principles/publications/. [71]

OECD (2018), *Developing Robust Project Pipelines for Low-Carbon Infrastructure*, OECD Publishing, Paris, https://doi.org/10.1787/9789264307827-en. [3]

OECD (2018), *Making Blended Finance Work for the Sustainable Development Goals*, OECD Publishing, https://doi.org/10.1787/9789264288768-en. [80]

OECD (2017), *Paris Collaborative on Green Budgeting*, OECD Publishing, Paris, https://www.oecd.org/environment/green-budgeting/. [68]

OECD (forthcoming), *Key Messages - EMnet Meeting - LAC Roundtable Consultation*, OECD Publishing, Paris. [14]

OECD (2022), *OECD Guidance on Transition Finance: Ensuring Credibility of Corporate Climate Transition Plans*, OECD Publishing, Paris, https://doi.org/10.1787/7c68a1ee-en. [90]

OECD et al. (2022), *Revenue Statistics in Latin America and the Caribbean 2022*, OECD Publishing, Paris, https://doi.org/10.1787/58a2dc35-en-es. [29]

OECD.Stat (2020), *Fossil Fuel Support - Detailed Indicators*, OECD Publishing, Paris, https://stats.oecd.org/Index.aspx?DataSetCode=FFS_INDICATOR_DETAILED%20#. [22]

OECD.Stat (2020), *Mobilisation*, https://stats.oecd.org/Index.aspx?DataSetCode=DV_DCD_MOBILISATION. [82]

OECD et al. (2021), *Latin American Economic Outlook 2021: Working Together for a Better Recovery*, OECD Publishing, Paris, https://doi.org/10.1787/5fedabe5-en. [17]

OECD/IEA (2021), *GHG Emissions from Fuel Combustion*, OECD Publishing/International Energy Agency, Paris, https://www.oecd-ilibrary.org/energy/data/iea-co2-emissions-from-fuel-combustion statistics_co2-data-en. [98]

Parry, I., S. Black and N. Vernon (2021), "Still Not Getting Energy Prices Right: A Global and Country Update of Fossil Fuel Subsidies", *IMF Working Paper*, No. 2021, International Monetary Fund, Washington, DC, https://www.imf.org/en/Publications/WP/Issues/2021/09/23/Still-Not-Getting-Energy-Prices-Right-A-Global-and-Country-Update-of-Fossil-Fuel-Subsidies-466004. [36]

Pekanov, A. and M. Schratzenstaller (2020), *The role of fiscal rules in relation with the green economy*, Economic Governance Support Unit and Policy Department A, European Parliament, Strasbourg, France, https://www.europarl.europa.eu/RegData/etudes/STUD/2020/614524/IPOL_STU(2020)614524_EN.pdf. [65]

Persaud, A. (2022), *A comprehensive, integrated, climate finance framework for the Earth*, https://cepr.org/voxeu/columns/comprehensive-integrated-climate-finance-framework-earth. [57]

Puig, J. and L. Salinardi (2015), "Argentina y los subsidios a los servicios públicos: un estudio", *Documento de Trabajo*, No. 183, Centro de Estudios, La Plata, Argentina, https://www.cedlas.econo.unlp.edu.ar/wp/wp-content/uploads/doc_cedlas183.pdf. [19]

Rentschler, J. and M. Bazilian (2017), "Policy Monitor – Principles for Designing Effective Fossil Fuel Subsidy Reforms", *Review of Environmental Economics and Policy*, Oxford University Press, Oxford, England, https://www.journals.uchicago.edu/doi/full/10.1093/reep/rew016. [20]

Responsible Investor (2022), *Colombia launches first LatAm Green Taxonomy, excludes nuclear and gas*, Private Equity International, London, https://www.responsible-investor.com/colombia-launches-first-latam-green-taxonomy-excludes-nuclear-and-gas/?msclkid=81aa500fbbe511ecbc469e7244b2fb77. [91]

S&P Global (2022), *World's 1st sovereign sustainability linked bond issued by Chile*, S&P Global Market Intelligence, London, https://www.spglobal.com/marketintelligence/en/news-insights/latest-news-headlines/world-s-1st-sovereign-sustainability-linked-bond-issued-by-chile-69226229. [46]

Santander (2021), *Responsible banking: Consolidated Annual Financial Report 2021*, Santander, Boston, https://www.santander.com/content/dam/santander-com/en/documentos/informe-financiero-anual/2021/ifa-2021-consolidated-annual-financial-report-en.pdf#page=15. [85]

SBFN/IFC (2021), *Accelerating Sustainable Finance Together: Global Progress Report of the Sustainable Banking and Finance Network*, Sustainable Banking and Finance Network/International Finance Corporation, New York/Washington, DC, https://sbfnetwork.org/wp-content/uploads/pdfs/2021_Global_Progress_Report_Downloads/SBFN_D003_GLOBAL_Progress_Report_29_Oct_2021-03_HR.pdf. [86]

Steele, P. and S. Pate (2020), *Tackling the triple crisis: Using debt swaps to address debt, climate and nature loss post-COVID-19*, International Institute for Environment and Development, London, https://pubs.iied.org/sites/default/files/pdfs/migrate/16674IIED.pdf. [53]

Tandon, A. (2021), *Transition finance: Investigating the state of play: A stocktake of emerging approaches and financial instruments*, OECD Environment Working Papers, No. 179, OECD Publishing, Paris, https://doi.org/10.1787/68becf35-en. [89]

Titelman, D. et al. (2022), *Fiscal Impact Estimates of a Net-Zero Emissions Transit for Major Hydrocarbon Producers in Latin America and the Caribbean*, Task Force on Climate, Development and the International Monetary Fund, https://www.bu.edu/gdp/files/2022/04/TF-WP-004-FIN.pdf. [30]

UN (2018), *Accelerating SDG 7 Achievement. Policy Brief 24. Energy Sector Transformation: Decentralized Renewable Energy for Universal Energy Access*, United Nations, New York, https://sustainabledevelopment.un.org/content/documents/17589PB24.pdf. [16]

UNDP (2022), *Building a common framework of sustainable finance taxonomies in Latin America and the Caribbean*, United Nations Development Programme – Latin America and the Caribbean, New York, https://www.undp.org/latin-america/press-releases/building-common-framework-sustainable-finance-taxonomies-latin-america-and-caribbean. [88]

Viguri et al. (2020), *Analysis of the external climate finance access and implementation. GCF, GEF, CIF and PCPF projects and programs by the Inter-American Development Bank*, Inter-American Development Bank, Washington, DC, https://publications.iadb.org/publications/english/document/Analysis-of-External-Climate-Finance-Access-and-Implementation-CIF-FCPF-GCF-and-GEF-Projects-and-Programs-by-the-Inter-American-Development-Bank.pdf. [73]

Vogt-Schilb, A. et al. (2019), "Cash transfers for pro-poor carbon taxes in Latin America and the Caribbean", *Nature Sustainability* 2, pp. 941–948, https://doi.org/10.1038/s41893-019-0385-0. [28]

World Bank (2022), *Carbon Pricing Dashboard*, World Bank, Washington, DC, https://carbonpricingdashboard.worldbank.org/. [32]

World Bank (2022), *Costa Rica Receives First Emission Reductions Payment from Forest Carbon Partnership Facility*, San José, https://www.worldbank.org/en/news/press-release/2022/08/16/-costa-rica-receives-first-emission-reductions-payment-from-forest-carbon-partnership-facility#:~:text=SAN%20JOS%C3%89%2C%20August%2012%2C%202022,degradation%E2%80%94commonly%20known%20as%20REDD%. [77]

World Bank (2021), *Super-sized Catastrophe Bond for Earthquake Risk in Latin America*, World Bank, Washington, DC, https://thedocs.worldbank.org/en/doc/192341554318525877-0340022019/original/casestudyfinancialproductsPacificAlliancefinal4.1.2019.pdf. [54]

World Bank (2021), *World Bank Catastrophe Bond Provides Jamaica $185 Million in Storm Protection*, World Bank, Washington, DC, https://www.worldbank.org/en/news/press-release/2021/07/19/world-bank-catastrophe-bond-provides-jamaica-185-million-in-storm-protection. [58]

Chapter 5

How to make it possible: Governing the green transition

The recovery from COVID-19 represents a unique opportunity to advance towards a greener development model in LAC. Citizens in the region show a high level of concern for environmental issues, suggesting that the green transition should be at the centre of a new social contract that reconnects society and institutions. To make the green transition possible, new institutions and policies to support those that will be temporarily affected by this transition are needed. A just green transition requires mechanisms to foster inclusive dialogue across all stakeholders and to build consensus around reforms, paying particular attention to overcoming the complex political economy of such a broad reform agenda. It is equally important to ensure that public institutions can work strategically and in close co-ordination, with a coherent, long-term view of making the green agenda a centrepiece of national development strategies.

The green transition is an opportunity to renew the social contract in LAC

LAC citizens are highly aware of the long-term risks of climate change and support for a green agenda is relatively broad

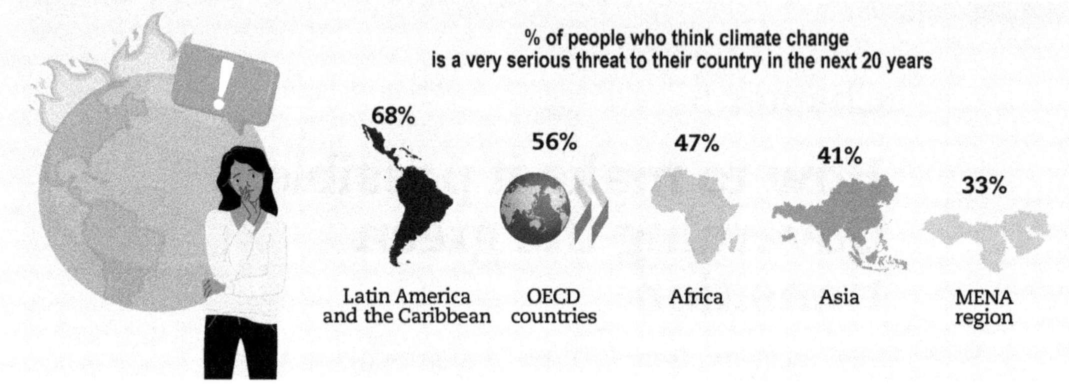

To make a green and just transition possible:

Involve and empower groups particularly vulnerable to climate change in environmental policy making, e.g. women, and local and indigenous communities

Bring the private sector on board: Over 60% of LAC firms need support to implement risk-based due diligence and the OECD RBC instruments

Consensus-building is key to overcome the barriers of the political economy of reform

Introduction

Moving the green transition forward is an immense task. In addition to the large financing needs associated with the green agenda (Chapter 4), green policies must be able to bring all interests on board so that the transition is possible, as well as fair, inclusive and sustainable. Taking into account the cross-cutting nature of the green agenda and the fact that it will leave winners and losers (at least temporarily), balancing the social, economic and institutional trade-offs arising from the transition and communicating them effectively is fundamental for its success. Coherent and inclusive policies are key to ensure an orderly green transition that accounts for its potential short-term challenges and costs, and supports vulnerable groups and sectors throughout the process.

This chapter analyses the institutional challenges posed by the green transition and presents policy options to address them. The first section takes as a starting point the fact that there is broad concern among citizens of Latin America and the Caribbean (LAC) around the seriousness of climate change and the importance of the green agenda. This section highlights that the green transition is not only an unavoidable agenda to address the existential threat of climate change but also an opportunity to restore dialogue and trust and improve well-being in the region. As such, green policies must be mainstreamed as part of a new social contract in LAC.

Moving the green agenda forward requires adequate, timely and transparent sharing of information as well as creating support for and avoiding resistance to policies that have strong and differentiated impacts across socio-economic groups, territories and generations. As such, the second section explores how to build consensus towards the green transition, making sure that all actors are effectively informed and involved in the process.

The third section argues that, for the green transition to be successful, policy instruments must be strengthened, with the coherence and long-term perspective of the agenda being fundamental. It analyses in detail the role of National Development Plans (NDPs) and other institutional mechanisms. The chapter concludes with key policy messages.

The green transition as the backbone of a new social contract

A new social contract that balances environmental sustainability with the needs of different socio-economic groups, territories and generations

The coronavirus (COVID-19) crisis exposed the weak foundations of the existing development model and aggravated the four structural development traps faced by LAC (OECD et al., 2021[1]). In pre-COVID-19 years, social grievances and higher aspirations for better living conditions already indicated that the pillars sustaining socio-economic progress through the years of bonanza (since the mid-2000s) needed rethinking. The rise of social discontent, marked by a wave of protests that started in 2019 and continued throughout 2020-22, confirmed the need to reach a new, overarching consensus and bridge the divide between society and public institutions.

These expressions of social discontent highlight the imperative for LAC countries to renew their social contract, to build back from the COVID-19 crisis in a more sustainable and inclusive way (OECD et al., 2021[1]). Social discontent is indeed partly driven by the impact of the crisis, but it also has a structural and multi-dimensional nature and can be explained by unmet aspirations for better jobs, quality public services, greater political representation and efforts to preserve the environment.

Given the widespread support for green policies in LAC, as well as the concern about climate change, the green transition can become the backbone of a new social contract that seeks to increase citizens' well-being. This new social contract should rethink the present development model from a multi-dimensional perspective, putting the green dimension at the centre. This would entail advancing towards more sustainable production and consumption strategies, stronger welfare systems adapted to the challenges of the green transition, and green financing for development models to underpin these efforts. As green policies can have asymmetric impacts across socio-economic groups, territories and generations, adopting an intersectional approach will be crucial in order to balance the various costs and benefits and to gain support – and avoid backlash – around the green agenda (Table 5.1).

Table 5.1. **Enhancing green dimensions of the social contract to improve people's well-being**

Improving people's well-being		Across		
		Socioeconomic groups (incl. income, gender, ethnic and racial)	**Territories**	**Generations**
Through	**Promoting sustainable production strategies**	Providing quality green jobs and investing in retraining programmes for "brown workers"	Defining transition strategies adapted to local skills and endowments (resource-rich vs. resource-poor communities)	Fostering green growth and sustainable resource management
	Strengthening social protection and public services to support the transition	Expanding the reach of social protection systems and social climate funds to support the transition from brown to green jobs	Ensuring wide territorial coverage and investing in flexible mitigation and adaption mechanisms to respond to shocks and extreme weather events	Better welfare systems to shield people from the adverse effects of climate change on health, income across time etc.
	Expanding sustainable financing for development	Fairer and stronger tax systems that discourage polluting and wasteful practices	Strengthening local financing and insurance mechanisms. Deepening local green markets	Sustainable debt management and impact investment

Source: Authors' elaboration.

The current global crisis may be understood as a "critical juncture" – that is, an exceptional time of severe crisis that redefines what is possible (ECLAC, 2020[2]). In effect, when facing extreme circumstances and tensions, many actors become more willing to change the status quo, thus opening windows of opportunity for social, economic and political change (Capoccia and Kelemen, 2007[3]; Weyland, 2008[4]). In the context of the COVID-19 crisis, the role of the state has been expanding, mostly through temporary interventions; a renewed social contract could respond to the need to adapt and strengthen state capacities in the medium and long term. This entails progressively building genuine welfare states, which in turn requires new social and fiscal pacts (ECLAC, 2022[5]; Arenas de Mesa, 2016[6]). Such welfare states must be adapted to the future, address the new risk structure, guarantee the broadening of the scope of rights, and urgently respond to the challenges of low productivity, social vulnerability and inequality, institutional weakness, and technological and climate change (ECLAC, 2022[5]).

Citizens' opinions about green policies: Where does the LAC region stand?

Latin Americans' concern about climate change is among the highest globally (Figure 5.1, Panel A) (Dechezleprêtre et al., 2022[7]; Ipsos, 2021[8]). On average, 68% of LAC citizens recognise climate change as a very serious threat to their country in the next 20 years vs. 56% in countries belonging to the Organisation for Economic Co-operation and Development (OECD), 47% in the African continent, 41% in Asia and the Pacific, and 33% in the Middle East and North Africa (MENA). Among the most concerning issues for LAC citizens are the depletion of natural resources, water pollution and deforestation; globally, on average, the top environmental priorities are global warming, air pollution and waste management (Ipsos, 2021[8]).

Nonetheless, when asked about the greatest risk to their safety in daily life, LAC citizens tend to be less concerned about the environment than those in OECD countries and on a par with those in the African continent (Figure 5.1, Panel B). Multiple factors could converge in framing long-term (e.g. 20 years) and short-term (e.g. daily life) concerns, including lack of exposure to or awareness about the immediate effects of climate change, which may indicate a greater need for awareness-raising campaigns. The presence in LAC of more important risks to safety in daily life, including violence and unsafe work environments, may also explain the relatively minor importance placed on the environment (Figure 5.1, Panel B). When asked about the most important problem in their country, less than 1% of LAC citizens mention environmental problems or global warming. They instead place greater importance on unemployment and the economy, although priorities vary across countries (Latinobarometro, 2021[9]). This raises the need to frame the green transition as part of a more comprehensive strategy (e.g. the United Nations 2030 Agenda) that encompasses economic, social and institutional objectives.

Overall, despite framing climate change and environmental issues as a longer-term (instead of immediate) problem, the majority of LAC citizens are willing to make sacrifices to protect the environment. A majority of citizens in Latin America (55.8%) think that the environment should be given priority, even if it causes slower economic growth and some loss of jobs, slightly above the global average of 53.8% (Figure 5.2).

Figure 5.1. **LAC citizens are highly aware of the long-term risks of climate change but tend to see the environment as a lesser risk in daily life**

Share of citizens who agree that climate change is a very serious threat in 20 years (Panel A) and share of citizens who mention the environment as the greatest risk to safety in daily life (Panel B), 2019

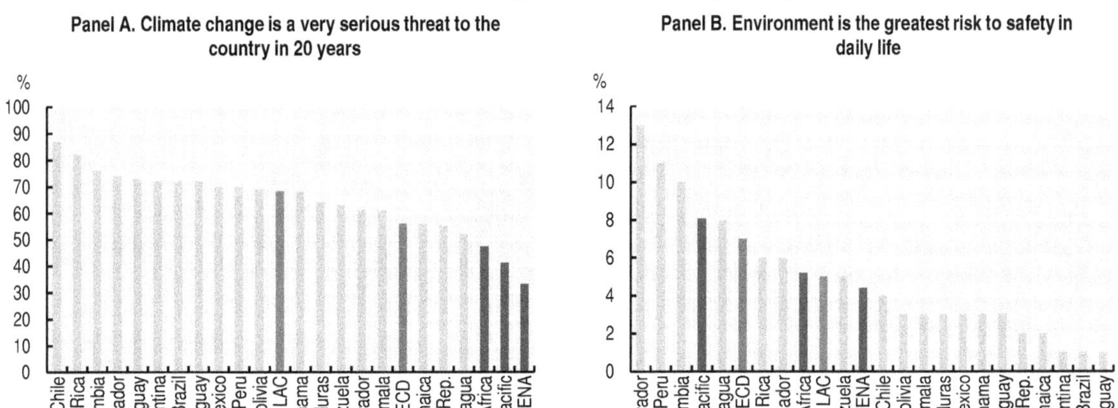

Notes: Question for Panel A: "Do you think that climate change is a very serious threat, a somewhat serious threat, or not a threat at all to the people in this country in the next 20 years? If you don't know, please just say so". Question for Panel B: "In your own words, what is the greatest source of risk to your safety in your daily life? Environment".

Source: (Lloyd's Register Foundation, 2020[10]).

StatLink https://stat.link/o2ezx6

Figure 5.2. **A majority of LAC citizens think that the environment should be given priority, even if it causes slower economic growth and some loss of jobs**

Share of citizens who prefer protecting the environment vs. economic growth, latest year available

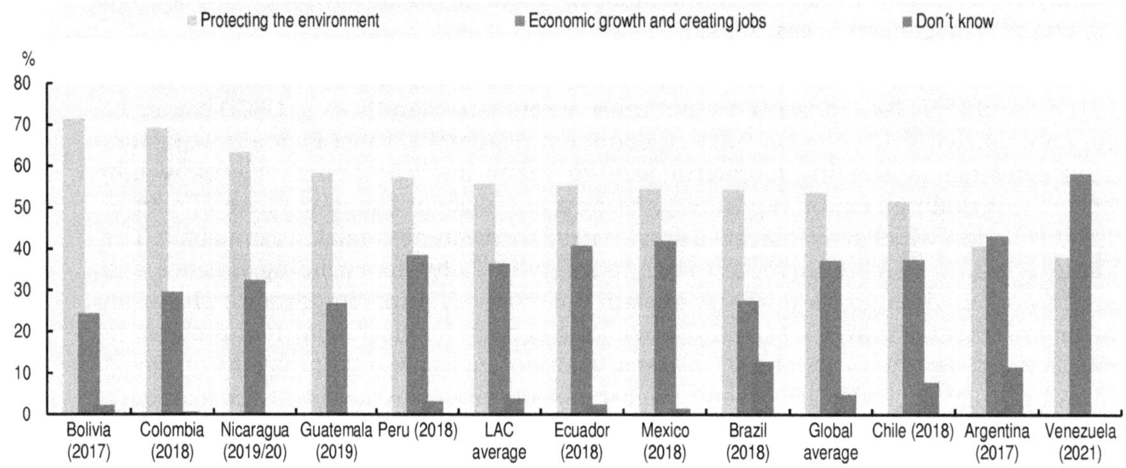

Note: Question: "Here are two statements people sometimes make when discussing the environment and economic growth. Which of them comes closer to your own point of view? A. Protecting the environment should be given priority, even if it causes slower economic growth and some loss of jobs; B. Economic growth and creating jobs should be the top priority, even if the environment suffers to some extent".

Source: (Inglehart et al., 2022[11]).

StatLink https://stat.link/vpfe7b

At the same time, in recent years, a greater share of LAC citizens is growing more dissatisfied with efforts to preserve the environment (Figure 5.3) and is calling for governments to act to fight climate change (Ipsos, 2020[12]). Among the regions analysed, LAC is the most dissatisfied with efforts to preserve the environment. The share of people satisfied with efforts to preserve the environment fell from 53% in 2012 to 42% in 2021 (Figure 5.3). The lowest levels of satisfaction with national preservation efforts were found in Brazil (23%) and Chile (19%); the greatest were found in some Central American countries, including Costa Rica (62%) and Guatemala (57%) (Gallup, 2022[13]). An SMS survey conducted in 13 Caribbean states in 2022 found that more than half of respondents (51%) believed that not enough is being done on climate action in their country (GeoPoll, 2022[14]). Overall support for the idea that the government and businesses – rather than individuals – should undertake the real efforts towards sustainability and environmental preservation is lower in Brazil, Chile, Colombia, Ecuador, Mexico, Paraguay and Peru than the global average (WIN, 2022[15]). Dissatisfaction with efforts to preserve the environment or the low levels of trust in government and firms in LAC may explain the preference for a more individualistic approach. As the green transition implies an all-of-society effort, this raises the need to deepen social cohesion and bridge the gap between citizens and institutions.

A further level of disaggregation helps in understanding which subgroups of citizens are more or less likely to be in favour of fighting climate change or supporting green policies. Based on the *AmericasBarometer 2016/17*, the most significant predictor of climate change concern in the LAC region is education, although wealth also plays a role. Worries about being affected by a natural disaster are almost as important as education in predicting climate change concern. It is therefore likely that attitudes about climate change may be shifting in the Caribbean in response to increased exposure to destructive natural disasters (Evans and Zeichmeister, 2018[16]).

Figure 5.3. **On average, LAC citizens are becoming more dissatisfied with efforts to preserve the environment in their country**

Share of citizens who are satisfied with efforts to preserve the environment in their country, LAC average, 2012-21

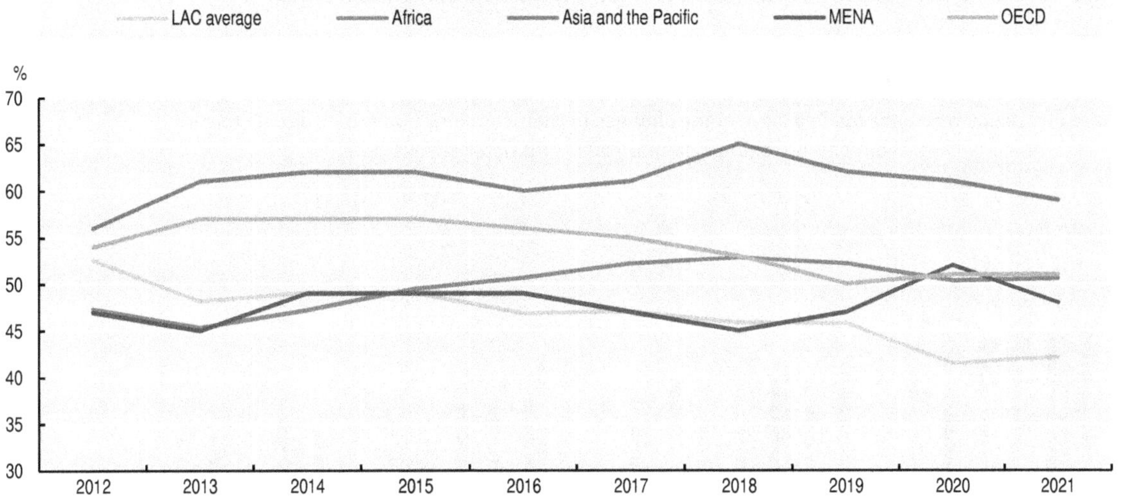

Note: Question: "In this country, are you satisfied or dissatisfied with efforts to preserve the environment?".
Source: (Gallup, 2022[13]).

StatLink https://stat.link/vco62l

Concern about climate change is consistent across the political spectrum. Interestingly, in the United States, identifying as a conservative is associated with a 25% decrease in climate change concern compared to political centrists. In LAC, there is almost no difference in level of concern between centrists and liberals, and only a small although statistically significant decrease in concern among conservatives (Evans and Zeichmeister, 2018[16]). The broad concern for the environment across the political spectrum is also confirmed by the representation of green parties in national legislatures, as in Brazil, Chile, Colombia and Mexico. These parties show mixed ideologies, with positions ranging from the far right to the left, especially in Brazil and Mexico (McBride, 2022[17]). Moreover, indigenous movements with an environmental agenda have also gained power in legislatures, most notably the Movement Toward Socialism party in Bolivia and the Pachakutik Movement in Ecuador (Rice, 2017[18]).

The COVID-19 crisis may have changed perceptions of climate change and support to fight it. A study conducted in February 2021 in 16 economies, including Brazil and Mexico, finds that the pandemic heightened concern about climate change. However, those who lost jobs or suffered income shocks due to COVID-19 were more reluctant to support green recovery policies that may require some short-term costs (Mohommad and Pugacheva, 2021[19]).

Income-driven considerations may therefore become more important for some parts of the population, despite high concern about climate change. Evidence from the 2008 global financial crisis shows that large, negative income shocks can affect the way people rank priorities towards a greater concern for jobs over climate change (Scruggs and Benegal, 2012[20]). A comparison of pre- and post-pandemic behavioural preferences shows that Latin Americans remain among the citizens most willing to change their behaviour out of concern about climate change. However, the changes to daily life caused by the pandemic resulted in decreasing concern about the environmental impact of products and services they buy or use (Figure 5.4). Protecting income and livelihoods in the near term is therefore key to sustaining support for climate-oriented green recovery policies (Mohommad and

Pugacheva, 2021[19]). It is particularly necessary to understand the needs and aspirations of the vulnerable. This subgroup has expanded as a consequence of the crisis; although it may share middle-class aspirations, its more unstable and precarious position makes it more prone to prioritise everyday issues over environmental considerations.

Figure 5.4. **Citizens' concern about the environmental impact of their consumption habits decreased during the COVID-19 pandemic**

Share of citizens who have made a lot or some changes in their consumption habits out of concern about climate change, January 2020 and October 2021

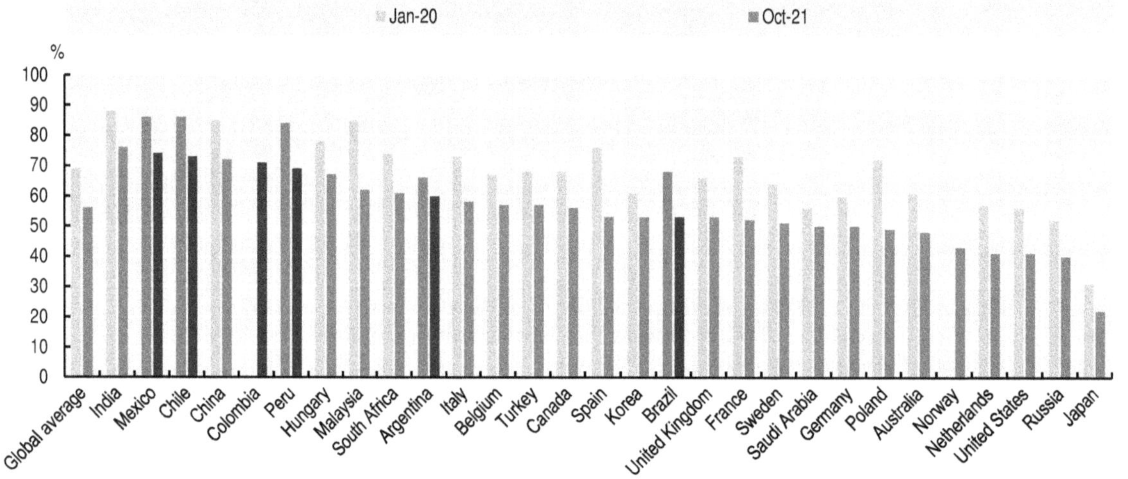

Note: Question: "Over the past few years, have you made any changes regarding the products and services you buy or use, specifically out of concern about climate change?".
Source: (Ipsos, 2021[21]).

StatLink https://stat.link/bqs576

Overall, this shows that the high concern about climate change and support for tackling environmental issues among LAC citizens could make the green transition the binding element of a wider social contract for the region. It is therefore key to consider the green transition as part of a comprehensive set of economic, social and institutional policies to advance towards a more inclusive, just and sustainable development agenda. However, as illustrated by the increasing discontent caused by the rising energy and fuel prices, these policies must also mitigate the short-term impacts of the transition on the most vulnerable groups in order to be politically viable (Section below on political economy).

Rising mistrust of public action and institutions may hamper the green transition

While Latin Americans may support a drive towards sustainability and a green economy in view of the risks posed by climate change and environmental degradation, they have little confidence in the efficiency and neutrality of public action and institutions. Even prior to the pandemic, rising mistrust was a common feature in LAC (OECD et al., 2019[22]; Maldonado Valera et al., 2021[23]). Trust in key public and political institutions – including the national police, the government, the judiciary, congress, electoral authorities and, at the lowest level, political parties – was persistently low and has tended to fall since the 2010s. In 2020, confidence was lowest in legislatures (20%) and political parties (13%) (Latinobarometro, 2021[9]). The share of the population with confidence in the government declined from 44% in 2011 to 38% in 2021. The share of people who think that corruption is widespread in the government has been consistently higher than 70% in the past decade, with the exception of 2020 (69%), and stood at 72% in 2021 (Figure 5.5).

Figure 5.5. **Low trust in institutions and the perception of widespread corruption are key structural problems in LAC**

Share of people with confidence in the government and share of people who think corruption is widespread in the government, LAC average, 2009-21

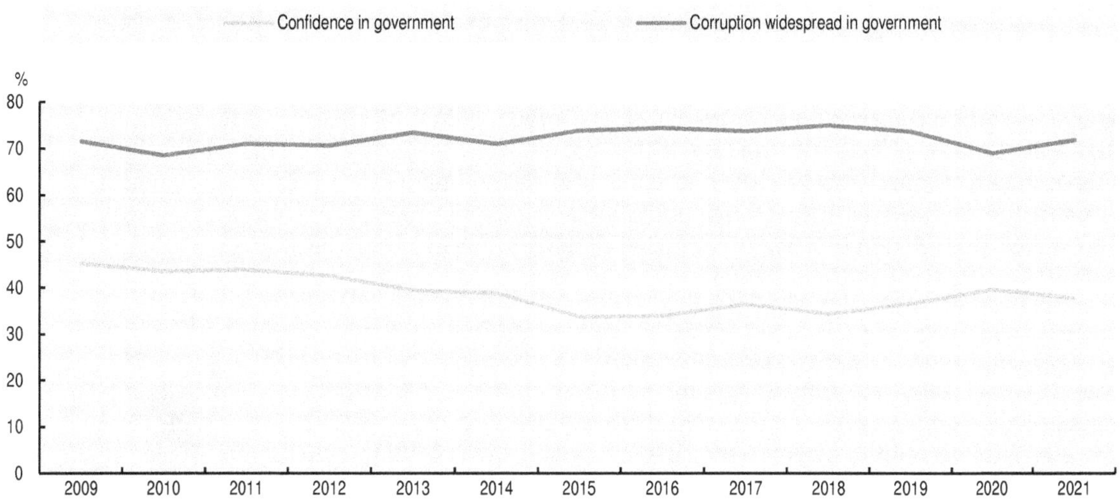

Note: Simple LAC average.

Source: (Gallup, 2022[13]).

StatLink https://stat.link/k04p1w

Trust is the cornerstone of democracy and public governance. Declining confidence in institutions can result in lower compliance with the law in general, as well as a decreasing observance of civic duties, such as paying taxes (OECD et al., 2019[22]; Rothstein, 2011[24]). Lower trust can also negatively affect the government's ability to implement reforms, thereby directly affecting citizens' well-being. This poses a challenge for defining a common policy agenda towards a green transition backed by a broad and long-term consensus. Low trust requires paying attention to the political economy of reforms and promoting institutional change, as well as specific measures to improve efficiency, transparency, accountability and the policy coherence of public institutions. In the face of a complex endeavour, institutions and public authorities in the region should leverage support for environmental policies to table a larger discussion around the pillars of the welfare state and the social contract. Growing concern about climate change and support for green recovery policies during the pandemic have built momentum for the discussion of renewed and sustainable social pacts centred around the environment and sustainability. The green agenda may therefore help build coalitions supporting a wider reform agenda in the region. Effective and evidence-based communication around the green agenda can also strengthen the credibility of government action and positively enhance key drivers of trust.

Political economy of the green transition: The need to build support and avoid resistance to change

In moving the green transition forward and making it the centrepiece of a new social contract in LAC, policy makers must be aware of the political economy issues that may enable or limit their efforts. A green transition involves a shift of resources among economic sectors and political constituencies, as well as institutional and policy changes, that may trigger the opposition of some interest groups (Arent et al., 2017[25]). Similarly, such a transition goes beyond formal government institutions and includes other stakeholders, such as informal networks, civil society, the private sector, specific lobbying groups, and a wealth of actors and structures operating at various levels, from local to international (Worker and Palmer, 2021[26]; Edenhofer et al., 2014[27]). These actors may have competing interests and political priorities, shifting public opinions, and different levels of exposure to climate policy.

Understanding these political economy dynamics can better position policy makers to anticipate the response to green policies and identify actions to support coalitions, shift incentives and amplify the contributions of non-state actors to advancing the green transition (Worker and Palmer, 2021[26]). In fact, vested interests and rent-seeking arrangements may make it extremely difficult to change the status quo. This is especially relevant in a context marked by low trust and deep intersecting inequalities, where powerful elites can greatly influence political decisions, hampering the building of broad consensus and implementation of inclusive social and fiscal pacts (OECD et al., 2021[1]).

A transparent governance and policy process can strengthen social cohesion and boost the prospects for building wider social and political consensus for sustainability. Stronger accountability mechanisms and citizen oversight can contribute to promoting integrity and avoiding policy capture. To boost trust in public policies, governments should develop mechanisms for recognition, participation and conflict resolution. More generally, it is crucial to improve the rule of law and the quality of democracy. These can be enhanced by implementing open and participatory policy decision-making processes; strengthening mechanisms for accountability and efficiency; and improving information and the quality of public debate by ensuring greater access to information systems and transparency bodies, as well as greater media openness and transparency (Maldonado Valera et al., 2022[28]). On this matter, the OECD Recommendation of the Council on Open Government includes provisions to promote citizen and stakeholder participation throughout the policy cycle, and greater transparency and access to information (OECD, 2017[29]). In addition, the OECD Guidelines for Citizen Participation Processes suggest a ten-step path to design, plan, and implement a citizen participation process, from the identification of a problem to solve to the evaluation of the process and the cultivation of a culture of participation. These Guidelines also suggests eight guiding principles that help ensure the quality of these processes, namely purpose, accountability, transparency, inclusiveness and accessibility, integrity, privacy, information, and evaluation (OECD, 2022[30]).

Building consensus for a green, inclusive and just transition

A green, inclusive and just transition requires institutional mechanisms to foster dialogue and build consensus around reforms. Otherwise, entrenched stakeholders may try to pull the process in opposite directions, with some promoting and others obstructing change (Arent et al., 2017[25]). Similarly, a green transition may bring not only benefits but also significant costs and negative impacts – across the board and for specific socio-economic groups – at least in the short term (Chapters 1, 3 and 4). Mitigating these short-term costs and communicating about them in a relevant, timely and transparent manner is important to bring the temporary losers on board, to strengthen the credibility of government action, and to generate perceptions of responsiveness and reliability among the population.

Policy makers must factor in these dynamics when reflecting on how to govern the green transition to make it possible and just. Among other considerations, it is important to identify and involve key stakeholders in the policy-making process from the beginning; to bring on board interest groups to avoid having them work against the climate action strategies; to understand the socio-political context to appropriately adjust the speed and scale of the transition; to devise a clear communication strategy, based on key outcomes, to shape the transition narrative; and to design a comprehensive set of policies to support people throughout the transition and avoid certain groups or sectors feeling disproportionately affected by it. As described below, these principles can be broadly summarised under four Cs: 1) concertation; 2) context; 3) communication; and 4) compensation (Cabutto, Nieto Parra and Vázquez Zamora, 2022[31]).

Concertation of the interests of all parties through inclusive and participatory processes

Initial loss of jobs, increased tax burden, transport and energy price increases, and more stringent regulations for some business activities may generate opposition to the green agenda, particularly if all interested parties are not involved in the decision-making process from the beginning. Identifying key stakeholders, as well as the institutions and networks through which they act, is therefore vital.

The complexities of the green transition imply that the public sector will need support from civil society, intermediary bodies and the private sector to move forward the agenda (Sections below on the role of civil society and the private sector). One of the most important short-term effects of the green transition involves employment contraction in some activities as the production structure shifts to more sustainable ones (Chapter 3). These transformations will require the involvement of trade unions, business associations, local leaders and non-governmental organisations (NGOs). Thus, the transition requires a full mobilisation not only of government but also of society.

A protected and promoted civic space is a prerequisite for such collaboration between civil society and governments and for a more inclusive green transition. Such a space is defined by the OECD as the set of legal, policy, institutional and practical conditions necessary for non-governmental actors to access information, express themselves, associate, organise and participate in public life (OECD, 2021[32]). To participate throughout policy- and decision-making cycles, evaluate results, express their views and provide oversight of government activities, citizens and civil society organisations (CSOs) need the guarantee – by law and in practice – of fundamental civil rights such as freedoms of expression, peaceful assembly, association and the right to privacy. Governments are thus encouraged to support a vibrant civic space as an enabling environment for citizens and non-governmental actors to fully exercise their democratic rights and actively engage on environmental issues (OECD, 2022[33]). Some practical actions may include the improvement of the enabling environment for CSOs by facilitating access to funding and simplifying the legal regime, as recommended in the OECD Open Government Review of Brazil (OECD, 2022[33]).

Engaging in transparent and participatory processes can have multiple benefits. First, it can help identify appropriate policies that safeguard the interests of all stakeholders. Given the cross-sectoral nature of the green transition, a multi-stakeholder process can help conciliate various interests and achieve a negotiated policy stance. Moreover, the process of engagement with key stakeholders can help sustain a high level of support for climate reform policies beyond short-term political cycles, even during political transitions, thereby preventing conflicts of interest or reform rollback when leadership changes (OECD et al., 2021[1]; UNECA, 2020[34]). For instance, past experiences with the phase out of fossil fuel subsidies show that successful reforms require extensive consultation in design and implementation (UNECA, 2020[34]).

To broaden the dialogue and boost the sense of ownership of agreements achieved, considerable effort must be made to give voice and influence to sectors and population groups that have been discriminated against or excluded, as well as to groups more vulnerable to shocks and emergencies. The state should maintain effective venues for dialogue and participation and have a decisive mediating role to ensure the interests of all actors are fully represented and not silenced by majority decisions. Citizen participation and social movements are significant in driving major changes and influencing the political agenda.

Deliberative processes are useful instruments to build consensus around policy challenges that require complex trade-offs and a long-term vision. A relatively recent example, directly connected to the green transition, is the 2019-20 French Citizens' Convention on Climate, which was set up as a direct response to social mobilisation in the country (OECD, 2020[35]). Although less common in LAC, climate assemblies are a representative deliberative process dealing exclusively with environmental issues. They involve a group of randomly selected citizens who are statistically stratified to make up a microcosm of society that deliberate based on evidence and information to provide policy recommendations to public authorities. Examples have taken place in Spain, the UK, Finland, France and Denmark (OECD, 2020[35]). In Brazil, a similar deliberative process took place in 2019 with the creation of the Citizen Council of Fortaleza, with 40 randomly-selected residents deliberating on solid waste management (Pogrebinschi, 2020[112]).

Protecting environmental defenders and local communities is a prerequisite of any real participatory process. Social conflicts associated with natural resources are increasing, and social environmental defenders can be at risk of physical harm. Between 2012 and 2020, 1 540 land and environmental defenders were murdered around the world, with LAC accounting for over two-thirds of killings, making it the region most affected by threats and attacks targeting human rights defenders and environmental activists (Figure 5.6) (Frontline Defenders, 2022[36]). These killings relate mainly to land use issues and the mining and extractive activities sector.

Figure 5.6. **Number of environmental defenders murdered in LAC and other countries, 2012-20**

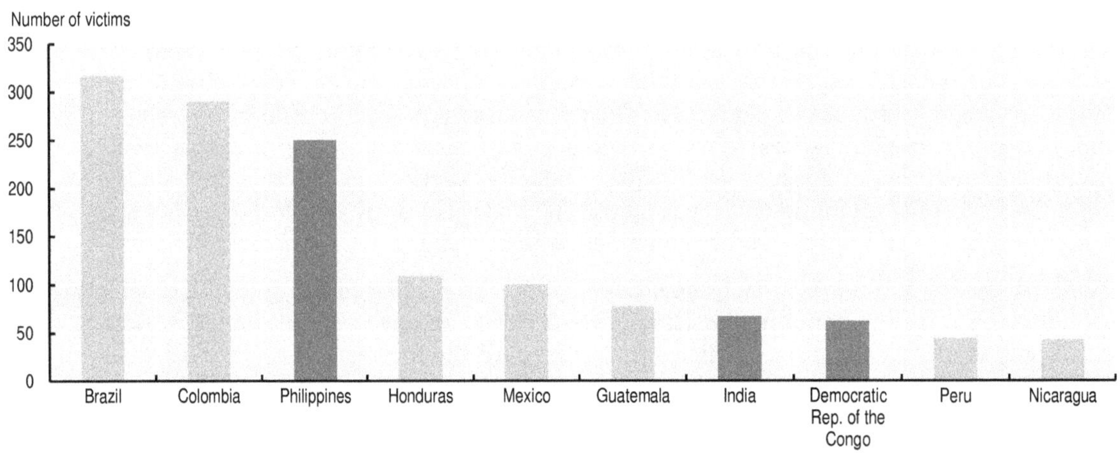

Source: (Global Witness, 2022[37]).

StatLink https://stat.link/qh95pn

Understanding the context to find the right policy pace and sequencing

There is no single blueprint for a successful green transition. Each country should balance local needs and priorities and adapt the speed and scale of the reform process to the socio-political context. For instance, fiscal policy is key to support the ambitious efforts needed to achieve the transition. However, as households and businesses are still recovering from the COVID-19 crisis, in the short term, governments may prefer to avoid increasing the fiscal burden and instead focus on policy options that may still contribute to strengthening public finances (e.g. intensifying the fight against tax evasion and avoidance or eliminating ineffective tax expenditures).

In the post-pandemic era, context-specific considerations are vital for the recovery. Well-designed green recovery plans (e.g. investments in renewable energy infrastructure, public and low-carbon transport options, clean-up of polluted sites) can be more acceptable than environmental tax reform (Vona, 2021[38]). For instance, investments in research and development (R&D) can foster innovation for the green transition and are also politically acceptable (OECD, 2019[95]). Depending on the context, policy makers may prefer to bundle reforms into a comprehensive package so that losses from one reform are compensated by gains from others, e.g. embedding income support schemes for poor households in wider energy subsidy reform packages. Or, if this is not possible, to reach specific agreements and incremental policy advances in areas with potential for accord (OECD et al., 2021[1]). For instance, by reacting early to a mix of climate change and geopolitical concerns, Sweden has strengthened the security and sustainability of its energy supply through a gradual energy transition based on consensus among the parties. Among others, it introduced a tax on CO_2 emissions and grants for heating networks powered by bioenergy since the 1990s. In 2003 it further introduced a system of green certificates that strengthened renewable electricity output by 13.3 TWh between 2003 and 2012 (Cruciani, 2016[114]).

Communicating effectively to build trust and favour consensus

In the context of polarised political discourse and rising mis- and dis-information, a commitment to evidence-based analysis and effective communication is imperative to frame a clear transition narrative and reach broad consensus (Matasick, Alfonsi and Bellantoni, 2020[39]). In particular, aversion to environmental policies may be due to misinformation and stereotypes, which often highlight the income costs of policies over their health benefits. Public communication[1] is a powerful tool, as ideas and narratives can shape what stakeholders see as the problem in climate policy and the potential solutions. Communicating results in an effective manner, both within the public administration and outside, can also improve the responsiveness and reliability of government action (OECD, 2021[40]). Transparent communication about the reasons behind potential delays in implementing specific plans or adjustments required during the course of action can help mitigate feelings of disillusionment.

Solid analysis and research by authoritative institutions, together with targets and indicators, are essential to build the case for action. This requires investing in data collection, building strong and independent national statistical offices, co-ordinating data publication and sharing across agencies, and committing to *ex post* evaluation. At present, data on the environment and climate change are still scarce globally. In particular, few statistical indicators have been developed to monitor oceans (United Nations Sustainable Development Goal [SDG] 14), sustainable consumption and production (SDG 12), gender equality (SDG 5) and sustainable cities and communities (SDG 11) (OECD, 2022[41]). In addition to this, out of the 231 unique indicators in the SDG framework, 114 have an environmental angle, but only 20 of those provide for gender-specific and/or sex disaggregation, evidencing a shortage of indicators to support gender-sensitive environmental policy making (Cohen and Shinwell, 2020[107]; OECD, 2021[108]).

Statistical capacity development for environmental and related issues is key to communicate the urgency of the green agenda to citizens and to make informed decisions. Investments should aim to facilitate data accessibility, comprehension and reuse in order to generate public value and have an impact on policy making (OECD, 2019[42]; OECD, 2018[43]; Grinspan and Worker, 2021[44]). Avoiding data fragmentation across agencies is also important. For instance, in Costa Rica, strong political support for open data from the Ministry of the Presidency helped establish institutional arrangements among agencies for continuous data provision and led to the development of the National System of Climate Change Metrics (SINAMECC), an online open platform for climate-related data (Grinspan and Worker, 2021[44]).

A number of issues pertaining to scientific communication and climate change communication more specifically should not be overlooked. These include the necessity to communicate complex data and scientific discoveries without turning people off; the importance of exhorting citizens to act against climate change without sending apocalyptic messages that may be disempowering; and the need to create empathic connections between climate change and citizens' everyday lives to avoid framing it as an abstract, isolated and faraway problem (Olano, n.d.[45]; Yale Climate Connections, 2017[46]).

Public institutions seeking to engage citizens on climate change action need to invest in the professionalisation of the public communication function, including on areas such as audience and behavioural insights. Indeed, the OECD Report on Public Communication finds that governments' focus on audience insights can be improved, with just 27% of Centres of Government (CoG) conducting this activity at least on a quarterly basis. Also, governments may not always be the best placed to circulate certain messages. Enlisting third-party messengers as vehicles for official information and developing partnerships with scientists or community leaders can help engage citizens in the green agenda in a more compelling manner, while amplifying the reach and trustworthiness of communication. The use of community messengers and influencers as relatable and trusted voices can also improve how inclusive communications are of diverse groups in society. Such people can help to identify under-served groups and the barriers to information. Choosing appropriate formats that can engage diverse audiences, on line (e.g. social media) and off line (OECD, 2021[40]) is also of the essence. More tailored messages are more likely to resonate than mainstream channels or content. By helping communicators account for the cognitive factors, barriers and biases shaping how people navigate an increasingly complex, crowded information ecosystem, the application of behavioural science can help engender more compelling communication (OECD, 2022[47]).

Compensating those who stand to lose to bring everybody on board

From the perspective of increasing the political acceptability of the green transition, governments will have to think about a comprehensive series of actions to support citizens and communities throughout the journey. The transition involves several monetary and non-monetary distributional effects, which policy makers should study carefully. In turn, policy actions undertaken will determine whether potential losers bear the costs of the transition, affecting the level and intensity of resistance to green policies.

To increase support for such policies, policy makers should distinguish between small and large distributional effects of climate policies and find the appropriate combination of revenue-recycling schemes (e.g. mechanisms that earmark income generated from carbon taxation to return it back to society), industrial and retraining policies, and compensation packages. When environmental policies are part of a broader political package, small distributional effects of climate policies may become almost irrelevant for political acceptability if offset by other policies (OECD et al., 2021[1]). For instance, bundling a carbon tax and policies promoting greater fiscal progressivity in a fiscal reform package may help reduce resistance to the potential income effect of the carbon tax (Vona, 2019[48]).

The argument is different when green policies have large distributional effects, most notably when resulting in job losses. Large distributional effects can also be spatially concentrated, as extractive or energy-intensive industries tend to be located in the same area, creating large constituencies that oppose green policies and influence the ideology of locally elected members of parliament (Vona, 2019[48]). In this case, the post-coal labour market transition in the Ruhr region in Germany shows that investing in long-term planning promoting industry diversification, active labour market policies and retraining facilities can help from both equity and political acceptability perspectives; by mitigating negative income effects and supporting the transition from brown to green jobs, such investment can decrease resistance to such transition policies (Chapter 3; Arora and Schroeder, 2022[113]).

Overall, environmental policies and resulting compensation schemes should aim to be fair, efficient and cost-effective. However, as aversion to specific types of policies may go beyond these considerations, calculations of political acceptability should also inform environmental policy making (Vona, 2021[38]). For instance, environmental tax reform raises stronger opposition than lump-sum redistribution or spending on green projects. This is due to the fact that upfront carbon compensation provides an immediate and direct benefit, while the benefits from lower labour taxation are harder to calculate, as they are indirect and uncertain.

Similarly, beyond compensation schemes (redistribution), and in view of creating a broad constituency in favour of green policies, general interventions aimed at tackling inequality (predistribution) may help expand the size of the middle class, increasing the number of citizens wealthy enough to care more about collective goods and the environment. In this regard, universal social protection and universal access to social services (notably education and health) are essential not only to compensate or render potential losses acceptable to large sectors of LAC societies but also to avoid deeper and/or new inequality gaps, as well as unintended poverty increases arising from a structural transition to sustainability.

Role of citizens and civil society, including the crucial role of women and local communities

Civil society participation can strengthen the outcome of a green transition envisioned as truly inclusive. Moreover, expanding quality spaces for civic participation can improve citizens' well-being. Realising that citizens have a role to play boosts civic engagement and a sense of belonging. If sustained over time, civic participation could also have a positive effect on social inclusion mechanisms by giving visibility to the needs of vulnerable groups and creating spaces for exchange. Civil society's participation throughout the cycle of green policies can play a vital role in the outcomes, as the inclusion of public views is closely linked with the sustainability of green projects (IDB, 2021[49]). An enabling environment, which encompasses a conducive legal and policy environment safeguarding freedom of association, is central to ensure that CSOs can operate in a free and autonomous manner and reach their full potential. On the contrary, active and effective participation in the policy cycle is hindered if CSOs are struggling to operate, arbitrarily dissolved or overburdened with disproportionate administrative obligations.

Civil society can bring new insights and innovative practices through local knowledge. It can also help anticipate emerging issues and support effective policy implementation by fostering trust among stakeholders. Making the different stages of the policy cycle accessible to members of communities directly affected by green policies also incentivises higher levels of transparency.

The OECD Recommendation of the Council on Open Government refers to stakeholders, grouping together both citizens and any interested and/or affected party such as civil society organisations (OECD, 2017[29]). Involving citizens and/or stakeholders is equally important, but their participation should not be treated identically. The line between these groups can be blurry and, in reality, is not always perfectly neat. Both groups can enrich public decisions, projects, policies, and services. Stakeholders can provide expertise and more specific input than citizens through mechanisms such as advisory bodies or experts' panels. Involving citizens can bring diversity by including rarely heard voices, it can help raise awareness and facilitate public learning about an issue and, in the medium to long term, it can strengthen democratic feelings and trust in institutions. The National Council of Environment (CONAMA) in Brazil, a national consultative and deliberative body involving government representatives and non-governmental stakeholders, is an example of participatory practice in environmental matters taking place in the region. Established in 1981, CONAMA has the authority to establish regulations for polluting activities and to carry out environmental impact studies for public and private projects (CONAMA, 2018[105]). Another example is the Climate Change online platform "BA Cambio Climatico" of the city of Buenos Aires that gathers open environmental data and promotes citizen participation for a greener city. The platform was co-developed with citizens and stakeholders through meetings, interviews, a hackathon, and eight roundtable discussions with over 600 inhabitants (OIDP, 2020[106]).

The regional Escazú Agreement[2] is an important tool to improve policy coherence and the transparency and accountability of national governance. The agreement aims to promote civic participation in environmental matters along three key dimensions: 1) access to environmental and climate information; 2) involvement in decision making; and 3) access to environmental justice. Enacted in 2018, the agreement entered into force in 2021 and has been signed by 24 countries and ratified by 12 LAC countries. It offers a benchmark and a guide for countries, giving legitimacy and support to these processes of change (Chapter 6; ECLAC, 2018[50]).

Women as key actors to address climate change

Given their over-exposure to the effects of climate change, the perspective of women can ensure that measures against climate change are inclusive and comprehensive. Women are more vulnerable to climate change impacts than men as they are more likely to be in poverty, have less access to basic human rights, such as free movement and land acquisition, and are more dependent on the natural resources which climate change threatens the most (Ward, 2022[117]). The Social Institutions and Gender Index (SIGI) 2019 shows that the level of discrimination in social institutions in LAC is highest in Haiti (39.9) and lowest in Colombia (15) (OECD, 2019[118]). Understanding their realities and including their life lessons throughout the policy cycle will help build more robust climate mitigation and adaptation actions and programmes. Moreover, by raising their visibility, women can become more relevant actors. This presence, such as their inclusion in vulnerability assessments, can have positive effects (IDB, 2021[49]; OXFAM, 2018[51]).

The processes of climate change adaptation and mitigation should be gender sensitive. Policies that consider gender sensitivities acknowledge these vulnerabilities and incorporate them throughout the policy cycle to produce stronger responses. A comprehensive green transition should also identify and recognise the potential women and girls have as agents of change. In LAC, women represented only 7% of environment ministers in 2021 (OECD, 2021[52]). Including women in environmental decision-making roles could increase not only the gender focus of environmental policies but also their effectiveness (OECD, 2021[52]; Strumskyte, Ramos Magaña and Bendig, 2022[109]). At a local level, women can be key actors in emergency agencies. For example, women in small villages of the Caribbean tend to know its members, making them better able to identify who is missing or hurt after a natural disaster, to organise shelters and to provide aid (IPCC, 2014[53]).

An integrated approach to gender equality and environmental sustainability – i.e. recognising the gender-environment nexus – could help to alleviate limitations to gender equality and women's economic empowerment and enhance women's role in environmental sustainability and green growth (OECD, 2022[110]). In agriculture, creating opportunities to increase women's ownership of rural land is a key priority, as it is often correlated with greater food security, greater bargaining power at household and community levels, greater economic independence, better child nutrition and lower levels of gender-based violence. While most women in LAC work the land, only 18% on average are land-owners. Although female land-ownership levels in LAC are higher than the global average of around 15%, high heterogeneity remains in the region, ranging from below 9% in Guatemala and Belize to over 30% in Peru (FAO/IFPRI, 2018[115]; Deere, Alvarado and Twyman, 2012[54]).

Promoting opportunities for young women and girls to study and to develop in green jobs will also help raise gender sensitivity and close the gender pay gap. Gender stereotyping in childhood and youth tends to discourage female presence in technical sectors that are critical for the green transition and the future of jobs, such as sciences, physics, computational molecular biology and digital technology (OECD, 2021[52]). Moreover, women and girls from local communities hold unique traditional knowledge and practices that could be key inputs for green jobs and the circular economy in general. A good example is the improvement they could bring to land use and farming, as they tend to know and employ ancient uses of plants and animals.

Community-driven development to engage local communities in social and environmental policy solutions

Locally led adaptation and mitigation efforts recognise that local communities are on the frontlines of climate change and are often best placed to identify solutions. It is therefore important to support the power of local communities to influence adaptation efforts. To avoid overburdening local partners, a more equitable distribution of power and resources towards local communities should take into account specific contexts, local cultures and interests at play. In turn, communities should be supported in the development of new capabilities (WRI, 2021[56]).

Local communities and actors rarely have a voice in the decisions that most affect them. Current top-down approaches should be replaced by new models in which local actors have greater voice, representation, power and resources to build resilience to climate change. Several LAC countries have recognised in their legal frameworks the right to Free, Prior and Informed Consent (FPIC) of indigenous peoples, which are also expressed in different forms in the ILO Indigenous and Tribal Peoples Convention (no.169), ratified by 15 LAC countries, and the 2007 UN Declaration on the Rights of Indigenous Peoples. The convention does not only include the right to prior consultation with indigenous peoples on any legislative or administrative measures that may affect them, but also their participation in the formulation, implementation and evaluation of plans and programmes. In Peru, the experience of the prior consultation with indigenous peoples on the Framework Law on Climate Change has resulted in the creation of the Indigenous Peoples' Platform to Address Climate Change for managing, formulating and following up on proposals on climate change mitigation and adaptation by indigenous groups (FIIAPP, 2021[116]).

Local communities should be engaged as partners in advancing the green transition. One way to link communities and marginalised groups with higher-level policy making is to provide flexible funding to local actors. This means according local actors a certain autonomy to adjust activities according to local priorities and changing circumstances. Updating climate finance implementation mechanisms and providing technical and financial assistance needed to achieve locally relevant and effective development impacts are some examples of ways to make this possible (WRI, 2021[56]; World Bank, 2021[57]). Community leaders can play a key role in designing and implementing investments in green programmes that correspond to their community's priorities.

Principles for locally led adaptation are intended to guide communities as they advance adaptation and mitigation programmes, seek funding, and undertake practices with increasing ownership by local partners. Three strategies can help: 1) putting design and funding in the hands of local actors; 2) enhancing institutional and technical capacity building; and 3) ensuring monitoring, evaluation and learning (WRI, 2021[56]).

Locally led does not mean locally isolated. Three aspects are important for the success of locally led green initiatives. First, local champions play a key role to advance reforms, maintaining a balance between local leadership and external support to ensure policy agendas are aligned with local goals. Second, an understanding of the local context and the political economy dynamics is needed to ensure the viability of policy reforms. Last, aligning the time horizons of policies with local priorities and capacities is relevant to avoid implementation gaps and find adaptable solutions that are seen as legitimate by local communities.

Role of the private sector: Responsible business conduct to support a green and just transition

The private sector is a key player to achieve sustainable and inclusive development and move the green agenda forward. As businesses are responsible for a significant share of environmental impacts, they are integral to a greener economy and society. The climate crisis highlights the central role of businesses and the need for firms to work with policy makers and other stakeholders to urgently adopt climate mitigation and adaptation actions. Public-private collaboration is also key for the creation of an enabling policy environment and an institutional framework that facilitates the twin digital and green transitions.

A number of policy-, science- and industry-led drivers have encouraged environmental action by the private sector (OECD, 2021[58]). Governments are increasingly using legislation and regulation as policy tools to promote more responsible business practices and sustainable financial flows. New legislation, particularly on human rights and environmental due diligence in global supply chains, has emerged in multiple jurisdictions in recent years.[3] For instance, in February 2022, the European Commission adopted a proposal for a directive on mandatory corporate sustainability due diligence with respect to human rights and environmental impacts (Chapter 6). Likewise, companies increasingly recognise their responsibility to address their social and environmental impact, enhance supply chain resilience and respond to sustainable consumption patterns.

Instruments for responsible business conduct

Responsible business conduct (RBC) sets out that all businesses, regardless of their legal status, size, ownership or sector, must avoid and address the negative impacts of their operations, supply chains and other business relationships while contributing to sustainable development in the countries where they operate (OECD, 2022[59]; OECD, 2019[60]). The *OECD Guidelines for Multinational Enterprises* (MNE Guidelines) and related OECD due diligence guidance[4] champion RBC to help companies rise to the challenge of transitioning to a green economy and preserving the world's greatest wealth: our natural ecosystems. The MNE Guidelines are the only multilaterally agreed and comprehensive code of RBC that governments have committed to promoting.

All governments adhering to the MNE Guidelines have the legal obligation to set up a National Contact Point (NCP) for RBC. NCPs are agencies established by governments to promote the MNE Guidelines, provide related due diligence guidance and handle non-judicial grievance mechanisms. Since 2000, close to 600 specific instances have been handled by NCPs in over 100 countries and territories (OECD, 2022[61]). To date, 51 governments have a NCP for RBC, including 8 in the LAC region (Argentina, Brazil, Chile, Colombia, Costa Rica, Mexico, Peru and Uruguay).

International RBC instruments can play a pivotal role in advancing the green transition in LAC by mainstreaming internationally agreed standards and safeguards into business decisions and actions (OECD, 2021[58]). For instance, Chapter VI of the MNE Guidelines (Environment) calls on enterprises to take account of the need to protect the environment and public health and safety, and generally to conduct their activities in a manner contributing to the wider goal of sustainable development. The environment chapter envisages that enterprises should have in place an environmental management system, with measurable objectives and targets, and should explore ways to improve environmental performance over the longer term. The guidelines also recommend that firms promote higher customer awareness of the environmental implications of using their products and services. The private sector is increasingly being held accountable on matters relating to RBC and the environment. As much as 24% of all specific instances submitted to NCPs made reference to provisions of the environment chapter of the MNE Guidelines (OECD, 2021[58]).

Risk-based due diligence is a key element of RBC. It refers to a process through which businesses identify, prevent and mitigate their actual and potential negative impacts and account for how those impacts are addressed. By implementing due diligence, businesses can not only mitigate the impacts of climate change and advance adaptation efforts but also be a major source of green finance and play a decisive role in making the green transition a driver of better jobs, greater respect for human rights, and increased integrity and trust. Carrying out due diligence is critical to ensure that business action on climate also takes into account social and human rights implications, possible trade-offs (or unforeseen adverse impacts across various risk areas as a result of action to address climate-related risk areas), and risk prioritisation considerations.

RBC practices in the region

As a response to the growing role firms play in preserving the environment and to increasing scrutiny by society, the private sector – globally and in LAC – has increased reporting on RBC and sustainability issues, as well as adoption of environmental management systems. The sustainability reporting rates of the largest 100 companies by revenue in LAC are among the highest in the world. It has increased by 6 percentage points, from 81% in 2017 to 87% in 2020, compared to 95% in North America and 85% in Western Europe. Mexico (100%), Brazil (85%), Argentina (83%), Colombia (83%) and Peru (81%) rank above the global average (77%), while Panama (60%), Costa Rica (56%) and Ecuador (31%) remain below (KPMG, 2020[62]). In the financial sphere, the number of signatories of the Principles for Responsible Investment doubled in emerging markets over 2019-21, with Brazil being among the emerging countries with the highest number of signatories (OECD, 2022[104]). Adoption of environmental management systems has gradually increased among firms in LAC, signalling greater awareness of sustainability criteria (Figure 5.7). However, growth in the number of ISO 14001-certified enterprises is mainly attributed to efforts made by firms in Argentina, Brazil, Colombia and Mexico (ECLAC, 2021[63]).

Figure 5.7. **Number of LAC firms adopting environmental management systems increased more than thirtyfold since 1999**

Number of ISO 14001-certified enterprises in LAC, 1999-2020

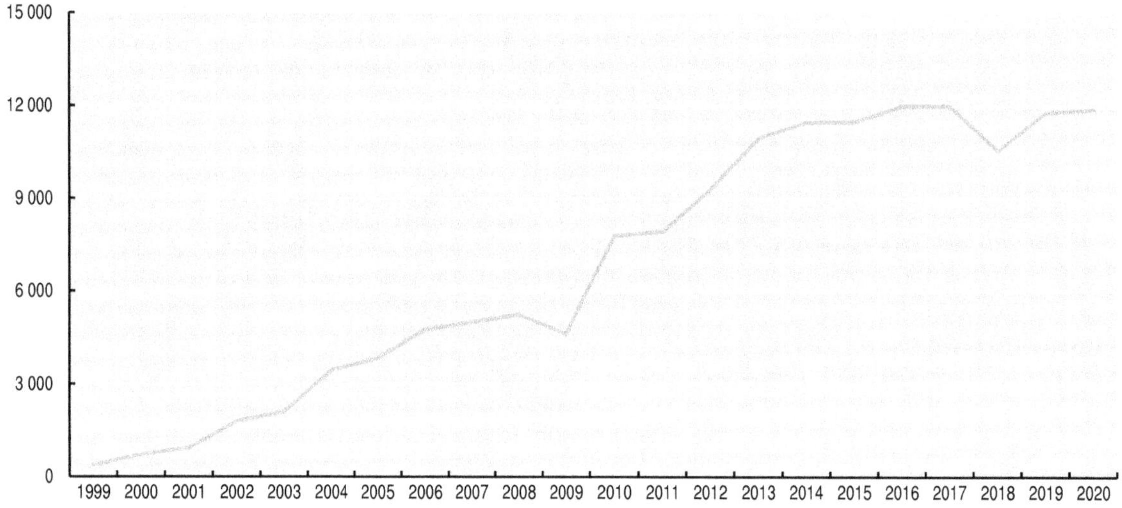

Note: This indicator provides information on the number of companies certified to ISO 14001. The figures include both private companies and public organisations. The international standard ISO 14001 is part of the family ISO 4000. This standard applies to any company, independent of its activity, size or country of operation, that is implementing an environmental management system on the basis of compliance with national legislation and continuous improvement of its performance.
Source: (ECLAC, 2021[63]).

StatLink https://stat.link/lgqfx8

According to the OECD 2021 Business Survey on RBC in LAC,[5] the majority of businesses surveyed have taken steps to manage social and environmental risks: 75% of respondents indicate they have in place a policy setting out RBC expectations and 55% report on their RBC-related practices. However, practical implementation of RBC remains a challenge in LAC countries. Only a minority of businesses seem to conduct risk assessments along the supply chain to minimise the negative impacts of their activities and maximise their positive contributions to sustainable development. Only 40% of respondents adopt a due diligence process when risks are identified, and only 21% take into account suppliers and business partners beyond Tier 1 in the supply chain. In addition, only 36% of businesses are familiar with the support that NCPs can provide to promote and facilitate RBC. Nevertheless, according to the survey, businesses are keen to fill the implementation gap for RBC and to address environmental, social and human rights issues in their supply chains. Over 60% indicate the need for further support and trainings to implement risk-based due diligence and the OECD RBC instruments (OECD, 2021[64]).

Key sectors linked to environmental impacts and climate change risks in LAC: Agriculture and extractives and minerals

Environmental considerations are of particular relevance for some main sectors of activity and trade in LAC, and thus support adoption of RBC practices and standards. For instance, the extractive and minerals sector plays a major socio-economic role in LAC. However, it causes significant environmental and social impacts, including cases of water, air and soil pollution, deforestation and loss of biodiversity. These impacts can also fuel social conflict. An increasing number of industry initiatives aim to address these issues. For instance, extractive companies in the region started to identify and assess environmental risks by tracking and disclosing data from mining sites (OECD, 2022[65]).

Through expansion of land use or increased yields, the agricultural sector can also contribute to environmental degradation, such as forest and biodiversity loss, soil degeneration, water pollution and overexploitation, and greenhouse gas (GHG) emissions. Many agricultural companies operating in the region are taking on these challenges, for example by using certification schemes and developing innovative technologies for sustainable production (OECD, 2022[66]).

To address the negative impacts in these key sectors, it is important that companies have strong internal due diligence mechanisms and policies, and that they integrate relevant environmental standards and international best practices into their operations. Environmental impacts from business activities are usually better managed and avoided when companies have created adequate spaces for dialogue, consultation and engagement with stakeholders throughout the supply chain due diligence process (OECD, 2017[67]). Stakeholder engagement can be more effective and better help identify and address risks when it is: 1) two-way, meaning that parties freely express opinions, share perspectives and listen to alternative viewpoints to reach mutual understanding, with some degree of shared decision-making power; 2) in "good faith", meaning that parties engage with the genuine intention of understanding how stakeholder interests are affected by enterprise activities; and 3) responsive, meaning that there is an active implementation of commitments agreed to by parties (OECD, 2017[67]). In both the agriculture and the extractives and minerals sectors in LAC, the private sector has made efforts to engage more effectively with local stakeholders and communities, for instance through multi-stakeholder initiatives and dialogues (OECD, 2022[66]; OECD, 2022[65]).

Promoting integrity and avoiding policy capture

Undue influence exerted by the private sector can act as a major barrier to advancing the green transition. Indeed, conduct by businesses in the political arena, or what is known as corporate political activity, can have an impact on policy direction and outcomes (Zinnbauer, 2022[68]). Some companies may have an interest in capturing policy making to delay or limit the changes it will bring about, with the ultimate objective of preserving the status quo and protecting their market power. For instance, power and energy are two sectors with strong vested interests in blocking potentially disruptive developments. Incumbents in these sectors often have significant political influence at the top levels of decision-making circles and are particularly prone to political corruption (Arent et al., 2017[25]). In LAC, there is a general widespread perception of policy capture by powerful elites; 73% of the population in 2020 believed that their country was run to the benefit of a few. In Argentina, Chile, El Salvador, Guatemala and Peru, the private sector was seen as holding the most power in the country, ahead of the government (Latinobarometro, 2021[9]).

Stronger regulations to promote integrity and accountability in the private sector are fundamental to a successful green transition. Lobbying and political finance regulations can ensure greater integrity. Relevant issues regarding lobbying include: strengthening the transparency and integrity of corporate interest representation; increasing the transparency of funding, membership composition and decision-making practices in the lobby target-setting adopted by lobbying and business associations; and supporting development and adoption of good practice principles, such as the 2010 OECD Principles for Transparency and Integrity in Lobbying. Several LAC countries including Argentina (2003), Colombia (2011), Chile (2014), Mexico (2010) and Peru (2003) have adopted lobbying laws or regulations. Chile, Colombia and Mexico have lobbyist registers; among these, Colombia does not impose sanctions for non-compliance. Only four countries (Argentina, Chile, Mexico and Peru) require that the agendas of public officials be made public and five countries (Argentina, Colombia, Costa Rica, Mexico and Peru) require the disclosure of names of members of permanent advisory bodies (OECD, 2020[69]).[6] Early observations confirm that countries with a regulatory framework to enhance the transparency of lobbying activities and policy making generally ensured a greater degree of accountability in policy decisions during the COVID-19 crisis (OECD, 2021[70]).

Political finance regulation is strongly regulated in LAC, with the exception of some countries in Central America (International IDEA, 2020[71]). However, an implementation gap persists: in 11 of 12 countries surveyed, cash contributions were still allowed in 2018, making it easier to circumvent political finance regulations. Moreover, digital technologies and social media are creating "grey areas" that make tracking digital advertisements for political parties and candidates more complex (OECD, 2021[70]).

As lobbying, political finance, corporate governance and, more broadly, RBC are increasingly connected, better co-operation is needed among experts, practitioners, regulators and advocates in these domains. Interlinking data streams through harmonised data frameworks and common identifiers is among measures that could help policy makers better monitor the influence of business in the policy-making process. The ability to triangulate disclosures made by politicians, businesses and lobbying associations on their activities could significantly increase transparency and accountability (Zinnbauer, 2022[68]).

A strategic view of the green transition: Preparing public institutions

Many LAC countries' constitutions such as Argentina, Colombia, Brazil, Bolivia, Ecuador, Paraguay, Peru and Venezuela already recognise the right to a healthy environment but, in practice, this may be difficult to enforce (UN, 2022[55]). Moving the green agenda forward will require institutions to evolve accordingly and to be ready to address the numerous challenges this will bring. A high level of co-ordination will be necessary to design and implement a systemic green transition (Chapters 2 and 3). An integrated vision of this transition will facilitate reaching consensus around green policies while reconciling diverging views and interests (Section on political economy). In practical terms, this integrated vision will be essential to co-ordinate actions across sectors and levels of government, and will provide a coherent approach that can guide actions and delivery mechanisms throughout the process.

Policy coherence and multi-level governance

Achieving a holistic green transition will require enhancing policy coherence for sustainable development across levels and sectors of government. Given the complex interconnections among economic, social and environmental challenges, policy coherence can help policy makers better understand the impacts and spillovers of policies.

Policy coherence has three main objectives (Soria Morales, 2018[72]). The first is to help foster synergies and minimise trade-offs across sectors. For instance, an increase in agricultural land use could undermine efforts in halting biodiversity loss and preserving terrestrial ecosystems. In contrast, fostering sustainable agricultural practices can help increase food sovereignty while supporting water use efficiency targets, especially as agriculture is the major user of the world's freshwater withdrawals.

The second is to help reconcile local, regional and national policy objectives with internationally agreed objectives and avoid fragmented responses. Most policies and objectives related to climate change and sustainable development are a responsibility shared across levels of government or are critically dependent on local actions. It is therefore important to foster co-ordination and alignment of objectives.

The third is to address the transboundary and long-term effects of policies. In particular, informed choices about sustainable development need to consider the long-term impact of policy decisions on the well-being of future generations.

Effective and inclusive institutional and governance mechanisms are important in addressing policy interactions (OECD, 2021[73]). The nature of wicked problems, such as climate change, demand a co-ordinated centre of government (CoG), able to provide horizontal and comprehensive responses that include as many key actors as possible. To ensure their long-term outcome, horizontal green policies should be enhanced across various ministries (e.g. planning, finance, environmental and social ministries) but also with key stakeholders inside and outside public institutions.

Policy co-ordination and coherence in green policies must also occur vertically, among the CoG, subnational governments, municipalities and their main stakeholders. This allows the CoG to oversee high-level co-ordination mechanisms and take an active role in aligning multi-department work plans with clear mandates and resources, ensuring that activities align with government priorities. Moreover, it could avoid policy duplication and overlap and respond to citizens' growing demands for better service delivery. It could also generate cost reductions and positive synergies, e.g. in dealing with issues such as service provision. For instance, water governance can be most effective when looking at hydrological instead of administrative boundaries (e.g. river basin management).

Beyond political short-termism: Linking policy instruments and long-term plans for the green transition in LAC

Climate change and the green transition are complex issues requiring substantial planning and investments. However, as the short-term costs of these actions could be more evident than their long-term benefits, governments are often constrained by public opinion and short-term political cycles, as well as by vested interests, which can derail policies to address these complex, longer-term challenges. It is therefore important that governments have a long-term vision to articulate their actions, such as National Development Plans (NDPs) and strategies, as well as Nationally Determined Contributions (NDCs), climate strategies, defined policies, regulations and sector plans to underpin their pledges (see Annex Table 5.A.1 in Annex 5.A1).

Nationally Determined Contributions: Overview of main priorities with regard to the green transition

LAC governments are aware of the requirement to transition to a greener society. This is why, in 2015, in accordance with the Paris Agreement, several LAC countries committed to developing their own NDC plans, i.e. climate action plans to cut emissions and adapt to climate change impacts. These documents provide information regarding mitigation targets, adaptation actions and economic diversification plans, with the primary objective of reducing GHG emissions to limit global warming to below 2°C – preferably to below 1.5°C – compared to pre-industrial levels by 2030. Countries also developed National Adaptation Plans (NAPs), which are flexible processes that build on each country's existing adaptation activities, helping integrate climate change into national decision making.

NDCs and NAPs allow countries to establish concrete targets, measures and policies allowing stakeholders from all sectors and institutions to contribute from their specific technical and budgetary capabilities and setting the basis for national climate action plans. These platforms also allow countries to co-ordinate whole-of-government approaches by committing to targets that are more ambitious and that help pivot away from imposed top-down targets towards self-determined, bottom-up pledges. The projected impact of NDCs and NAPs can be aggregated to show where LAC governments stand in relation to climate goals and to highlight institutional and financial needs at the ground level. Article 13 of the Paris Agreement also establishes an Enhanced Transparency Framework (ETF) under which countries are required to communicate progress in the implementation of their NDCs.

Even though early adoption and implementation of the NDC framework has led to some ambitious decarbonisation plans in LAC countries, as well as to considerable advances in the energy and transport sectors, there is still a long way to go in terms of implementation. As some countries submit the second iteration of their NDCs, the main challenge is to transform these ambitious objectives into measurable results (Cárdenas, Bonilla and Brusa, 2021[74]). Out of the selected 14 LAC countries analysed in this section, only the Dominican Republic, Panama and Peru have proposed or developed a national monitoring system to track the effective implementation of their commitments. For instance, Panama's National Climate Transparency Platform aims to facilitate the collection, management and dissemination of climate-related data in a consultative and transparent manner (Wetlands International, 2021[75]).

Of the countries analysed, 12 had submitted an update of their NDCs by April 2022 (Table 5.2), primarily strengthening their targets and monitoring methodology. Updates from Ecuador and Uruguay are pending; as they submitted their first NDCs in 2019 and 2017, respectively, they are still under the five-year renewal commitment. Most countries in the Caribbean have submitted updated NDCs, apart from the Bahamas and St. Vincent and the Grenadines. Trinidad and Tobago submitted its first NDC in 2018 and thus remains in the five-year renewal commitment.

Regarding total GHG emission reduction targets committed to by LAC countries in their NDCs, comparison is not straightforward, given the difference in baseline scenarios and reference years employed. Costa Rica's 2020 NDC update is among the few that are rated 2°C compatible (CAT, 2020[76]). By contrast, despite a 50% GHG emission reduction target by 2030 with respect to 2005 (Figure 5.9), Brazil's 2020 and 2022 NDC updates allow higher emissions than the 2016 NDC (Unterstell and Martins, 2022[77]; CAT, 2022[78]). Mexico's 2020 NDC update also decreased its emission reduction ambition with respect to the 2016 NDC (CAT, 2020[79]).

While Argentina, Brazil, Colombia, Costa Rica and Panama established only unconditional targets, the majority of countries also set conditional targets – i.e. commitments that directly depend on the delivery of international funding (Chapters 4 and 6) (Figure 5.8). The latter is especially true for some heavily-indebted Caribbean countries that are currently struggling with the negative effects of climate change. For instance, the Barbados' 2021 NDC update commits to a GHG emission reduction target of 35% relative to business-as-usual emissions in 2030 without international support (unconditional), which could increase up to 70% with international support (conditional). Similarly, given that the intensity and frequency of climate change impacts are above the ability of the country to adapt, the sectoral targets set in the NDC of Antigua and Barbuda are entirely contingent upon receiving international support for technology transfer, capacity building and financial resources, with an estimated cost of around USD 1-1.7 billion (United States dollar) (UNFCCC, 2021[80]).

Figure 5.8. **For many LAC countries, achieving the most ambitious emission reduction targets depends on external funding**

Total GHG emission reduction targets commitments in LAC NDCs by 2030

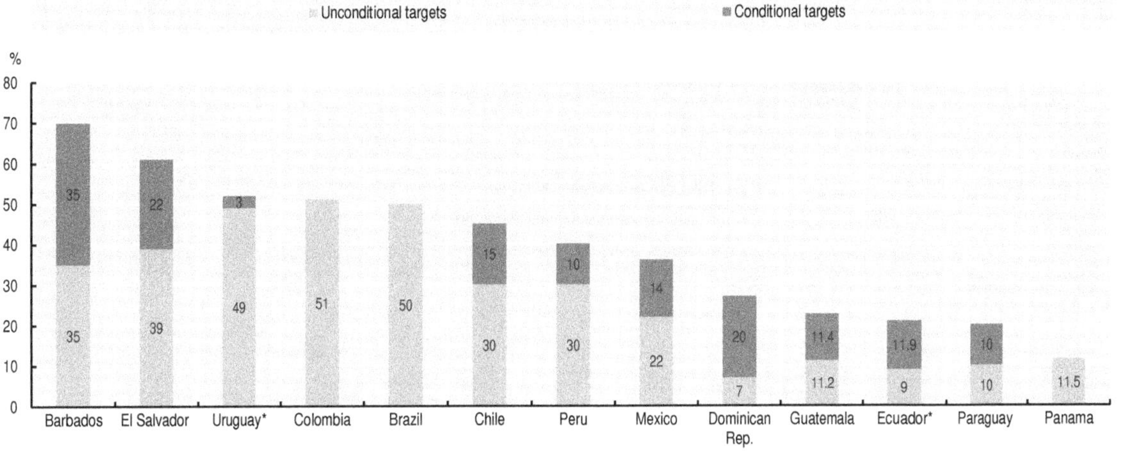

*Ecuador and Uruguay targets refer to 2025, not 2030.

Notes: Total GHG emission reduction targets correspond to the sum of unconditional and conditional targets. Argentina and Costa Rica did not officially set a relative target. Antigua and Barbuda did not officially set an economy-wide GHG emission reduction target in its updated NDC.

Source: Authors' elaboration based on countries' NDCs.

StatLink https://stat.link/e1qn4l

Countries are free to decide the NDC's reference point (e.g. base year), scope and coverage, as well as the timeframe for implementation and methodologies used; these elements may therefore differ across countries. Six LAC countries set targets referencing a base year; six others used a dynamic business-as-usual scenario (Table 5.2). The latter is the most common across LAC GHG emission reduction targets, having the advantage of giving countries more flexibility as mitigation information changes over time. Nevertheless, constant updates also raise questions about countries' level of ambition, with no guarantee that the same or a higher level of ambition will be maintained for the next NDC. This can also have direct implications for policy uncertainty, as green public policy plans need to be adapted every time targets change.

Moving towards a static baseline target could provide more certainty for policy makers and non-state actors, considering that the target's progress would be easier to track (Vaidyula and Hood, 2018[81]). It would also allow for greater accountability in terms of monitoring the progress of implementation of commitments. However, it is important that this fixed target align with the best understanding of emissions. For instance, Costa Rica's 2020 NDC update does not consider a relative target with reference to a base year but only an absolute level of net emissions by 2030 (9.11Mt CO_2e [carbon dioxide equivalent]) and an absolute maximum budget of 106.53 Mt CO_2e net emissions for the period 2021-30 (UNFCCC, 2020[82]). Similarly, Argentina's 2021 NDC update commits only to an absolute target, applying to all sectors of the economy, equivalent to 349 Mt CO_2e by 2030 (UNFCCC, 2021[83]).

While the focus of most NDCs is on adaptation policies (Table 5.2), in practice countries tend to invest more in mitigation (Buchner et al., 2021[84]), evidencing the gap between political ambitions and actual investment. Eight LAC countries have a clearer tendency for adaptation measures rather than mitigation in their NDCs, three opt for a mixed approach, and two have a stronger focus on mitigation goals. However, these commitments contrast with actual climate-related primary investments made by private and public institutions. Over 2019/20, investments in climate mitigation averaged USD 28 billion, while investments in climate adaptation stood at USD 4.5 billion, and investments with multiple objectives stood at USD 2.5 billion (Buchner et al., 2021[84]).

Table 5.2. **NDCs in LAC countries**

Country	NDC update (yes/no), year of publication	GHG emission reduction target (reference to base year vs. BAU scenario)	Adaptation vs. mitigation focus
Argentina	Yes, 2021	Absolute emissions limit for 2030	Mixed
Brazil	Yes, 2020	Base year, 2005	Mitigation
Chile	Yes, 2020	Base year, 2016 and absolute emissions limit for 2030	Mixed
Colombia	Yes, 2020	BAU and absolute emissions limit for 2030	Mixed
Costa Rica	Yes, 2020	Absolute emissions limit for 2030	Mitigation
Dominican Republic	Yes, 2020	BAU	Adaptation
Ecuador	No, 2019	Base year, 2008 and 2010	Adaptation
El Salvador	Yes, 2022	BAU	Adaptation
Guatemala	Yes, 2022	Base year, 2016	Adaptation
Mexico	Yes, 2020	BAU	Adaptation
Panama	Yes, 2020	BAU	Mixed
Paraguay	Yes, 2021	BAU and absolute emissions limit for 2030	Adaptation
Peru	Yes, 2020	Base year	Adaptation
Uruguay	No, 2017*	Base year, 1999	Adaptation

*An update of the Uruguay NDC is planned for November 2022.

Note: BAU = Business as usual. The adaptation vs. mitigation focus is based on the number of specific targets defined within LAC countries' NDCs under each category. A focus on adaptation means the number of adaptation-specific targets exceeds the number of mitigation-specific targets included in the NDC or vice versa.

Source: Authors' elaboration based on countries' NDCs: (UNFCCC, 2021[85]).

Within the mitigation lens, LAC countries prioritise policy measures in sectors such as agriculture, energy (transport and mobility), forestry and land use/land-use change, with one of the primary objectives being to shift public transport systems from fossil fuel to renewable energy (Figure 5.9, Panel A). Costa Rica, for example, developed its own National Decarbonisation Plan 2020-2050, which aims to convert the public transport system to electric power. It also seeks to accelerate and scale up actions to transform agricultural sector activities that produce the most emissions (UNFCCC, 2021[85]). Regarding adaptation, the focus is on agriculture, health and water resources, with the aim of developing countries' agriculture frontier, strengthening adequate use of the land, containing deforestation and promoting implementation of environmental health indicators associated with climate change and the health status of the population. For instance, Uruguay developed the Programme of General Measures for Mitigation and Adaptation, which includes adaptation measures for the agriculture, biodiversity and health sectors and for coastal, water and fishing resources. A common objective across the region is to develop disaster risk management programmes across all sectors to adapt further and to build resilience to climate change impacts; however, only 6% of adaptation targets in current NDCs are directed towards disaster risk management (Figure 5.9, Panel B).

Figure 5.9. **Sector distribution of NDC targets of 14 selected LAC countries**

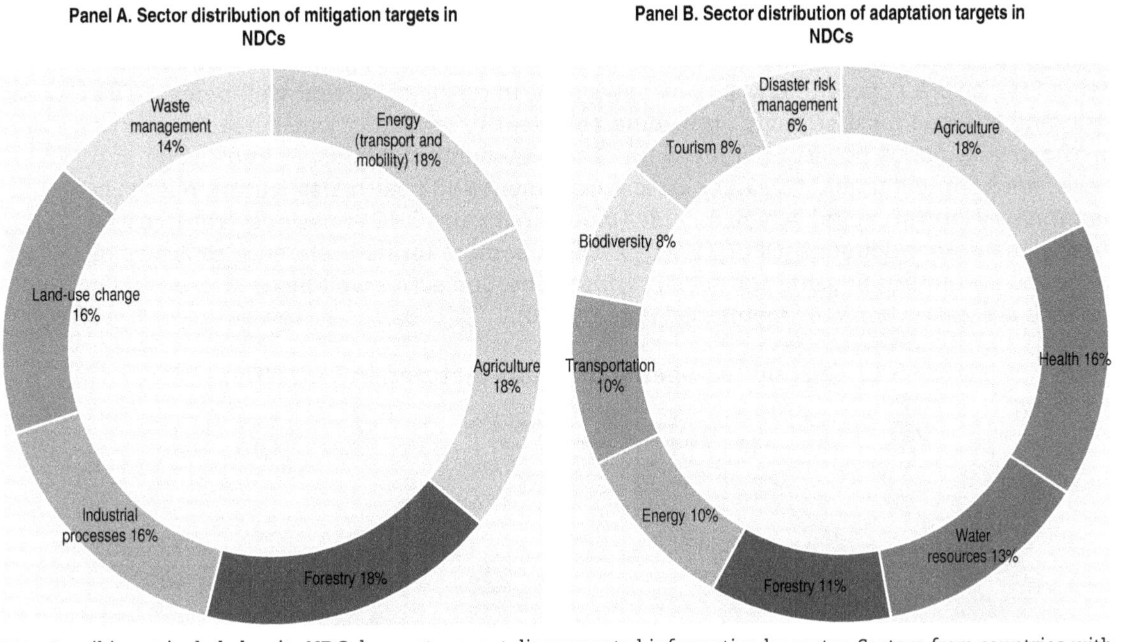

Note: Brazil is not included as its NDC does not present disaggregated information by sector. Sectors from countries with mixed approaches are considered in both figures.

Source: Authors' elaboration based on countries' NDCs: (UNFCCC, 2021[85]).

StatLink https://stat.link/ot9ixw

Institutional mechanisms to ensure coherent policy making: National Development Plans and beyond

As the green transition involves complex and interconnected policies, NDPs can help link short-term policies with longer-term objectives and support consistent implementation over time. These long-term frameworks can also provide a clear and comprehensive logic to evaluate policy spillovers and set cross-cutting objectives to orient actions across levels and sectors of government (Box 5.1). The NDPs can also help guide a whole-of-government approach to raising societal well-being by incorporating a multi-dimensional perspective (OECD, 2021[86]).

Among the most important attributes that can contribute to the effectiveness of these long-term frameworks are: 1) clear goals and indicators to define priorities, allocate financial resources, monitor progress and identify gaps; 2) a solid legal framework or high-level commitment to give the framework greater credibility and authority; 3) a link with the national budget, allowing concrete assessment of policy feasibility; 4) a bottom-up approach, subnational engagement and participatory planning processes to enhance legitimacy; 5) a multi-dimensional lens to reflect the multifaceted nature of countries' development challenges and better consider social, environmental and economic goals, as well as inclusion and sustainability issues; and 6) a monitoring and evaluation (M&E) system, independent from political cycles and government turnover, which is key for assessing policy impacts and implementation and for enabling learning and improvement over time (Box 5.1) (OECD et al., 2019[22]; OECD, 2021[86]).

In particular, by adopting a multi-dimensional approach to public policy, particularly in the agenda-setting stage, NDPs can focus government attention on priority areas to raise citizens' well-being and sustainability. While the 2030 Agenda is the most notable example of a multi-dimensional societal progress framework used by LAC countries,

NDPs also play an important role in multi-dimensional agenda setting. Besides promoting alignment across development agendas, linking NDPs with the 2030 Agenda promotes longer frameworks, which are key for sustainable goals (OECD, 2021[86]). Using text-mining analysis, a review of the intensity with which the different dimensions of the OECD Well-Being Framework appear in LAC NDPs shows that, in terms of current well-being, NDPs still tend to focus most strongly on income and wealth, reflecting widespread concerns in the region about poverty (Figure 5.10). The well-being domains of knowledge and skills, environmental quality, safety, civic engagement and health also feature relatively commonly. Natural capital (which includes biodiversity and GHG emissions) features less clearly than other domains (OECD, 2021[86]). There is more limited reference to issues of work and job quality, housing and social connections. Subjective well-being and work-life balance are least commonly mentioned (Figure 5.10).

Figure 5.10. **Current well-being domains featured in LAC NDPs**

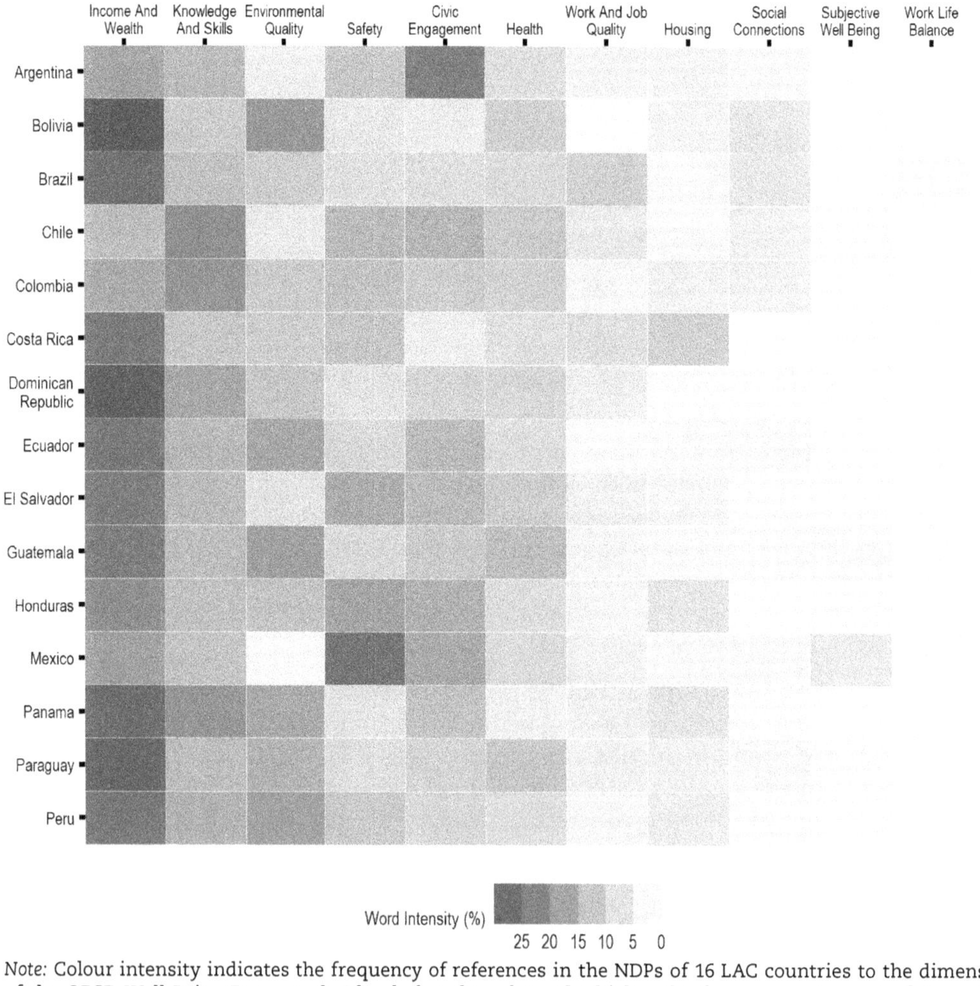

Note: Colour intensity indicates the frequency of references in the NDPs of 16 LAC countries to the dimensions of the OECD Well-Being Framework. The darker the colour, the higher the frequency. The sum of the relative frequencies across all dimensions in a country's NDP is 100. Each country's text data come from the latest NDP (or equivalent) approved by the end of 2020.

Source: OECD (2021[86]).

Box 5.1. Role of the United Kingdom Climate Change Act in providing a long-term framework for emissions reduction

The United Kingdom's (UK) pioneering 2008 Climate Change Act was the world's first comprehensive, long-term and legally binding framework law to address climate change mitigation and adaptation. It is credited with having succeeded in reducing emissions, withstanding political opposition and transforming climate governance in the United Kingdom. The Act sets a legally binding long-term emission reduction target and requires the identification of interim targets every five years.

Beyond its legal framework, the Act has endured thus far thanks to the associated institutions of monitoring, transparency and public accountability. A series of semi-structured interviews with 33 high-level UK policy makers found that the most important features of the Act have been the long-term emissions target; an empowered independent advisory body; mandatory five-year carbon budgets; mandatory government reporting to parliament and the public; and an iterative five-year adaptation planning cycle to ensure learning and flexibility (Fankhauser, Averchenkova and Finnegan, 2018[87]). In particular, the role of an independent expert body – the Climate Change Committee – in advising on carbon budgets and reporting on progress to parliament has helped create the norm of parliamentary and public scrutiny of climate change progress. Moreover, as its mandate extends beyond parliamentary elections, the direction of UK action on climate change has remained focused on the long-term target (Worker and Palmer, 2021[26]; OECD, 2021[88]).

Various mechanisms can help ensure coherent and consistent planning for the green transition. Adequate incentives can help promote active adoption of green policies in areas of government that are not necessarily specialised in the environment. In the short and medium term, the use of existing and pragmatic tools for policy coherence can help create more opportunities for policy makers to reach compromise and implement green policies. These procedures can be adopted all along the public policy cycle and in close co-operation with various jurisdictions, granting spaces for exchange and joint work while ensuring coherent planning and implementation.

In the first place, *ex ante* analysis using well-being indicators is useful to assess the costs and benefits of a policy before its implementation (OECD, 2018[89]). If no social, economic or environmental counter effects are identified during the cost-benefit analysis, policy makers can proceed with the design of the green policy using several tools. Environmental checklists can help ensure that no basic considerations are left unattended. The government of Japan has designed models for various economic activities and sectors that might serve as reference (Japan International Cooperation Agency, n.d.[90]). The opinions and approval of civil society and local communities are crucial for the successive steps of the green policy cycle and can be expressed through opinion surveys (D'Arcangelo et al., 2022[91]) or local meetings. The National Environmental Certification Service for Sustainable Investments of Peru offers a diverse range of mechanisms for citizen participation. Hearing citizen views and including them throughout the policy cycle may prevent future setbacks and strengthen policy design. Peer learning and benchmarking among local governments can also help improve policy design and avoid potential conflicts.

Once a policy is legitimated, several tools can guide implementation. Among them, collaboration agreements between the CoG and the local government implementing the policy can help clarify its main objectives and different stages, including key elements for the success of green policies that tend to be forgotten, such as civil society participation or compensation measures. To ensure that local bureaucracies have the necessary tools and information to implement the policy, CoGs can incentivise the creation of executive units at the municipal government. This way, a focal point is identified and can be subject to monitoring but also profit from trainings and advice from the national counterpart for capacity building. For instance, executive units can be trained in the basic requirements for a sustainable use of land, making sure that no area will be assigned an activity that will put the environment at risk. The National Institute for Federalism and Municipal Development of Mexico provides several examples of collaboration agreements to strengthen co-ordination between the CoG and municipalities and to develop capacity building for local bureaucracies. In case this collaboration does not suffice and the executive unit faces delays or unforeseen events, technical assistance can be requested. The territorial strategic plan of Argentina, for example, foresees technical assistance for provincial governments to ensure that diagnostics, strategies and instruments are well implemented (ECLAC, 2017[92]). Use of environmental communication can help explain in simpler terms the technical aspects of a green policy while keeping citizens informed of its goals and evolution, ensuring transparency (Aparicio Cid, 2016[93]). Where and how this information will be communicated is key, as imperfect information is higher among the most disadvantaged groups (Vona, 2021[38]).

Given the complexity of green policies and the multi-level co-ordination they require, the M&E phase is crucial to identify potential setbacks and improvements. The government of Germany has developed a set of environmental targets and indicators, together with a data collection system, that could prove useful for the design of green M&E systems (OECD, 2020[94]). Unlike traditional M&E systems, which are only accessible by the direct authorities implementing the policies, a digital dashboard of monitoring and control could allow easier exchange of information among levels of government. By implementing an accessible system of control, CoGs can follow the evolution of the policy without having to be physically in every jurisdiction, while providing constant guidance to the local executive units.

These institutional mechanisms can strengthen green policies by improving communication and co-operation channels among relevant actors; promoting and disseminating tools and practices for capacity building; promoting inter-jurisdictional exchange activities to foster mutual learning and co-operation; strengthening the capacities of intermediary and local bureaucracies; and reducing the risks of policy failures or judicialisation.

Government as a strategic actor to advance the green transition

When it comes to moving the green agenda forward, governments are expected to play active roles beyond being regulators and providing defining frameworks (Table 5.3). Governments have various tools at their disposal that can send clear signals to markets and individuals to help shape their behaviour.

First, governments can lead by example by investing in R&D or establishing ambitious targets for reducing the carbon footprint of public-sector activities. Among other measures, they can make more efficient and electrify their transport fleets. Governments can also send clear messages to businesses and citizens through their roles as regulator, consumer and investor. Last, governments can accompany and support communities and firms during the transition by temporarily absorbing and mitigating some of the short-term costs and distributional effects. Among other actions, they can offer reskilling programmes for workers employed in extractive activities to support their transition to greener jobs. They can also provide subsidies to encourage businesses to adopt more sustainable practices (Table 5.3).

Table 5.3. **Governments, in their diverse roles, can apply a range of tools to shape the green transition**

	Roles of government in the green transition				
	Leader	**Regulator**	**Consumer**	**Investor**	**Facilitator**
Tools and policies	Green budgeting; awareness-raising campaigns; emission reduction targets for public activities; public investment in R&D; open data and statistical development to anticipate future needs	Sustainable labelling and standards; land use mapping and spatial planning for sustainable land management; binding GHG reduction targets; energy efficiency standards; due diligence standards for RBC	Green supply chains and green public procurement	Earmarked funding; pre-investment analysis with clear sustainability and impact criteria; taxonomy for sustainable activities	Reskilling and trainings; green subsidies and incentives; feed-in tariffs; environmental taxes; social climate funds; multi-stakeholder dialogues and platforms
Examples	The Ministry of Environment of Panama developed a Climate Change Labelling Manual for Public Investment Projects serves as a tool to identify, classify, weight and mark relevant expenditures to address the climate crisis in the government's budget system. In Uruguay, the Ministries of Environment and Finance, along with the national digital government agency (AGESIC) and the Inter-American Development Bank, co-hosted a hackathon to promote data reutilization related to climate change, including climate finance data.	Rule No.2 of 4 June 2014, issued by the Secretary of Logistics and Information Technology in the Ministry of Planning, Budget, and Management of Brazil, states that new federal public building projects must be developed to ENCE* Class A project standards, and retrofitting work must obtain the ENCE Class A label for individual lighting and air conditioning systems.	The municipality of Mendoza (Argentina) uses its public procurement system to enable a triple-impact economy (economic, social and environmental) by allowing procuring agencies to prioritise goods and services from companies certified as B Corporations (i.e. complying with environmental, social and governance criteria).	The Ministry of Finance of Colombia developed a green taxonomy for mobilising public and private finance for climate change.	In Costa Rica, the Phytosanitary State Service established a Voluntary Certification in Good Agricultural Practices by which farmers can participate in capacity-building activities on good agricultural practices relative to workers' health and the environment; receive support for implementing such practices on their farms; and be audited to verify compliance with requirements of the certification.

*ENCE is the National Energy Conservation Label.
Source: Authors' elaboration.

In particular, governments can use various tools to stimulate the innovation ecosystem at the heart of the green transition. A more sustainable development model depends not only on technological innovation but also on social, economic, behavioural and business model innovations (OECD, 2019[95]; Söderholm, 2020[96]) (Box 5.2). The mix of technologies used in production and consumption processes needs to change radically in the next decades to achieve the 2030 Agenda and reach the targets set in the Paris Agreement. Several advances are evident across sectors. For instance, electric vehicles, solar photovoltaics and onshore wind are now mature technologies. However, further research is needed for cost-effective and scalable solutions to improve water quality, alleviate air pollution and effectively manage solid waste.

Investment in R&D remains low and government-driven in LAC (Chapter 3). Another problem in LAC is that the private sector has little incentive to invest in socially innovative R&D, owing to long time horizons and uncertainty relative to commercial viability.

To counter this inertia, governments can play a key role in lifting barriers to accelerate innovation and in creating the right environment to support the transition. Truly transformative change can arise only if governments commit to systemic change (Chapters 2 and 3) and implement environment-friendly and mission-driven policies that change technologies, regulations, price signals and social attitudes (OECD, 2019[95]; OECD, 2021[98]).

For this reason, policy alignment across sectors and levels of government (Section on policy coherence), coupled with a clear vision and a long-term commitment by government to ensure policy certainty (Section on long-term planning), is key to support a well-functioning innovation ecosystem and to encourage the private sector to make risky long-term investments towards the green transition. Moreover, governments can use public funds to leverage further private investments in innovative green projects (Chapter 4) and establish dedicated platforms where governments, businesses and researchers can co-operate on key transition issues.

Box 5.2. **Social innovation in LAC to support the transition**

Social innovation refers to the design and implementation of new solutions that deal with socio-economic and environmental problems; it implies conceptual, process, product or organisational change. Unlike technological innovation, social innovation can include technologies only if they contribute to solving social challenges and improve people's well-being (OECD, n.d.[99]).

To tap the full potential of social innovation, an enabling policy framework is needed, to support public, non-profit and private actors to co-construct and implement socially innovative solutions (OECD, n.d.[99]). NGOs have been the driving force behind social innovation initiatives and are essential for their implementation in LAC. However, without sufficient state support at local, subnational and national levels, these social innovations lack scalability (Chatham House, 2020[100]; SI-DRIVE, 2015[101]).

Some governments in LAC, including in Argentina, Colombia and Chile, have embedded social innovation programmes within government ministries. In the early 2000s, the city of Medellín (Colombia), introduced an innovative cable car system to provide transport into the city for the poorest people living in hillside communities; it is a successful example of a government-backed social innovation initiative that integrated education, social programmes and participatory budgets (IDB, 2016[102]; Chatham House, 2020[100]). Universities are also relevant actors in promoting social innovation in LAC, as exemplified by the Social Innovation Lab of the Pontificia Universidad Católica de Chile and the Latin American Social Innovation Network (LASIN) of universities in LAC and Europe.

Beyond alliances and government backing, funding is another important element for social innovation initiatives (SI-DRIVE, 2015[101]). For example, Potencial, a group of social entrepreneurs in Chile, developed a product that retained 97% of particulate emissions produced by domestic stoves, which are widely used by low-income households for heating. Thanks to public funding by Impacta Energía 2016 of the Government Laboratory and the Ministry of Energy of Chile, and Capital Semilla 2017 from the Production Development Corporation (CORFO), Potencial was able to advance in prototyping and testing. However, the project now needs further private investment to be scaled up (Soto Espinace, 2019[103]).

Key policy messages

LAC citizens are among the most concerned about climate change globally. This concern is also consistent across the political spectrum. The COVID-19 crisis may have further changed perceptions of climate change and support to fight it, opening a "critical juncture" during the recovery phase. Governments could leverage this high level of concern to make the green agenda the binding element of a new, sustainable social contract for the region. Box 5.3 highlights the most critical and salient considerations for governing a just, inclusive and sustainable green transition.

Box 5.3. **Key policy messages**

Understanding climate policy and the political economy factors that may enable or limit reform is crucial for effective policy formulation and implementation. Among others, four elements, or 4 Cs, are important:

- *Concertation*: involving key stakeholders from the beginning and throughout all the policy-making cycle to reach a common agenda.
- *Context*: adjusting the speed and scale of the transition to the national and international socio-political context.
- *Communication*: implementing an evidence-based, inclusive and compelling communication strategy based on concrete facts and key outcomes, as well as audience and behavioural insights, to shape the transition narrative and rally diverse segments of the population to action.
- *Compensation*: putting in place policies to support the most vulnerable groups along the path to transition while also helping populations cope with the irreversible and differentiated effects of ever more frequent and extreme weather events.

The multitude of actors, sectors and constituencies affected by the green agenda highlights the need to build consensus and establish a shared platform for negotiation. Participatory policy-making processes are even more important, to avoid powerful elites capturing policies in their interests or derailing reforms. In particular, governments need to:

- Foster the participation of citizens, civil society and local communities in order to make the most of their knowledge and develop inclusive and transparent policies with a special focus on sectors and population groups that have been historically excluded, as well as groups more vulnerable to shocks.
- Include women, and local indigenous communities in immediate response efforts and promote their participation in the decision-making process, to broaden the scope of dialogue and increase ownership of agreements achieved.
- Encourage businesses to ensure expectations relating to climate action are met and that environmental impacts are identified, addressed and mitigated as part of supply chain due diligence processes. In particular, governments could provide further support and training to LAC firms to implement risk-based due diligence and the OECD Responsible Business Conduct instruments.

To effectively steer society towards a more sustainable development path, public institutions must become more efficient and work strategically to:

- Invest greater efforts in ensuring an integrated approach to balance economic, social and environmental trade-offs and take advantage of policy spillovers among these fields.
- Align objectives across levels and sectors of government to promote a whole-of-government approach and ensure coherent implementation of the green agenda.
- Link policy instruments with long-term strategies, particularly National Development Plans, for consistent policy implementation over time. The Enhanced Transparency Framework established under the Paris Agreement is a key tool to track progress on climate action.
- Make full use of the various tools at the disposal of governments, ranging from regulation to investment and financing, and leverage their role as strategic market agents in order to send clear signals to businesses and to shape consumer behaviour (e.g. through green public procurement or ambitious sustainability criteria for state-owned enterprises).

Notes

1. Public communication is understood as the government function to deliver information, listen and respond to citizens in the service of the common good and of democratic principles. It is considered distinct from political communication, the legitimate but partisan communication conducted by elected officials, political parties and figures that supports personal, party or electoral objectives.
2. Officially the Regional Agreement on Access to Information, Public Participation and Justice in Environmental Matters in Latin America and the Caribbean.
3. For more information, see the OECD Due Diligence Policy Hub: https://mneguidelines.oecd.org/due-diligence-policy-hub.htm.
4. The MNE Guidelines (last update in 2011) represent the most comprehensive set of government-backed recommendations on RBC available and have become the key reference point for responsible business, together with the International Labour Organization Tripartite Declaration of Principles concerning Multinational Enterprises and Social Policy and the UN Guiding Principles on Business and Human Rights. The *OECD Due Diligence Guidance for Responsible Business Conduct* (2018) provides practical support to enterprises on the implementation of the due diligence process in six stages. In addition, sectoral due diligence guidance instruments have been issued in the extractive, garment, agricultural and financial sectors. For more information, see http://mneguidelines.oecd.org/guidelines/.
5. The survey was disseminated in Argentina, Brazil, Chile, Colombia, Costa Rica, Ecuador, Mexico, Panama and Peru. It was open to responses from companies headquartered in or doing business in these countries for a period of nine weeks (23 November 2020 to 31 January 2021). In total, 501 company representatives across the target countries and a broad range of industries responded. For more information, see https://mneguidelines.oecd.org/survey-business-rbc-latin-america-caribbean.htm.
6. Data are drawn from the 2018 OECD Questionnaire on Public Integrity in Latin America, covering 12 countries. Respondents were predominantly senior officials in central government, supreme audit institutions and electoral commissions.

References

Aparicio Cid, R. (2016), "Environmental communication: conceptual approaches for an emerging field", *Comunicación y Sociedad*, Vol. 25, pp. 209-235, https://doi.org/10.32870/cys.v0i25.4427. [93]

Arenas de Mesa, A. (2016), *Sostenibilidad fiscal y reformas tributarias en América Latina*, LC/G.2688-P, Economic Commission for Latin America and the Caribbean, Santiago, https://www.cepal.org/es/publicaciones/40624-sostenibilidad-fiscal-reformas-tributarias-america-latina. [6]

Arent, D. et al. (2017), *The Political Economy of Clean Energy Transitions*, Oxford University Press, Oxford, UK, https://doi.org/10.1093/oso/9780198802242.001.0001. [25]

Arora, A. and H. Schroeder (2002), "How to avoid unjust energy transitions: insights from the Ruhr region", *Energy, Sustainability and Society*, Vol. 12 (19), pp. 345-5, https://doi.org/10.1186/s13705-022-00. [113]

Buchner, B. et al. (2021), *Global Landscape of Climate Finance 2021*, Climate Policy Initiative, http://www.climatepolicyinitiative.org/publication/global-landscape-of-climate-finance-2021/. [84]

Cabutto, C., S. Nieto Parra and J. Vázquez Zamora (2022), *Hacia un contrato social renovado en tiempos de Covid-19 para América Latina: ¿por qué? ¿qué incluir?, ¿cómo llevarlo a cabo?*, Blog, Foco Economico, https://dev.focoeconomico.org/2022/01/19/hacia-un-contrato-social-renovado-en-tiempos-de-covid-19-para-america-latina-por-que-que-incluir-como-llevarlo-a-cabo/. [31]

Capoccia, G. and R. Kelemen (2007), "The study of critical junctures: theory, narrative, and counterfactuals in historical institutionalism", *World Politics*, Vol. 59/3, pp. 341-369, https://doi.org/10.1017/S0043887100020852. [3]

Cárdenas, M., J. Bonilla and F. Brusa (2021), *Climate Policies in Latin America and the Caribbean: Success Stories and Challenges in the Fight Against Climate Change*, Inter-American Development Bank, Washington, DC, https://publications.iadb.org/publications/english/document/Climate-policies-in-latin-america-and-the-caribbean.pdf. [74]

CAT (2022), *Brazil: Targets*, Climate Action Tracker, Climate Analytics/NewClimate Institute, Berlin, https://climateactiontracker.org/countries/brazil/targets/. [78]

CAT (2020), *Climate Target Update Tracker: Costa Rica*, Climate Action Tracker, Climate Analytics/NewClimate Institute, Berlin, https://climateactiontracker.org/climate-target-update-tracker/costa-rica/. [76]

CAT (2020), *Climate Target Update Tracker: Mexico*, Climate Action Tracker, Climate Analytics/NewClimate Institute, Berlin, https://climateactiontracker.org/climate-target-update-tracker/mexico/. [79]

Cohen, G. and M. Shinwell (2020), "How far are OECD countries from achieving SDG targets for women and girls?: Applying a gender lens to measuring distance to SDG targets", *OECD Statistics Working Papers*, No. 2020/02, OECD Publishing, Paris, https://doi.org/10.1787/17a25070-en. [107]

CONAMA (2018), *O que é o CONAMA?*, Conselho Nacional do Meio Ambiente, Brasília, http://conama.mma.gov.br/o-que-e-o-conama. [105]

Cruciani, M. (2016), *The Energy Transition in Sweden*, Etude de l'Ifri, French Institute of International Relations, Paris, www.ifri.org/en/publications/notes-de-lifri/energy-transition-sweden. [114]

Chatham House (2020), *The Circular Economy in Latin America and the Caribbean: Opportunities for Building Resilience*, Research Paper, Chatham House, London, http://www.chathamhouse.org/2020/09/circular-economy-latin-america-and-caribbean/02-lac-institutions-and-policies. [100]

D'Arcangelo, F. et al. (2022), "A framework to decarbonise the economy", *OECD Economic Policy Papers*, No. 31, OECD Publishing, Paris, https://doi.org/10.1787/4e4d973d-en. [91]

Dechezleprêtre, A. et al. (2022), "Fighting climate change: International attitudes toward climate policies", *OECD Economics Department Working Papers* 1714, https://doi.org/10.1787/3406f29a-en. [7]

Deere, C., G. Alvarado and J. Twyman (2012), "Gender Inequality in Asset Ownership in Latin America: Female Owners vs. Household", *Development and Change*, Wiley & Sons Inc., Hoboken, New Jersey, pp. 505-530, https://doi.org/10.1111/j.1467-7660.2012.01764.x. [54]

ECLAC (2022), *Social Panorama of Latin America 2021*, LC/PUB.2021/17-P, Economic Commission for Latin America and the Caribbean, Santiago, https://hdl.handle.net/11362/47718. [5]

ECLAC (2021), *CEPALSTAT*, Economic Commission for Latin America and the Caribbean, Santiago, https://statistics.cepal.org/portal/cepalstat/. [63]

ECLAC (2020), *Social Panorama of Latin America 2020*, LC/PUB.2021/2-P/Rev.1, Economic Commission for Latin America and the Caribbean, Santiago, http://hdl.handle.net/11362/46687. [2]

ECLAC (2017), *Planificación para el desarrollo en América Latina y el Caribe: enfoques, experiencias y perspectivas*, United Nations Economic Commission for Latin America and the Caribbean, Santiago, www.cepal.org/es/publicaciones/42139-planificacion-desarrollo-america-latina-caribe-enfoques-experiencias. [92]

ECLAC (2018), *Regional Agreement on Access to Information, Public Participation and Justice in Environmental Matters in Latin America and the Caribbean*, Economic Commission for Latin America and the Caribbean, Santiago, https://repositorio.cepal.org/bitstream/handle/11362/43583/1/S1800428_en.pdf. [50]

Edenhofer, O. et al. (2014), *IPCC, 2014: Climate Change 2014: Mitigation of Climate Change. Contribution of Working Group III to the Fifth Assessment Report of the Intergovernmental Panel on Climate Change*, Cambridge University Press, Cambridge. [27]

Evans, C. and E. Zeichmeister (2018), *Education and Risk Assessments Predict Climate Change Concerns in Latin America and the Caribbean*, Latin American Public Opinion Project, Vanderbilt University, Nashville, https://www.vanderbilt.edu/lapop/insights/IO929en.pdf. [16]

Fankhauser, S., A. Averchenkova and J. Finnegan (2018), *10 Years of the UK Climate Change Act*, Centre for Climate Change Economics and Policy/Grantham Research Institute on Climate Change and the Environment, London School of Economics and Political Science, London, https://www.lse.ac.uk/granthaminstitute/wp-content/uploads/2018/03/10-Years-of-the-UK-Climate-Change-Act_Fankhauser-et-al.pdf. [87]

FAO/IFPRI (2018), *The gender gap in land rights*, Food and Agriculture Organization of the United Nations, Rome, https://www.fao.org/documents/card/en/c/I8796EN/. [115]

FIIAPP (2021), *Plataforma de los Pueblos indígenas para enfrentar el Cambio Climático (PPICC) de Perú*, Fundación Internacional y para Iberoamérica de Administración y Políticas Públicas, https://euroclimaplus.org/apoyo-a-la-constitucion-de-la-plataforma-de-los-pueblos-indigenas-para-enfrentar-el-cambio-climatico-ppicc. [116]

Frontline Defenders (2022), *Global Analysis 2021*, https://www.frontlinedefenders.org/en/resource-publication/global-analysis-2021-0. [36]

Gallup (2022), *Gallup World Poll*, Gallup Inc., Washington, DC, https://ga.gallup.com/. [13]

GeoPoll (2022), *GeoPoll Report: Climate Change Perceptions in the Caribbean*, GeoPoll, Denver, USA, https://www.geopoll.com/blog/climate-change-perceptions-caribbean/#Interactive_Data_Dashboard. [14]

Global Witness (2022), *Number of lethal attacks against defenders*, Global Witness, London, http://www.globalwitness.org/en/campaigns/environmental-activists/numbers-lethal-attacks-against-defenders-2012/. [37]

Grinspan, D. and J. Worker (2021), *Implementing Open Data Strategies for Climate Action: Suggestions And Lessons Learned for Government and Civil Society Stakeholders*, World Resources Institute, Washington, DC, https://doi.org/10.46830/wriwp.19.00093. [44]

IDB (2021), *10 key points on climate change impacts, opportunities and priorities for Latin America and the Caribbean*, Inter-American Development Bank, Washington, DC, https://blogs.worldbank.org/latinamerica/10-key-points-climate-change-impacts-opportunities. [49]

IDB (2016), *Study of Social Entrepreneurship and Innovation Ecosystems in the Latin American Pacific Alliance Countries, Country Analysis: Colombia*, Technical Note N. IDB-TN-1150, Multilateral Investment Fund (IADB)/Fundación Ecología y Desarrollo, Washington, DC/Zaragoza, Spain, https://publications.iadb.org/publications/english/document/Study-of-Social-Entrepreneurship-and-Innovation-Ecosystems-in-the-Latin-American-Pacific-Alliance-Countries-Country-Analysis-Colombia.pdf. [102]

Inglehart, R. et al. (2022), *World Values Survey: Round Seven*, JD Systems Institute, Madrid, https://www.worldvaluessurvey.org. [11]

International IDEA (2020), *Political Finance Database (database)*, International Institute for Democracy and Electoral Assistance, Stockholm, https://www.idea.int/data-tools/data/political-finance-database. [71]

IPCC (2014), *Climate Change 2014: Synthesis Report. Contribution of Working Groups I, II and III to the Fifth Assessment Report of the Intergovernmental Panel on Climate Change*, Intergovernmental Panel on Climate Change, Geneva, https://epic.awi.de/id/eprint/37530/1/IPCC_AR5_SYR_Final.pdf. [53]

Ipsos (2021), *Climate Change & Consumer Behaviour: A Global Advisor Survey*, Ipsos, Paris, https://www.ipsos.com/sites/default/files/ct/news/documents/2021-10/Climate-Change-and-Consumer-Behaviour-2021.pdf. [21]

Ipsos (2021), *Earth Day 2021*, Ipsos Global Advisor, Ipsos, Paris, https://www.ipsos.com/sites/default/files/ct/news/documents/2021-04/G%40%20Earth%20Day%202021.pdf. [8]

Ipsos (2020), *How Does the World View Climate Change and Covid-19*, Ipsos, Paris, http://www.ipsos.com/sites/default/files/ct/news/documents/2020-04/earth-day-2020-ipsos.pdf. [12]

Japan International Cooperation Agency (n.d.), *Table. List of Environmental Checklists*, Japan International Cooperation Agency, Tokyo, https://www.jica.go.jp/english/our_work/social_environmental/guideline/pdf/checklist.pdf. [90]

KPMG (2020), *The KPMG Survey of Sustainability Reporting 2020*, KPMG, Amstelveen, Netherlands, https://assets.kpmg/content/dam/kpmg/xx/pdf/2020/11/the-time-has-come.pdf. [62]

Latinobarometro (2021), *Latinobarometro 2020*, Corporación Latinobarómetro, Santiago, https://www.latinobarometro.org/latContents.jsp?Idioma=0&CMSID=Datos&Idioma=0&CMSID=Datos. [9]

Lloyd's Register Foundation (2020), *World Risk Poll*, powered by Gallup, Lloyd's Register Foundation, London, https://wrp.lrfoundation.org.uk/explore-the-poll/. [10]

Maldonado Valera, C. et al. (2022), *Social cohesion and inclusive social development in Latin America: A proposal for an era of uncertainties*, Project Documents, LC/TS.2021/133/Rev.1, Economic Commission for Latin America and the Caribbean, Santiago, https://repositorio.cepal.org/handle/11362/47884. [28]

Maldonado Valera, C. et al. (2021), *Panorama de la cohesión social en América Latina y el Caribe*, Documentos de Proyectos, LC/TS.2021/205, Economic Commission for Latin America and the Caribbean, Santiago, https://www.cepal.org/es/publicaciones/47667-panorama-la-cohesion-social-america-latina-caribe. [23]

Matasick, C., C. Alfonsi and A. Bellantoni (2020), "Governance responses to disinformation: How open government principles can inform policy options", *OECD Working Papers on Public Governance*, No. 29, OECD Publishing, Paris, https://doi.org/10.1787/d6237c85-en. [39]

McBride, J. (2022), *How Green-Party Success Is Reshaping Global Politics*, Council on Foreign Relations, New York, http://www.cfr.org/backgrounder/how-green-party-success-reshaping-global-politics. [17]

Mohommad, A. and E. Pugacheva (2021), *Impact of COVID-19 on Attitudes to Climate Change and Support for Climate Policies*, WP/22/23, International Monetary Fund, Washington, DC, https://www.imf.org/-/media/Files/Publications/WP/2022/English/wpiea2022023-print-pdf.ashx. [19]

OECD (2022), *About Responsible Business Conduct*, OECD Publishing, Paris, https://mneguidelines.oecd.org/about.htm#:~:text=A%20key%20element%20of%20RBC,how%20those%20impacts%20are%20addressed. [59]

OECD (2022), "Business Insights on Emerging Markets 2022", *OECD Emerging Markets Network*, OECD Development Centre, Paris, www.oecd.org/dev/oecdemnet.htm. [104]

OECD (2022), *Empowering Women in the Transition Towards Green Growth in Greece*, OECD Publishing, Paris, https://doi.org/10.1787/a9eacee6-en. [110]

OECD (2022), *Database of specific instances*, OECD Publishing, Paris, https://mneguidelines.oecd.org/database/. [61]

OECD (2022), *OECD Guidelines for Citizen Participation Processes*, Working Party on Open Government, OECD Publishing, Paris. [30]

OECD (2022), *Open Government Review of Brazil: Towards an Integrated Open Government Agenda*, OECD Public Governance Reviews, OECD Publishing, Paris, https://doi.org/10.1787/3f9009d4-en. [33]

OECD (2022), *Public communication trends after COVID-19: Innovative practices across the OECD and in four Southeast Asian countries*, Working Party on Open Government, OECD Publishing, Paris. [47]

OECD (2022), *Responsible Business Conduct in the Agriculture Sector in Latin America and the Caribbean*, OECD Publishing, Paris, https://mneguidelines.oecd.org/responsible-business-conduct-in-the-agriculture-sector-in-latin-america-and-the-caribbean%20.pdf. [66]

OECD (2022), *Responsible Business Conduct in the Extractive and Minerals*, OECD Publishing, Paris, https://mneguidelines.oecd.org/responsible-business-conduct-in-the-extractive-and-minerals-sector-in-latin-america-and-the-caribbean.pdf. [65]

OECD (2022), *The Short and Winding Road to 2030: Measuring Distance to the SDG Targets*, OECD Publishing, Paris, https://doi.org/10.1787/af4b630d-en. [41]

OECD (2021), *Civic Space Scan of Finland*, OECD Public Governance Reviews, OECD Publishing, Paris, https://doi.org/10.1787/f9e971bd-en. [32]

OECD (2021), *Gender and the Environment: Building Evidence and Policies to Achieve the SDGs*, OECD Publishing, Paris, https://doi.org/10.1787/3d32ca39-en. [52]

OECD (2021), *How's Life in Latin America?: Measuring Well-being for Policy Making*, OECD Publishing, Paris, https://doi.org/10.1787/2965f4fe-en. [86]

OECD (2021), *Implementing the OECD Recommendation on Policy Coherence for Sustainable Development – Guidance Note COM/DCD/DAC/GOV/PGC*, OECD Publishing, Paris, https://www.oecd.org/gov/pcsd/pcsd-guidance-note-publication.pdf. [73]

OECD (2021), *Lobbying in the 21st Century: Transparency, Integrity and Access*, OECD Publishing, Paris, https://doi.org/10.1787/c6d8eff8-en. [70]

OECD (2021), *OECD Business Survey Results on Responsible Business Conduct in Latin America and the Caribbean*, OECD Publishing, Paris, https://mneguidelines.oecd.org/oecd-business-survey-results-on-responsible-business-conduct-in-latin-america-and-the-caribbean.pdf. [64]

OECD (2021), *OECD Report on Public Communication: The Global Context and the Way Forward*, OECD Publishing, Paris, https://doi.org/10.1787/22f8031c-en. [40]

OECD (2021), *Public Sector Innovation Facets: Mission-Oriented Innovation*, OECD Publishing, Paris, https://oecd-opsi.org/wp-content/uploads/2021/10/OECD-Innovation-Facets-Brief-Mission-Oriented-Innovation-2021.pdf. [98]

OECD (2021), "Recent progress in mapping the gender-environment nexus", in *Gender and the Environment: Building Evidence and Policies to Achieve the SDGs*, OECD Publishing, Paris, https://doi.org/10.1787/129a4345-en. [108]

OECD (2021), *The role of OECD instruments on responsible business conduct in progressing environmental objectives*, OECD Publishing, Paris, http://mneguidelines.oecd.org/The-role-of-OECD-instruments-on-responsible-business-conduct-in-progressing-environmental-objectives.pdf. [58]

OECD (2021), *The United Kingdom's pioneering Climate Change Act*, OECD International Programme for Action on Climate, OECD Publishing, Paris, https://www.oecd.org/climate-action/ipac/practices/the-united-kingdom-s-pioneering-climate-change-act-c08c3d7a/. [88]

OECD (2020), *Government at a Glance: Latin America and the Caribbean 2020*, OECD Publishing, Paris, https://doi.org/10.1787/13130fbb-en. [69]

OECD (2020), *Innovative Citizen Participation and New Democratic Institutions: Catching the Deliberative Wave*, OECD Publishing, Paris, https://doi.org/10.1787/339306da-en. [35]

OECD (2020), *Policy Framework on Sound Public Governance: Baseline Features of Governments that Work Well*, OECD Publishing, Paris, https://doi.org/10.1787/c03e01b3-en. [94]

OECD (2019), *Innovation and Business/Market Opportunities associated with Energy Transitions and a Cleaner Global Environment*, prepared by the OECD as input for the 2019 G20 Ministerial Meeting on Energy Transitions and Global Environment for Sustainable Growth, OECD Publishing, Paris, https://www.oecd.org/g20/summits/osaka/OECD-G20-Paper-Innovation-and-Green-Transition.pdf. [95]

OECD (2019), *Responsible Business Conduct in LAC: OECD scope and activities*, OECD Publishing, Paris, https://mneguidelines.oecd.org/RBC-LAC-scope-and-activities.pdf. [60]

OECD (2019), *Social Institutions and Gender Index (SIGI) Policy Simulator*, OECD Publishing, Paris, https://sim.oecd.org/Default.ashx?lang=En&ds=SIGI&d1c=ame&cs=ame. [118]

OECD (2019), *The Path to Becoming a Data-Driven Public Sector*, OECD Digital Government Studies, OECD Publishing, Paris, https://doi.org/10.1787/059814a7-en. [42]

OECD (2018), *Cost-Benefit Analysis and the Environment: Further Developments and Policy Use*, OECD Publishing, Paris, https://doi.org/10.1787/9789264085169-en. [89]

OECD (2018), *Open Government Data Report: Enhancing Policy Maturity for Sustainable Impact*, OECD Digital Government Studies, OECD Publishing, Paris, https://doi.org/10.1787/9789264305847-en. [43]

OECD (2017), *OECD Due Diligence Guidance for Meaningful Stakeholder Engagement in the Extractive Sector*, OECD Publishing, Paris, https://doi.org/10.1787/9789264252462-en. [67]

OECD (2017), *Recommendation of the Council on Open Government*, OECD/LEGAL/0438, OECD Publishing, Paris, https://legalinstruments.oecd.org/en/instruments/OECD-LEGAL-0438. [29]

OECD (n.d.), *Social Innovation*, OECD Publishing, Paris, https://www.oecd.org/regional/leed/social-innovation.htm. [99]

OECD et al. (2019), *Latin American Economic Outlook 2019: Development in Transition*, OECD Publishing, Paris, https://doi.org/10.1787/g2g9ff18-en. [22]

OECD et al. (2021), *Latin American Economic Outlook 2021: Working Together for a Better Recovery*, OECD Publishing, Paris, https://doi.org/10.1787/5fedabe5-en. [1]

OIDP (2020), *Cocreación de la plataforma Buenos Aires Cambio Climático*, Observatorio Internacional de la Democracia Participativa, Barcelona, www.oidp.net/es/practice.php?id=1326. [106]

Olano, M. (n.d.), *Communicating the Climate Crisis*, Climate XChange, Boston, https://climate-xchange.org/communicating-the-climate-crisis/#speak. [45]

OXFAM (2018), *The Weight of Water on Women: The Long Wake of Hurricane María in Puerto*, OXFAM, Oxford, England, https://www.oxfamamerica.org/explore/research-publications/research-backgrounderwash-gender-report-puerto-rico. [51]

Pogrebinschi, T. (2017), *LATINNO Dataset*, WZB, Berlin, https://latinno.net/es/case/3409/. [112]

Rice, R. (2017), "Indigenous Political Representation in Latin America", *Oxford Research Encyclopedia of Politics*, Oxford University Press, Oxford, England, https://doi.org/10.1093/acrefore/9780190228637.013.243. [18]

Rothstein, B. (2011), *The Quality of Government: Corruption, Social Trust, and Inequality in International Perspective*, University of Chicago Press, Chicago, https://doi.org/10.1111/gove.12146. [24]

Scruggs, L. and S. Benegal (2012), "Declining public concern about climate change: Can we blame the great recession?", *Global Environmental Change*, Elsevier, Amsterdam, pp. 505–515, https://doi.org/10.1016/j.gloenvcha.2012.01.002. [20]

SI-DRIVE (2015), *Social Innovation Strategies – Regional Report D 3.6*, Social Innovation: Driving Force of Social Change (SI_DRIVE), Dortmund, Germany, https://www.si-drive.eu/wp-content/uploads/2018/03/D3.6-SI-DRIVE-Global-Region-Report-2015.pdf. [101]

Söderholm, P. (2020), "The green economy transition: the challenges of technological change for sustainability", *Sustainable Earth*, Springer Nature, Berlin, https://doi.org/10.1186/s42055-020-00029-y. [96]

Soria Morales, E. (2018), *Why is policy coherence essential for achieving the 2030 Agenda?*, United Nations System Staff College, Turin, Italy, http://www.unssc.org/news-and-insights/blog/why-policy-coherence-essential-achieving-2030-agenda. [72]

Soto Espinace, R. (2019), "MPzero: Sustainable, Affordable and Clean Heating Available to Everyone", *Social Innovations Journal*, Social Innovations Partners, Bryn Mawr, PA, https://socialinnovationsjournal.org/editions/issue-53/75-disruptive-innovations/2924-mpzero-sustainable-affordable-and-clean-heating-available-to-everyone. [103]

Strumskyte, S., S. Ramos Magaña and H. Bendig (2022), "Women's leadership in environmental action", *OECD Environment Working Papers*, No. 193, OECD Publishing, Paris, https://doi.org/10.1787/f0038d22-en. [109]

UN (2022), *Constitucionalismo Ambiental en América Latina: Recomendaciones del Sistema de las Naciones Unidas para Chile*, United Nations, Santiago, https://chile.un.org/sites/default/files/2022-03/Constitucionalismo%20ambiental_21Marzo_OK_0.pdf. [55]

UNECA (2020), *Political economy of a green economy: transition in Africa*, United Nations Economic Commission for Africa, Addis Ababa, https://hdl.handle.net/10855/43792. [34]

UNFCCC (2021), *Actualización de la meta de emisiones netas de Argentina al 2030*, Presidency of Argentina, Buenos Aires, https://unfccc.int/sites/default/files/NDC/2022-05/Actualizacio%CC%81n%20meta%20de%20emisiones%202030.pdf. [83]

UNFCCC (2021), *Antigua and Barbuda: Updated Nationally Determined Contribution*, Government of Antigua and Barbuda, Ministry of Health, Wellness and Environment, https://unfccc.int/sites/default/files/NDC/2022-06/ATG%20-%20UNFCCC%20NDC%20-%202021-09-02%20-%20Final.pdf. [80]

UNFCCC (2021), *NDC Registry*, United Nations Framework Convention on Climate Change, Bonn, Germany, https://www4.unfccc.int/sites/NDCStaging/Pages/Home.aspx. [85]

UNFCCC (2020), *Contribución Nacionalmente Determinada 2020: Costa Rica*, United Nations Framework Convention on Climate Change, Bonn, Germany, https://unfccc.int/sites/default/files/NDC/2022-06/Contribucio%CC%81n%20Nacionalmente%20Determinada%20de%20Costa%20Rica%20 2020%20-%20Versio%CC%81n%20Completa.pdf. [82]

Unterstell, N. and N. Martins (2022), *NDC: Analysis of the 2022 update submitted by the Government of Brazil*, Analysis Report, Instituto Talanoa, Rio de Janeiro, Brazil, http://www.institutotalanoa.org/documentos. [77]

Vaidyula, M. and C. Hood (2018), "Accounting for baseline targets in NDCs: Issues and options for guidance", *OECD/IEA Climate Change Expert Group Papers*, No. 2018/02, OECD Publishing, Paris, https://www.oecd.org/env/cc/Accounting_for_baselines_targets_in_NDCS.pdf. [81]

Vona, F. (2021), "Managing the distributional effects of environmental and climate policies: The narrow path for a triple dividend", *OECD Environment Working Papers*, No. 188, OECD Publishing, Paris, https://doi.org/10.1787/361126bd-en. [38]

Vona, F. (2019), "Job losses and political acceptability of climate policies: why the 'job-killing' argument is so persistent and how to overturn it", *Climate Policy*, Taylor & Francis, Abingdon, UK, https://www.tandfonline.com/doi/abs/10.1080/14693062.2018.1532871?journalCode=tcpo20. [48]

Ward, K. (2022), "Gender Equality for a Sustainable Tomorrow: How closing the STEM gap can support climate action", *The Forum Network*, OECD Publishing, Paris, www.oecd-forum.org/posts/gender-equality-for-a-sustainable-tomorrow-how-closing-the-gender-gap-can-support-climate-action. [117]

Wetlands International (2021), *Development of the National Framework for Climate Transparency of Panama*, Wetlands International, Wageningen, Netherlands, https://lac.wetlands.org/wp-content/uploads/sites/2/dlm_uploads/2021/03/CBIT.TdRs-Programming-PSS-Senior.pdf. [75]

Weyland, K. (2008), "Toward a New Theory of Institutional Change", *World Politics*, Cambridge University Press, Cambridge, UK, pp. 281-314, https://doi.org/10.1353/wp.0.0013. [4]

WIN (2022), *WIN World Survey (WWS): 2021 Edition*, Worldwide Independent Network of Market Research (WIN), Milan, https://winmr.com/win-world-survey/. [15]

Worker, J. and N. Palmer (2021), "A Guide to Assessing the Political Economy of Domestic Climate Change Governance", Working Paper, World Resources Institute, Washington, DC, https://www.wri.org/research/guide-assessing-political-economy-domestic-climate-change-governance. [26]

World Bank (2021), *Promoting Climate Change Action in Latin America and the Caribbean*, World Bank Group, Washington, DC, https://reliefweb.int/report/world/promoting-climate-change-action-latin-america-and-caribbean (accessed on 6 April 2022). [57]

WRI (2021), *50 Organizations Committed to Locally Led Adaptation. Now What?*, World Resources Institute, Washington, DC, https://www.wri.org/insights/enacting-principles-locally-led-adaptation. [56]

Yale Climate Connections (2017), *Climate communication: Do apocalyptic messages work?*, Yale School of the Environment, Yale University, New Haven, Connecticut, https://yaleclimateconnections.org/2017/08/climate-communication-do-apocalyptic-messages-work/. [46]

Zinnbauer, D. (2022), *Corporate Political Responsibility: Mobilizing the Private Sector for Political Integrity*, International IDEA, Stockholm, https://www.idea.int/sites/default/files/publications/corporate-political-responsibility.pdf. [68]

Annex 5.A. Recent environmental legislation in LAC

Table 5.A.1. Recent environmental legislation in LAC

Country	Regulation
Argentina	Law of Minimum Budgets for Climate Change Mitigation and Adaptation No. 27.520 (2020): establishes the minimum budgets to guarantee adequate actions, instruments and strategies for adaptation and mitigation of climate change under the terms of Article 41 of the National Constitution. Law for the Implementation of Comprehensive Environmental Education No. 27.621 (2021): dictates general guidelines for environmental training and constitutes the basis for guiding the design of training itineraries in the framework of environmental training for those who work in the public service.
Brazil	Decree N° 11.075 (2022): establishes procedures for preparing Sectorial Plans for the Mitigation of Climate Change and creates the National System for the Reduction of Greenhouse Gas Emissions. It creates a carbon market focused on the export of credits. Federal Decree No. 11.043 (2022): establishes the National Solid Waste Plan (Planares) that brings guidelines, strategies, actions and targets to modernise solid waste management to put into practice the objectives set out in the National Solid Waste Policy – Law No. 12.305 of 2010. Climate Change Law 12.187 (2010): establishes Brazil's National Climate Change Policy, including its principles, objectives, guidelines and instruments.
Chile	Framework Law on Climate Change N.21.455 (2022): establishes climate governance, powers and obligations of state agencies for climate action and sets the goal of carbon neutrality and resilience by 2050. It recognises principles of non-regression and progressivity and incorporates principles of an ecosystem approach, equity and climate justice, territoriality, transparency and citizen participation. Law No. 21.364 (2021):creates the National Disaster Prevention and Response System (SINAPRED) and the National Disaster Prevention and Response Service (SENAPRED) and replaces the National Emergency Office (ONEMI) with the National Disaster Prevention and Response Service.
Colombia	Law N.° 2169 (2021): promotes low-carbon development by establishing minimum goals and measures in terms of carbon neutrality and climate resilience. It also establishes the target of fully implementing the Colombian green taxonomy by 2030. Law 2111 (2021): criminalises deforestation, creates five other environmental crimes and applies more rigorous penalties and sanctions to five existing crimes in the Colombian Penal Code. Law No. 1931 (2018): establishes Guidelines for Climate Change Management in Colombia.
Costa Rica	Decree 43443-MINAE (2022): reforms the Regulations to the Mining Code, with the aim of ordering the processing of concessions for the mining community and the way in which the eventual closure of mining projects should be carried out to avoid negative environmental impacts. Decree No. 41091 (2018): imparts the National Policy on Adaptation to Climate Change, seeking to articulate efforts on adaptation to climate change through effective and informed citizen participation.
Dominican Republic	General Law on Integrated Waste Management and Co-processing 225-20 (2020): aims to prevent generation of waste and establish the legal regime for its integral management. Law on Environmental Education and Communication 94-20 (2020): aims to include environmental education in different levels, cycles, grades, modalities and stages of the school system and higher education in public and private education centres to raise environmental awareness across society. Law Establishing the Payment of Environmental Services 44-18 (2018): ensures the conservation, preservation, restoration and sustainable use of ecosystems through a general framework for compensation and remuneration of environmental services to guarantee environmental or ecosystem services.
Ecuador	Decree No. 59 (2021): establishes an institutional framework to fight climate change, creating the Ministry of Environment, Water and Ecological Transition. Organic Law for a Special Regimen in the Galápagos (2015): ratifies the government's commitment to protect the natural heritage of the Galapagos Islands under technical, scientific and sustainable development criteria.
El Salvador	General Law on Water Resources (2021): recognises the human right to water and sanitation as components of the right to an adequate standard of living. It establishes that water may not be privatised under any condition and creates the Salvadoran Water Authority (ASA). Law on Integral Waste Management and Promotion of Recycling (2020): aims to achieve sanitary and environmentally safe use and disposal of waste to protect human health and the environment and to promote a circular economy. Prohibition of Metallic Mining Law (2017): aims to prohibit metal mining in the soil and subsoil of the territory of the Republic.
Guatemala	Regulation 164 (2021): regulates integral management of solid waste to protect human health and the environment. National Environmental Education Policy 189 (2017): seeks to develop awareness-raising environmental education programmes with a multicultural and equity approach. Framework Law for Climate Change (Decree 7-2013): aims to prevent, plan for and respond in an urgent, adequate, co-ordinated and sustained manner to the impacts of climate change.

Country	Regulation
Mexico	NOM-001-SEMARNAT (2021): establishes permissible limits for pollutants in wastewater discharges into receiving bodies owned by the nation. It updates NOM-001-SEMARNAT-1996, renewing technical aspects that were no longer complied with due to the passage of time. General Law on Climate Change (2018): establishes provisions to address the adverse effects of climate change and to regulate actions for mitigation and adaptation to climate change. Energy Transition Law (2015): establishes guidelines for the use of renewable energies and the financing of the energy transition.
Panama	Law No. 287 (2022): recognises nature as a subject of rights, as well as the obligations of the state and all people to ensure respect and protection of these rights. Decree No.107 (2021): prohibits export of timber extracted from natural forests and water reservoirs in logs. Decree No. 34 (2018): establishes the legal and institutional framework for climate change in Panama. It also establishes 2030 climate goals and a 2030-50 transparency framework.
Paraguay	Law No.6911 (2022): declares the Tobatí stream in the Department of Cordillera a protected wild area under public domain with the category of protected landscape management. Decree No.7017 (2022): regulates Law 3239/2007 on water resources of Paraguay. Law No.5875 (2017): establishes the general regulatory framework for planning for and responding to the impacts of climate change in an urgent, co-ordinated and sustained manner.
Peru	Decree No.023 (2021): approves the National Climate Change Policy towards 2030. Law No.30754 (2018): establishes the Framework Law on Climate Change for the planning, articulation, execution, monitoring, evaluation, reporting and dissemination of public policies for the comprehensive management of climate change.
Uruguay	Decree No.135/021 (2021): establishes air quality objectives and maximum emission limits. Law No.19.889 Art. 291 (2020): creates the Ministry of Environment. Decree No. 310 (2017): approves the National Climate Change Policy and the first NDC.

Note: The table presents the most recent and relevant legislation adopted until May 2022 in the 14 LAC countries analysed.

Source: Authors' elaboration.

Chapter 6

International partnerships for a green and just transition

Regional and international co-operation is necessary to ensure the successful implementation of climate change mitigation and adaptation policies. Many countries in LAC are natural resource-intensive exporters with rich biodiversity, placing them as important actors in climate negotiations. However, they also face an export structure biased towards primary sectors, a condition that makes them vulnerable to new international environmental standards and regulations. This chapter argues that, to transition to a sustainable model of development, LAC governments need a strong convening power and to enhance a unified voice in multilateral environmental agendas. This will help to better illustrate the region's particularities and improve alignment of national policies and internationally established environmental goals. In addition, the chapter looks at how governments will also have to face the green economy's impact on trade. Policies will need to take into account the additional costs imposed on exports in the medium term, along with the effects of international environmental standards and regulations. Finally the chapter shows how regional co-ordination and further co-operation among LAC countries, and sub-regional grouping will be the way forward for advancing on renewable energies and green transition policies in the LAC region.

Regional and international co-operation is crucial for the success of a green and just transition in LAC

The LAC region plays an active but fragmented role in multilateral efforts for advancing the green transition

The region partakes in over **10 coalitions** in climate negotiations

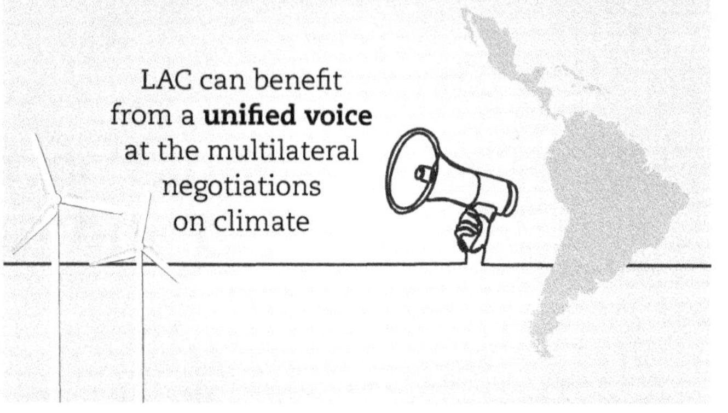

LAC can benefit from a **unified voice** at the multilateral negotiations on climate

Using a comprehensive tool-mix, international co-operation can facilitate LAC's active efforts towards the green transition

LAC holds more than 60% of the world's identified lithium reserves...

... and can play a big role in the push for international low-carbon hydrogen

The region could become one of the world's leading providers of carbon credits, and be among the first to develop a regional carbon market

Latin America reached nearly 40% of the market-share in 2020

International partnerships can help the LAC region navigate the trade-offs of the green transition on trade

International partnerships, in co-ordination with governments, private sector and civil society, can support LAC countries in adapting to new international environmental standards and regulations.

Introduction

The challenges of achieving the low-carbon development targets and a green transition cannot be confronted only at the national level. To transition to a sustainable development model, the LAC region needs a strong convening power and to enhance a unified voice in multilateral environmental agendas. Notwithstanding its fragmented position on climate issues, the region has been active in international climate negotiations and has strongly supported ambitious environmental treaties. Even so, commitments made by LAC countries at the international level do not always translate into national actions, indicating that national policies are in need of further alignment with internationally established environmental goals.

Transitioning to a greener and more sustainable economy has substantial implications for numerous aspects of the region's development path. In particular, a green economy's impact on trade is a concern better addressed earlier than later. Policies will need to take into account the additional costs to exports and the effects of international environmental standards and regulations in the short and long term. In addition, LAC countries can harness international trade in both goods and services to facilitate the transition to a circular economy.

Addressing global challenges, such as climate change, biodiversity loss or pollution, and promoting a green transition as part of coronavirus (COVID-19) recovery are key objectives of the European Union (EU) internally and with respect to key EU partners. The EU's strategic leadership on the green transition has major implications for trading partners, modalities of international co-operation and transition requirements, as stated in the European Green Deal (hereafter, "the Green Deal"). These pose a challenge to the LAC region but, above all, offer an opportunity to make the most of multilateral efforts to bring forth a green and just transition. Boosting international partnerships, in particular with the European Union as part of the "Global Gateway", can help LAC adapt to the new environmental trade norms and regulations.

The LAC community can benefit from the renewed impetus of international partners to pursue a green agenda, transforming commitments into action through harmonised use of international co-operation instruments, following the Development in Transition approach. To ensure a sustainable development the region needs to ensure policy coherence across policy sectors and levels (local, regional and international), harness joint efforts for capacity building and technology transfer on energy sources, and co-ordinate common regulations and norms for carbon markets.

The rest of this chapter is organised as follows. First, it analyses LAC's positioning in climate negotiations at the multilateral level and the extent to which the region could benefit from a strengthened regional agenda and better alignment of national strategies with international commitments. Second, it assesses, through a focus on trade, how international partnerships can soften the impact of the green transition in the LAC region while helping the region grasp its opportunities. This section analyses the impacts of the green transition on trade in terms of expected additional costs on exports in the medium term and the effects of international green norms and regulations on the region's imports and exports. This includes an analysis of the EU Green Deal, one of the most prominent climate policies, and its effects on LAC through possible impacts on trade and potential avenues for co-operation. Third, the chapter stresses the importance of international partnerships in facilitating LAC efforts to translate multilateral commitments into concrete progress in advancing the green transition, primarily by implementing a harmonised and integrated use of international co-operation instruments and working towards common norms, standards and regulations. The chapter concludes by providing preliminary key policy messages for further consideration and implementation.

Aligning multilateral efforts with national strategies for a green and just transition

Through numerous environmental fora and summits, the international climate agenda has evolved to include additional priorities and frameworks. From its global nature, all countries are called to participate in the efforts for reducing carbon dioxide (CO_2) emissions. Although the implications of CO_2 reduction targets in human development standards of developing countries remain a matter of debate, it is clear that it is possible to decouple economic development from CO_2 emissions (Figure 6.1). LAC can reach higher Human Development Index (HDI) levels while meeting its low emissions targets. Green transition policy experiences exist within partners that are already reducing their levels of emissions. At the same time, in this global agenda, LAC countries can play a predominant role in climate negotiations by sharing experiences of sustainable development with other regions of similar and lower development levels. Climate change has shown that continuing a path of exponential growth of CO_2 emissions is no longer an option.

Figure 6.1. **CO_2 emissions per capita vs. HDI**

Time series 1995-2019

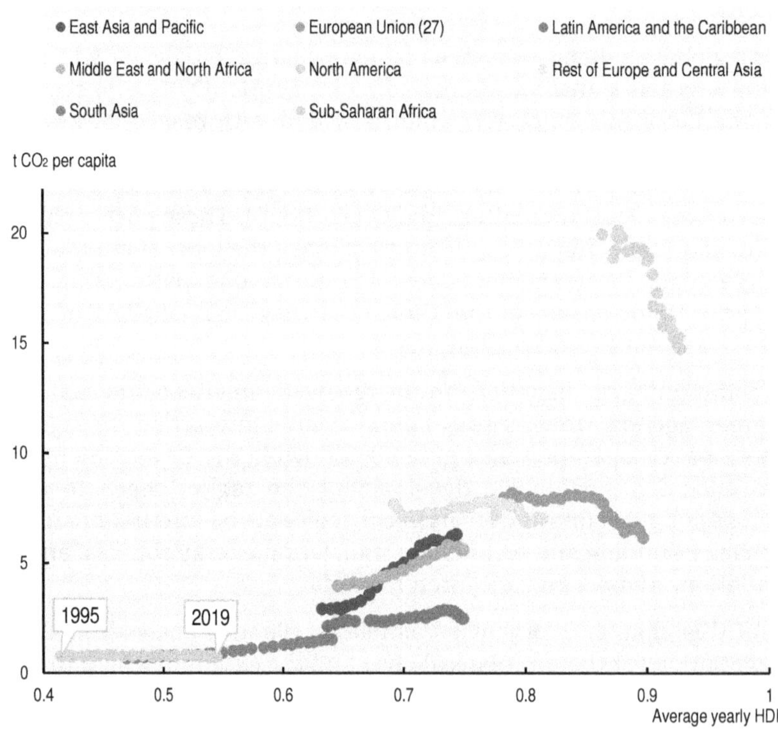

Note: Climate Watch Historical CO_2 Emissions excluding LUCF.

Source: Authors' calculations based on (Climate Watch, 2022[1]) and (UNDP, 2022[2]).

StatLink https://stat.link/fsamjk

From the isolated goal of tackling climate change, international actors have directed their efforts towards green growth initiatives while ensuring that natural assets continue to provide the resources and environmental services on which well-being relies (OECD, 2011[3]). By recognising the need for a fundamental shift to an economic system that is less damaging to the environment, leaders have moved the discussion towards bringing forward the green transition (Box 6.1). This notion expands the scope of environmental action to include an economy in which growth in income and employment is driven by public and private investments that reduce CO_2 emissions and pollution, enhance energy and resource efficiency, and prevent the loss of biodiversity and ecosystem services (UNEP, 2011[4]).

The COVID-19 recovery has been a push to advance the green transition. Recovery packages have not, however, met the level of the ambition. If not consistently combined with structural reforms, these programmes are a "flash in the pan" rather than a socio-environmental transformation (Burger, Kristof and Matthey, 2020[5]).

Efforts to bring forth the green transition can have challenging effects for the labour force if not properly co-ordinated. Policy makers need to ensure that the transition is not only green but also just – in other words, that it balances environmental sustainability with the needs of the people, that are most negatively affected. The Paris Climate Agreement (hereafter, "the Paris Agreement") takes into account the just transition of the workforce and the importance of creating decent work and quality jobs. The development of social dialogue, expansion of social protection, securing of rights in the workplace and creation of employment are key aspects of a fair and inclusive transition (ILO, 2015[6]). The Solidarity and Just Transition Silesia Declaration, adopted at the 24th Conference of the Parties (COP24) to the UN Framework Convention on Climate Change (UNFCCC) in 2018, proved a significant step in highlighting the importance of addressing the vulnerability of labour markets in carbon-intensive sectors.

Box 6.1. **From climate negotiations to a green and just transition**

The signing of the United Nations Framework Convention on Climate Change (UNFCCC) by 154 nations at the Rio Earth Summit in 1992 marked the beginning of multilateral climate negotiations. Aiming for the "stabilisation of greenhouse gas (GHG) concentrations in the atmosphere at a level that would prevent dangerous anthropogenic interference with the climate system", the convention categorised parties according to various commitments and established the "common but differentiated responsibilities and respective capabilities" principle. In 1997, participating parties adopted the Kyoto Protocol (entered into force in 2005), which set internationally binding emissions reduction targets based on a rigid interpretation of the aforementioned principle.

In the Copenhagen Accord of 2009, countries recognised the need to limit global warming to below 2°C to prevent dangerous climate change but were unable to agree on a clear path towards a legally binding treaty. A consensus was reached in 2015, with a new treaty – the Paris Agreement – and came into force in November 2016. It commits all parties to limit global warming to "well below 2°C".

In 2021, the Glasgow Climate Pact strengthened ambitions across three pillars of collective climate action. On adaptation, parties recognised "the importance of the global goal on adaptation for the effective implementation of the Paris Agreement" and "welcomed the launch of the comprehensive two-year Glasgow–Sharm el-Sheikh work programme on the global goal on adaptation". On mitigation, parties agreed to "establish a work programme to urgently scale up mitigation ambition and implementation in this critical decade". Last, on climate finance, parties reaffirmed their pledge to reach the target of committing USD 100 billion per year to climate action in developing and vulnerable countries.

Bringing forth the green transition requires a whole-of-society approach, capable of addressing the needs of workers. While the Paris Agreement contains some elements related to these needs and priorities, the Solidarity and Just Transition Silesia Declaration, adopted at COP24 in 2018, provides further directions on how to advance a green and just transition.

Source: (Andersen, 2015[7]), (Climate Strategies, 2020[8]).

LAC has lacked a unified voice in the international arena, as the region negotiates within multiple coalitions

The fragmented positioning of LAC countries in the international arena for climate negotiation constitutes a missed opportunity for the region. Advancing common priorities and objectives could be easier and more effectively done when backed by the entire LAC community. Greater regional co-operation might favour policy implementation and co-ordination, particularly due to the opportunity to further align national plans with global environmental goals.

The lack of a unified voice in the context of environmental conferences can be attributed to the existing fragmentation of LAC regional integration and often reflects subregional economic ties. In contrast to Europe, where integration has revolved around the European Union, regionalism in LAC encompasses a plurality of groups, which are often competing, complementary and overlapping in function and membership and reflect distinct waves of regionalism throughout the continent (Ruano and Saltalamacchia, 2021[9]). At the core of the complexity and plurality of LAC organisations is the region's state resistance to the establishment of supranational organs and the delegation of competences (Nolte, 2021[10]). In addition, a primary focus on tangible economic benefits through regional co-operation and the lack of a deeper philosophical or ideological commitment to the integration process have resulted in regional projects that often lack long-term perspectives (Pastrana, 2013[11]).

While the LAC region has a reputation for active participation in international climate negotiations, throughout much of its history, there have been few active overall Latin American coalitions operating in climate change talks. Even though the Central American Integration System and the Caribbean Community (CARICOM) have proved longstanding actors in climate negotiations and regional environmental initiatives, their involvement has mostly been limited to operational roles rather than in political negotiations. Similarly, the United Nations regional groups, such as the Group of Latin America and the Caribbean, have not been active in substantive negotiation (Watts and Depledge, 2018[12]).

Subregional coalitions with diverse and fragmented narratives have characterised the positioning of LAC in climate negotiations. The first subregional coalition was the Bolivarian Alliance for the People of Our America (ALBA) at COP15 in Copenhagen (2009). This group has adopted a hard-line approach, focusing on climate justice, equity and an uncompromising interpretation of the principle of "common but differentiated responsibilities and respective capabilities", with an emphasis on industrial countries' historical responsibility for global warming (Marzano Franco, 2016[13]). While often criticised for its radical positions, the coalition has been impactful in strengthening developing nations' voice in climate negotiations and in reminding developed nations of their obligations under the convention.

The Independent Association of Latin America and the Caribbean (AILAC) – established in 2012 and composed of Chile, Colombia, Costa Rica, Guatemala, Honduras, Panama, Paraguay and Peru – presents itself as a "third way" in the North-South discussion, its primary role being to build bridges among negotiating groups. While partly prompted as a response to ALBA, the group has had significantly more impact on climate negotiations. Its bridge-building language was a major factor in framing the fight against climate change not as a divisive but as a common endeavour, thus ultimately contributing to adoption of the Paris Agreement and celebrated by many as an opportunity to break down the historic North-South divide (Watts and Depledge, 2018[12]).

Argentina, Brazil, Uruguay (A-B-U) is a regional coalition formed in 2016 among three countries with very strong historical, economic and political ties. This coalition identified climate change adaptation as a strategic agenda item and established a framework of adaptation principles, which were adopted in their majority by the Group of 77 (G77) and the People's Republic of China (hereafter, "China") (Lorenzo Arana, 2020[14]).

The remaining groups in LAC are not based on regional criteria, are sometimes not limited to countries of the region, and often overlap. The notable exception is Mexico, whose participation is primarily limited to the Environmental Integrity Group (EIG) (Figure 6.2).

Figure 6.2. **LAC countries' participation in selected climate-related international coalitions**

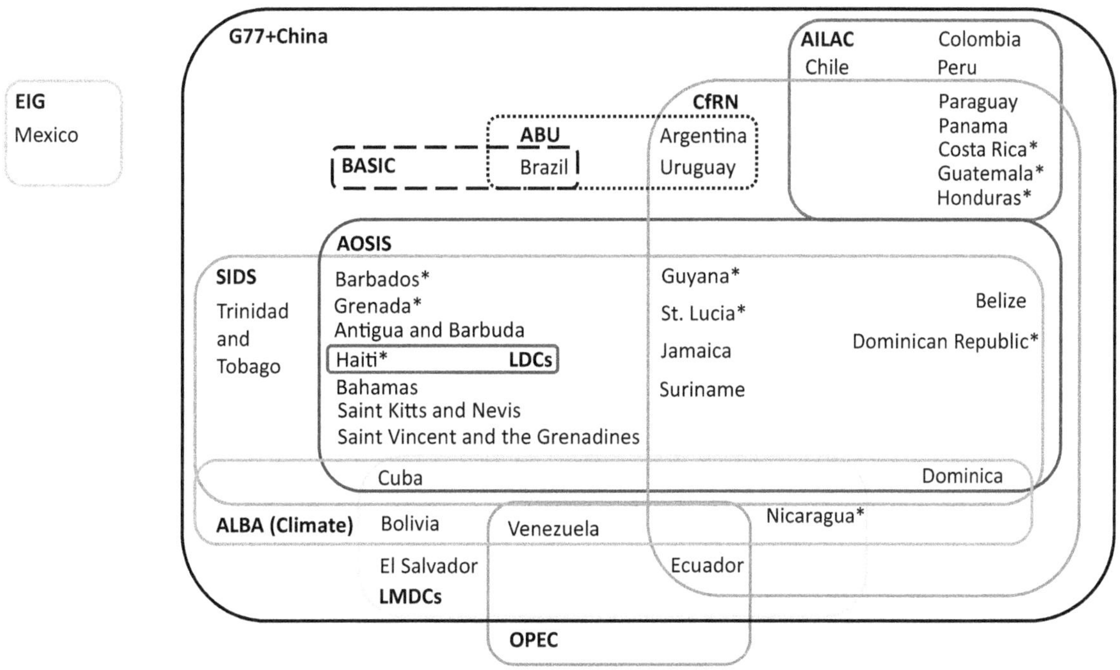

Note: *Members of the Climate Vulnerable Forum (CVF). A-B-U: Argentina, Brazil and Uruguay. AILAC: Independent Alliance of Latin America and the Caribbean. ALBA: Bolivarian Alliance for the Peoples of Our America. AOSIS: Alliance of Small Island States. EIG: Environmental Integrity Group. CfRN: Coalition for Rainforest Nations. LDCs: Least Developed Countries. LMDCs: Like-minded Developing Countries. OPEC: Organisation of the Petroleum Exporting Countries. SIDS: Small Islands Developing States. Non-exhaustive coalitions in the region; some coalitions relate to the environment as part of a broader agenda. Brazil has been part of the LMDCs in the past.

Source: Authors' elaboration based on (Delgado Pugley, 2021[15]); (Klöck et al., 2020[16]); (Watts and Depledge, 2018[12]).

Preventing fragmentation of environmental policies is crucial to harness the full potential of the green transition. As LAC's fragmented voice in climate negotiations in many ways constitutes a missed opportunity, future efforts will need to prioritise enhanced policy dialogue and a regional environmental agenda. To avoid further tensions, this common agenda would need to be based on a bare minimum consensus among LAC countries and gradually expand to more ambitious plans and strategies that are ultimately integrated into national policies.

Finally, while the green transition's overarching goals can be subject to contestation, a science-based policy approach that mobilises technical expertise and capacities can prove an effective tool in overcoming regional organisations' political paralysis. While only a national initiative, the establishment in Chile of an independent scientific body that formulates science-based recommendations not only provided the most advanced sectoral perspectives on policy and technology but also identified overlap with other agendas (IDB, 2021[17]). Similar policies at the regional level could help overcome the limitations of regional initiatives and enhance efforts for technical co-operation.

Strategic international institutional arrangements have helped advance the green agenda in LAC

Despite political fragmentation at the regional level, LAC countries demonstrate exceptional adherence to environmental goals and the green agenda within the framework of international conventions and fora through commitments, plans and strategies established at the national level. Regardless of their approach to climate negotiations, LAC countries' participation in strategic institutional arrangements at the international level has advanced the green transition. Commitments reached are a prominent way of identifying not only LAC countries' dedication to environmental goals but also their specific priorities regarding sustainable development. In particular, numerous LAC countries have set ambitious goals for reaching net zero emissions, with an overwhelming majority aiming for 2050. Furthermore, within the context of the Paris Agreement mandates, all LAC countries have submitted Nationally Determined Contributions (NDCs). In fact, 24 have submitted a second or updated NDC, some of them with notable improvements from one document to the next (IDB, 2021[17]), demonstrating their firm commitment to ratcheting up efforts in the fight against climate change (Chapter 5) (NDC Partnership, 2022[18]).

As the region hosts about 50% of the planet's biodiversity, efforts for its protection and preservation remain a top priority. Ever since the Convention on Biological Diversity (CBD) came into force in 1993, LAC countries have been undertaking the challenge of sustaining their natural heritage without losing sight of economic development. As all LAC countries are parties to the convention, they have been submitting ambitious National Biodiversity Strategies and Action Plans. Most countries are party to the CBD's supplementary agreements, ensuring that biodiversity strategies are included in all planning and activities affecting national biodiversity. As the Aichi Biodiversity Targets, aimed at the year 2020, have gone largely unmet, policy makers – especially those in LAC countries – have an enormous responsibility to boost biological diversity efforts in the next and coming decades.

While LAC is well known for its rainforests, about one-quarter of the region is actually desert and drylands. LAC countries are also party to the United Nations Convention to Combat Desertification (UNCCD), established in 1996 to support the incorporation of long-term national action programmes into international co-operation and partnership arrangements to combat desertification and mitigate the effects of droughts. As the convention enjoys strong political support in the region and as most policy makers are in favour of mainstreaming relevant goals into national policies, 22 LAC countries are currently participating in the UNCCD Land Degradation Neutrality Target Setting Programme (UNCCD, 2022[19]).

With the purpose of facilitating the sharing of experiences, including both successes and lessons learned, 32 countries of the region have presented Voluntary National Reviews (VNRs) to the High-level Political Forum on Sustainable Development (HLPF). Of these countries, 16 have submitted more than one VNR; some have submitted a third (UN DESA, 2020[20]). More recently, COP26 demonstrated (once again) LAC commitment to a green transition. The region's overwhelming participation in pledges to end deforestation and cut methane emissions is a case in point (Table 6.1).

Table 6.1. **LAC international commitments on climate change, selected LAC countries**

	COP26 commitments				
	Glasgow's Leaders' Declaration on Forests and Land Use (target: to end deforestation)	**Global Coal to Clean Power Transition Statement** (target: to quit coal)	**Global Methane Pledge** (target: to cut methane emissions)	**Net zero target date**	**Updated or second NDC**
Argentina	Yes	No	Yes	2050	Yes
Brazil	Yes	No	Yes	2060	Yes
Chile	Yes	Yes	Yes	2050	Yes
Colombia	Yes	No	Yes	2050	Yes
Costa Rica	Yes	No	Yes	2050	Yes
Dominican Republic	Yes	No	Yes	2050	Yes
Ecuador	Yes	Yes	Yes	2050	No
El Salvador	Yes	No	Yes	No target set	Yes
Guatemala	Yes	No	Yes	No target set	Yes
Mexico	Yes	No	Yes	No target set	Yes
Panama	Yes	No	Yes	2050	Yes
Paraguay	Yes	No	No	No target set	Yes
Peru	Yes	No	Yes	2050	Yes
Uruguay	Yes	No	Yes	2050	No

Source: Authors' elaboration based on (UN, 2021[21]); (UN, 2021[22]); (European Commission, 2021[23]); (NDC Partnership, 2022[18]).

These institutional commitments show the extent to which LAC countries have been advancing their national plans and strategies, in the context of international agreements. These commitments, however, are frequently regarded as a matter of international relations in the region and thus are not always integrated into domestic planning bodies. This lack of integration between the environmental agenda and domestic plans or local governments is reflected in the lack of consistency across numerous policies or sectors related to environmental goals (IDB, 2021[17]). Advancing the green transition requires going beyond setting priorities and objectives and calls for concrete efforts to align national and international agendas and even find consensus on a regional environmental agenda. Firm political commitment on all these fronts will prove key in advancing from *de jure* to *de facto* implementation of the aforementioned plans and goals.

The implementation of these multilateral agendas ought, indeed, to be the region's focus in the coming years. Achievements in terms of a green transition, however, require more than a clear setting of priorities. They require an alignment of national and international agendas, firm financial commitments, and agreement on common rules at the regional level. It is also a matter of ensuring that spillovers are addressed and accounted for. International co-operation is fundamental to help LAC governments put such agendas in motion.

LAC countries are involved in several international partnerships pertinent to facilitating their green transition. These partnerships include efforts ranging from climate change mitigation to energy transition, sustainable mobility, forest and landscape conservation, and sustainable agricultural practices. In these initiatives, the LAC region has co-operated with multiple partners with the purpose of aligning multilateral commitments with national actions for a green transition. International initiatives include those established between Argentina and the European Commission through the EUROCLIMA+ programme. Other multilateral initiatives in the area of forest rehabilitation and conservation have been carried out with the support of the Green Climate Fund (GCF): USD 82 million have been disbursed to support South American efforts. In the same area, the United States Agency for International Development (USAID) has been a key partner for Brazil in achieving conservation projects in the Amazon. In other areas, such as renewable energies and improvements in the institutional framework of the national climate change plan, Brazil's partnership with the European Investment Bank (EIB) and the Deutsche Gesellschaft für Internationale Zusammenarbeit (German Agency for International Cooperation [GIZ]) stands out.

Regional and subregional initiatives can function as drivers of the green transition

While the participation of LAC countries in climate negotiations is characterised by a variety of positions, initiatives taken regionally or sub-regionally highlight the potential of LAC's role in advancing the green agenda. From South-South Co-operation (SSC) and Triangular Co-operation (TrC) to geographically specific agreements, these initiatives have been functioning as major drivers of the green transition and show promising potential for upcoming environmental challenges.

The LAC region now has the Escazú Agreement, the region's first treaty on environmental matters and the world's first to include provisions for human rights defenders in environmental matters. It is the only legally binding agreement stemming from the United Nations Conference on Sustainable Development (Rio+20) and rooted in the tenets of Principle 10 of 1992 Rio Declaration. The Escazú Agreement aims to promote access to information, public participation and justice in environmental matters. By linking global and national frameworks, the agreement sets regional standards, fosters capacity building – particularly through SSC – and lays the foundations for a supporting institutional architecture. It also offers tools for improved policy and decision making (Box 6.2) (ECLAC, 2018[24]).

Box 6.2. A landmark environmental agreement in LAC: The Escazú Agreement

The Regional Agreement on Access to Information, Public Participation and Justice in Environmental Matters in Latin America and the Caribbean, also known as the Escazú Agreement, is the first environmental agreement adopted in LAC and the world's first treaty to include provisions relating to the protection of environmental human rights defenders. The agreement entered into force in April 2021; to date, it has been signed by 24 countries and ratified by 13.

It contains provisions similar to those of the Aarhus Convention, signed by 39 member states of the United Nations Economic Commission for Europe (UNECE), which grants citizens the right to access information and to participate in the processes of environmental decision making, as well as the right to obtain redress in the event that their rights are not respected. The Escazú text was adapted to LAC's specific needs and contains, for instance, provisions for the protection of environmental human rights defenders and people or groups in vulnerable situations.

The agreement constitutes an important opportunity to promote responsible business conduct (RBC) and to address social and environmental impacts. It might also be a critical tool to generate certainty and stability in investments by bringing together states, companies and societies to optimise the outcomes of sustainable investments, which is crucial for Latin America's green transition. It can certainly also be an important opportunity to raise companies' awareness of the necessity of involving human rights defenders in due diligence processes.

Source: Authors' elaboration based on ECLAC (2018[24]).

The Declaration for the Conservation of the Marine Corridor of the Eastern Tropical Pacific, announced at the COP26 by Colombia, Costa Rica, Ecuador and Panama, is a good example of a cross-border and subregional initiative for the conservation and sustainable use of the region's biodiversity. The area will span 500 000 km^2 and will connect the

Galapagos Islands in Ecuador, Malpelo Island in Colombia, and Cocos Islands and Coiba Island in the territorial waters of Costa Rica and Panama. The declaration includes the definition of an adequate model for the protection and management of these protected areas. If it is to be sustained, the support of civil society, international co-operation organisations and the private sector will be important (Climate Tracker, 2021[25]).

With the Amazon rainforest spanning nine countries and representing over half of the planet's remaining rainforests, regional mechanisms for preserving this particular ecosystem are a region-wide and global concern. The Amazon Co-operation Treaty Organization has been on the front line of these efforts (ACTO, 2022[26]). The Leticia Pact is another prominent example. This pact has fostered a particularly good balance of financing, regulations and capacity building. Among other efforts, the signing countries have committed to exchange and implement experiences in the comprehensive management of fires; exchange information to improve the capabilities of monitoring the region's climate, biodiversity, water and hydro-biological resources; and develop education and awareness-raising activities on the role and function of the Amazon.

Caribbean countries face particular challenges when it comes to climate change (Box 6.3). Conscious of their limitations regarding their financial, human and infrastructure resources for confronting the effects of climate change, these countries have been implementing efforts towards a green transition not only through government entities but also through multilateral institutions. CARICOM is a prime example of a multilateral organisation working both to push for ambitious strategies regarding mitigation initiatives and to recognise the adaptation needs of the countries most vulnerable to climate change. Its strategic plan for 2015-19 emphasised environmental management and the protection of the region's natural assets across all sectors of development, as well as empowering the community in its preparedness to cope with and manage the effects of natural disasters, both manmade and resulting from climate change (CARICOM, 2014[27]).

Box 6.3. **Climate vulnerability and the Caribbean**

Climate change poses a serious threat to all Caribbean nations. According to the Intergovernmental Panel on Climate Change (IPCC), average temperatures in the region have increased by 0.1°C to 0.2°C per decade over the past three decades. Rainfall patterns have shifted, with the number of consecutive dry days expected to increase. Additionally, sea level rise has occurred at a rate of about 2 cm to 4 cm per decade over the past 33 years, a trend that presents risks to the region's freshwater resources and to its largely coastal population, which is dependent on tourism and agriculture. Accordingly, 16 out of 39 members of the Small Islands Developing States (SIDS) belong to the Caribbean region. This all comes in combination with Caribbean countries' minimal contribution to global GHG emissions.

This stark vulnerability calls for adapted approaches when it comes to building specific partnerships with these countries. The Multidimensional Vulnerability Index has been fundamental in the efforts to reopen discussion on eligibility criteria for concessional financing by addressing the structural challenges often faced by SIDS, including those related to their remoteness, economic concentration and dependence on external flows, such as remittances, foreign direct investment (FDI) and tourism revenues. The index is composed of a set of indicators related to economic, financial, environmental and geographic vulnerability and can therefore function as a critical tool for measuring Caribbean countries' needs and priorities. Recent efforts are in the process of expanding the scope of vulnerability. In particular, the Caribbean Development Bank is continuing discussion on vulnerability measures by piloting a new concept of the Recovery Duration Adjuster, which incorporates aspects of both vulnerability and resilience.

Box 6.3. **Climate vulnerability and the Caribbean** *(cont.)*

CARICOM has been among the regional organisations placing significant emphasis on the vulnerability of Caribbean countries when it comes to climate change and on the crucial need to push for more initiatives related to adaptation. Within its integrated strategic priorities, the organisation includes environmental resilience, the primary goal of which is to strengthen the community's information gathering, monitoring and analysis infrastructure to reduce countries' vulnerability to the risk of disasters and the effects of climate change and to guarantee effective management of the natural resources of its member states.

Sources: (IDB, 2014[28]); (CARICOM, 2014[27]); (Laguardia Martínez, 2017[29]); (CARICOM, 2017[30]); (UNDP, 2021[31]); (Dowrich-Phillips, 2022[32]).

LAC has also made important efforts in fighting climate change through SSC and TrC (together, SSTC) within Ibero-America. In particular, the share of the environment sector in total SSTC initiatives in Ibero-America has doubled in the last decade, reaching 8.4% of total SSTC in 2020. In the past year, the environment ranked third in terms of number of initiatives, behind only the traditionally most significant sectors, such as health and agriculture, in terms of volume.

It is also noteworthy that the objective of moving towards sustainable development has been mainstreamed in many additional sectors of activity, especially since the approval of the United Nations 2030 Agenda for Sustainable Development. For instance, of 155 initiatives implemented in the agricultural sector between 2019 and 2020, approximately 23% focused explicitly on progress towards sustainable practices within their respective titles or objectives, and 10% focused on the need to adapt to the effects of climate change. In short, almost one out of every three SSTC initiatives (30%) in the agricultural and livestock sector defines key dimensions of environmental improvement as priority objectives (based on *Sistema Integrado de Datos de Iberoamérica sobre Cooperación Sur-Sur y Triangular of SEGIB*).

The various agreements and initiatives reached at regional and subregional levels highlight the commitment and methods of LAC countries in advancing the green agenda beyond the limitations created by political fragmentation in climate negotiations. They are a strong driver of the region's environmental objectives and an important tool for firmly embedding climate policy in LAC growth strategies.

Seizing opportunities and softening the impacts of the green transition for trade in LAC

Inevitably, advancing the green transition goes beyond treaties and multilateral initiatives. It entails ramifications for various aspects of the region's development model, among which trade relations hold a distinctive position. Through proper regulations and policies, the region stands to reap substantial benefits related to the green agenda. LAC countries can take the opportunity to face the entrenched productivity trap and move towards a transformation of its production model.

In particular, much-needed measures will imply additional costs for both LAC imports and exports, at least in the medium term. LAC countries will need to prepare for such challenges and search for effective methods to tackle them early on. The establishment of international green standards and regulations will pose challenges and opportunities for the region's green development, affecting in particular the structure of LAC exports.

Such standards and regulations will boost LAC efforts towards the creation of a circular economy, which will require enhanced co-ordination at the regional level.

Measures will impose additional costs on LAC imports in the medium term

NDCs provide a wealth of information on how each country intends to achieve its goals in terms of climate change mitigation and adaptation. The implementation of NDC commitments is closely linked to the use of so-called "environmental goods and services".[1] These are products manufactured or services rendered for the main purposes of: 1) preventing or minimising pollution, degradation or natural resources depletion; 2) repairing damage to air, water, waste, noise, biodiversity and landscapes; and 3) carrying out other activities related to environmental protection or resource management, such as measurement and monitoring, control, research and development (R&D), education, training, information and communication (UN et al., 2014[33]).

NDCs offer benchmarks for understanding their impact on local markets and the strategies that would help LAC countries face the evolution of international regulations and standards. Linking LAC countries' NDC commitments to their likely impact on import flows in the coming years can reveal possible trade implications and shed light on policy recommendations.

Of the six types[2] of NDC measures affecting expenditure on manufactured imports, three are assumed to lower the cost: 1) reduction of tariff and non-tariff barriers for renewable energy technology; 2) explicit mention of a policy measure's intention to reduce dependence on imported fuel; and 3) encouragement of technology transfer from advanced economies. The other three types would raise the cost of imports: 1) ban on the importation of old or energy-inefficient goods; 2) imposition of new standards and labelling requirements; and 3) renewable energy development (Saalfield, forthcoming[34]).

Banning the importation of old or energy-inefficient goods was mentioned in 36% of the region's NDCs, imposing new domestic standards was mentioned in 42%, and renewable energy development was mentioned in 85%. The first two measures might reasonably be expected to increase the average cost of goods such as vehicles and appliances, as these measures would lock out older, cheaper substitutes and push consumers towards newer, more expensive alternatives. Both measures are more common in Caribbean NDCs, where higher medium-term fuel savings hasten the payback timeline on products such as electric vehicles (EVs) (Table 6.2).

The third measure, renewable energy development, is common across the region. Renewable energy technologies are considered essential to meeting national emissions reduction targets, so it is not surprising that a large majority of NDCs commit to scaling related infrastructure in the coming decade. While the region's commitment to renewables is commendable, infrastructure development will likely continue to be led by foreign firms and rely on foreign components, contributing to short-term import expenditure. Establishing local clean technology sectors will require time, financing, technology transfer and the development of a relatively specialised workforce. The region may find opportunities in other areas, for example in the emerging green hydrogen market (ECLAC, 2021[36]). Even then, the establishment of a substantial green hydrogen industry would be capital-intensive and thus likely to contribute to import expenditure in the short to medium term.

Table 6.2. **Country tally of NDC measures affecting short-term import expenditure in LAC**

	1 Block energy-inefficient imports	2 Impose new domestic standards	3 Develop renewable energy	4 Reduce trade barriers for efficient imports	5 Reduce dependence on imported fuel	6 Encourage technology transfer
Latin America						
Argentina	✓		✓			✓
Bolivia			✓			✓
Brazil						
Chile			✓			✓
Colombia	✓	✓	✓			✓
Costa Rica		✓	✓			✓
Ecuador			✓			
El Salvador			✓			✓
Guatemala			✓			✓
Honduras			✓			✓
Mexico			✓			✓
Nicaragua			✓			
Panama		✓	✓			✓
Paraguay	✓		✓	✓		✓
Peru						
Uruguay		✓	✓			✓
Venezuela	✓	✓	✓			✓
Caribbean						
Antigua	✓	✓	✓		✓	✓
Bahamas	✓	✓	✓	✓	✓	✓
Barbados		✓		✓		✓
Belize	✓	✓	✓			
Cuba			✓	✓		✓
Dominica	✓		✓		✓	
Dominican Republic	✓	✓	✓	✓		✓
Grenada		✓	✓		✓	
Guyana		✓	✓	✓		
Haiti	✓		✓			✓
Jamaica	✓					
Saint Kitts and Nevis			✓			✓
Saint Lucia			✓			✓
Saint Vincent and the Grenadines		✓	✓	✓	✓	✓
Suriname	✓	✓	✓			✓
Trinidad and Tobago			✓			

Source: Nationally Determined Contributions, accessed via (UNFCCC, 2022[35]).

On the savings side, reducing barriers for energy-efficient imports was mentioned in 21% of the NDCs, reducing dependence on imported fuel was mentioned in 15%, and requests for technology transfer were found in 70%. Reducing trade barriers such as tariffs has the shortest time horizon of the three cost-saving policies, as it can lower the cost of energy-efficient imports almost immediately. However, reducing import tariffs may further entrench the region's pattern of high import dependence. Last, in theory, technology transfer has great potential to foster import substitution but it has proved difficult to implement.

Implementation of NDC commitments is closely linked to the use of so-called environmental goods and services. While there is no internationally agreed list of environmental goods, probably the most comprehensive list available is the Combined List of Environmental Goods (CLEG) compiled by the OECD (Sauvage, 2014[37]). The CLEG includes 248 products at the 6-digit level of the World Customs Organization's Harmonized Commodity Description and Coding System (HS).[3]

When envisaging the transition to a greener and more sustainable development path, its implications for the region's external position must be carefully considered. In fact, the external restriction is a longstanding constraint on development in LAC. Since 1990, the region registered a current account surplus only between 2003 and 2007, in the context of the so-called commodity super-cycle (Chapter 1). Using the CLEG as a benchmark, LAC has consistently posted a deficit in its trade in environmental goods over the last two decades (Figure 6.3, Panel A). Three-quarters of the region's imports come from China, the United States and the European Union, while intraregional imports account for just 5% of total expenditure (Figure 6.3, Panel B). Moreover, regional export capacity is highly concentrated: just one country (Mexico) accounted for 84% of regional exports of environmental goods between 2018 and 2020.

Figure 6.3. **LAC trade balance in environmental goods**

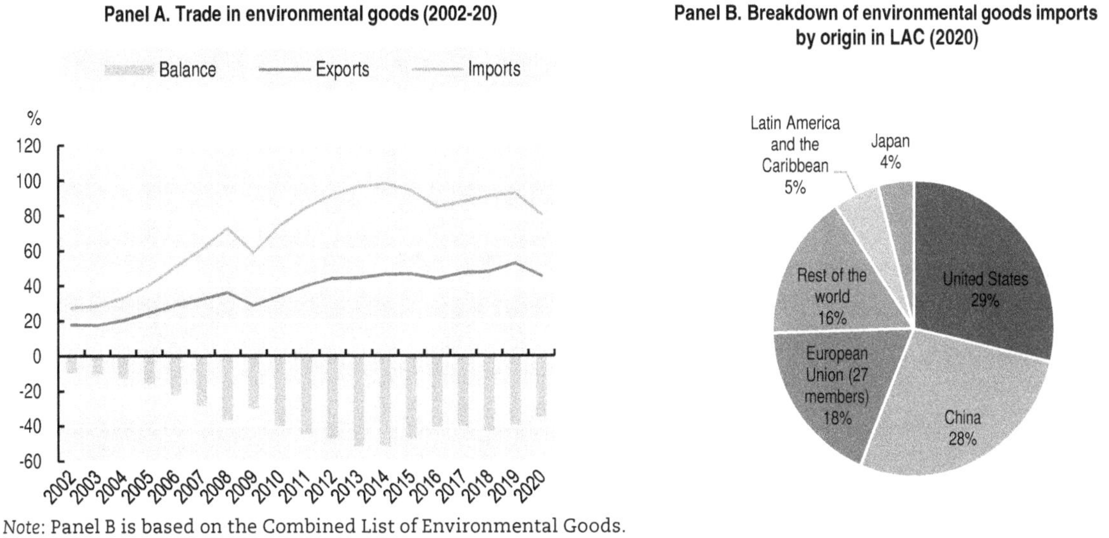

Note: Panel B is based on the Combined List of Environmental Goods.
Source: Authors' elaboration using data from (UN Comtrade, 2020[38]).

StatLink https://stat.link/owfzdj

As a whole, the green transition will likely involve a substantial import expenditure for LAC over the coming years (possibly decades), putting strain on the region's balance of payments. In particular, continuing imports in the renewables sector, on which LAC NDCs place substantial emphasis, can exacerbate the region's existing trade deficit in environmental goods and lead to higher import costs.

International co-operation can play a substantial role in seizing the opportunities and softening the impacts of the transition for the region's imports. Promoting regional production capabilities in the renewables sector will be crucial not only to avoid excessive imports but also from a political economy perspective, to strengthen local coalitions in favour of the green transition. Such initiatives can function as a significant driver of buy-in for the green agenda and can prove a decisive factor in the creation of jobs, thus ensuring both a green and just transition. While it is incumbent on the countries to implement policies aimed at developing local capabilities, development partners and alliances will also play a crucial role in channelling financial resources through investments, digital tools and technical know-how towards that effort.

LAC exports need to prepare to face evolving environmental standards and regulations in international trade: The LAC-European Union case

Globally, some of the most important trade partners for the LAC region are China, the United States and the European Union. The three actors contribute the most to GHG emissions and represent important shares of the global economy. The three have led important strategies in the past decades to fight climate change and reach net zero levels. Each follows a different strategy.

The European Union has positioned itself as a global leader on environment. With the adoption of the European Green Deal, climate policy is now firmly embedded in the EU's new growth strategy. This ambitious agenda aims to decouple economic growth from resource use. The European Union and its member states have leveraged their climate leadership and the power of the European single market to engage other countries in raising the ambition of their climate policies. LAC citizens consider the European Union to be the world leader on environment protection (59%), well above the United States (12%) and China (9%) (Figure 6.4).

Figure 6.4. **Leadership across diverse sectors among three global players**

Which country do you consider the world leader in each of the following aspects?

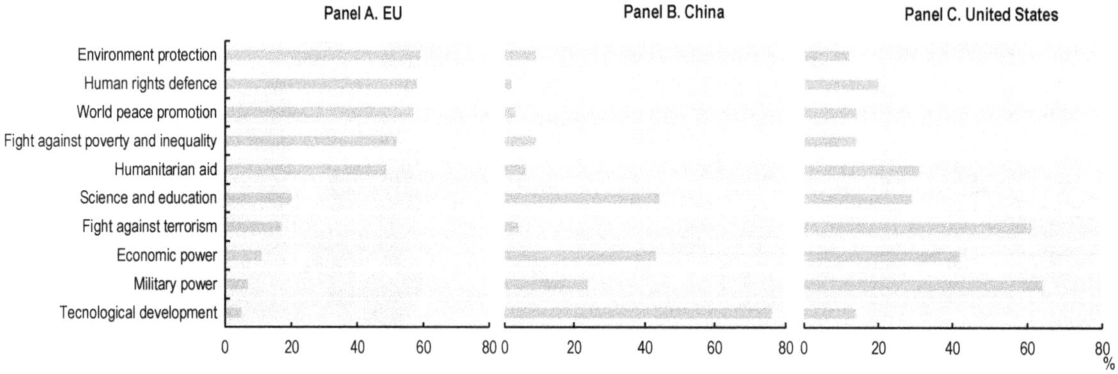

Source: (Latinobarómetro/Nueva Sociedad/Friedrich-Ebert-Stiftung, 2022[41]).

StatLink https://stat.link/i9leo4

Required changes or adaptations will be a cost in the short and medium term but could trigger LAC to transform development models based on the green transition. The EU's strategic leadership on the green transition and the European Green Deal's implications for trading partners are a case in point. While such standards imply considerable trade-offs for LAC countries, proper preparation and co-ordination can ensure that the region advances with both its green and its development objectives.

Why will the new norms and regulations of the Green Deal affect LAC exports to the European Union?

The possible effects in LAC of the new standards and regulations of the Green Deal arise from the trade balance between the regions. LAC remains a key supplier of agribusiness products and raw materials to the EU. In 2021, the EU27 was the destination of 8.9% (EUR 89.9 billion) of total exports from LAC. The top five exporting countries were Brazil

(EUR 30.9 billion or 34.4%), Mexico (EUR 13.9 billion or 15.5%), Argentina (EUR 8.34 billion or 9.3%), Chile (EUR 6.56 billion or 7.3%) and Peru (EUR 5.66 billion or 6.3%).

Vegetables are among the top products exported from LAC to the EU at a value of EUR 17.3 billion (19.4%). The second-most important are mineral products. The third-most significant are prepared foodstuffs, beverages, spirits, vinegar and tobacco, representing EUR 11.9 billion (13.3%) (Figure 6.5). Food and feed products are heavily regulated in the EU market, with EU food regulations applying to both imported and EU-made products, while compliance with EU sanitary, phytosanitary and technical requirements varies across LAC countries.

Figure 6.5. **Share of LAC exports to EU27 in 2021**

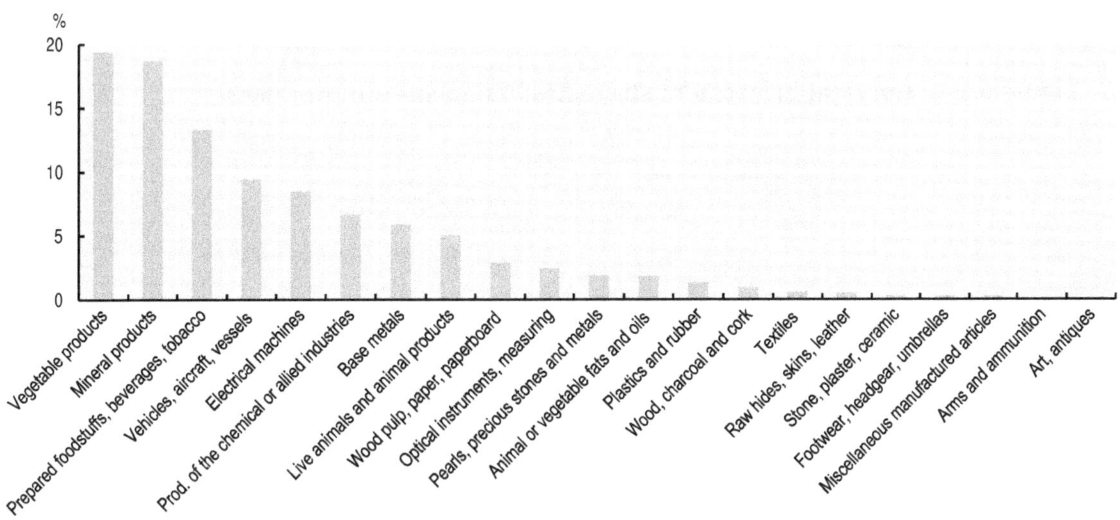

Note: Exports by Harmonized Commodity Description and Coding System sections.
Source: Authors' elaboration based on (International Trade Center, 2022[42]).

StatLink https://stat.link/xzalc6

Another element to consider in LAC-EU trade relations is that the similarity of productive structures of some LAC economies makes them natural competitors in the EU common market. The fact that many LAC countries rely on some few export commodities, such as bananas or coffee, entails high competition in reaping the benefits of EU markets.

While the Green Deal is, above all, a transformative agenda for Europe, it also has important implications for the European Union as a global actor and for its relations with other regions like LAC. The Green Deal focuses on eight areas: climate, energy, agriculture, industry, environment and oceans, transport, finance and regional development, and research and innovation. Dozens of interconnected regulations will be reviewed in the coming years, and multiple strategies and action plans have been established to fulfil the carbon-neutral goal. LAC will face important trade-offs from this institutional arrangement and from the need for continued trade and co-operation with the European Union.

How will Green Deal strategies affect LAC exports and potentially boost both the green transition and strategic autonomy?

The Green Deal policies with the largest potential impacts on LAC countries are those focused on producing healthier food and feed and on the implementation of new standards of sustainability. Given LAC's concentration in exporting agricultural goods and raw materials to the EU, policies such as the proposal for a regulation on deforestation-free products, the Fit for 55 Package, the New Circular Economy Action Plan, the Farm to Fork Strategy and the Biodiversity Strategy for 2030 will potentially have impacts for LAC that need to be considered.

The Green Deal includes diverse investment strategies (Figure 6.6). One-third of the EUR 1.8 trillion investments from the NextGenerationEU Recovery Plan and the European Union's Multiannual Financial Framework 2021-2027 will finance the Green Deal (European Commission, 2019[43]). The New Social Climate Fund invests in energy efficiency, new heating and cooling systems, cleaner mobility and a socially fair transition (European Commission, 2021[44]). The Just Transition Fund, the European Regional Development Fund and the European Social Fund Plus all support eligible territories through territorial just transition plans. InvestEU attracts private investments to help regions find new sources of growth, including renewable energies (European Commission, 2020[45]).

Figure 6.6. **Green Deal investment strategies: Financial compromises**

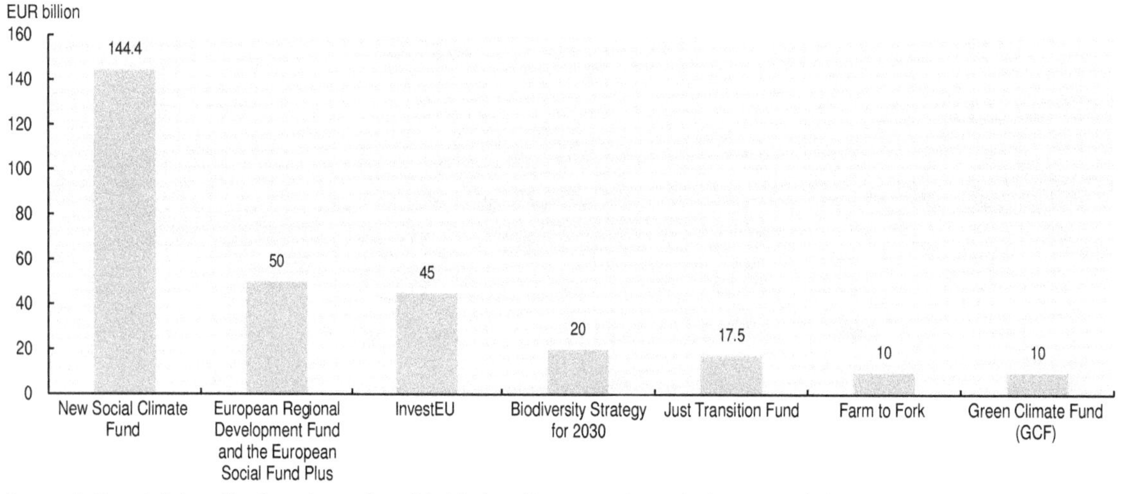

Source: Authors' elaboration based on selected initiatives (European Commission, 2022[46]); (European Commission, 2020[47]); (European Commission, 2020[45]); (Green Climate Fund, 2020[48]); (European Commission, 2019[43]); (European Commission, 2022[49]).

StatLink https://stat.link/q8j7tm

The Biodiversity Strategy for 2030 will enlarge EU-protected areas with high biodiversity and climate value on land and at sea, and will establish measures to restore degraded ecosystems, especially those with superior potential to capture and store carbon (European Commission, 2022[46]). The research and innovation component of the Farm to Fork Strategy will focus on food, bioeconomy, natural resources, agriculture, fisheries, aquaculture and the environment. Knowledge transfer will be essential, too. The European Union will collaborate with third countries to support a global move towards sustainable food systems (European Commission, 2020[47]).

The GCF supports the implementation of climate mitigation and adaptation policies. Projects cover several areas, such as forest conservation and rehabilitation in Argentina, adoption of digital agricultural production and climate adaptation technologies in Colombia, light rail in Costa Rica, cooling in El Salvador, and installation of greenhouses, micro-tunnel facilities and hydrometeorology in Guatemala, among others (Green Climate Fund, 2022[50]).

The Neighbourhood, Development and International Co-operation Instrument-Global Europe (NDICI-Global Europe) is the new financial mechanism to support third countries.

It merges several former EU external financing instruments, unifying grants, blending, and guarantees to overcome long-term development challenges and contribute to achieving international commitments, such as under the 2030 Agenda and the Paris Agreement. The programme finances three areas: geographic programmes, thematic programmes and global challenges. There is an additional emerging challenges and priorities cushion. Moreover, the EU can promote public and private investment in support of sustainable development through the European Fund for Sustainable Development Plus (EFSD+) (European Commission, 2021[51]).

The NDICI-Global Europe has a budget of EUR 79.5 billion for 2021-27, with LAC being assigned EUR 3.3 billion, equivalent to 5.62% of the amount allocated to geographic programmes. The areas of co-operation are: 1) good governance, democracy, the rule of law and human rights, including gender equality; 2) poverty eradication, human development and the fight against inequality and discrimination; 3) migration, forced displacement and mobility; 4) environment and climate change; 5) inclusive and sustainable economic growth and decent employment; 6) peace, stability and conflict prevention; and 7) partnership (European Parliament/European Council, 2021[52]).

In terms of the inclusive and sustainable economic growth and decent employment component, the instrument seeks to: 1) support microfinance, skill and competency development; 2) improve application of international labour standards, gender equality and living wages; 3) reduce the risk of exclusion and marginalisation of specific groups; 4) promote fair taxation and redistributive public policies; 5) strengthen social protection systems; 6) improve the business environment and investment climate; 7) foster an enabling policy environment for economic development, particularly for small and medium-sized enterprises (SMEs); 8) encourage accountability, mandatory due diligence and human rights commitments; 9) promote internal economic, social and territorial cohesion between urban and rural areas; 10) facilitate the development of creative industries and a sustainable tourism sector; 11) diversify sustainable and inclusive agricultural and food value chains; 12) strengthen value addition, regional integration, competitiveness and fair trade; and 13) strengthen multilateralism and co-operation in science, technology, research, digitalisation, open data, big data, artificial intelligence and innovation (European Parliament/European Council, 2021[52]).

Delving into the details of the Green Deal policies and their impact on LAC, the Directive on mandatory corporate sustainability due diligence for large companies with respect to human rights and environmental impacts in their supply chains is a far-reaching proposal adopted by the EU. Both EU and non-EU companies with activities in the EU should adopt due diligence processes along their value chains. The proposal aims to create legal certainty and a level playing field for businesses, as well as transparency for consumers and investors, and offers enhanced protection of human and environmental rights, in accordance with international conventions (European Commission, 2022[53]). It also points to a growing global recognition of RBC standards, which includes the OECD Guidelines for Multinational Enterprises and related guidance on supply chain due diligence in addressing environmental threats, dependencies and adverse impacts. At the EU member state level, France and Germany have led due diligence regulations. Belgium and the Netherlands are also laying out their own plans (WEF, 2022[54]).

The proposal for regulation on deforestation-free products is one of the most recent and ambitious initiatives to promote sustainable consumption. It focuses on halting deforestation associated with the production of soy, beef, palm oil, wood, cocoa and coffee and the products derived from them. Only legally deforestation-free products will be allowed on the EU market. In addition, mandatory due diligence standards will be required in marketing, and countries will be assessed according to their level of deforestation risk.

Deforestation monitoring will be carried out with geolocation systems. Food supply chains will be subject to stricter surveillance (European Commission, 2021[55]). EU initiatives on general corporate due diligence and on deforestation offer the potential for LAC countries and companies to lead on the implementation of due diligence processes and RBC to gain competitive advantages in global value chains and to strengthen the export economy.

Fit for 55 is a package of proposals aimed at ensuring a just, competitive and green transition. Investments in a low-carbon economy can stimulate economic growth and employment, accelerate the transition to clean energy, and increase long-term competitiveness. Fit for 55 strengthens existing regulations and presents new initiatives on climate, energy and fuels, transport, buildings, land use and forestry to reach an emissions reduction of 55% by 2030. Important proposals within Fit for 55 include ReFuelEU Aviation and FuelEU Maritime, which promote the use of sustainable fuels in the aviation and maritime sectors as a complement to the EU Emissions Trading System (EU ETS) by incentivising fuel suppliers to blend in increasing levels of sustainable fuels and incentivising the use of low-carbon synthetic fuels (electrofuels) (European Commission, 2021[56]). These energy transition initiatives will require the development of and large investments in new technologies, as well as the transformation of consumption patterns of governments, businesses and end-consumers. These adjustments will not be immediate, hence the importance of protection systems to help mitigate the possible negative effects. In terms of foreign trade, implementation of the new policies could result in further increases in freight rates, which are already at record highs, especially for container cargo (UNCTAD, 2021[57]).

Another key proposal is the evolution of the EU ETS towards a Carbon Border Adjustment Mechanism (CBAM), which aims to combat carbon leakage by identifying and tracing the emissions involved in EU imports (Box 6.4).

Box 6.4. **Regulations to achieve carbon neutrality: The CBAM in the European Union**

The EU ETS was the world's first international emissions trading scheme to fight climate change. It limits the amount of GHG emissions that industrial installations in specific sectors can emit. Emission allowances must be purchased on the EU ETS market, and some free allowances are distributed to prevent carbon leakage.

From 2023 onwards, the CBAM will gradually complement the EU ETS. The CBAM will be based on a system of certificates to cover the emissions involved in products imported by the European Union. Initially, it will only apply to a number of selected products with a high risk of carbon leakage: iron and steel, cement, fertilisers, aluminium, and power generation. To ensure its conformity with World Trade Organization (WTO) rules, the rollout of the CBAM must take place in tandem with the withdrawal of free allowances allocated to European producers of the same products under the EU ETS. A reporting system for covered products will be in use from 2023 onwards. A transition period will be put in place between 2023 and 2025. The free subsidies will be phased out gradually from 2026, and importers will start paying a financial adjustment. Products manufactured in the European Union and those imported from elsewhere will receive the same treatment.

Box 6.4. Regulations to achieve carbon neutrality: The CBAM in the European Union *(cont.)*

This mechanism targets direct GHG emissions of the production of semi-elaborated and elaborated products, not raw materials. After the transition, the European Commission will evaluate whether the mechanism will be expanded to new products and sectors even further down the value chains. In addition, the CBAM will generate additional revenue, estimated at above EUR 2.1 billion for 2030. However, there is a risk that EU companies will relocate their carbon-intensive production to other countries with lower standards or that more carbon-intensive imports will replace European products.

The leading LAC exporters of iron and steel (in tonnes) to the EU27 in 2019 were Brazil, Mexico and Venezuela. Colombia led exports in cement, Trinidad and Tobago in fertilisers, and Venezuela in aluminium. Countries will be required either to match EU standards or to pay the financial adjustment. Other costs for exporting countries, such as the implementation of emissions traceability systems, could also result from the CBAM scheme.

Source: Authors' elaboration based on (European Commission, 2021[58]); (European Commission, 2021[59]); (Bellora and Fontagné, 2022[60]).

Last, the New Circular Economy Action Plan, one of the main components of the Green Deal, lays the foundation for a cleaner and more competitive EU. It focuses on the most resource-intensive sectors with high circularity potential including food, water, nutrients, packaging, plastics, textiles, construction, buildings, batteries, vehicles, and electronics and information and communications technology. Its scope involves the whole life cycle of products: design, sustainable consumption, waste disposal/management, etc. (European Commission, 2022[61]). At the EU level, a new comprehensive Strategy for a Sustainable Built Environment will be adopted as part of the EU Circular Economy Action Plan to promote circularity principles throughout the life cycle of buildings. At the international level, the plan proposes the formation of a Global Circular Economy Alliance to discuss the potential of an international agreement on natural resource management.

In addition, the EU will collaborate with third countries to support a global move towards sustainable food systems, implementation of animal welfare, reduction of the use of pesticides and the fight against antimicrobial resistance. Moreover, it will assist small-scale farmers in meeting the standards and accessing markets (European Commission, 2020[62]). These strategies, together with the areas prioritised in the NDICI-Global Europe, will be crucial for LAC to meet the challenges of the new legislation planned or being implemented (Table 6.3).

Table 6.3. **Summary of sectors affected by Green Deal policies, and challenges for LAC**

Green Deal regulations	Products/supply chain	Challenges for LAC
Farm to Fork: new stricter organic standards for organic production.	Foodstuffs and feedstuffs	Replacement of agrochemicals with organic inputs and obtaining of organic certifications. These production transformations generate higher production costs.
Carbon tax due to CBAM.	Iron and steel, cement, fertilisers, aluminium and power generation	Technology transfer and investments required to achieve a greener production transition that minimises use of coal.
Proposal for a regulation on deforestation-free products.	Soy, beef, palm oil, wood, cocoa, coffee and their and derived products	Implementation of traceability and due diligence systems in supply chains. Operators should collect the geographic co-ordinates of land where products were produced.
Proposal for a revision of the Sustainable Use of Pesticides Directive.	Foodstuffs and feedstuffs	Adjustment of production supply to new requirements that could raise production costs.
Proposal for a revision of the Feed Additives Regulation to reduce the environmental impact of livestock farming.	Foodstuffs and feedstuffs	Adjustment of production supply to new requirements that could raise production costs.
Evaluation and revision of the existing animal welfare legislation, including on animal transport and slaughter.	Foodstuffs and feedstuffs	Adjustment of production supply: 1) to new requirements that could raise production costs; and 2) to the waste control plans that have been approved.
New mandatory requirements to reduce (over)packaging.	Food (i.e. fresh and processed fruits and vegetables, including juices and wines)	Change from plastic to biodegradable packaging that raises production costs. Adoption of new technologies according to new requirements.
New mandatory requirements for recycled content and special attention for microplastics, as well as biobased and biodegradable plastics.	Foodstuffs	Change from plastic to biodegradable packaging that raises production costs. Adoption of new technologies according to new requirements.
New legislative initiative on re-use to replace single-use packaging, tableware and cutlery with reusable products in food services.	Foodstuffs	Increased packaging requirements that raise production costs. Adoption of new technologies according to new requirements.
A new EU Strategy for textiles to strengthen competitiveness and innovation in the sector and boost the EU market for textile re-use.	Textiles	Adoption of new technologies according to new requirements with higher production costs.
A comprehensive Strategy for a Sustainably Built Environment promoting circularity principles for buildings and construction.	Construction materials	Adoption of new technologies according to new requirements with higher production costs.

Source: Authors' elaboration based on (European Commission, 2020[62]) and (European Commission, 2020[63]).

Impacts and opportunities of the Green Deal for LAC food production

Food production remains a significant challenge for LAC competitiveness in the EU market. Current EU consumption trends lean towards more responsible, sustainable (organic) and fair consumption. EU organic farming objectives restrict the use of agrochemicals and require the development of sustainable practices in planting, processing, transport, distribution and consumption. However, it is costly for countries to adapt their agricultural practices to organic production when approved pesticides are modified or their use limited. The substitution of chemical agro-inputs with biological agro-inputs directly affects the cost structure of products, requiring time for adaptation. Moreover, authorised fertilisers and pesticides may not work efficiently to control pests, especially those of the tropics. Additionally, given the distance between the two regions, transport times reduce product shelf life for both organic and conventional products.

The implementation of the Green Deal will demand food chain traceability, and there will be additional controls on the use of antibiotics, hormones, biologically active substances, feed additives and chemical residues, as well as on animal welfare, organic production, cold chain and labelling, among others.

In particular, with the Farm to Fork Strategy and the Biodiversity Strategy for 2030, the European Union aims to reduce the use and risk of chemical pesticides and the use of more hazardous pesticides by 50% by 2030. In addition, the new regulations propose to minimise the use of fertilisers by at least 20% and the sale of antimicrobials for farm animals and aquaculture by 2030, and to achieve 25% of total arable land under organic farming by 2030 (European Commission, 2020[47]). Countries often struggle to realise the infrastructure and resources needed to comply with all the requirements, even while their exports are accepted in other destinations.

Currently, there is strict control of pesticide residues in plant and animal products intended for human consumption, or Maximum Residue Levels. More than 300 fresh products are approved for certain pesticides and maximum quantities of pesticides (European Commission, 2022[64]). To export animals and their products to the European Union, export countries must have: the approval of veterinary medicine residues; monitoring plans for 13 classes of bovine, ovine/caprine, porcine, equine, poultry, aquaculture, milk, eggs, rabbit, wild game, farmed game, honey and casings products and by-products; and certification establishments producing such products (European Commission, 2022[65]). The approval of residue plans requires laboratory tests and analyses. Sanitary admissibility and compliance with EU technical requirements are heterogeneous across LAC countries. Most countries (16) have achieved approval of residue plans in aquaculture, followed by honey (11) and casings (7). Some countries have achieved sanitary admissibility for most products, including Argentina (12 approved plans), Chile (9) and Uruguay (8) (European Commission, 2022[65]).

Private standards used in the EU tend to be even more restrictive than public standards, e.g. those required by retailers from their suppliers for agri-food products. This can add further pressure on third-country exporters to invest in compliance with increasingly stringent standards, introducing an additional barrier such that only companies with significant financial capacities can make the adaptations in the short term. This takes place irrespective of national capacities and institutional arrangements, such as the existence of accredited laboratories that can verify compliance.

Organic certification costs can also become a barrier to trade if not properly addressed with efficient adaptation processes. Regulation instruments, such as heavy compliance costs (especially for small farmers), copious documentation and the inherent difficulties of tropical lands requiring powerful agrochemicals (which often lack sustainable substitutes) are exacerbated to the extent that producers have split systems to serve different markets.

Mutual recognition of equivalent standards in the case of organic products can be a useful option when LAC and the EU have very similar but not equal legislation. These agreements already exist between the EU and largest organic produce exporters. Chile has been recognised as an equivalent third country in organic products since 2018 (European Council, 2017[66]). Under the new framework of organic production, EU Regulation 2018/848 of 30 May 2018, the European Commission was authorised to open negotiations with Argentina and Costa Rica, among other countries, with a view to concluding agreements on trade in organic products (European Council, 2021[67]).

There is also the risk at the end of the value chain that the market will not recognise the additional costs involved in the organic production process, especially in the case of commodities that tend to have standard prices. This can be a risk for smallholders, not only for the adoption of organic practices but also for the adoption of any standard that requires significant investments. This is the case of the proposal of the Directive on corporate sustainability due diligence, which although planned to apply to large companies, may potentially affect the most vulnerable groups in its implementation if no transition support mechanisms are foreseen.

Likewise, the Farm to Fork Strategy proposes that farmers receive a fair price for their products. The EU has several policy tools for its farmers to achieve this objective, such as payments under the Common Agricultural Policy or competition law to enable groups of farmers to negotiate prices collectively. However, Farm to Fork has not so far included mechanisms to guarantee fair income for non-EU farmers that supply EU consumers, or to compensate the costs of the transition, potentially affecting the weakest actors in the supply chain.

Seizing the opportunities and softening the impacts of the Green Deal for LAC exports

Overall, international green regulations, including the Green Deal, are expected to guide the prioritisation of upcoming government agendas in LAC. However, in this policy-making process, it is crucial that trade partners communicate their concerns about draft legislation to make visible potential impacts and design tailored co-operation projects that help mitigate the effects on sectors that could potentially be affected. Moreover, new legislation should offer ample transition periods and flexibility to allow smooth adjustments, as well as tools to cope with the new requirements. Last, it is important to develop programmes for in-depth and direct technical assistance to emerging regions, including LAC. Developing tailor-made productive transformation programmes to adapt the exportable supply to the new requirements and to provide facilities and increased resources for the implementation of large projects (e.g. infrastructure, science and technology) requires significant investments in the long term.

The European Union's ambition to promote sustainable development as part of its external actions has also been part of the institutional framework of Association Agreements (AAs). Moving forward, and although the EU demonstrates an awareness of the challenges posed by the Green Deal to both European and non-European economies, the actualisation of AAs could contribute to confirming and materialising the EU's commitment to supporting the needed adjustments by LAC countries (Box 6.5).

Additional efforts need to be implemented at the national level. For instance, establishing national sustainability roadmaps is a key step for facing international regulations, such as under the Green Deal. LAC countries should strengthen their institutions and create public-private co-ordination mechanisms among agencies that regulate trade, agricultural, industrial, environmental, energy, planning, and science and technology policies and the most representative private-sector companies with an interest in investing in and working with the European market.

Adapting to Green Deal standards would also improve national standards in LAC countries, thus accelerating the achievement of sustainable economic development goals. LAC has an opportunity to foster a productive transformation towards cleaner technologies and to add higher value to exports. One benefit for LAC of adapting to the Green Deal will be the implementation of sustainability standards that meet EU requirements, allowing the region's exports to preserve and possibly expand their share in the European market.

Box 6.5. Revamping Association Agreements (AAs)

The European Union has integrated sustainable development chapters into the trade agreements signed in the last decade to assist the just transition among its trading partners. The AAs between the European Union, its member states and countries of the LAC region are a case in point. While in practice they often remain limited to their trade-related components and can be perceived as mere free trade agreements, AAs are also intended to promote and enhance (political) policy dialogue and development co-operation among contracting parties. Countries including Chile, Colombia, Mexico and Peru, or even regions, such as Central America, have been among the actors benefitting from such initiatives and frameworks for co-operation. Negotiations for modernising the agreement with Chile and efforts to renew Mexico's AA can be considered significant steps in accelerating LAC objectives in bringing forth the green transition.

Nevertheless, the AAs' most ambitious environmental goals also signal their greatest risks. Increased regulations and higher environmental standards can lead to additional barriers to trade and, therefore, to new waves of protectionism. In order for both LAC and EU to contribute effectively to the international agreements on sustainable development, AAs will need to further emphasise the importance of interconnecting the trade pillars with the other two pillars of dialogue and co-operation.

The AAs may be a relevant instrument for bi-regional dialogue. If their full potential is harnessed and a comprehensive approach is adopted, AAs are capable of creating a shared space for policy dialogue, of advancing regulatory convergence and productive transformation aimed at changing the current economic model, and of reconstructing the region's social contract. More importantly, they offer an opportunity to advance the ecological transition to decarbonisation and sustainability.

Sources: (Gómez Arana, 2021[68]); (Rodríguez Díaz and Sanahuja, 2021[69]).

The productive adaptations required to continue and expand LAC exports under new international regulations will not be automatic and will require important investments in the short term. Therefore, new regulations, such as under the Green Deal, will certainly have an effect on upcoming government agendas in LAC and will present opportunities to put in place policies to foster transformative sustainable development. The speed and depth of the reforms will depend on each government's capacity to articulate a systemic green model. International co-operation can help accelerate this transition and reduce its impacts.

Scaling up the circular economy in LAC: The role of trade and regional co-ordination

International trade can facilitate the transition to a circular economy by providing the technology and scale required to undertake efficiently relevant activities such as recycling, refurbishment, remanufacturing and valorisation of residues and waste as an input for other industries (Chapter 3). Trade may also extend the life cycle of final products and materials and facilitate their reincorporation into production cycles, as well as generate demand for products and business models that support the transition to a resource efficient and circular economy. Moreover, trade in services may promote new business models that facilitate the sharing of under-utilised products through digital platforms and technologies.

The exported volume of economy goods has been increasing over the last decade, both globally and regionally. Globally, the main export category corresponds to products for recycling, especially waste and scrap metal. By contrast, almost 90% of circular economy exports in LAC correspond to products for valorisation, mostly of residues from soybean oil extraction (Figure 6.7). These are mainly exported to Southeast Asian countries to produce animal and fish feed. At present, the Harmonised Commodity Description and Coding System, used by all countries to compile their trade statistics, does not include separate codes for some circular economy goods. Therefore, trade in circular economy goods is not fully captured by international statistics.

Figure 6.7. **Exported volume of circular goods by category, averages, 2017-19**

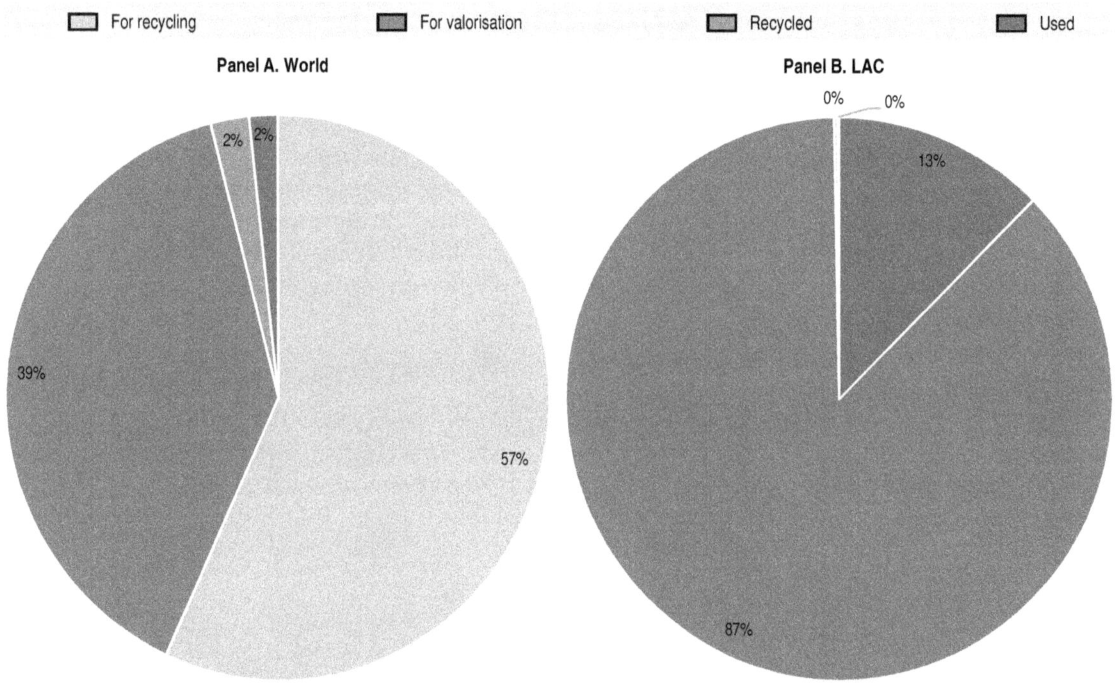

Source: (ECLAC, 2021[70]), based on (CEPII, 2021[71]); (International Trade Analysis Database, 2021[72]).

StatLink https://stat.link/zg1u2f

The potential contribution of trade to the transition to a circular economy in LAC depends critically on its co-ordination with national and international efforts. The reduction of tariff and non-tariff barriers for trade in goods and services that contribute to circularity at each stage of production and consumption, especially at the end of life phase of the goods, is crucial. Greater granularity in national and international trade classifications will be needed to identify circular economy goods more precisely. Regional integration bodies may also play a key role in harmonising standards and promoting joint solutions, such as the European Union in its Circular Economy Action Plans (from 2015 and 2020). Section 7 of the 2020 Circular Economy Action Plan considers to explore how the European Union can succeed by defining a "Safe Operating Space" in which the use of various natural resources does not exceed certain local, regional or global thresholds and the environmental impact remains within planetary limits. Here, new sustainable models will open up business and employment opportunities with key partners around the world. Last, the promotion of standards and certifications that endorse the circularity of processes could help firms adopt sustainable practices.

Several countries in Latin America are defining standards and formulating strategies for the circular economy, including several aspects related to international trade. These centre around key export sectors, circular firms and products, market access, and the promotion of FDI or international financing for circular economy projects, especially for SMEs. Other efforts relate to establishing registers of circular suppliers, which facilitates export promotion. Multiple countries are including circularity criteria in their green public procurement strategies. Member countries of the Pacific Alliance are co-ordinating parts of their national circular economy strategies, for example, in relation to the sustainable management of plastics (ECLAC, 2021[70]).

Several international initiatives are supportive of LAC potential to develop a circular economy. For instance, the International Trade Centre's GreenToCompete initiative focuses on assisting small businesses in developing countries, including in LAC, by providing information about green opportunities and innovations. The programme is divided into three areas: climate resilience, circularity and biodiversity. The circular economy raises global ambitions to reduce pollution, emissions and waste while increasing prosperity. The programme encourages businesses and value chains to implement circular practices to reduce production costs, increase productivity and boost innovation by developing and commercialising new products and services (International Trade Centre, 2022[73]).

Harmonised use of international co-operation instruments to facilitate design and implementation of a green policy agenda in LAC

In recognising the increasing linkages across domestic and global dynamics, international co-operation and partnerships appear essential for realising the full potential of the needed reforms that will contribute to a greener development model in LAC.

International co-operation can act as a facilitator of LAC's will to transform international commitments into concrete national and regional policy measures that advance a green transition (OECD et al., 2019[74]). Mission-driven partnerships, along with balanced use of international co-operation tools, can provide capacity building, align standards and ensure the coherence of national and international efforts (as per the Development in Transition narrative), providing the right mix to enhance LAC efforts towards the green transition (OECD et al., 2021[75]).

Facilitating the design and implementation of a green policy agenda can trigger change of the region's development model. This should at least include increased efforts in policy coherence, aligning national and international priorities; capacity building for creating and adapting to new skills; technology transfer as a method for adding value to the region's production structures; and tapping into LAC's global potential by advancing towards common regulations and standards.

A harmonised use of these international co-operation instruments can ensure that LAC countries adopt an integrated approach to the green transition, placing equal emphasis on each of these priorities and taking advantage of possible synergies in the implementation of the green agenda. The region's existing strengths and weaknesses – particularly in relation to institutional capacities and arrangements that are capable of utilising these instruments – vary from one country to the next. The promotion of a common framework for enhanced co-operation, and a space for policy dialogue and the sharing of experiences based on each country's strong points, advancements and lessons learned, could help in facing the green transition's trade-offs.

Policy coherence can improve alignment of national plans and strategies with international commitments and help manage externalities

Operationalising a sustainable and inclusive recovery is not to be taken for granted. At the national level, the interactions among inclusiveness, the green and digital transitions, productivity, and resilience often imply trade-offs. The design and implementation of green policies require intense co-ordination in order to exploit available synergies across all political areas and to avoid negative externalities that transcend distinct policy sectors and national boundaries (Chapter 5). Increased efforts in policy coherence as an instrument for international partnerships can foster a green and just transition by accounting for the complexities of policy impacts at various levels of government and the interactions among regional, national and international actors.

The 2030 Agenda signalled the moment for going beyond the traditional principle of Policy Coherence for Development to promote a broader, more ambitious vision, reformulated as Policy Coherence for Sustainable Development, as stated in SDG 17.14. This approach aims to integrate the dimensions of sustainable development through both domestic and international policy making and to advance integrated implementation of the 2030 Agenda by: 1) fostering synergies and maximising benefits across economic, social and environmental policy areas; 2) balancing domestic policy objectives with internationally recognised Sustainable Development Goals (SDGs); and 3) addressing the transboundary and long-term impacts of policies, including those likely to affect developing countries (OECD, 2019[76]). An example is the OECD International Programme for Action on Climate (IPAC), which supports countries' progress towards net zero GHG emissions and more resilient economies by 2050. Through regular monitoring, policy evaluation, and feedback on results and good practices, IPAC helps countries strengthen and co-ordinate their climate action. It complements and supports the UNFCCC and the Paris Agreement monitoring frameworks. Moreover, the Framework to Decarbonise the Economy aims to design and implement country-specific decarbonisation strategies with a mix of cost-effective, comprehensive, inclusive and acceptable policies. Both IPAC and the Framework for Decarbonising the Economy are part of the OECD-led Horizontal Project on Climate and Economic Resilience, which is intended as a holistic tool to support policy makers (OECD, 2022[77]).

Even though the importance of enhancing policy coherence for the green transition domestically cannot be overstated, its impact can remain rather limited if not aligned with similar efforts in other countries. Bringing forth the green transition on a global scale requires an immense amount of co-ordination among national leaders and measurable efforts for policy integration. The SDGs, while offering a unique opportunity for the establishment of a common framework and guidance for sustainable development, still require concrete domestic policies and initiatives that are aligned with recognised global sustainable development objectives and goals.

Mainstreaming the SDGs at national and local levels is subject to the specificities of country contexts and the particularities of development needs across borders. Efforts vary, from SDG budgetary and legal initiatives to SDG localisation efforts and establishing the goals as national priorities. Notable regional initiatives include the efforts of Colombia, Cuba, Guatemala, Mexico and Paraguay to localise the SDGs, with some countries already conducting Voluntary Local Reviews and Voluntary Subnational Reviews. Mexico has also been carrying out studies on quantifying the benefits of climate action through the implementation of both the 2030 Agenda and the Paris Agreement, with the purpose of reducing implementation costs, avoiding work duplication, and identifying possible opportunities and synergies (Secretaría de Economía, 2021[78]). The VNRs submitted to the HLPF, while not comprehensive, offer a notion of some of the methods used in the region to mainstream SDGs domestically.

LAC countries are already directly referencing the various forms of international and regional co-operation and frameworks that allow for greater policy coherence (Table 6.4), but the challenge of implementing these nationally remains. Thus far, the Paris Agreement is the framework most widely used to benchmark progress in the development of climate change mitigation policies and energy transition plans.

Table 6.4. **Methods for mainstreaming SDGs in LAC**

As reported by VNRs of 2019-21

	Considering and identifying SDG interlinkages	SDG-specific plans, strategies and policies	Integrating SDGs into NDPs and strategic frameworks	SDG localisation	Coherence with international and regional co-operation and frameworks
2019					
Chile		✓			✓
Guatemala					✓
Guyana	✓		✓		✓
Saint Lucia	✓		✓		✓
2020					
Argentina	✓	✓		✓	✓
Barbados					
Costa Rica		✓	✓	✓	✓
Ecuador	✓		✓	✓	✓
Honduras	✓			✓	
Panama		✓			
Peru		✓	✓		✓
Saint Vincent and the Grenadines					
Trinidad and Tobago	✓	✓	✓		✓
2021					
Antigua and Barbuda	✓		✓		
Bahamas					
Bolivia	✓		✓		
Colombia	✓			✓	✓
Cuba	✓		✓	✓	✓
Dominican Republic	✓		✓		✓
Guatemala	✓			✓	✓
Nicaragua			✓		✓
Mexico	✓		✓	✓	✓
Paraguay	✓			✓	✓
Uruguay	✓		✓		✓

Source: Authors' elaboration based on (GIZ, 2021[79]); (GIZ, 2020[80]); (GIZ, 2019[81]).

When time is of the essence, as is the case with climate change and biodiversity loss, measurements of progress and results are crucial. The United Nations Environment Programme index on SDG 17.14.1 can be considered the most relevant indicator for measuring national policy coherence for sustainable development (UNEP, 2022[82]). However, the index has remained on the sidelines, its use limited to a few examples of one-off research. LAC countries need to take greater advantage of such tools in order to identify relative progress and regress, and to collect measurable policy lessons.

The role of regional institutions and organisations is key in balancing national policy objectives with international sustainable development goals. Regional actors can promote internationally recognised environmental goals by accounting for the political, social and environmental particularities of their regions, thus acting as a direct link in aligning international sustainability goals with national policies. The articulation of strategies for sustainable development and the analysis of key policy interlinkages in topics related to trade, technology transfer and financing for development can prove major contributors in advancing both the SDGs and the Paris Agreement's mandates. Existing regional structures can promote the integration of the SDGs and disaster risk reduction measures into national and territorial planning and strengthen countries' statistical capacity to support more effective evidence-based policy making and SDGs measurements (UN DESA,

2020[20]). The Regional Knowledge Management Platform for the Sustainable Development Goals, or SDG Gateway, is a promising demonstration of the region's efforts in monitoring the progress and needs of countries in implementing the 2030 Agenda (UN, 2020[83]). The OECD had a steering role in promoting hands-on initiatives for aligning the Sustainable Development Goals with national policies, in line with the Recommendation of the Council on Policy Coherence for Sustainable Development (OECD, 2019[76]). For instance, in 2020 it accompanied the national government in Colombia, through a peer-to-peer exercise involving Spain and Sweden as peers, in establishing a platform for multi-stakeholder dialogue on the SDGs implementation. This exercise build on the SDG investor Map for Colombia, an UNDP tool to align the private sector to SDG financing in Colombia.

In an increasingly interdependent world, countries' actions and policies can have positive or negative effects on other countries' ability to achieve the SDGs. If ignored, these international spillovers may result in one country achieving SDGs at the cost of another or in missed positive synergies. Interdependence becomes even more crucial when considering that 97 SDG targets (57% of all 169 targets) entail transboundary elements (OECD, 2019[84]). Externalities, both positive and negative, need to be understood, measured and meticulously managed in order to avoid situations where one country's achievements are counteracted or neutralised by the transboundary policies implemented in third countries.

Concerning positive externalities, LAC countries seem to have a significant head start at the global level. By taking into consideration the environmental and social impacts embodied in trade, combined with the economic, financial and security dimensions of spillovers, the region scored 95.1/100 in the International Spillover Index, meaning that it has a minimal effect on the ability of other countries to achieve the SDGs, in comparison with the 70.1 OECD average (Sachs et al., 2021[85]). Similarly, the Global Commons Stewardship Index measures how countries are affecting key components of the environmental system, such as climate change, biodiversity and land-use change, both within their borders and through impacts involved in trade and consumption. The index can prove a useful tool for measuring regions' environmental progress. LAC's population-weighted average spillover impact reaches 69/100, while the OECD is only at 32/100. No region scored as low as LAC in proportional terms, once again demonstrating the minimal negative externalities attributed to the region (SDSN/Yale Center for Environmental Law & Policy/Center for Global Commons at the University of Tokyo, 2021[86]).

While the significance of LAC's comparatively positive policy spillover cannot be overstated, their contribution is undermined by large negative spillovers generated by high-income and OECD countries. Data on the direct effects of these countries' policies on developing countries – and LAC countries specifically – are still lacking. But unsustainable supply chains, for instance, driven by trade in timber, palm oil, rubber, coffee, soy and other commodities, have led to disastrous deforestation and biodiversity loss (Sachs et al., 2021[85]).

Even though OECD members have made progress in implementing and raising awareness of the 2030 Agenda domestically, it has often come at the expense of adapting national policies to support developing countries in making progress towards the SDGs or, in other words, developing partner-friendly policies. Comprehensive strategies that balance a broad range of objectives and challenges at home frequently forget or neglect the effects of their policies on developing countries. To increase the positive and the avoid negative impacts of sustainable policies, OECD countries need to work with developing countries to enhance evidence-based dialogue founded on quality reporting and assessments of policies. They also need strengthen efforts to co-ordinate development co-operation with all development partners by aligning with partner country indicators, synchronising planning cycles with partner countries' cycles and using sector co-ordination structures with the aim of ensuring cross-sector coherence (OECD/EC-JRC, 2021[87]).

LAC would benefit from enhanced efforts in capacity building and technology transfer

Capacity building and technology transfer for developing renewable energies

Building national and local capacities would allow the formulation, monitoring, implementation and evaluation of environmental policies supporting LAC's green transition. Nevertheless, developing the necessary capacities goes beyond enhancing capabilities at the policy level; it requires multifaceted strategies that address all stakeholders and sectors. It is critical that the region encourages the development of strategic projects at the business and academic levels in order to mitigate and adapt to the effects of climate change. Technology transfer related to smart mobility, renewable energies and a circular economy are also key in international efforts to switch to green and sustainable development models.

International and regional partnerships seeking to develop capacity building and technology transfer exist in LAC concerning two interrelated streams: new renewable energies, such as wind and hydrogen; and production of minerals, such as lithium, conducive to the production of renewable energy. Already, a "Latin American Green Deal" is being proposed, given the need for a major rethinking of trade and investment strategies in the region (Lebdioui, 2022[88]). The region can further learn from the European Union (as mentioned in the previous section) and build its own deal considering the region's needs on trade, investments and economies of scale. International co-operation programmes could play a role, mainly those that include a component of technical assistance, as they can strengthen LAC resilience in the long term. There is a lot of room to expand them, including by further involving the private sector. Regional coordination appears as well as the key to unlock the potential of renewable energies in LAC.

While LAC's great natural reserve is highly vulnerable to the effects of climate change, it also has significant alternative energy potential due to its hydrographic, wind and mineral wealth. Thanks to state-led investment in hydropower and major escalation of onshore wind and photovoltaic solar power development, renewables make up 33% of LAC's total energy supply, well above the global average of 13% (Chapter 2) (Sistema de Informacion energetica de Latinoamerica y el Caribe (SieLAC), 2020[89]).

Latin America accounts for 60% of all identified global lithium reserves, most located in Argentina, Bolivia and Chile, the so-called "Lithium Triangle" (Chapter 3). Other countries, such as Brazil, Mexico and Peru, also appear to hold important lithium resources. However, it requires significant investment, as most of the profit of the lithium industry comes from a long value chain, with extracting and exporting countries unlikely to realise any significant gains. A sustainable governance perspective is needed to ensure that investment is available and that gains are distributed in ways that improve the well-being of societies, especially local communities. Any opportunity that lithium offers needs to be reinvested in a more ambitious and long-term technological transformation that allows societies to overcome productivity traps (López-Calva, 2022[90]).

A new lithium partnership between Argentina, Bolivia, Chile and Mexico is already being explored, to co-operate on exploration, exploitation and the development of new technologies (Domínguez, 2022[91]). In fact, joint initiatives from producer countries have great potential as demand for lithium continues to spike. In order to support the global energy transition and achieve local economic development objectives, there needs to be a rapid increase in production. For that to be possible, a multi-stakeholder approach must be adopted to ensure that LAC countries do not repeat past errors in extractive industries, meaning that there is a need for consultations with local communities and transparency with regards to environmental impacts. Producer countries have an opportunity to join efforts, share experiences and build partnerships for developing technologies that allow sustainable exploitation of this resource.

Similarly, embracing the potential of green hydrogen production within LAC countries can ease the pressures of energy shortages, increase prosperity, and reduce the risk of climate-related loss and damage. Several initiatives are underway. Chile and Germany have already published roadmaps for the development of their domestic markets and are looking to co-operate with other countries to optimise the demand and supply of low-carbon hydrogen, thus jointly contributing to the decarbonisation of global economies. Chile's strategy in particular focuses on promoting co-operation among industry, academia and technical centres, encouraging the construction of R&D roadmaps by public and private sectors for solving local implementation challenges, and creating work groups with public companies to accelerate adoption of green hydrogen in their activities and supply chains (Government of Chile, 2020[92]). Colombia has undertaken similar actions recently, publishing its own roadmap for the production of both green and blue hydrogen, specifically focusing on regulating the production process, researching and developing production technologies at a local level, and establishing competitive prices (IDB, 2021[93]).

Latin America can benefit significantly from early participation in global initiatives and fora on hydrogen production and use and from seeking opportunities for international collaboration. In particular, establishing permanent regional structures for co-operation would allow a co-ordinated regional approach while maintaining national autonomy (IEA, 2021[94]). For instance, the International Hydrogen Energy Centre, launched in 2021, aims to develop hydrogen energy globally and attract experimental R&D funding, with a significant emphasis on capacity building through appropriately designed training programmes, training in innovative policy formulation and the modernisation of local industry to meet specific development needs (UNIDO, 2021[95]).

With LAC's relatively low-density population and large distances between cities and towns, wind power also offers substantial potential for supplying the energy needs of inhabitants across the region. While Argentina, Brazil and Mexico currently dominate the sector, more LAC countries are joining the wind market as a result of the expert knowledge provided by the Latin American Task Force of the Global Wind Energy Council. The wind park developed in Guajira, Colombia in 2022, with the co-operation of ISAGEN – a private energy company – and the Spanish company Grupo Elecnor, is the largest and first of many upcoming wind projects planned for the country (Anderson, 2022[96]).

Partnerships on capacity building and technology transfer for overcoming environmental challenges

The European Union can be a strategic ally to transfer technology and develop innovation towards a green transition. In particular, LAC countries can take advantage of the European Union's strategic leadership on the green transition by building on its expertise, experience and technologies related to environmental goals. Whether this concerns policies and practices regarding public actors or developing the capacities of private companies and stakeholders, the shared values of LAC and EU countries in facing environmental challenges can function as a unifying force for transferring knowledge and know-how from one region to the other.

Initiatives carried out in this regard vary in scope and expected results. The EUROCLIMA+ programme supports intra-regional dialogue and co-operation on climate issues within Latin America and the promotion of bi-regional dialogue between Latin America and the European Union in international fora (EUROCLIMA+, 2022[97]). The AL-INVEST Verde programme has expanded its scope from past initiatives focused on growth and jobs. It aims to promote sustainable growth and job creation by supporting the transition to a low-carbon, resource-efficient and more circular economy in LAC through a focus on innovation and digitalisation. The LIFE programme supports the Green Deal by promoting the implementation of energy efficiency and small-scale renewable energies, with the European Commission approving an investment package in 2021 of more than EUR 290 million for 132 new LIFE projects (European Commission, 2021[98]). Among the Commission's latest announcements on green co-operation with LAC are the European Fund for Sustainable Development Plus (EFSD+), which aims to leverage more than EUR 12 billion of public and private investment for the region, and a new edition of EUROCLIMA+, with EUR 140 million in support to LAC (European Commission, 2022[99]).

The Responsible Business Conduct in Latin America and the Caribbean (RBCLAC) project, implemented since 2019 by the OECD together with the International Labour Organization (ILO), the Office of the United Nations High Commissioner for Human Rights and the European Union, supports governments, businesses and stakeholders in their efforts to promote RBC in line with international standards (OECD, 2022[100]). Among other efforts, the RBCLAC project carries out regional and national activities to strengthen the capacity of businesses and governments in order to mainstream environmentally and socially responsible business and corporate due diligence in key sectors. The *Eco-Eficiencia Empresarial* programme demonstrates the initiative attributed to the Central American region along similar lines. The programme originated within the Business Alliance for Development of Costa Rica but has now expanded to El Salvador, Guatemala, Honduras, Nicaragua and Panama. It identified a significant lack of knowledge among member companies on managing their environmental impact. The programme offers a collaborative platform aimed at orienting and strengthening the capacities of businesses, facilitating the exchange of good practices and training and the formation of alliances on priority issues related to sustainable development (ILO, 2021[101]).

In addition, there is growing recognition of the potential of SSTC to facilitate technology development and transfer for climate action in developing countries. SSTC on climate technologies are taking place in and among all geographical regions and cover all priority areas outlined in developing countries' NDCs and NAPs. While commonly initiated at the national level, the implementation of SSTC projects usually involves various stakeholders, including local government, civil society organisations (CSOs), research and training institutions, and the private sector (Chapter 3) (UNFCCC, 2021[102]).

International co-operation can prove effective in accelerating the green transition through the use of new digital tools and technologies to develop new skills and productive methods (Chapter 3). Bilateral co-operation between the European Union and Brazil created a partnership for Scientific and Technological Cooperation which allowed calls for joint research and innovation in the domains of water management and agriculture, among others. Triangular co-operation has also helped in promoting the adoption of new technologies to cope with new standards of production and regulations. Peru benefited from a co-operation initiative with Germany and Brazil and created the Environmental Technology Centre (CTA) to cope with the urgent need for environmental technology experts generated by new environmental regulations and production criteria (OECD, 2020[40]).

Overall, adopting an approach that focuses on building capacities and enhancing technology transfer at all levels of government and in the private sector can ensure that the benefits of sustainable energy sources are exploited and that they foster the well-being of local societies. The region boasts high-quality universities and research centres, specialised CSOs, and peasant and indigenous communities with strong territorial roots. These capacities, if properly developed and strengthened through international partnerships, can prove a key contribution to a green and just transition (ILO, 2021[101]).

Common approaches represent opportunity to develop a regional carbon market

Protecting the environment and delivering on climate action is also a matter of policy co-ordination at regional and global levels. Common regulations can derive from, for instance, agreements for protecting the oceans or ecological zones, common import standards, international carbon prices and classification systems of economic activities, or green taxonomies.

Market mechanisms are commonly used as national drivers for fostering a greener development model. In fact, economic levers are among those most used to reach net-zero emissions. Carbon pricing is a prominent example of an effective way countries can reduce emissions. Explicit carbon pricing can be implemented domestically through carbon taxes and emissions trading systems. While such explicit forms of carbon pricing are relatively rare in developing countries, almost every country has experience with fuel excise taxes. Fuel excise taxes are economically and administratively similar to carbon taxes, and could be reformed to better align with the climate costs of fossil fuel use (OECD, 2021[107]).

International carbon markets allow countries that struggle to meet the emissions reduction targets stated in their NDCs – or are in search of less expensive emissions cuts – to purchase emissions reductions from nations that have already cut their emissions by more than the amount pledged. In ideal conditions, carbon markets are capable of generating win-win situations whereby both countries meet their climate commitments and additional finance is provided to the country generating emissions reductions. Poor design and construction of regulations can leave countries "off the hook" in making meaningful contributions by hindering ambition for increased emissions cuts, not ensuring additional emissions reductions and double counting emissions reductions (UNFCCC, 2015[103]).

Estimates show that setting fossil fuel prices that reflect their true cost would cut global CO_2 emissions by over one-third. In fact, efficient fuel pricing by 2025 would reduce global CO_2 emissions to 36% below baseline levels. Efficient fuel pricing can raise substantial revenues as well: in the case of LAC, revenues could reach around 2% of GDP (Parry, Black and Vernon, 2021[104]). Over half of the Group of Twenty (G20) nations have in place a form of carbon pricing at the national level, be it a carbon tax or an ETS. Some LAC countries have also advanced the agenda, as many are considering or have implemented a type of carbon tax or ETS (Chapter 4) (Table 6.5). They can also draw on experience with fuel excise taxes and fossil fuel subsidy reform (OECD, 2021[107]).

Table 6.5. **Carbon pricing**

Carbon taxes and ETSs in LAC

	Carbon tax	ETS
Argentina	Implemented	..
Brazil	Under consideration	Under consideration
Chile	Implemented	Under consideration
Colombia	Implemented	Under consideration
Mexico	Implemented	Implemented
Uruguay	Under consideration	..

Note: Latest data available: April 2021.

Source: (World Bank, 2021[105]).

Article 6 of the Paris Agreement sets the stage for a common regulation on carbon market mechanisms. However, it can be considered one of the most complex and controversial aspects of the global accord. Exactly six years after the Paris Agreement, COP26 brought forth an agreement on a Global Carbon Market Mechanism, largely completing the Paris Agreement Article 6 Rulebook. Key decisions were made concerning the approval process and issuance of credits, the eligibility of projects and activities to be included, how to deal with legacy projects and credits under the Kyoto Protocol's Clean Development mechanism, and the making of corresponding adjustments to host states' emissions accounts (Clifford Chance, 2021[106]). While progress has been made, critics argue that the language used falls short of fully realising the regulations needed to bring market forces to bear as strongly as possible on emissions reduction.

For this reason, carbon markets under proper regulation offer an outstanding opportunity for LAC countries' development financing. Under this new structure, and given its unique ability to offer green projects and nature-based solutions at cheaper marginal costs than elsewhere, LAC countries can position themselves as the world's largest providers of carbon credits (Arbache, 2021[107]). At the same time, policy makers will need to ensure that gains from the development of the carbon credit market do not come at the cost of increased land grabs and other forms of detrimental consequences. An emphasis on proper regulation and transparency is key to minimise possible risks, specifically those related to a just transition.

The region has great potential to offer carbon credits derived from nature-based solutions, such as the forest removal and emissions avoided by preventing deforestation. Some initiatives are already taking shape in the region, such as the Latin America and the Caribbean Carbon Market Initiative (ILACC), which adopts a comprehensive approach (Box 6.6).

Another economic benefit of accelerating the carbon market in the region is the encouragement of companies to comply with environmental regulations and compete in the international context of growing threats to access concessional and non-concessional resources. This acceleration is particularly important for the competitiveness of agro-exports as evidenced by the Farm to Fork programme.

Box 6.6. **LAC's quest for a regional carbon market: The ILACC**

A promising way to position LAC in the global carbon market is through regional integration of national initiatives (with the potential of creating a regional market) that allows not only economies of scale and cost reduction but also an important pipeline of projects.

The creation of a regional market could position the LAC region as a leader in the new international scenario. The objective of the ILACC is to promote global competitiveness of the supply of carbon credits generated in the region, expanding the impacts on job creation, income, development of value chains, technologies, clusters, green business products and the fight against poverty.

Three critical factors – infrastructure, technical training and a product pipeline – require immediate attention to promote the development of the regional carbon market. The first is infrastructure. This means having an entire services platform that enables, in practice, the correct and adequate functioning of the regional market, including a basic legal framework, monitoring and control, and data collection and compilation systems, as well as the full set of services necessary for the carbon market value chain to function: taxonomy, standards, certifications and legal services, among many others. The second is technical training. The market, still in formation, will require many qualified professionals for the conception, development, execution and management of green projects. Generally speaking, these skills are not available in the quantity and degree of specialisation needed. Without these skilled workers, it will be difficult for a market to gain the trust of players or develop a pipeline of projects with the speed and quality required. The third is a pipeline of diversified and quality projects. For the regional market to establish itself as an international hub, it will be necessary to attract demand through a timely and diversified supply of credits from conservation, agroforestry, ecological restoration, energy and many other projects.

Source: (CAF, 2022[108]).

Another key element regarding common regulations from which the LAC region stands to benefit is classification systems of the sustainability of economic activities, or green taxonomies. Such regulations can offer investors clearer standards, reinforce transparency and commitments to green, social and sustainability-linked activity, and mitigate the risk of greenwashing. Although there is currently no specific transition taxonomy in LAC, a few countries (including Brazil, Colombia, Chile, the Dominican Republic and Mexico) are in the process of developing sustainable or green finance taxonomies. Other LAC countries are considering similar actions or carrying out relevant research through their co-operation with the International Finance Corporation, the IDB and the GIZ. If properly implemented, such initiatives can play a promising role in protecting the environment and delivering on climate action (Chapter 4).

Tools from policy coherence to capacity building, technology transfer and common standards can play crucial roles in adopting a multidimensional and integrated approach to advancing the green agenda. Nevertheless, their utility and potential are nullified if they are not implemented within the context of international partnerships. In view of the global nature of the climate crisis, international co-operation remains a crucial – if not the only – means of fostering a green and just transition.

Overall, international co-operation is needed for the transition to a greener and sustainable economy, but the transition's impact on the labour force and the most vulnerable populations should not be left on the sidelines. LAC population is highly vulnerable to the effects of climate change due to its high dependence on agribusiness and natural resource exploitation. For a green transition to be just, well designed and multi-stakeholder, international partnerships should be considered not only key to mitigating the effects of climate change but also a strategy for transforming LAC development and production models, strengthening the social structure, and smoothing labour market adaptation as a consequence of international regulations. Bringing forth a green economy is itself a challenge; a multi-stakeholder approach that takes into consideration citizens' needs and priorities can ultimately function as a multiplier force for gaining popular support and contributing to the final push for a green and just transition.

Key policy messages

International partnerships are crucial to facilitate LAC countries' efforts towards a green transition. They can contribute to: 1) overcoming current implementation shortcomings of multilateral agreements and instruments; 2) aligning national and regional priorities with multilateral sustainability goals; and 3) promoting technology transfer and capacity building, including through policy dialogues and sharing experiences in peer learning. International partnerships can also play a role in helping LAC countries address trade challenges when facing new international green norms and regulations (Box 6.7).

Box 6.7. Key policy messages

With regards to regional and international partnerships:

- Regional integration and co-operation can enhance the effectiveness of sustainable policies by addressing LAC's existing fragmentation and lack of a unified voice at the multilateral level, and by supporting the alignment of national green strategies with global environmental goals.
- Ratify and implement the Escazú Agreement. It represents a unique opportunity to enhance the protection of environmental human rights defenders.

Box 6.7. **Key policy messages** *(cont.)*

- International partnerships should promote further dialogues across regions on green transition policy experiences. LAC countries can play a predominant role in this global agenda by sharing experiences of sustainable development with other regions of similar and lower development levels.
- International partnerships can contribute to softening spillover effects caused by green transition efforts within and outside the LAC region – mainly in trade channels, labour markets, and consumption and production patterns.
- International partnerships can support new policies needed to address the substantial expenditure on exports. Promoting regional production capabilities in the renewables sector, with financial resources and technical know-how, will be crucial not only to avoid excessive costs but also from a political economy perspective, to strengthen local coalitions in favour of the green transition.
- International partnerships can also support LAC countries to address the effects on LAC exports of newly established international environmental standards and regulations.
- Ensure the coherence of domestic green policy objectives with internationally recognised sustainable development goals. Approaches should take into account the transboundary impacts of policies, the articulation of strategies at the local, regional, national and global levels.
- International partnerships need to put more emphasis on capacity building and technology transfer to develop renewable energy sources in LAC.
- Regional partnerships within LAC are the key to unlock the region's potential on renewable energies.
- Accelerate a regional carbon market under proper regulation.

With regards to trading partners:

- Trade partners need to communicate their concerns about draft legislation to make visible any potential impacts and to design tailored co-operation projects that help mitigate the effects on sectors potentially affected.
  - New legislation should offer transition periods to allow smooth adjustments.
  - Increase investments in developing tailor-made productive transformation programmes to adapt exportable supply to the new requirements and to provide facilities and increased resources for the implementation of large projects (e.g. infrastructure, science and technology).
  - Accompany measures, such as funding, technical co-operation and capacity building, to help trading partners adopt more sustainable practices, especially in the most vulnerable countries.
- The actualisation of AAs between the European Union and LAC countries, taking into better account the impacts of the green transition, along with further integration across the political, trade and co-operation pillars for the purpose of supporting LAC's green transition, remains crucial.
- Harness international trade in both goods and services to facilitate the transition to a circular economy.

With regards to LAC governments:

- Adapt to green regulations including Green Deal standards. This would improve national standards in LAC, thus fostering opportunities for materialising a productive transformation towards cleaner technologies and adding value to exports.

Box 6.7. **Key policy messages** *(cont.)*

- Establish national roadmaps for environmental sustainability and climate change adaptation to new norms and regulations, in particular those affecting LAC.
- Support efforts to strengthen institutions that favour public-private co-ordination mechanisms, with a view to promote investments in LAC related to the green transition.
- Promote sustainable product industries, such as biodegradable plastics, smart materials and bio-inputs, in order to support the transition of traditional sectors to the new green sectors.

Notes

1. The first type of environmental goods and services in the EGSS is environmental specific services.

   These services comprise environmental protection and resource management products that are "characteristic" or typical of those activities. Hence, environmental specific services are environmental protection and resource management specific services produced by economic units for sale or own use. Examples of environmental specific services are waste and wastewater management and treatment services, and energy and water-saving activities.

   Consistent with the definition of environmental protection and resource management activities, environmental specific services are those services that have the main purpose of:

   (a) Preventing or minimising pollution, degradation or natural resource depletion (including the production of energy from renewable sources);

   (b) Treating and managing pollution, degradation and natural resource depletion;

   (c) Repairing damage to air, soil, water, biodiversity and landscapes;

   (d) Carrying out other activities such as measurement and monitoring, control, research and development, education, training, information and communication related to environmental protection or resource management.

   The second type of environmental goods and services is environmental sole-purpose products. Environmental sole-purpose products are goods (durable or non-durable) or services whose use directly serves an environmental protection or resource management purpose and that have no use except for environmental protection or resource management. Examples of these products include catalytic converters, septic tanks (including maintenance services), and the installation of renewable energy production technologies (e.g. solar panels).

   The third type of environmental goods and services is adapted goods. Adapted goods are goods that have been specifically modified to be more "environmentally friendly" or "cleaner" and whose use is therefore beneficial for environmental protection or resource management. For the purposes of the EGSS, adapted goods are either:

   (a) "Cleaner" goods, which help to prevent pollution or environmental degradation because they are less polluting at the time of their consumption and/or scrapping, compared with equivalent "normal" goods. Equivalent normal goods are goods that provide similar utility except for the impact on the environment. Examples include mercury-free batteries and cars or buses with lower air emissions;

   (b) "Resource-efficient" goods, which help to prevent natural resource depletion because they contain fewer natural resources in the production stage (e.g. recycled paper and renewable energy, heat from heat pumps and solar panels); and/or in the use stage (e.g. resource efficient appliances and water-saving devices such as tap filters).

   The fourth type of goods and services is environmental technologies. Environmental technologies are technical processes, installations and equipment (goods), and methods or knowledge (services), whose technical nature or purpose is environmental protection or resource management. Environmental technologies can be classified as either:

   (a) End-of-pipe (pollution treatment) technologies, which are mainly technical installations and equipment produced for measurement, control, treatment and restoration/correction of pollution, environmental degradation, and/or resource depletion. Examples include sewage treatment plants, equipment for measuring air pollution, and facilities for the containment of high-level radioactive waste;

(b) Integrated (pollution prevention) technologies, which are technical processes, methods or knowledge used in production processes that are less polluting and less resource-intensive than the equivalent "normal" technology used by other producers. Their use is less environmentally harmful than that of relevant alternatives.

2. The six types of NDC measures affecting expenditure on manufactured imports considered in this chapter are:
   1) Banning the importation of old or energy-inefficient goods (such as used cars). Because such bans favour new, higher-tech versions that tend to be more expensive, this type of measure is generally assumed to increase the average cost of the product group and contribute positively to short-term import expenditure. Four of 17 Latin American NDCs and 8 of 16 Caribbean NDCs include commitments to such bans (almost all refer to banning the importation of older or less efficient vehicles). In the case of five Caribbean countries, this measure is proposed to reduce dependence on imported fuel.
   2) Imposing new domestic standards. This measure refers to the introduction of standards and labelling requirements for goods sold or used domestically, such as home appliances. Again, favouring newer, higher-tech goods over older, less-efficient substitutes is assumed to increase the average cost of goods in the product group and contribute positively to short-term import expenditure. Five of 17 Latin American NDCs and 8 of 16 Caribbean NDCs include commitments to new domestic standards. These are mostly efficiency standards for vehicles and durable goods or regulations on specific refrigerants found in appliances such as refrigerators and air conditioners.
   3) Renewable energy development. 15 out of 17 Latin American NDCs and 13 of 16 Caribbean NDCs include commitments on renewable energy development. As most of the world's major renewable energy developers are based in advanced economies, LAC countries have historically relied on FDI to execute renewable energy projects. As of 2020, four of the five biggest renewable energy developers active in the region were European (Smith, 2020[106]). The components for solar, wind and geothermal energy projects are mostly imported from abroad via local subsidiaries or project developers, thus contributing to the region's overall import bill. Until local manufacturing capabilities are substantially scaled up, renewable energy development will continue to require substantial imports in the coming years.
   4) Reducing trade barriers for energy-efficient imports. This measure refers to the reduction or removal of tariff and non-tariff trade barriers, allowing specific environmental goods to enter the domestic market at lower prices. This measure is assumed to reduce overall import expenditure, particularly if demand for the goods is relatively inelastic. Only 1 of 17 Latin American NDCs and 6 of 16 Caribbean NDCs include commitments to reduce trade barriers for energy-efficient imports. For example, Guyana's NDC notes that "legislation has been enacted to remove import duty and tax barriers for the imports of renewable energy equipment, compact fluorescent lamps, and LED lamps to incentivize and motivate energy-efficient behaviour." The Bahamas, Saint Lucia and Saint Vincent and the Grenadines committed to reducing import duties on low-emitting vehicles.
   5) Reducing dependence on imported fuel. Lower dependence on imported fuel is a strong economic co-benefit of climate change mitigation in the region's energy sectors, with potential to slash significantly overall import expenditure. While 5 of the 16 Caribbean NDCs commit to policies explicitly intended to reduce expenditure on imported fuel, to date, none of the 17 Latin American NDCs do. This is particularly significant for island states, where fuel costs are generally higher.
   6) Technology transfer. The UNFCCC encourages the transfer of technology and intellectual property from developed to developing countries, so references in NDCs are often *pro forma* and short on details. In principle, technology transfer might allow LAC manufacturers to produce more environmental goods domestically, thus reducing expenditure on clean technology from abroad. Thirteen of 17 Latin American NDCs and 10 of 16 Caribbean NDCs express interest in receiving technology transfer. Most countries cite their developing status and historical lack of responsibility for climate change and request financial and technological assistance from the international community to meet their emissions goals.
3. The World Customs Organization's Harmonized System (HS) uses code numbers to define products. A code with a low number of digits defines broad categories of products; additional digits indicate sub-divisions into more detailed definitions. Six-digit codes are the most detailed definitions that are used as standard.

References

ACTO (2022), *Understand the importance of ACTO*, Amazon Cooperation Treaty Organization, Brasília, Brazil, http://otca.org/en/about-us/ (accessed on 5 July 2022). [26]

Andersen, S. (2015), "International Climate Negotiations: Top-down, Bottom-Up or a Combination of Both?", *The International Spectator*, Vol. 50, No. 1, Taylor & Francis, Abingdon-on-Thames, UK, pp. 15-30, https://doi.org/10.1080/03932729.2014.997992 (accessed on 21 June 2022). [7]

Anderson, U. (2022), *Gobierno inauguró Guajira 1, el primer parque eólico que aportará 20 MW de energía*, Grupo La República, Lima, https://larepublica.co/economia/gobierno-inaugurara-hoy-el-parque-eolico-guajira-1-el-mas-grande-de-colombia-3289884 (accessed on 30 June 2022). [96]

Arbache, J. (2021), *Moving towards a Latin American carbon market*, Development Bank of Latin America, Caracas, https://www.caf.com/en/knowledge/views/2021/12/moving-towards-a-latin-american-carbon-market/ (accessed on 5 July 2022). [107]

Bellora, C. and L. Fontagné (2022), *The EU in Search of a WTO-compatible Carbon Border Adjustment Mechanism*, VoxEU, London, https://voxeu.org/article/search-wto-compatible-carbon-border-adjustment-mechanism (accessed on 28 June 2022). [60]

Burger, A., K. Kristof and A. Matthey (2020), *The Green New Consensus*, Umweltbundesamt, Dessau-Roßlau, Germany, https://www.umweltbundesamt.de/sites/default/files/medien/376/publikationen/thenewgreenconsensus_englisch_bf.pdf. [5]

CAF (2022), *Value Proposition: Latin America and Caribbean Initiative for the Development of the Carbon Market (ILACC)*, Development Bank of Latin America, Caracas, https://scioteca.caf.com/handle/123456789/1916. [108]

CARICOM (2017), *Caribbean Community Environmental and Natural Resources Policy Framework*, The Secretariat of the Caribbean Community, Turkeyen, https://caricom.org/Att.%20I%20-%20DRAFT%20FINAL%20CARICOM%20ENV%20%20NAT%20RESOURCE%20POLICY.pdf (accessed on 18 July 2022). [30]

CARICOM (2014), *Strategic Plan for the Caribbean Community 2015-2019: Repositioning CARICOM. Vol. 1 – The Executive Plan*, Caribbean Community Secretariat, Turkeyen, https://caricom.org/wp-content/uploads/STRATEGIC-PLAN-2016_opt.pdf (accessed on 18 July 2022). [27]

CEPII (2021), *Centre for International Prospective Studies and Information*, Centre d'Études Prospectives et d'Informations Internationales, Paris, http://cepii.fr/CEPII/en/welcome.asp (accessed on 11 July 2022). [71]

Clifford Chance (2021), *COP26: Article 6 Rulebook for the New Global Carbon Market Mechanism Agreed*, Clifford Chance, London, https://www.cliffordchance.com/content/dam/cliffordchance/briefings/2021/11/cop26-article-6-rulebook-for-the-new-global-carbon-market-agreed.pdf. [106]

Climate Strategies (2020), *Implementing Just Transition after COP24*, Climate Strategies, London, https://climatestrategies.org/wp-content/uploads/2019/01/Implementing-Just-Transition-after-COP24_FINAL.pdf (accessed on 6 July 2022). [8]

Climate Tracker (2021), *The new marine "mega reserve" in Galápagos Island, explained*, Climate Tracker, Sydney, Australia, https://climatetracker.org/the-declaration-for-the-conservation-of-the-marine-corridor-of-the-eastern-tropical-pacific-explained/ (accessed on 5 July 2022). [25]

Climate Watch (2022), *Historical CO2 Emissions*, https://www.climatewatchdata.org/data-explorer/historical-emissions?page=1 (accessed on 17 August 2022). [1]

Delgado Pugley, D. (2021), *Latin America before the COP26: Positions and perspectives*, Fundación Carolina, Madrid, https://doi.org/10.33960/issn-e.1885-9119.DT58. [15]

Domínguez, P. (2022), *Hay comunicación con Bolivia, Argentina y Chile para crear asociación de litio: AMLO*, Milenio, Monterrey, Mexico, https://www.milenio.com/politica/amlo-busca-crear-asociacion-litio-bolivia-argentina-chile (accessed on 30 June 2022). [91]

Dowrich-Phillips, L. (2022), *CDB devises new metric to make small states eligible for financing*, Loop News, https://caribbean.loopnews.com/content/cdb-devises-new-metric-make-small-states-eligible-financing-3 (accessed on 19 July 2022). [32]

ECLAC (2021), *International Trade Outlook for Latin America and the Caribbean: Pursuing a Resilient and Sustainable Recovery*, Economic Commission for Latin America and the Caribbean, Santiago, https://www.cepal.org/en/publications/47536-international-trade-outlook-latin-america-and-caribbean-2021-pursuing-resilient. [70]

ECLAC (2021), *Latin America and the Caribbean has All the Right Conditions to Become a Renewable Energy Hub with Great Potential in Green Hydrogen*, Economic Commission for Latin America and the Caribbean, Santiago, https://www.cepal.org/en/news/latin-america-and-caribbean-has-all-right-conditions-become-renewable-energy-hub-great (accessed on 6 July 2022). [36]

ECLAC (2018), *Regional Agreement on Access to Information, Public Participation, and Justice in Environmental Matters in Latin America and the Caribbean*, Economic Commission for Latin America and the Caribbean, Santiago, https://www.cepal.org/en/escazuagreement (accessed on 2022). [24]

EUROCLIMA+ (2022), *El Salvador y la Unión Europea presentan su plan de acción frente al cambio climático*, EUROCLIMA+, Brussels, https://www.euroclima.org/actualidad-ec/noticias-es/1695-el-salvador-y-la-union-europea-presentan-su-plan-de-accion-frente-al-cambio-climatico (accessed on 29 June 2022). [97]

European Commission (2022), *Biodiversity Strategy for 2030*, European Commission, Brussels, https://environment.ec.europa.eu/strategy/biodiversity-strategy-2030_en (accessed on 5 July 2022). [46]

European Commission (2022), *Circular economy action plan*, European Commission, Brussels, https://environment.ec.europa.eu/strategy/circular-economy-action-plan_en#:~:text=The%20new%20action%20plan%20announces,for%20as%20long%20as%20possible (accessed on 29 June 2022). [61]

European Commission (2022), *EU legislation on MRLs*, European Commission, Brussels, https://ec.europa.eu/food/plants/pesticides/maximum-residue-levels/eu-legislation-mrls_en (accessed on 18 May 2022). [64]

European Commission (2022), *EU-Latin America & Caribbean Leaders' Meeting: Joining Forces for a Sustainable post-COVID Recovery*, Eurpean Commission, Brussels, https://ec.europa.eu/commission/presscorner/detail/hu/ip_21_6541 (accessed on 29 June 2022). [99]

European Commission (2022), *Health and consumer protection for animal and plant product*, European Commission, Brussels, https://trade.ec.europa.eu/access-to-markets/en/content/health-and-consumer-protection-animal-and-plant-product (accessed on 18 May 2022). [65]

European Commission (2022), *Just and sustainable economy: Commission lays down rules for companies to respect human rights and environment in global value chains*, European Commission, Brussels, https://ec.europa.eu/commission/presscorner/detail/en/ip_22_1145 (accessed on 5 July 2022). [53]

European Commission (2022), *Just Transition funding sources*, https://ec.europa.eu/info/strategy/priorities-2019-2024/european-green-deal/finance-and-green-deal/just-transition-mechanism/just-transition-funding-sources_en (accessed on 17 August 2022). [49]

European Commission (2021), *Carbon Border Adjustment Mechanism: Questions and Answers*, European Commission, Brussels, https://ec.europa.eu/commission/presscorner/detail/en/qanda_21_3661 (accessed on 5 July 2022). [59]

European Commission (2021), *Communication from the Commission to the European Parliament, the Council, the European Economic and Social Committee and the Committee of the Regions Empty. 'Fit for 55': delivering the EU's 2030 Climate Target on the way to climate neutrality*, European Commission, Brussels, https://eur-lex.europa.eu/legal-content/En/TXT/HTML/?uri=CELEX:52021DC0550&from=EN (accessed on 4 July 2022). [56]

European Commission (2021), *European Green Deal: Commission adopts new proposals to stop deforestation, innovate sustainable waste management and make soils healthy for people, nature and climate*, European Commission, Brussels, https://ec.europa.eu/commission/presscorner/detail/en/ip_21_5916 (accessed on 5 July 2022). [55]

European Commission (2021), *European Green Deal: Commission proposes transformation of EU economy and society to meet climate ambitions*, European Commission, Brussels, https://ec.europa.eu/commission/presscorner/detail/en/IP_21_3541 (accessed on 1 July 2022). [44]

European Commission (2021), *Launch by the United States, the European Union, and Partners of the Global Methane Pledge to Keep 1.5C Within Reach*, European Commission, Brussels, https://ec.europa.eu/commission/presscorner/detail/en/statement_21_5766 (accessed on 5 July 2022). [23]

European Commission (2021), *LIFE Programme: More than €290 million in EU funding for nature, environment and climate action projects**, European Commission, Brussels, https://ec.europa.eu/commission/presscorner/detail/en/ip_21_6178 (accessed on 30 June 2022). [98]

European Commission (2021), *Neighbourhood, Development and International Cooperation Instrument – Global Europe (NDICI – Global Europe)*, European Commission, Brussels, https://ec.europa.eu/neighbourhood-enlargement/funding-and-technical-assistance/neighbourhood-development-and-international-cooperation-instrument-global-europe-ndici-global-europe_en (accessed on 15 July 2022). [51]

European Commission (2021), *Proposal for a Regulation of the European Parliament and of the Council establishing a carbon border adjustment mechanism*, European Commission, Brussels, https://ec.europa.eu/info/sites/default/files/carbon_border_adjustment_mechanism_0.pdf. [58]

European Commission (2020), *Communication from the Commission to the European Parliament, the Council, the European Economic and Social Committee and the Committee of the Regions. A Farm to Fork Strategy for a fair, healthy and environmentally-friendly food system*, European Commision, Brussels, https://eur-lex.europa.eu/legal-content/EN/TXT/HTML/?uri=CELEX:52020DC0381&from=EN (accessed on 19 July 2022). [62]

European Commission (2020), *Financing the green transition: The European Green Deal Investment Plan and Just Transition Mechanism*, European Commission, Brussels, https://ec.europa.eu/regional_policy/en/newsroom/news/2020/01/14-01-2020-financing-the-green-transition-the-european-green-deal-investment-plan-and-just-transition-mechanism (accessed on 4 July 2022). [45]

European Commission (2020), *From farm to fork: Our food, our health, our planet, our future*, European Commission, Brussels, https://ec.europa.eu/commission/presscorner/detail/en/fs_20_908 (accessed on 5 July 2022). [47]

European Commission (2020), *New Circular Economy Action Plan shows the way to a climate-neutral, competitive economy of empowered consumers*, European Commission, Brussels, https://ec.europa.eu/newsroom/growth/items/671357/en (accessed on 30 June 2022). [63]

European Commission (2019), *A European Green Deal*, European Commission, Brussels, https://ec.europa.eu/info/strategy/priorities-2019-2024/european-green-deal_en (accessed on 1 July 2022). [43]

European Council (2021), *Council Decision (EU) 2021/1345 of 28 June 2021 authorising the opening of negotiations with Argentina, Australia, Canada, Costa Rica, India, Israel, Japan, New Zealand, South Korea, Tunisia and the United States...*, European Council, Brussels, https://eur-lex.europa.eu/eli/dec/2021/1345/oj (accessed on 1 July 2022). [67]

European Council (2017), *Council Decision (EU) 2017/436 of 6 March 2017 on the signing, on behalf of the European Union, of the Agreement between the European Union and the Republic of Chile on trade in organic products*, European Council, Brussels, https://eur-lex.europa.eu/legal-content/EN/TXT/?uri=CELEX%3A32017D0436 (accessed on 1 July 2022). [66]

European Parliament/European Council (2021), *Regulation (EU) 2021/947 of the European Parliament and of the Council of 9 June 2021 establishing the Neighbourhood, Development and International Cooperation Instrument – Global Europe, amending and repealing Decision No 466/2014/EU...*, European Parliament/European Council, Brussels, https://eur-lex.europa.eu/eli/reg/2021/947/oj (accessed on 15 July 2022). [52]

GIZ (2021), *Partners for Review: 2021 Voluntary National Reviews. A snapshot of trends in SDG reporting*, Deutsche Gesellschaft für Internationale Zusammenarbeit, Bonn, Germany, https://www.partners-for-review.de/wp-content/uploads/2021/11/P4R-2021-VNR-Analysis_FINAL.pdf. [79]

GIZ (2020), *2020 Voluntary National Reviews. A Snapshot of Trends in SDG Reporting*, Deutsche Gesellschaft für Internationale Zusammenarbeit, Bonn, Germany, https://www.partners-for-review.de/wp-content/uploads/2020/12/P4R-2020-VNR-Analysis_FINAL-Nov.-2020.pdf. [80]

GIZ (2019), *Partners for Review: Voluntary National Reviews submitted to the 2019 High-level Political Forum for Sustainable Development. A Comparative Analysis*, Deutsche Gesellschaft für Internationale Zusammenarbeit, Bonn, Germany, https://www.partners-for-review.de/wp-content/uploads/2020/12/P4R-Analysis-VNRs-2019.pdf. [81]

Gómez Arana, A. (2021), *Interregionalismo y Acuerdos de Asociación UE-América Latina y el Caribe*, Fundación Carolina, Madrid, https://doi.org/10.33960/issn-e.1885-9119.DT54. [68]

Government of Chile (2020), *National Green Hydrogen Strategy: Chile, a clean energy provider for a carbon neutral planet*, Ministry of Energy, Government of Chile, Santiago, https://energia.gob.cl/sites/default/files/national_green_hydrogen_strategy_-_chile.pdf. [92]

Green Climate Fund (2022), *Project Portfolio*, Green Climate Fund, Incheon, South Korea, https://www.greenclimate.fund/projects (accessed on 1 July 2022). [50]

Green Climate Fund (2020), *Initial Resource Mobilisation*, https://www.greenclimate.fund/about/resource-mobilisation/irm (accessed on 17 August 2022). [48]

IDB (2021), *Climate Policies in Latin America and the Caribbean: Success Stories and Challenges in the Fight against Climate Change*, Inter-American Development Bank, New York, https://doi.org/10.18235/0003239. [17]

IDB (2021), *Hoja de Ruta del Hidrógeno en Colombia*, Inter-American Development Bank, New York, https://cyted.org/es/noticias/hoja-de-ruta-del-hidr%C3%B3geno-en-colombia. [93]

IDB (2014), *Small Island States*, Inter-American Development Bank, New York, https://www.iadb.org/en/ove/climate-change-caribbean-small-island-states#:~:text=As%20developing%20economies%20relying%20on,disasters%20identified%20by%20the%20Intergovernmental (accessed on 18 July 2022). [28]

IEA (2021), *Hydrogen in Latin America: From near-term opportunities to large-scale deployment*, International Energy Agency, Paris, https://iea.org/reports/hydrogen-in-latin-america. [94]

ILO (2021), *Just transition and green jobs in the framework of biodiversity protection in Latin America and the Caribbean*, International Labour Organization, Geneva, Switzerland, https://ilo.org/americas/publicaciones/WCMS_831480/lang--es/index.htm. [101]

ILO (2015), *Guidelines for a just transition towards environmentally sustainable economies and societies for all*, International Labour Organization, Geneva, Switzerland, https://ilo.org/wcmsp5/groups/public/@ed_emp/@emp_ent/documents/publication/wcms_432859.pdf. [6]

International Trade Analysis Database (2021), *BACI: International Trade Database at the Product-Level*, Centre d'Études Prospectives et d'Informations Internationales, http://cepii.fr/CEPII/en/bdd_modele/bdd_modele_item.asp?id=37 (accessed on 11 July 2022). [72]

International Trade Center (2022), *Trade Map Statistics*, International Trade Center, Geneva, Switzerland, https://trademap.org/ (accessed on 18 May 2022). [42]

International Trade Centre (2022), *GreenToCompete*, International Trade Centre, Geneva, Switzerland, https://greentocompete.org/ (accessed on 20 July 2022). [73]

Klöck, C. et al. (2020), *Coalitions in the Climate Change Negotiations*, Routledge, London, https://doi.org/10.4324/9780429316258. [16]

Laguardia Martínez, J. (2017), "The Environmental Governance in the Caribbean of the CARICOM for the Management of Climate Change", *Entretextos*, Vol. 27, https://revistasacademicas.iberoleon.mx/index.php/entretextos/article/view/260. [29]

Latinobarómetro/Nueva Sociedad/Friedrich-Ebert-Stiftung (2022), *Latin America – European Union: views, agendas and expectations*, Latinobarómetro/Nueva Sociedad/Friedrich-Ebert-Stiftung, Santiago/Buenos Aires/Bonn, Germany, https://data.nuso.org/ (accessed on 19 May 2022). [41]

Lebdioui, A. (2022), *Latin American Trade in the Age of Climate Change: Impact, Opportunities, and Policy Options*, LSE Latin American and the Caribbean Centre, https://www.lse.ac.uk/lacc/assets/documents/CH-LSE-Latin-American-Trade-in-the-Age-of-Climate-Change-AMENDS.pdf (accessed on 17 August 2022). [88]

López-Calva, L. (2022), *Lithium in Latin America: A new quest for "El Dorado"?*, United Nations Development Programme – Latin America and the Caribbean, New York, https://undp.org/latin-america/blog/lithium-latin-america-new-quest-el-dorado (accessed on 29 June 2022). [90]

Lorenzo Arana, I. (2020), "Argentina, Brazil, and Uruguay (A-B-U)", *Negotiating Climate Adaptation*, pp. 8-18, https://doi.org/10.1007/978-3-030-41021-6. [14]

Marzano Franco, K. (2016), *Tackling Climate Change in Latin America*, Konrad-Adenauer-Stiftung, Bonn, Germany, https://www.kas.de/c/document_library/get_file?uuid=ba43934b-d004-4ca5-4519-58ce8a3dbd98&groupId=252038. [13]

NDC Partnership (2022), *NDC Content*, NDC Partnership, Washington, DC, https://ndcpartnership.org/climate-tools/ndcs (accessed on 5 July 2022). [18]

Nolte, D. (2021), "From the summits to the plains: The crisis of Latin American Regionalism", *Latin American Policy*, Vol. 12, No. 1, Wiley Online Library, Hoboken, NJ, pp. 181-192, https://doi.org/10.1111/lamp.12215. [10]

OECD (2022), *Meeting of the OECD Council at Ministerial Level, Paris, 9-10 June 2022*, OECD Publishing, Paris, https://www.oecd.org/mcm/2022-MCM-Key-Issues-Paper-EN.pdf. [77]

OECD (2022), *Promoting Responsible Business Conduct in Latin America and the Caribbean*, OECD Publishing, Paris, https://mneguidelines.oecd.org/rbclac.htm (accessed on 2 June 2022). [100]

OECD (2021), *Taxing Energy Use for Sustainable Development: Opportunities for energy tax and subsidy reform in selected developing and emerging economies*, OECD Publishing, Paris, https://www.oecd.org/tax/tax-policy/taxing-energy-use-for-sustainable-development.pdf [107]

OECD (2019), *Measuring Distance to the SDG Targets 2019: An Assessment of Where OECD Countries Stand*, OECD Publishing, Paris, https://doi.org/10.1787/a8caf3fa-en. [84]

OECD (2019), *Recommendation of the Council on Policy Coherence for Sustainable Development*, OECD Publishing, Paris, https://oecd.org/gov/pcsd/recommendation-on-policy-coherence-for-sustainable-development-eng.pdf. [76]

OECD (2011), *Towards Green Growth*, OECD Green Growth Studies, OECD Publishing, Paris, https://doi.org/10.1787/9789264111318-en. [3]

OECD et al. (2021), *Latin American Economic Outlook 2021: Working Together for a Better Recovery*, OECD Publishing, Paris, https://doi.org/10.1787/5fedabe5-en. [75]

OECD et al. (2020), *Latin American Economic Outlook 2020: Digital Transformation for Building Back Better*, OECD Publishing, Paris, https://doi.org/10.1787/e6e864fb-en. [40]

OECD et al. (2019), *Latin American Economic Outlook 2019: Development in Transition*, OECD Publishing, Paris, https://doi.org/10.1787/g2g9ff18-en. [74]

OECD/EC-JRC (2021), *Understanding the Spillovers and Transboundary Impacts of Public Policies: Implementing the 2030 Agenda for More Resilient Societies*, OECD Publishing, Paris, https://doi.org/10.1787/862c0db7-en. [87]

Parry, I., S. Black and N. Vernon (2021), *Still Not Getting Energy Prices Right: A Global and Country Update of Fossil Fuel Subsidies*, IMF Working Paper, No. WP/21/236, International Monetary Fund, Washington, DC, https://imf.org/-/media/Files/Publications/WP/2021/English/wpiea2021236-print-pdf.ashx. [104]

Pastrana, E. (2013), "Why regionalism has failed in Latin America: lack of stateness as an important factor for failure of sovereignty transfer in integration projects", *Contexto Internacional*, Vol. 35, No. 2, Pontifícia Universidade Católica do Rio de Janeiro, Brazil, pp. 443-469, https://doi.org/10.1590/S0102-85292013000200005. [11]

Rodríguez Díaz, J. and J. Sanahuja (2021), *The Mercosur-European Union Agreement: scenarios and options for strategic autonomy, productive transformation and social and ecological transition*, Fundación Carolina, Madrid, https://doi.org/10.33960/AC_20.2021. [69]

Ruano, L. and N. Saltalamacchia (2021), "Latin American and Caribbean Regionalism during the Covid-19 Pandemic: Saved by Functionalism?", *The International Spectator*, Vol. 56, No. 2, Taylor & Francis Online, Abingdon-on-Thames, UK, pp. 93-113, https://tandfonline.com/doi/full/10.1080/03932729.2021.1900666. [9]

Saalfield, J. (forthcoming), *Potential Trade Implications of Latin America's Climate Commitments Under the Paris Agreement*, International Trade Series, Economic Commission for Latin America and the Caribbean, Santiago. [34]

Sachs, J. et al. (2021), *Sustainable Development Report 2021: The Decade of Action for the Sustainable Development Goals*, Cambridge University Press, Cambridge, UK, https://doi.org/10.1017/9781009106559. [85]

Sauvage, J. (2014), *The Stringency of Environmental Regulations and Trade in Environmental Goods*, OECD Trade and Environment Working Papers, https://doi.org/10.1787/5jxrjn7xsnmq-en. [37]

SDSN/Yale Center for Environmental Law & Policy/Center for Global Commons at the University of Tokyo (2021), *Global Commons Stewardship Index 2021*, The Sustainable Development Solutions Network/Yale Center for Environmental Law & Policy/Center for Global Commons at the University of Tokyo, New York/New Haven, CT/Tokyo, https://resources.unsdsn.org/global-commons-stewardship-index-2021. [86]

Secretaría de Economía (2021), *Informe Nacional Voluntario 2021, Agenda 2030 en México*, Ministry of Economy, Government of Mexico, Mexico City, https://economia.gob.mx/files/gobmx/agenda2030/INV2021_F4.pdf (accessed on 5 July 2022). [78]

Sistema de Informacion energetica de Latinoamerica y el Caribe (SieLAC) (2020), *Estadística Energética [database]*, https://sielac.olade.org/default.aspx. [89]

Smith, S. (2020). "Rystad Energy: Latin America set for renewable energy boom". *Energy Global*, 11 September [online] https://www.energyglobal.com/special-reports/11092020/rystad-energy-latin-america-set-for-renewable-energy-boom/. [106]

UN (2021), *Glasgow Leaders' Declaration on Forests and Land Use*, UN Climate Change Conference UK 2021, UN, New York, https://ukcop26.org/glasgow-leaders-declaration-on-forests-and-land-use/ (accessed on 5 July 2022). [21]

UN (2021), *Global Coal to Clean Power Transition Statement*, UN Climate Change Conference UK 2021, UN, New York, https://ukcop26.org/global-coal-to-clean-power-transition-statement/ (accessed on 5 July 2022). [22]

UN (2020), *2030 Agenda in Latin America and the Caribbean*, United Nations General Assembly, New York, https://agenda2030lac.org/en (accessed on 1 July 2022). [83]

UN Comtrade (2020), *UN Comtrade Database*, United Nations International Trade Statistics Database, United Nations Statistics Division, New York, https://comtrade.un.org/ (accessed on 11 July 2022). [38]

UN, et al. (2014), *System of Environmental Economic Accounting 2012— SEEA Central Framework*. United Nations, https://unstats.un.org/unsd/envaccounting/seearev/seea_cf_final_en.pdf, (accessed on 6 Oct 2022). [33]

UN DESA (2020), *Economic Commission for Latin America and the Caribbean (ECLAC)*, United Nations Department of Economic and Social Affairs, New York, https://sdgs.un.org/un-system-sdg-implementation/economic-commission-latin-america-and-caribbean-eclac-24520 (accessed on 1 July 2022). [20]

UNCCD (2022), *Annex II: Latin America and the Caribbean (LAC)*, United Nations Convention to Combat Desertification, Bonn, Germany, https://unccd.int/convention/regions/annex-iii-latin-america-and-caribbean-lac (accessed on 29 June 2022). [19]

UNCTAD (2021), *Review of Maritime Transport 2021*, United Nations Conference on Trade and Development, Geneva, Switzerland, https://unctad.org/webflyer/review-maritime-transport-2021. [57]

UNDP (2022), *Human Developmen Report Data Center*, https://hdr.undp.org/data-center/documentation-and-downloads (accessed on 12 August 2022). [2]

UNDP (2021), *Towards a Multidimensional Vulnerability Index*, Discussion Paper, United Nations Development Programme, New York, https://undp.org/publications/towards-multidimensional-vulnerability-index (accessed on 19 July 2022). [31]

UNEP (2022), *Methodology for SDG-indicator 17.14.1: Mechanisms in Place to Enhance Policy Coherence for Sustainable Development*, United Nations Environment Programme, Nairobi, https://wedocs.unep.org/20.500.11822/38262 (accessed on 5 July 2022). [82]

UNEP (2011), *Towards a Green Economy: Pathways to Sustainable Development and Poverty Eradication*, United Nations Environment Programme, Nairobi, https://unep.org/resources/report/towards-green-economy-pathways-sustainable-development-and-poverty-eradication-10. [4]

UNFCCC (2022), *NDC Registry*, United Nations Framework Convention on Climate Change, Bonn, Germany, https://unfccc.int/NDCREG (accessed on 11 July 2022). [35]

UNFCCC (2021), *Potential of South-South and triangular cooperation on climate technologies for advancing the implementation of NDCs and NAPs*, United Nations Framework Convention on Climate Change, Bonn, Germany, https://unfccc.int/ttclear/tec/brief9.html#SScompilation. [102]

UNFCCC (2015), *Paris Agreement*, United Nations Framework Convention on Climate Change, Bonn, Germany, https://unfccc.int/files/meetings/paris_nov_2015/application/pdf/paris_agreement_english_.pdf (accessed on 12 April 2022). [103]

UNIDO (2021), *Supporting the establishment and development of the International Hydrogen Energy Centre*, United Nations Industrial Development Organization, Vienna, https://unido.org/sites/default/files/files/2021-09/Supporting%20the%20establishment%20and%20development%20of%20the%20International_0.pdf. [95]

Watts, J. and J. Depledge (2018), "Latin America in the climate change negotiations: Exploring the AILAC and ALBA coalitions", *WIREs Climate Change*, Wiley Interdisciplinary Reviews, Wiley-Blackwell, Hoboken, NJ, https://doi.org/10.1002/wcc.533. [12]

WEF (2022), *How mandatory human rights and environmental due diligence can create a sustainable future*, World Economic Forum, Cologny, Switzerland, https://weforum.org/agenda/2022/02/good-corporate-governance-and-a-sustainable-future-the-role-of-mandatory-human-rights-and-environmental-due-diligence/ (accessed on 30 June 2022). [54]

World Bank (2021), *Carbon Pricing Dashboard*, World Bank Group, Washington, DC, https://carbonpricingdashboard.worldbank.org/map_data. [105]

Country notes

- Argentina
- Brazil
- Chile
- Colombia
- Costa Rica
- Dominican Republic
- Ecuador
- El Salvador
- Guatemala
- Mexico
- Panama
- Paraguay
- Peru
- Uruguay

READER'S GUIDE

The statistical tables follow key areas identified in the *Latin American Economic Outlook (LEO)*: 1) socio-economic dimension; 2) citizens' perceptions and institutions; 3) productivity and innovation; 4) environment and the green transition; and 5) fiscal position.

Latin America and the Caribbean (LAC) average is a simple average of the largest set of LAC countries for which data are available.

Organisation for Economic Co-operation and Development (OECD) average is a simple average of the largest set of all OECD member countries, as of May 2022, for which data are available.

Countries for which data are not available for both years of comparison have been excluded from the averages to ensure comparability between years. Exceptions to this are mentioned in the notes.

Applying the same criteria as in previous LEO editions, data selection prioritises comparability across LAC countries and shows the latest comparable data available at the report's publication date.

Social dimension

Extreme poverty:[1] refers to the percentage of the population whose average per-capita income is below the extreme poverty line, as specified by the United Nations Economic Commission for Latin America and the Caribbean. Method of computation: "n" is defined as the total number of persons and "i" is the number of people whose per-capita income is below the extreme poverty line; the percentage of people living in extreme poverty is expressed as I=i/n (known as "headcount index"). The average per-capita income (yPC) is calculated by dividing the total income of each household by the number of people forming it. Data from ECLAC (2022[1]), *Statistical Database and Publications*, https://statistics.cepal.org/portal/cepalstat/dashboard.html.

Poverty:[1] refers to the percentage of the population whose average per-capita income is below the poverty line, as specified by the United Nations Economic Commission for Latin America and the Caribbean. Method of computation: "n" is the total number of people and "p" is the number of people whose per-capita income is below the poverty line; the percentage of people living in poverty is expressed as P=p/n. This indicator includes people under the extreme poverty line, by definition. The average per-capita income (yPC) is calculated by dividing the total income of each household by the number of people forming it. Data from ECLAC (2022[1]), *Statistical Database and Publications*, https://statistics.cepal.org/portal/cepalstat/dashboard.html.

Gini index:[2] measures the extent to which the distribution of income (or, in some cases, consumption expenditure) among individuals or households within an economy deviates from a perfectly equal distribution. A Gini index of zero represents perfect equality, while an index of 100 represents perfect inequality. Data from World Bank (2022[2]), *World Bank Open Data*, https://data.worldbank.org/indicator/SI.POV.GINI.

Share of internet users:[3] measures people with access to the internet as a percentage of the total population. Data from International Telecommunication Union (2022[3]), *Global ICT Statistics*, www.itu.int/en/ITU-D/Statistics/Pages/stat/default.aspx.

Share of total population in informal households overall and by quintile:[4] provides the distribution of the total population living in informal households overall and by quintile. An informal household has all of its workers in informal work. Quintiles are based on monthly total household consumption or income. Data are from OECD (2021[4]), *Key Indicators of Informality based on Individuals and their Households* (KIIbIH) database, https://stats.oecd.org/Index.aspx?DataSetCode=KIIBIH_B6 and https://stats.oecd.org/Index.aspx?DataSetCode=KIIBIH_B7.

Health expenditure: refers to the level of current health expenditure as a percentage of gross domestic product (GDP). Estimates of current health expenditures include healthcare goods and services consumed each year. This indicator does not include capital health expenditures, such as buildings, machinery, information technology, and stocks of vaccines for emergencies or outbreaks. Data from World Bank (2022[4]), *World Bank Open Data*, https://data.worldbank.org/indicator/SH.XPD.CHEX.GD.ZS.

SIGI index: measures discrimination against women in social institutions (e.g. formal and informal laws, social norms, and practices). Lower values indicate lower levels of discrimination in social institutions: the SIGI ranges from 0% for no discrimination to 100% for very high discrimination. Data from OECD (2022[5]), *Social Institutions and Gender Index (SIGI) Data*, www.genderindex.org/data/.

PISA score in science: measures the mean score in science performance as measured by the Programme for International Student Assessment (PISA) for each country. Scientific performance measures the scientific literacy of a 15-year-old in the use of scientific knowledge to identify questions, acquire new knowledge, explain scientific phenomena and draw evidence-based conclusions about science-related issues. Data from OECD (2022[6]), *Science performance (PISA) indicator*, https://data.oecd.org/pisa/science-performance-pisa.htm#indicator-chart.

Productivity and innovation

Labour productivity: measures output per employed person as a percentage of United States output per employed person (in 2021 international dollars, converted using purchasing power parity). Data from the Conference Board (2022[7]), *Total Economy Database*, www.conference-board.org/data/economydatabase/total-economy-database-productivity.

High-technology exports:[5] measures exports of products with high research and development (R&D) intensity as a percentage of total manufactured exports. Data from World Bank (2022[2]), *World Bank Open Data*, https://data.worldbank.org/indicator/TX.VAL.TECH.MF.ZS.

Research and development expenditure:[6] measures capital and current expenditures as a percentage of GDP in the four main sectors: business enterprise, government, higher education and private non-profit. R&D covers basic research, applied research and experimental development. Data from World Bank (2022[2]), *World Bank Open Data*, https://data.worldbank.org/indicator/GB.XPD.RSDV.GD.ZS.

Citizens' perception and institutions

Indicators measure the share of the population:[7]

- satisfied with efforts to preserve the environment
- with confidence in the national government
- that thinks corruption is widespread throughout the government
- satisfied with the education system
- satisfied with the availability of quality health care

as a percentage of the adult population. Data from Gallup (2022[8]), *Global Datasets for Public Use*, www.gallup.com/analytics/318875/global-research.aspx.

Environment and the green transition

Loss of natural and semi-natural vegetated land: measures tree cover, grassland, wetland, shrub land and sparse vegetation converted to any other land cover type as a percentage of the total natural and semi-natural vegetated land. The denominator used is the "stock" of natural and semi-natural land measured in square kilometres at the start of the period. Data from OECD (2019[9]), *Environment Database - Intact Forest Landscapes*, https://stats.oecd.org/Index.aspx?DataSetCode=INTACT_FOREST_LANDSCAPES#.

Greenhouse gas (GHG) emissions per capita: measures GHG emissions per capita, excluding land-use change and forestry (LUCF). GHGs include carbon dioxide (CO_2), methane, nitrous oxide and F-gases (fluorinated gases), which are converted and reported in tonnes of carbon dioxide equivalent (t CO_2e). Data from Climate Watch (2022[10]), *Climate Watch Historical GHG emissions*, www.climatewatchdata.org/ghg-emissions. Authors chose the Climate Analysis Indicators Tool (CAIT) as the data source because it is the most comprehensive on Climate Watch and includes all sectors and gases, and data is up to date until 2019. Climate Watch Historical GHG Emissions data are derived from several sources. Fuel combustion data are from OECD/IEA (2021[11]), *CO_2 emissions from Fuel Combustion*, www.oecd-ilibrary.org/energy/data/iea-co2-emissions-from-fuel-combustion-statistics_co2-data-en. Agriculture data are from FAO (2022[12]), *FAOSTAT Emissions*, www.fao.org/food-agriculture-statistics/data-release/data-release-detail/en/c/1304919/.

Air pollution as exposure to PM2.5:[8] refers to the percentage of the population exposed to more than 10 micrograms per cubic metre (μg/m^3) and is expressed as annual averages. Fine particulate matter (PM2.5) refers to a range of air pollutants that pose the greatest health risk globally, affecting more people than any other pollutant. Chronic exposure to PM2.5 considerably increases the risk of respiratory and cardiovascular diseases in particular. Data from OECD (2022[13]), *Air pollution exposure indicator*, https://data.oecd.org/air/air-pollution-exposure.htm.

Contribution of renewables to total primary energy supply:[9] renewables include the primary energy equivalent of hydro (excluding pumped storage), geothermal, solar, wind, tide and wave sources. Energy derived from solid biofuels, biogasoline, biodiesels, other liquid biofuels, biogases and the renewable fraction of municipal waste is also included. Data updated until 2019, from OECD (2022[14]), *Renewable energy indicator*, https://data.oecd.org/energy/renewable-energy.htm.

Marine protected areas:[10] measures the marine protected area as a percentage of a country's total marine area. Data from World Bank (2022[2]), *World Bank Open Data*, https://data.worldbank.org/indicator/ER.MRN.PTMR.ZS. and United Nations Environment World Conservation Monitoring Centre (2022[15]), *World Database on Protected Areas*, www.protectedplanet.net/en/search-areas?geo_type=country.

Fiscal position

Environmentally related tax revenue:[11] measures the revenues from environmentally related taxes as a percentage of GDP. It includes taxes on GHGs, fuel taxes, taxes on road use, forestry taxes and revenue from auctioned permits of emission trading systems for GHGs. Data from OECD (2022[16]), *Revenue Statistics in Latin America and the Caribbean 2022*, https://stats.oecd.org/Index.aspx?DataSetCode=ERTR.

Total tax revenues:[12] measures total tax revenues as a percentage of GDP. Data from OECD (2022[16]), *Revenue Statistics in Latin America and the Caribbean 2022*, https://stats.oecd.org/Index.aspx?DataSetCode=RS_GBL.

Share of VAT (value added tax):[12] measures VAT as a percentage of GDP. Data from OECD (2022[16]), *Revenue Statistics in Latin America and the Caribbean 2022*, https://stats.oecd.org/Index.aspx?DataSetCode=RS_GBL.

Share of PIT (personal income tax):[12] measures taxes on the income, profits and capital gains of individuals as a percentage of GDP. Data from OECD (2022[16]), *Revenue Statistics in Latin America and the Caribbean 2022*, https://stats.oecd.org/Index.aspx?DataSetCode=RS_GBL.

Share of CIT (corporate income tax):[12] measures taxes on the income, profits and capital gains of corporations as a percentage of GDP. Data from OECD (2022[16]), *Revenue Statistics in Latin America and the Caribbean 2022*, https://stats.oecd.org/Index.aspx?DataSetCode=RS_GBL.

Perception of tax evasion: measures the share of the population that claims to have heard of people who paid less taxes than they should have as a percentage of the adult population. Data from Latinobarómetro (2020[17]), *Latinobarómetro 2020*, www.latinobarometro.org/latOnline.jsp.

Social expenditure:[13] measures public social spending as a percentage of GDP. The main social policy areas are as follows: old age, survivors, incapacity-related benefits, health, family, active labour market programmes, unemployment, housing, and other social policy areas. For OECD countries, data refer to *Social expenditure* from OECD (2022[18]), https://stats.oecd.org/Index.aspx?DataSetCode=SOCX_AGG. For LAC countries, data refer to *Social public expenditure* from ECLAC (2022[19]), https://statistics.cepal.org/portal/cepalstat/dashboard.html?lang=en&indicator_id=3127&area_id=411.

Debt service: measures debt service as a percentage of tax revenue. Debt service is calculated as general government primary lending/borrowing minus general government net lending/borrowing. Authors calculations based on data from IMF (2022[20]), *World Economic Outlook Database*, www.imf.org/en/Publications/WEO/weo-database/2022/April and OECD (2022[16]), *Revenue Statistics in Latin America and the Caribbean 2022*, https://stats.oecd.org/Index.aspx?DataSetCode=RS_GBL.

Notes

1. Poverty and extreme poverty: All data are national-level data, except for Argentina, for which only urban-level data are available, wherefore it is excluded from the LAC averages. For the OECD and LAC averages in 2016, data for Chile is from 2015. For the LAC average in 2020, data for Honduras and Panama are from 2019.
2. Gini index: For the OECD and LAC averages in 2016, data for Chile is from 2015. For the LAC average in 2020, data for Honduras, Panama and El Salvador are from 2019.
3. Share of internet users: For the LAC average in 2020, data for Belize is from 2019.
4. Informality indicators: For the averages of informality by quintiles: in 2009, data for Argentina, Costa Rica, Colombia, Mexico and Peru are from 2010 and data for Uruguay is from 2008. For the averages in 2018, data for Chile is from 2017 and data for Costa Rica is from 2019. For the averages of total informality: in 2009, data for Colombia, Mexico, Argentina Costa Rica and Peru are from 2010. For the averages in 2018, data for Chile is from 2017 and data for Costa Rica is from 2019.
5. High-technology exports: For the LAC average in 2020, data for Antigua and Barbuda, Honduras, Suriname and St. Vincent and the Grenadines are from 2019.

6. Research and development expenditure: For the OECD average in 2016, data for Australia, Switzerland and New Zealand are from 2017. For the OECD and LAC averages in 2019, data for Costa Rica is from 2018.
7. Indicators measure the share of the population: For the LAC average in 2016, data for Jamaica is from 2017. For the OECD and LAC averages in 2021, data for Belgium, Chile, El Salvador and Israel are from 2020.
8. Air pollution as exposure to PM2.5: For the LAC average in 2019, data for Jamaica is from 2018.
9. Contribution of renewables to total primary energy supply: For the OECD and LAC averages in 2020, data for Costa Rica is from 2019.
10. Marine protected areas: For the LAC average in 2016, data for Haiti is from 2017.
11. Environmentally related tax revenue: For the OECD and LAC averages in 2020, data for Costa Rica is from 2019.
12. Tax revenues: For the OECD average of total tax revenues in 2020, data for Australia and Japan is from 2019. For the OECD averages of the shares of VAT/PIT/CIT in 2020, data for Australia, Greece and Japan are from 2019.
13. Social expenditure: Social spending is defined according to the OECD SOCX methodology, www.oecd.org/social/soc/SOCX_Manuel_2019.pdf. For the LAC average in 2020, data for Cuba is from 2019. For the OECD average in 2019, data for Canada, Colombia, Costa Rica, New Zealand and Switzerland are from 2018.

References

Climate Watch (2022), *Historical GHG Emissions*, World Resources Institute, Washington, DC, https://www.climatewatchdata.org/ghg-emissions. [10]

ECLAC (2022), *CEPALSTAT: Public social expenditure according to the classification of the functions of government (as a percentage of GDP)*, Economic Commission for Latin America and the Caribbean, Santiago, https://statistics.cepal.org/portal/cepalstat/dashboard.html?lang=en&indicator_id=3127&area_id=411. [19]

ECLAC (2022), *CEPALSTATS.Statistical Databases and Publications*, https://statistics.cepal.org/portal/cepalstat/dashboard.html?lang=en. [1]

FAO (2022), *FAOSTAT Emissions*, Food and Agriculture Organization, Rome, https://www.fao.org/food-agriculture-statistics/data-release/data-release-detail/en/c/1304919/. [12]

Gallup (2022), *Global Datasets for Public Use*, Gallup, Washington, DC, https://www.gallup.com/analytics/318923/world-poll-public-datasets.aspx. [8]

IMF (2022), *World Economic Outlook Database*, https://www.imf.org/en/Publications/WEO/weo-database/2022/April. [20]

International Telecommunication Union (2022), *Statistics*, International Telecommunication Union, Geneva, https://www.itu.int/en/ITU-D/Statistics/Pages/stat/default.aspx. [3]

Latinobarómetro (2020), *Análisis Online 2020: Se las arregló para pagar menos impuesto del que debía*, Latinobarómetro, Santiago, https://www.latinobarometro.org/latOnline.jsp. [17]

OECD (2022), *Air pollution exposure* (indicator), https://doi.org/10.1787/8d9dcc33-en (accessed on 24 October 2022). [13]

OECD (2022), *Renewable energy* (indicator), https://doi.org/10.1787/aac7c3f1-en (accessed on 24 October 2022). [14]

OECD (2022), *Science performance (PISA)* (indicator), https://doi.org/10.1787/91952204-en (accessed on 24 October 2022). [6]

OECD (2022), *Social Expenditure - Aggregated data*, OECD Publishing, Paris, https://stats.oecd.org/Index.aspx?DataSetCode=SOCX_AGG. [18]

OECD (2022), *Social Institutions and Gender Index (SIGI) Data*, OECD Publishing, Paris, https://www.genderindex.org/data/. [5]

OECD (2021), *OECD.Stat: Total population in informal households by quintile (Percentage of total population living in informal households in the subgroup)*, OECD Publishing, Paris, https://stats.oecd.org/Index.aspx?DataSetCode=KIIBIH_B7. [4]

OECD (2019), *OECD.Stat: Intact Forest Landscapes*, OECD Publishing, Paris, https://stats.oecd.org/Index.aspx?DataSetCode=INTACT_FOREST_LANDSCAPES. [9]

OECD et al. (2022), *OECD.Stat: Revenue Statistics in Latin America and the Caribbean 2022*, OECD Publishing, Paris, https://stats.oecd.org/Index.aspx?DataSetCode=ERTR. [16]

OECD/IEA (2021), *GHG Emissions from Fuel Combustion*, OECD Publishing/International Energy Agency, Paris, https://www.oecd-ilibrary.org/energy/data/iea-co2-emissions-from-fuel-combustion-statistics_co2-data-en. [11]

The Conference Board (2022), *Total Economy Database - Data*, The Conference Board, Inc., New York, https://www.conference-board.org/data/economydatabase/total-economy-database-productivity. [7]

UNEP-WCMC (2022), *World Database on Protected Areas*, United Nations Environment World Conservation Monitoring Centre, Cambridge, UK, https://www.protectedplanet.net/en/search-areas?geo_type=country. [15]

World Bank (2022), *World Bank Open Data*, World Bank Group, Washington, DC, https://data.worldbank.org/. [2]

ARGENTINA

1. Recent trends

Poverty in urban areas in Argentina increased from 21.5% in 2016 to 34.4% in 2020, surpassing the Latin America and the Caribbean (LAC) average of 26.3%, partly owing to the impact of the COVID-19 pandemic. Extreme poverty in urban areas also increased in that period from 2.9% to 6.3% but remains below the LAC average (8.7%). Inequality remained relatively stable. The Gini index increased from 42.0 in 2016 to 42.3 in 2020, remaining below the LAC average (45.3). Regarding environmental indicators, in 2019, greenhouse gas (GHG) emissions per capita were 8.2 tonnes of carbon dioxide equivalent (t CO_2e), higher than the averages for LAC (6.3) and slightly below for countries belonging to the Organisation for Economic Co-operation and Development (OECD) (9.1). That year, the share of population exposed to air pollution levels that pose risks to human health (PM2.5 at more than 10 µg/m^3) was 98.5%, higher than 95.4% for LAC and 61.0% for the OECD. The marine protected area of Argentina accounted for 11.8% of its territorial waters in 2021, compared to 7.3% for LAC and 18.6% for the OECD. On the fiscal side, environmentally related tax revenue was 1.8% of gross domestic product (GDP) in 2020, above LAC (1.0%) and below the OECD (2.1%). Total tax revenues as a percentage of GDP (29.4%) remains higher than the LAC average (21.9%) but below the OECD average (33.5%).

2. Long-term development policies for a green transition

Following the General Environment Law of 2002, Argentina approved laws on waste management (Law No. 25916 of 2004), water management (Law No. 25688 of 2002) forest protection (Law No. 26331 of 2007), and glaciers (Law No. 26639 of 2010). In 2019, Argentina promoted the Law on Minimum Standards for Adaptation and Mitigation of Global Climate Change and created the National Climate Change Cabinet which is developing the National Response Plan.

In terms of **mitigation**, the National Energy and Climate Change Action Plan promotes biofuels, renewable energies and energy efficiency to reduce the energy sector's GHG emissions. In addition, Argentina is committed to reducing deforestation and degradation through its REDD+ (Reducing Emissions from Deforestation and forest Degradation) strategy. The National Plan of Sustainable Mobility sets goals for a sustainable vehicle fleet and an electric charging network. The 266/2022 ministerial resolution created the *Programa de Movilidad Integral no motorizada* to implement sustainable mobility policies. **Adaptation** policies include the Plan GIRSU for improved urban solid waste treatment. The federal plan *Casa Común* promotes green projects with a social impact. To make the green transition socially inclusive, the *Potenciar Empleo Verde* programme supports sustainable economic actors that create decent jobs.

Regarding **international partnerships within the region**, Argentina ratified the Escazú Agreement in 2021 to enhance public participation in decision making and access to justice in environmental matters. Through MERCOSUR, it co-ordinates environmental policies and participates in a project with the National Metrology Institute of Germany to develop energy efficiency standards for household appliances. Argentina is working with 13 Iberoamerican countries in the project *H2Transel* to develop the production of hydrogen. **Beyond LAC**, Argentina is working with EUROCLIMA+ on forest management, electric mobility and energy efficiency. Argentina is part of the Partnership for Action on Green Economy which supports national efforts in the transition to green economies. Co-financed with the Green Climate Fund (GCF), the country is scaling up investments by small and medium-sized enterprises in renewable energy and energy efficiency. The GCF approved REDD+ result-based payments for the Period 2014-16. Since 2021, Argentina has been part of the international Pathways PtX initiative, led by the German government, to promote sustainable hydrogen markets.

In terms of **green finance**, the Ministry of Economy is developing a roadmap to issue sovereign green, social and sustainability (GSS) bonds. In 2019, the National Securities Commission presented guidelines for the issuance of GSS Marketable Securities and created a Sustainable Finance Programme. It also approved a special regime for sustainable collective investment products (General Resolution No. 885 of 2021), three guidance, advisory and educational documents on sustainable finance (General Resolution No. 896 of 2021), and a simplified regime for the issuance of social impact bonds (General Resolution No. 940 of 2022). The National Bank of Argentina is developing its first sustainable financing lines, and the Central Bank recently became a full member of the Network for Greening the Financial System. In 2020, the Ministry of Economy created the *Mesa Técnica de Finanzas Sostenibles (MTFS)* as a permanent forum to strengthen sustainable finance in Argentina and develop a national sustainable finance strategy. Within the MTFS framework, a joint statement was signed by the regulators of the banking, insurance and capital market sectors to promote sustainable finance and advance in the analysis of climate-related financial risks.

Key indicators – Argentina

	Argentina		LAC		OECD	
Social						
	2016	**2020**	**2016**	**2020**	**2016**	**2020**
Extreme poverty	2.9	6.3	8.1	8.7	N/A	N/A
Poverty	21.5	34.4	25.9	26.3	N/A	N/A
Share of Internet users (% of population)	71.0	85.5	53.7	68.1	81.8	88.0
	2016	**2020**	**2016**	**2020**	**2016**	**2019**
Gini index	42.0	42.3	46.3	45.3	34.9	34.2
	2010	**2018**	**2009**	**2018**	**2009**	**2018**
Share of total population in informal households (%)	35.7	36.8	43.4	36.3	N/A	N/A
Share of total population in informal households, upper-income quintile (%)	18.2	N/A	24.5	13.6	N/A	N/A
Share of total population in informal households, lower-income quintile (%)	73.7	N/A	70.4	72.0	N/A	N/A
	2016	**2019**	**2016**	**2019**	**2016**	**2019**
Health expenditure (% of GDP)	9.0	9.5	6.5	6.8	8.7	8.8
SIGI index	N/A	N/A	N/A	25.4	N/A	17.5
	2015	**2018**	**2015**	**2018**	**2015**	**2018**
PISA score in science	432	404	411	407	489	487
Productivity and innovation						
	2016	**2021**	**2016**	**2021**	**2016**	**2021**
Labour productivity (% of the United States)	40.7	34.8	29.0	26.6	69.9	67.2
	2016	**2020**	**2016**	**2020**	**2016**	**2020**
High-tech exports (% of manufactured exports)	9.0	6.9	8.4	7.2	16.5	16.2
	2016	**2019**	**2016**	**2019**	**2016**	**2019**
R&D expenditures (% of GDP)	0.6	0.5	0.3	0.4	1.8	2.0
Citizens' perceptions and institutions						
	2016	**2021**	**2016**	**2021**	**2016**	**2021**
Share of population satisfied with efforts to preserve the environment (%)	43.9	35.0	46.2	42.0	55.3	51.2
Share of population with confidence in national government (%)	42.0	32.7	34.2	39.4	40.3	46.3
Share of population that thinks corruption is widespread throughout government (%)	77.7	75.4	74.5	70.0	59.0	54.9
Share of population satisfied with the education system (%)	57.1	52.0	64.9	54.3	67.0	66.8
Share of urban population satisfied with the availability of quality health care (%)	50.8	51.0	49.2	48.9	68.2	69.5
Environment and the green transition						
	2004-19		**2004-19**		**2004-19**	
Loss of natural and semi-natural vegetated land (%)	0.9		1.5		1.3	
	2016	**2019**	**2016**	**2019**	**2016**	**2019**
GHG emissions per capita excluding LUCF (t CO_2e)	8.6	8.2	6.3	6.3	9.6	9.1
Air pollution – exposure to PM2.5 (annual average exposure to more than 10 µg/m^3, % of population)	99.1	98.5	95.5	95.4	61.3	61.0
	2016	**2020**	**2016**	**2019**	**2016**	**2020**
Contribution of renewables to total primary energy supply (%)	7.3	8.5	34.2	33.4	19.7	22.9
	2016	**2021**	**2016**	**2021**	**2016**	**2021**
Marine protected areas (% of territorial waters)	4.0	11.8	2.5	7.3	16.5	18.6
Fiscal position						
	2016	**2020**	**2016**	**2020**	**2016**	**2020**
Environmentally related tax revenue (% of GDP)	2.1	1.8	1.1	1.0	2.4	2.1
Total tax revenues (% of GDP)	30.7	29.4	22.2	21.9	33.6	33.5
Share of VAT (% GDP)	7.1	6.9	5.8	5.6	6.6	6.7
Share of PIT (% GDP)	2.1	2.3	2.1	2.2	7.8	8.3
Share of CIT (% of GDP)	2.9	2.8	3.3	3.4	2.9	2.7
Perception of tax evasion (%)	N/A	34.7	N/A	27.3	N/A	N/A
Debt service (% of total tax revenue)	6.2	8.2	11.3	13.0	5.6	5.2
	2016	**2020**	**2016**	**2020**	**2016**	**2019**
Social expenditure (% of GDP)	14.2	17.2	11.5	13.6	20.0	19.9

Note: See the Reader's Guide for definitions and sources.

StatLink https://stat.link/rjizf8

BRAZIL

1. Recent trends

Poverty in Brazil decreased from 20.9% in 2016 to 18.4% in 2020, contrary to Latin America and the Caribbean (LAC) average, which increased slightly from 25.9% to 26.3% in the same period. Brazil's trend can be explained by the wide fiscal support (with cash transfers and tax packages) during 2020 to alleviate the economic impact of the COVID-19 pandemic. Extreme poverty decreased in that period, from 5.3% to 5.1% and remains below the LAC average (8.7%). The population living in completely informal households decreased from 37.7% in 2009 to 29.7% in 2018, compared to the LAC average of 36.3% in 2018. Regarding environmental indicators, in 2019, total greenhouse gas (GHG) emissions per capita were 5.0 tonnes of carbon dioxide equivalent (t CO_2e), lower than the averages for LAC (6.3) and for countries belonging to the Organisation for Economic Co-operation and Development (OECD) (9.1). That year, the share of the population exposed to air pollution levels that pose risks to human health (PM2.5 at more than 10 $\mu g/m^3$) was 81.7%, lower than 95.4% for LAC and higher than 61.0% for the OECD. The marine protected areas accounted for 26.8% of Brazil's territorial waters in 2021, compared to 7.3% for LAC and 18.6% for the OECD. On the fiscal side, environmentally related tax revenue was 0.7% of gross domestic product (GDP) in 2020, lower than LAC (1.0%) and the OECD (2.1%). Total tax revenues as a percentage of GDP (31.6%) remain higher than the average for LAC (21.9%), but below the OECD (33.5%).

2. Long-term development policies for a green transition

Brazil plays a key role in the green transition, as it is home to about 60% of the world's rainforest area. Two of the country's key sustainable policies are the National Policy on Climate Change (2009), which is currently being updated, and the National Programme for Green Growth, which aims to promote green jobs.

Brazil's **mitigation** actions build on key sectoral plans, notably in agriculture, energy, and forest protection. To record offset and trade GHG emissions, a National System for the Reduction of Greenhouse Gas Emissions (SINARE) was established in 2022. Brazil launched the updated National Plan for Adaptation and Low Carbon Emission in Agriculture (ABC+ 2020-2030), which promotes new technologies and sustainable practices in agricultural systems. Additionally, the National Bioinputs Program (2020) fosters the use of bio-based products in agriculture. Regarding energy, the National Energy Plan 2050 places a strong emphasis on expanding hydropower capacity and includes actions regarding the oil and gas sectors. Brazil also takes a sectoral approach to **adaptation**, guided by the National Adaptation Plan to Climate Change (2016), which defines 11 priority sectors.

Brazil's **international partnerships** focus mainly on forest conservation. **Within** LAC, Brazil participates in the Amazon Cooperation Treaty Organization (1995), which launched an integrated water management programme in the Amazon Basin. In 2020, Brazil joined the *TerrAmaz* programme with the Amazonian countries to monitor deforestation and promote sustainable agricultural practices, with funding from the French Development Agency (AFD). **Beyond** the region, the Partnership for the Conservation of Amazon Biodiversity and the United States Agency for International Development (USAID) promote private-sector engagement through the Partnership Platform for the Amazon. The Development Bank of Minas Gerais (DBMG), in partnership with the European Investment Bank (EIB), finances a series of climate actions and renewable energy projects.

Concerning **green finance**, Brazil created the National Fund on Climate Change (2009) and the Amazon Fund (2008), to provide funding for Brazil's REDD+ (Reducing Emissions from Deforestation and Forest Degradation) strategy adopted in 2016. Brazil's National Development Bank for Economic and Social Development (BNDES) partnered with the Climate Bond Initiative to attract international investments for sustainable projects and improved banking environmental, social and governance indicators. The BNDES was the first Brazilian bank to issue a green bond in the international capital market in 2017, and the first to issue green financial bills in 2020 in the domestic market. In 2021, the BNDES launched the Sustainability Bond Framework and created its own Sustainable Taxonomy. The value of the Brazilian sustainable bond market grew 177% from 2020 to 2021. Brazil's voluntary carbon market is expected to grow rapidly thanks to the SINARE, which provides a market infrastructure, and the National Payment Policy for Environmental Services (2021), which encourages its use.

Key indicators – Brazil

	Brazil		LAC		OECD	
Social						
	2016	**2020**	**2016**	**2020**	**2016**	**2020**
Extreme poverty	5.3	5.1	8.1	8.7	N/A	N/A
Poverty	20.9	18.4	25.9	26.3	N/A	N/A
Share of Internet users (% of population)	60.9	81.3	53.7	68.1	81.8	88.0
	2016	**2020**	**2016**	**2020**	**2016**	**2019**
Gini index	53.3	48.9	46.3	45.3	34.9	34.2
	2009	**2018**	**2009**	**2018**	**2009**	**2018**
Share of total population in informal households (%)	37.7	29.7	43.4	36.3	N/A	N/A
Share of total population in informal households, upper-income quintile (%)	23.8	10.7	24.5	13.6	N/A	N/A
Share of total population in informal households, lower-income quintile (%)	71.3	70.5	70.4	72.0	N/A	N/A
	2016	**2019**	**2016**	**2019**	**2016**	**2019**
Health expenditure (% of GDP)	9.2	9.6	6.5	6.8	8.7	8.8
SIGI index	N/A	21.2	N/A	25.4	N/A	17.5
	2015	**2018**	**2015**	**2018**	**2015**	**2018**
PISA score in science	401	404	411	407	489	487
Productivity and innovation						
	2016	**2021**	**2016**	**2021**	**2016**	**2021**
Labour productivity (% of the United States)	26.6	25.6	29.0	26.6	69.9	67.2
	2016	**2020**	**2016**	**2020**	**2016**	**2020**
High-tech exports (% of manufactured exports)	16.0	11.4	8.4	7.2	16.5	16.2
	2016	**2019**	**2016**	**2019**	**2016**	**2019**
R&D expenditures (% of GDP)	1.3	1.2	0.3	0.4	1.8	2.0
Citizens' perceptions and institutions						
	2016	**2021**	**2016**	**2021**	**2016**	**2021**
Share of population satisfied with efforts to preserve the environment (%)	37.8	23.1	46.2	42.0	55.3	51.2
Share of population with confidence in national government (%)	26.4	32.7	34.2	39.4	40.3	46.3
Share of population that thinks corruption is widespread throughout government (%)	78.5	71.1	74.5	70.0	59.0	54.9
Share of population satisfied with the education system (%)	48.3	51.7	64.9	54.3	67.0	66.8
Share of urban population satisfied with the availability of quality health care (%)	30.7	42.2	49.2	48.9	68.2	69.5
Environment and the green transition						
	2004-19		**2004-19**		**2004-19**	
Loss of natural and semi-natural vegetated land (%)	0.8		1.5		1.3	
	2016	**2019**	**2016**	**2019**	**2016**	**2019**
GHG emissions per capita excluding LUCF (t CO_2e)	5.2	5.0	6.3	6.3	9.6	9.1
Air pollution – exposure to PM2.5 (annual average exposure to more than 10 µg/m^3, % of population)	76.3	81.7	95.5	95.4	61.3	61.0
	2016	**2020**	**2016**	**2019**	**2016**	**2020**
Contribution of renewables to total primary energy supply (%)	42.6	47.8	34.2	33.4	19.7	22.9
	2016	**2021**	**2016**	**2021**	**2016**	**2021**
Marine protected areas (% of territorial waters)	1.7	26.8	2.5	7.3	16.5	18.6
Fiscal position						
	2016	**2020**	**2016**	**2020**	**2016**	**2020**
Environmentally related tax revenue (% of GDP)	0.9	0.7	1.1	1.0	2.4	2.1
Total tax revenues (% of GDP)	32.2	31.6	22.2	21.9	33.6	33.5
Share of VAT (% GDP)	6.6	7.0	5.8	5.6	6.6	6.7
Share of PIT (% GDP)	2.6	3.0	2.1	2.2	7.8	8.3
Share of CIT (% of GDP)	3.2	2.7	3.3	3.4	2.9	2.7
Perception of tax evasion (%)	N/A	24.5	N/A	27.3	N/A	N/A
Debt service (% of total tax revenue)	20.2	13.2	11.3	13.0	5.6	5.2
	2016	**2020**	**2016**	**2020**	**2016**	**2019**
Social expenditure (% of GDP)	17.3	22.5	11.5	13.6	20.0	19.9

Note: See the Reader's Guide for definitions and sources.

StatLink https://stat.link/inwf0s

CHILE

1. Recent trends

Poverty in Chile increased from 13.7% in 2015 to 14.2% in 2020, below the Latin America and the Caribbean (LAC) average of 26.3%, partly owing to the impact of the COVID-19 pandemic. Extreme poverty increased in that period from 1.8% to 4.5%, below the LAC average (8.7%). The population living in totally informal households decreased from 23.5% in 2009 to 18.2% in 2017, compared to the LAC average 36.3% in 2018. Regarding environmental indicators, in 2019, greenhouse gas (GHG) emissions per capita were 6.0 tonnes of carbon dioxide equivalent (t CO_2e), lower than the averages for LAC (6.3) and countries belonging to the Organisation for Economic Co-operation and Development (OECD) (9.1). That year, the share of the population exposed to air pollution levels that pose risks to human health (PM2.5 at more than 10 $\mu g/m^3$) was 98.6%, higher than 95.4% for LAC and 61% for the OECD. The marine protected area of Chile accounted for 41.3% of its territorial waters in 2021, compared to 7.3% for LAC and 18.6% for the OECD. On the fiscal side, environmentally related tax revenue was 1.4% of gross domestic product (GDP) in 2020, above LAC (1.0%) and below the OECD (2.1%). Total tax revenues as a percentage of GDP (19.3%) remain lower than the averages for LAC (21.9%) and the OECD (33.5%).

2. Long-term development policies for a green transition

Chile has committed to adopting greener policies in various sectors, primarily energy as it accounts for 77% of total GHG emissions. In 2021, Chile established the Long-Term Climate Strategy, which lays the foundations for environmental actions in the coming years.

With respect to Chile's **mitigation** measures, the National Strategy of Green Hydrogen aims to develop 5 GW of capacity for electrolysis in 2025, produce the cheapest green hydrogen in the world and be among the three main exporters by 2040. Law No. 20.571 enables households with solar systems to generate their own energy, with surplus sold in the energy market. The Residential Ministry grants subsidies to vulnerable households to implement renewable energy systems. In the transport sector, the National Strategies of Sustainable Mobility and Electro Mobility aim to improve the use of energy resources, time, and road and urban space. In 2021, new electric and low-emission buses were introduced in the public transport system, and the *Mi Taxi Eléctrico* programme was launched. As a circular economy initiative, the National Strategy of Organic Wastes 2040 aims to increase the valorisation rate of organic wastes at the municipal level from 1% to 66%. To increase accountability, the *Huella Chile* programme was created to promote the quantification and reporting of corporate GHG emissions. Among Chile's **adaptation** policies, the Adaptation Plan for Fishery and Aquaculture adopts a systemic approach to increase the resilience of marine ecosystems and coastal communities.

Regarding **international partnerships, within** the region, as a member of the Pacific Alliance, Chile participates in the Working Group of Finance and Sustainable Development, which organises regional co-operation for the adoption of environmental, social and governance criteria. **Beyond** LAC, Chile is part of *MinSus*, a programme for the sustainable management of mineral resources, in partnership with the Economic Commission for Latin America and the Caribbean (ECLAC) and the German Agency for Economic Cooperation (GIZ). The National Sustainable Mobility Strategy, which sets the vision for mobility and urban planning in 2050, was developed with aids from EUROCLIMA+, the GIZ, and the French Development Agency (AFD). A National Preparation Committee for Green Taxonomy was created in co-operation with the Inter-American Development Bank (IDB) and the Climate Bonds Initiative.

Regarding **green finance**, between 2019 and 2022, Chile presented its frameworks for green, social, and sustainable (GSS) bonds, which are linked to key performance indicators. Since 2020, a fiscal framework is being designed by the Finance Ministry, with assistance from the Nationally Determined Contributions (NDC) Support Programme of the United Nations Development Programme (UNDP), to measure the effectiveness of public and private green investments. In 2019, Chile was the first country in the region to issue green bonds; in 2022, it was first in the world to issue bonds linked to sustainability, which currently constitute 28.7% of the public debt. The *Mesa Público-Privada de Finanzas Verdes* initiative co-ordinates the public and private sectors for capacity building towards climate change.

	Key indicators – Chile					
	Chile		LAC		OECD	
Social						
	2015	**2020**	**2016**	**2020**	**2016**	**2020**
Extreme poverty	1.8	4.5	8.1	8.7	N/A	N/A
Poverty	13.7	14.2	25.9	26.3	N/A	N/A
Share of Internet users (% of population)	83.6	88.3	53.7	68.1	81.8	88.0
	2015	**2020**	**2016**	**2020**	**2016**	**2019**
Gini index	44.4	44.9	46.3	45.3	34.9	34.2
	2009	**2017**	**2009**	**2018**	**2009**	**2018**
Share of total population in informal households (%)	23.5	18.2	43.4	36.3	N/A	N/A
Share of total population in informal households, upper-income quintile (%)	20.0	9.4	24.5	13.6	N/A	N/A
Share of total population in informal households, lower-income quintile (%)	32.8	34.9	70.4	72.0	N/A	N/A
	2016	**2019**	**2016**	**2019**	**2016**	**2019**
Health expenditure (% of GDP)	8.5	9.3	6.5	6.8	8.7	8.8
SIGI index	N/A	36.1	N/A	25.4	N/A	17.5
	2015	**2018**	**2015**	**2018**	**2015**	**2018**
PISA score in science	447	444	411	407	489	487
Productivity and innovation						
	2016	**2021**	**2016**	**2021**	**2016**	**2021**
Labour productivity (% of the United States)	42.9	46.5	29.0	26.6	69.9	67.2
	2016	**2020**	**2016**	**2020**	**2016**	**2020**
High-tech exports (% of manufactured exports)	8.5	15.8	8.4	7.2	16.5	16.2
	2016	**2019**	**2016**	**2019**	**2016**	**2019**
R&D expenditures (% of GDP)	0.4	0.3	0.3	0.4	1.8	2.0
Citizens' perceptions and institutions						
	2016	**2020**	**2016**	**2021**	**2016**	**2021**
Share of population satisfied with efforts to preserve the environment (%)	31.1	23.5	46.2	42.0	55.3	51.2
Share of population with confidence in national government (%)	20.4	16.2	34.2	39.4	40.3	46.3
Share of population that thinks corruption is widespread throughout government (%)	82.5	80.4	74.5	70.0	59.0	54.9
Share of population satisfied with the education system (%)	46.0	41.8	64.9	54.3	67.0	66.8
Share of urban population satisfied with the availability of quality health care (%)	35.5	33.4	49.2	48.9	68.2	69.5
Environment and the green transition						
	2004-19		**2004-19**		**2004-19**	
Loss of natural and semi-natural vegetated land (%)	0.8		1.5		1.3	
	2016	**2019**	**2016**	**2019**	**2016**	**2019**
GHG emissions per capita excluding LUCF (t CO_2e)	5.9	6.0	6.3	6.3	9.6	9.1
Air pollution – exposure to PM2.5 (annual average exposure to more than 10 µg/m³, % of population)	98.9	98.6	95.5	95.4	61.3	61.0
	2016	**2020**	**2016**	**2019**	**2016**	**2020**
Contribution of renewables to total primary energy supply (%)	27.1	29.7	34.2	33.4	19.7	22.9
	2016	**2021**	**2016**	**2021**	**2016**	**2021**
Marine protected areas (% of territorial waters)	12.6	41.3	2.5	7.3	16.5	18.6
Fiscal position						
	2016	**2020**	**2016**	**2020**	**2016**	**2020**
Environmentally related tax revenue (% of GDP)	1.2	1.4	1.1	1.0	2.4	2.1
Total tax revenues (% of GDP)	20.1	19.3	22.2	21.9	33.6	33.5
Share of VAT (% GDP)	8.3	8.0	5.8	5.6	6.6	6.7
Share of PIT (% GDP)	1.8	2.0	2.1	2.2	7.8	8.3
Share of CIT (% of GDP)	4.2	4.7	3.3	3.4	2.9	2.7
Perception of tax evasion (%)	N/A	14.1	N/A	27.3	N/A	N/A
Debt service (% of total tax revenue)	1.4	2.7	11.3	13.0	5.6	5.2
	2016	**2020**	**2016**	**2020**	**2016**	**2019**
Social expenditure (% of GDP)	16.1	20.2	11.5	13.6	20.0	19.9

Note: See the Reader's Guide for definitions and sources.

StatLink https://stat.link/s5wcaj

COLOMBIA

1. Recent trends

Poverty in Colombia increased from 30.9% in 2016 to 39.8% in 2020, above the Latin America and the Caribbean (LAC) average of 26.3%, partly owing to the impact of the COVID-19 pandemic. Extreme poverty increased in that period from 12.0% to 19.2% and remains above the LAC average (8.7%). The population living in completely informal households decreased from 62.6% in 2010 to 52.2% in 2018, compared to the LAC average of 36.3% in 2018. Regarding environmental indicators, in 2019 greenhouse gas (GHG) emissions per capita were 3.7 tonnes of carbon dioxide equivalent (t CO_2e), lower than the averages for LAC (6.3) and for countries belonging to the Organisation for Economic Co-operation and Development (OECD) (9.1). That year, the share of the population exposed to air pollution levels that pose risks to human health (PM2.5 at more than 10 µg/m^3) was 99.3%, higher than 95.4% for LAC and 61.0% for the OECD. The marine protected area of Colombia accounted for 17.2% of its territorial waters in 2021, compared to 7.3% for LAC and 18.6% for the OECD. On the fiscal side, environmentally related tax revenue was 0.6% of gross domestic product (GDP) in 2020, below LAC (1.0%) and the OECD (2.1%). Total tax revenues as a percentage of GDP in 2020 (18.7%) was below averages for LAC (21.9%) and the OECD (33.5%).

2. Long-term development policies for a green transition

Colombia transformed the plan for the implementation and monitoring of its nationally determined contributions (NDCs) into Law No. 2169 of Climate Action (2021). The long-term Climate Strategy of Colombia E2050 aims for the net zero scenario by 2050. The Green Growth Policy (CONPES 3934 of 2018) is an inter-sectoral plan to boost green economic productivity and sustainable use of natural resources.

Within Colombia's **mitigation** measures, the Comprehensive Strategy for Deforestation Control and Forest Management commits to reduce deforestation to zero in 2030. *Vida Manglar* is a market-driven mangrove conservation project in Cispata, that entered the carbon market in 2021. Law No. 1964 of 2019 promotes the use of electric vehicles, by means of tax exemptions and targeted subsidies. Among upcoming policies, the tax reform presented in Congress (in 2022) proposes a tax for single-use plastics, as well as other environmental tariffs related to the energy matrix. Regarding **adaptation** policies, the National Plan of Adaptation to Climate Change aims to increase Colombia's capacity to respond to the impacts of climate change. The Strategy for the Strengthening of the Business Sector in Climate Risk Management to Maintain Competitiveness helps businesses to adopt a climate risk management strategy. Colombia has 25 Comprehensive Climate Change Sectorial Management Plans, which allow ministries and territorial entities to identify, orient and evaluate mitigation and adaptation measures.

Regarding **international partnerships within** the region, Colombia is member of the Pacific Alliance, which runs various environmental and climate change-related projects. Colombia is part of the Amazon Cooperation Treaty Organization and is currently participating in the integrated water management programme in the Amazon Basin with Brazil and Peru. It is also part of the *TerrAmaz* programme to monitor deforestation and promote sustainable agricultural practices, with support from the French Development Agency (AFD). Colombia ratified the *Escazú* Regional Agreement on access to information, public participation and justice in environmental matters. **Beyond** LAC, Colombia created the Global Blue Carbon Coalition in 2022 to accelerate investments in coastal carbon sinks, in partnership with Australia, Costa Rica, France, and Conservation International.

Regarding **green finance**, in 2021, Colombia was the first country in LAC to issue green bonds (*TES verdes 2031*) in the domestic market and to publish its Green Taxonomy in 2022. The National Strategy for Climate Finance includes climate change criteria in economic and financial planning and evaluation and helped create a platform for information on climate finance. The Corridor of Climate Financing is an intermediation system to connect climate actions with resources and support. It is being used to create the Climate Finance Accelerator with the United Kingdom through the UK PACT, and to support the National Development Banks in the design of green innovative instruments with the German Agency for Economic Cooperation (GIZ). Colombia is designing the National Program of Tradable Emissions Quotas, which will allow the sale and auction of emission rights and should be implemented by 2030.

Key indicators – Colombia

	Colombia		LAC		OECD	
Social						
	2016	**2020**	**2016**	**2020**	**2016**	**2020**
Extreme poverty	12.0	19.2	8.1	8.7	N/A	N/A
Poverty	30.9	39.8	25.9	26.3	N/A	N/A
Share of Internet users (% of population)	58.1	69.8	53.7	68.1	81.8	88.0
	2016	**2020**	**2016**	**2020**	**2016**	**2019**
Gini index	50.6	54.2	46.3	45.3	34.9	34.2
	2010	**2018**	**2009**	**2018**	**2009**	**2018**
Share of total population in informal households (%)	62.6	52.2	43.4	36.3	N/A	N/A
Share of total population in informal households, upper-income quintile (%)	48.3	21.6	24.5	13.6	N/A	N/A
Share of total population in informal households, lower-income quintile (%)	68.1	91.9	70.4	72.0	N/A	N/A
	2016	**2019**	**2016**	**2019**	**2016**	**2019**
Health expenditure (% of GDP)	7.5	7.7	6.5	6.8	8.7	8.8
SIGI index	N/A	15.0	N/A	25.4	N/A	17.5
	2015	**2018**	**2015**	**2018**	**2015**	**2018**
PISA score in science	416	413	411	407	489	487
Productivity and innovation						
	2016	**2021**	**2016**	**2021**	**2016**	**2021**
Labour productivity (% of the United States)	25.1	27.8	29.0	26.6	69.9	67.2
	2016	**2020**	**2016**	**2020**	**2016**	**2020**
High-tech exports (% of manufactured exports)	10.2	9.9	8.4	7.2	16.5	16.2
	2016	**2019**	**2016**	**2019**	**2016**	**2019**
R&D expenditures (% of GDP)	0.3	0.3	0.3	0.4	1.8	2.0
Citizens' perceptions and institutions						
	2016	**2021**	**2016**	**2021**	**2016**	**2021**
Share of population satisfied with efforts to preserve the environment (%)	40.2	37.2	46.2	42.0	55.3	51.2
Share of population with confidence in national government (%)	26.1	28.9	34.2	39.4	40.3	46.3
Share of population that thinks corruption is widespread throughout government (%)	86.0	80.4	74.5	70.0	59.0	54.9
Share of population satisfied with the education system (%)	63.8	51.1	64.9	54.3	67.0	66.8
Share of urban population satisfied with the availability of quality health care (%)	45.1	46.0	49.2	48.9	68.2	69.5
Environment and the green transition						
	2004-19		**2004-19**		**2004-19**	
Loss of natural and semi-natural vegetated land (%)	0.5		1.5		1.3	
	2016	**2019**	**2016**	**2019**	**2016**	**2019**
GHG emissions per capita excluding LUCF (t CO_2e)	3.7	3.7	6.3	6.3	9.6	9.1
Air pollution – exposure to PM2.5 (annual average exposure to more than 10 µg/m³, % of population)	99.3	99.3	95.5	95.4	61.3	61.0
	2016	**2020**	**2016**	**2019**	**2016**	**2020**
Contribution of renewables to total primary energy supply (%)	23.9	25.0	34.2	33.4	19.7	22.9
	2016	**2021**	**2016**	**2021**	**2016**	**2021**
Marine protected areas (% of territorial waters)	2.1	17.2	2.5	7.3	16.5	18.6
Fiscal position						
	2016	**2020**	**2016**	**2020**	**2016**	**2020**
Environmentally related tax revenue (% of GDP)	0.8	0.6	1.1	1.0	2.4	2.1
Total tax revenues (% of GDP)	19.1	18.7	22.2	21.9	33.6	33.5
Share of VAT (% GDP)	4.8	5.4	5.8	5.6	6.6	6.7
Share of PIT (% GDP)	1.1	1.3	2.1	2.2	7.8	8.3
Share of CIT (% of GDP)	4.9	4.6	3.3	3.4	2.9	2.7
Perception of tax evasion (%)	N/A	25.5	N/A	27.3	N/A	N/A
Debt service (% of total tax revenue)	10.1	14.0	11.3	13.0	5.6	5.2
	2016	**2020**	**2016**	**2020**	**2016**	**2019**
Social expenditure (% of GDP)	12.5	15.5	11.5	13.6	20.0	19.9

Note: See the Reader's Guide for definitions and sources.

StatLink https://stat.link/i0gty9

COSTA RICA

1. Recent trends

Poverty in Costa Rica increased from 16.5% in 2016 to 19.4% in 2020, partly owing to the impact of the COVID-19 pandemic, remaining below the Latin America and the Caribbean (LAC) average of 26.3%. Extreme poverty slightly decreased over that period from 4.2% to 4.0% and remained below the LAC average (8.7%). The population living in completely informal households remained constant at 26.9% in 2010 and 27.0% in 2019, below the LAC average of 36.3% in 2018. Regarding environmental indicators, in 2019, total greenhouse gas (GHG) emissions per capita were 3.1 tonnes of carbon dioxide equivalent (t CO_2e), lower than the averages for LAC (6.3) and countries belonging to the Organisation for Economic Co-operation and Development (OECD) (9.1). That year, the share of the population exposed to air pollution levels that pose risks to human health (PM2.5 at more than 10 $\mu g/m^3$) was 99.9%, higher than 95.2% for LAC and 61.0% for the OECD. The marine protected area accounted for 2.7% of its territorial waters at the beginning of 2021, compared to 7.3% for LAC and 18.6% for the OECD. In December 2021, the government signed a decree that expands the protected marine area of the *Area de Conservación Marina Coco* to 161 129 km^2, thereby increasing the national marine protected area to 30% of territorial waters. On the fiscal side, environmentally related tax revenue was 2.3% of gross domestic product (GDP) in 2019, above both LAC (1.0% in 2020) and the OECD (2.1% in 2020). Total tax revenue as a percentage of GDP (22.9%) remains higher than the average for LAC (21.9%), but lower than the OECD (33.5%).

2. Long-term development policies for a green transition

Costa Rica is the only tropical country in the world that has managed to reverse deforestation, largely due to the Payments for Ecosystem Services programme and a System of Conservation Areas that covers 26.2% of the nation's territory. The National Plan for Decarbonisation 2018-2050 aims to reduce emissions to net zero by 2050. The National Bioeconomy Strategy (2020) provides a framework to integrate the productive and environment sectors. The Implementation Plan for the National REDD+ (Reducing Emissions from Deforestation and forest Degradation) Strategy sets an operational framework for forest conservation for the period 2018 to 2024.

Within its **mitigation** measures, in 2019, Costa Rica amended the Decree 36693-MINAET, which introduces a national moratorium on oil exploitation. The Ministry of Finance and the Ministry of Environment and Energy (MINAE) are designing a Green Fiscal Reform that prioritises lines according to their environmental impacts as well as a reform to the selective consumption tax based on environmental criteria. Together with the Economic Commission for Latin America and the Caribbean (ECLAC), they are also studying a reform to the fuel tax and the creation of a tax on solid wastes. The MINAE, in collaboration with the German Agency for Economic Cooperation (GIZ), designed a plan for the production and consumption of green hydrogen. Regarding its **adaptation** policies, Costa Rica approved the National Adaptation Policy 2018-2030, which establishes a roadmap to strengthen the country's resilience to the impacts of climate change.

Regarding **international partnerships, within** the region, Costa Rica holds the current presidency of the Forum of Ministers of Environment of LAC, which discusses the region's role in the face of environmental planetary crisis. **Beyond** LAC, Costa Rica is part of a consortium that created in 2022 the Global Blue Carbon Coalition to accelerate investments in coastal carbon sinks. The MINAE is leading various environmental projects with funding from the Global Environment Facility and implemented by the United Nations Development Programme.

Regarding **green finance**, the General Securities Superintendency authorized three green bonds. *Acción Clima II* and the *Políticas Climáticas Verticalmente Integradas* project, managed by GIZ, did a diagnosis of the potential financing sources to implement climate actions in Costa Rica's municipalities, which contributed to the development of the Climate Finance Taxonomy. The government is working on the publication of the *Guía Oficial de Acceso a los recursos del Fondo Verde del Clima en Costa Rica*, to orient public or private actors interested on accessing green finance through the Green Climate Fund (GCF). The General Superintendence of Financial Entities developed a methodology to capture climate information on credit operations managed by supervised financial institutions.

	Key indicators – Costa Rica					
	Costa Rica		LAC		OECD	
Social						
	2016	**2020**	**2016**	**2020**	**2016**	**2020**
Extreme poverty	4.2	4.0	8.1	8.7	N/A	N/A
Poverty	16.5	19.4	25.9	26.3	N/A	N/A
Share of Internet users (% of population)	65.9	80.5	53.7	68.1	81.8	88.0
	2016	**2020**	**2016**	**2020**	**2016**	**2019**
Gini index	48.7	49.3	46.3	45.3	34.9	34.2
	2010	**2019**	**2009**	**2018**	**2009**	**2018**
Share of total population in informal households (%)	26.9	27	43.4	36.3	N/A	N/A
Share of total population in informal households, upper-income quintile (%)	7.7	6.1	24.5	13.6	N/A	N/A
Share of total population in informal households, lower-income quintile (%)	62.9	75.6	70.4	72.0	N/A	N/A
	2016	**2019**	**2016**	**2019**	**2016**	**2019**
Health expenditure (% of GDP)	7.3	7.3	6.5	6.8	8.7	8.8
SIGI index	N/A	27.9	N/A	25.4	N/A	17.5
	2015	**2018**	**2015**	**2018**	**2015**	**2018**
PISA score in science	420	416	411	407	489	487
Productivity and innovation						
	2016	**2021**	**2016**	**2021**	**2016**	**2021**
Labour productivity (% of the United States)	38.8	38.4	29.0	26.6	69.9	67.2
	2016	**2020**	**2016**	**2020**	**2016**	**2020**
High-tech exports (% of manufactured exports)	18.4	15.7	8.4	7.0	16.5	16.2
	2016	**2018**	**2016**	**2019**	**2016**	**2019**
R&D expenditures (% of GDP)	0.4	0.4	0.3	0.4	1.8	2.0
Citizens' perceptions and institutions						
	2016	**2021**	**2016**	**2021**	**2016**	**2021**
Share of population satisfied with efforts to preserve the environment (%)	52.2	59.7	46.2	42.0	55.3	51.2
Share of population with confidence in national government (%)	27.9	30.7	34.2	39.4	40.3	46.3
Share of population that thinks corruption is widespread throughout government (%)	75.5	76.7	74.5	70.0	59.0	54.9
Share of population satisfied with the education system (%)	82.1	70.8	64.9	54.3	67.0	66.8
Share of urban population satisfied with the availability of quality health care (%)	65.2	70.8	49.2	48.9	68.2	69.5
Environment and the green transition						
	2004-19		**2004-19**		**2004-19**	
Loss of natural and semi-natural vegetated land (%)	1.1		1.5		1.3	
	2016	**2019**	**2016**	**2019**	**2016**	**2019**
GHG emissions per capita excluding LUCF (t CO_2e)	3.1	3.1	6.3	6.3	9.6	9.1
Air pollution – exposure to PM2.5 (annual average exposure to more than 10 µg/m³, % of population)	99.9	99.9	95.5	95.2	61.3	61.0
	2016	**2020**	**2016**	**2019**	**2016**	**2020**
Contribution of renewables to total primary energy supply (%)	50.4	50.5	34.2	33.4	19.7	22.9
	2016	**2021**	**2016**	**2021**	**2016**	**2021**
Marine protected areas (% of territorial waters)	0.8	2.7	2.5	7.3	16.5	18.6
Fiscal position						
	2016	**2019**	**2016**	**2020**	**2016**	**2020**
Environmentally related tax revenue (% of GDP)	2.3	2.3	1.1	1.0	2.4	2.1
	2016	2020	2016	2020	2016	2020
Total tax revenues (% of GDP)	23.5	22.9	22.2	21.9	33.6	33.5
Share of VAT (% GDP)	4.4	4.5	5.8	5.6	6.6	6.7
Share of PIT (% GDP)	1.3	1.5	2.1	2.2	7.8	8.3
Share of CIT (% of GDP)	2.3	1.9	3.3	3.4	2.9	2.7
Perception of tax evasion (%)	N/A	33.8	N/A	27.3	N/A	N/A
Debt service (% of total tax revenue)	**11.8**	**20.4**	**11.3**	**13.0**	**5.6**	**5.2**
	2016	2020	2016	2020	2016	2019
Social expenditure (% of GDP)	11.9	12.3	11.5	13.6	20.0	19.9

Note: See the Reader's Guide for definitions and sources.

StatLink https://stat.link/xs5zcu

DOMINICAN REPUBLIC

1. Recent trends

Poverty in Dominican Republic decreased from 26.7% in 2016 to 21.8% in 2020, below the Latin America and the Caribbean (LAC) average of 26.3%. Extreme poverty decreased in that period from 7.0% to 5.6%, also below the LAC average (8.7%). The Gini index decreased from 45.7 in 2016 to 39.6 in 2020, below the LAC average (45.3). Regarding environmental indicators, in 2019 greenhouse gas (GHG) emissions per capita were 3.9 tonnes of carbon dioxide equivalent (t CO_2e), lower than the averages for LAC (6.3) and countries belonging to the Organisation for Economic Co-operation and Development (OECD) (9.1). That year, the share of the population exposed to air pollution levels that pose risks to human health (PM2.5 at more than 10 $\mu g/m^3$) was 100%, higher than 95.4% for LAC and 61.0% for the OECD. The marine protected area of the Dominican Republic accounted for 18.0% of its territorial waters in 2021, compared to 7.3% for LAC and 18.6% for the OECD. On the fiscal side, environmentally related tax revenue was 1.4% of gross domestic product (GDP) in 2020, above LAC (1.0%) but below the OECD (2.1%). Total tax revenue as a percentage of GDP (12.6%) remains much lower than the averages for LAC (21.9%) and the OECD (33.5%).

2. Long-term development policies for a green transition

The Dominican Republic ranks as one of the most vulnerable countries to climate change in the world. The National Council for Climate Change, established in 2008, is responsible for formulating, implementing, and enforcing climate change policies and projects. The Strategic Plan for Climate Change 2011-2030 aims to decarbonise the economy and improve capacity to adapt to climate change. The National Policy for Climate Change (Decree 269-15, approved in 2016) is a transversal policy to manage climate variability, incorporate adaptation within the National Development Strategy 2030 and favour low-carbon development. More recently, the Climate Change and Gender Action Plan of 2018 aims to integrate gender equality into mitigation and adaptation policies.

Within the Dominican Republic's **mitigation** efforts, the Ministry of Environment is implementing the *Plan de Movilidad Urbana Sostenible del Gran Santo Domingo*, financed by the European Union, and developed under the MobiliseYourCity initiative. The plan includes the creation of the first Urban Sustainable Mobility Observatory in 2022. In 2018, the Dominican Republic set a ten-year public-private association with the Blue Finance to co-administer the marine protected area of *Arrecifes del Sureste*, with financing from the Sustainable Ocean Fund. Among the country's **adaptation** policies, the National Adaptation Plan for Climate Change 2015-2030 aims to reduce the country's vulnerabilities. The Technological Institute of Santo Domingo is working with the United States Agency for International Development (USAID) to engage local communities and academic institutions to build resilience within urban environments. The Ministry of Agriculture launches contingency plans every year to act to minimise negative impacts of climate change events.

Regarding **international partnerships within** the region, the country is working on capacity-building programmes for air quality management, environmental impact assessment, and solid waste and marine litter management within the Dominican Republic-Central America-United States Free Trade Agreement. It also takes part in an ecosystem-based adaptation project in the Central American Dry Corridor and the Arid Zones of the Dominican Republic, with support from the Green Climate Fund (GCF), the Central American Bank for Economic Integration (CABEI), the United Nations Environment Programme (UNEP) and the Food and Agriculture Organization (FAO). **Beyond** LAC, the Dominican Republic is the third country to have signed (in 2021) an agreement with the World's Bank Forest Carbon Partnership Facility to receive payments for actions to reduce GHG emissions from deforestation. The National Strategy to Strengthen Human Resources and Skills to Advance towards Green Development, with Low Emissions and Climate Resilience (2012) was created through a partnership with the United Nations Climate Change Learning Partnership.

Concerning **green finance**, the public-private *BanReservas* bank created the Program of Green Finance, which promotes the financing of sustainable energy and agroecologist projects. In 2022, the Ministry of Environment, the Securities Superintendence of the Dominican Republic, and the International Finance Corporation (IFC) signed an agreement to develop the Dominican Republic Green Taxonomy.

Key indicators – Dominican Republic	Dominican Republic		LAC		OECD	
Social						
	2016	**2020**	**2016**	**2020**	**2016**	**2020**
Extreme poverty	7.0	5.6	8.1	8.7	N/A	N/A
Poverty	26.7	21.8	25.9	26.3	N/A	N/A
Share of Internet users (% of population)	63.9	76.9	53.7	68.1	81.8	88.0
	2016	**2020**	**2016**	**2020**	**2016**	**2019**
Gini index	45.7	39.6	46.3	45.3	34.9	34.2
	2010	**2018**	**2009**	**2018**	**2009**	**2018**
Share of total population in informal households (%)	N/A	N/A	43.4	36.3	N/A	N/A
Share of total population in informal households, upper-income quintile (%)	N/A	N/A	24.5	13.6	N/A	N/A
Share of total population in informal households, lower-income quintile (%)	N/A	N/A	70.4	72.0	N/A	N/A
	2016	**2019**	**2016**	**2019**	**2016**	**2019**
Health expenditure (% of GDP)	5.8	5.9	6.5	6.8	8.7	8.8
SIGI index	N/A	18.2	N/A	25.4	N/A	17.5
	2015	**2018**	**2015**	**2018**	**2015**	**2018**
PISA score in science	332	336	411	407	489	487
Productivity and innovation						
	2016	**2021**	**2016**	**2021**	**2016**	**2021**
Labour productivity (% of the United States)	29.6	30.8	29.0	26.6	69.9	67.2
	2016	**2020**	**2016**	**2020**	**2016**	**2020**
High-tech exports (% of manufactured exports)	4.5	8.8	8.4	7.2	16.5	16.2
	2016	**2019**	**2016**	**2019**	**2016**	**2019**
R&D expenditures (% of GDP)	N/A	N/A	0.3	0.4	1.8	2.0
Citizens' perceptions and institutions						
	2016	**2021**	**2016**	**2021**	**2016**	**2021**
Share of population satisfied with efforts to preserve the environment (%)	60.3	55.3	46.2	42.0	55.3	51.2
Share of population with confidence in national government (%)	54.9	57.5	34.2	39.4	40.3	46.3
Share of population that thinks corruption is widespread throughout government (%)	64.9	60.1	74.5	70.0	59.0	54.9
Share of population satisfied with the education system (%)	81.0	71.8	64.9	54.3	67.0	66.8
Share of urban population satisfied with the availability of quality health care (%)	56.8	65.1	49.2	48.9	68.2	69.5
Environment and the green transition						
	2004-19		**2004-19**		**2004-19**	
Loss of natural and semi-natural vegetated land (%)	1.7		1.5		1.3	
	2016	**2019**	**2016**	**2019**	**2016**	**2019**
GHG emissions per capita excluding LUCF (t CO_2e)	3.6	3.9	6.3	6.3	9.6	9.1
Air pollution – exposure to PM2.5 (annual average exposure to more than 10 µg/m^3, % of population)	100	100	95.5	95.4	61.3	61.0
	2016	**2020**	**2016**	**2019**	**2016**	**2020**
Contribution of renewables to total primary energy supply (%)	11.4	11.0	34.2	33.4	19.7	22.9
	2016	**2021**	**2016**	**2021**	**2016**	**2021**
Marine protected areas (% of territorial waters)	9.1	18.0	2.5	7.3	16.5	18.6
Fiscal position						
	2016	**2020**	**2016**	**2020**	**2016**	**2020**
Environmentally related tax revenue (% of GDP)	1.8	1.4	1.1	1.0	2.4	2.1
Total tax revenues (% of GDP)	13.1	12.6	22.2	21.9	33.6	33.5
Share of VAT (% GDP)	4.6	4.4	5.8	5.6	6.6	6.7
Share of PIT (% GDP)	1.2	1.3	2.1	2.2	7.8	8.3
Share of CIT (% of GDP)	2.0	2.0	3.3	3.4	2.9	2.7
Perception of tax evasion (%)	N/A	27.9	N/A	27.3	N/A	N/A
Debt service (% of total tax revenue)	19.4	25.8	11.3	13.0	5.6	5.2
	2016	**2020**	**2016**	**2020**	**2016**	**2019**
Social expenditure (% of GDP)	7.7	12.3	11.5	13.6	20.0	19.9

Note: See the Reader's Guide for definitions and sources.

StatLink https://stat.link/kmaytv

ECUADOR

1. Recent trends

Poverty in Ecuador increased from 24.3% in 2016 to 30.6% in 2020, above the Latin America and the Caribbean (LAC) average of 26.3%, partly owing to the impact of the COVID-19 pandemic. Extreme poverty increased in that period from 7.5% to 10.8% and remains above the LAC average (8.7%). The Gini index increased from 45.0 in 2016 to 47.3 in 2020, close to the LAC average (45.3). Regarding environmental indicators, in 2019, greenhouse gas (GHG) emissions per capita were 4.2 tonnes of carbon dioxide equivalent (t CO_2e), lower than the averages for LAC (6.3) and countries belonging to the Organisation for Economic Co-operation and Development (OECD) (9.1). That year, the share of the population exposed to air pollution levels that pose risks to human health (PM2.5 at more than 10 µg/m^3) was 100%, higher than 95.4% for LAC and 61.0% for the OECD. The marine protected area of Ecuador accounted for 13.3% of its territorial waters in 2021, compared to 7.3% for LAC and 18.6% for the OECD. In 2022, Ecuador created *Hermandad*, a new 60 000 km^2 marine protected area, thereby increasing the national marine protected area to 19.2% of territorial waters. On the fiscal side, environmentally related tax revenue was 0.3% of gross domestic product (GDP) in 2020, below LAC (1.0%) and OECD averages (2.1%). Total tax revenues as a percentage of GDP (19.1%) remain lower than the averages for LAC (21.9%) and the OECD (33.5%).

2. Long-term development policies for a green transition

Ecuador is particularly exposed to natural hazards associated with the *El Niño* phenomenon. It is also home to the Galápagos Islands' unique biodiversity. Ecuador was the first LAC country to decree the ecological transition a state policy (Decree No. 59 of 2021). A National Climate Change Strategy 2012-2025 is currently in place and a National Plan for the Transition to Decarbonisation (2021) defines a sectoral roadmap. The Plan to Create Opportunities 2021-2025 promotes investment, conservation, and sustainable use of natural resources.

Within its **mitigation** efforts, Ecuador focuses on energy, transport, environmental conservation, and the circular economy. The Energy Efficiency National Plan 2016-2035 sets sectoral objectives and the Energy Master Plan 2016-2025 emphasises electricity generation from renewable sources. The National Electromobility Strategy for Ecuador (2021) was developed with the Inter-American Development Bank (IDB). In 2021, the government created two new protected areas (*Mazán* and *Taita Imbabura*), three water protection areas and a National Park. To promote circular economy, Ecuador approved a Pact for the Circular Economy (2019), a Law for the Circular Economy (2021), and the *Libro Blanco de Economía Circular de Ecuador* (2021). The government has made regulatory efforts to prevent environmental damage streaming from oil and mining sectors. Regarding **adaptation**, a National Adaptation Plan is currently under development in co-operation with the Green Climate Fund (GCF) and the United Nations Development Programme (UNDP). Current adaptation projects focus on reducing flood risk in semi-urban coastal areas (2020), adaptation to impacts on water resources (2018), and an integrated landscape management plan for ecosystem services in the Andes (2020).

With its **international partnerships**, Ecuador focuses on adaptation of rural areas and communities, and several economic opportunities. **Within** LAC, Ecuador aims to improve the adaptive capacity of Awá and Afro-descendant communities, in co-operation with Colombia and the World Food Programme (WFP) (2018), and of poor and vulnerable people in the Andean ecosystems, in co-operation with Bolivia and Peru (2020). Colombia, Costa Rica, Ecuador, and Panama agreed to construct a marine protected corridor along the Pacific. **Beyond** LAC, Ecuador is working on an Economic Recovery Strategy, supported by EUROCLIMA+, and promotes sustainable tourism, in collaboration with the German Import Promotion Desk. Ecuador receives support from the European Union to reactivate its food industry (*CREA: Camino a la Reactivación del Ecuador Agroalimentario*) and promote sustainable non-traditional exports (*NEXT Ecuador*). Ecuador is implementing a project for ecosystem rehabilitation, green agriculture and tourism, and renewable energies in the Galápagos Islands, with support from the GCF.

Regarding **green finance**, Ecuador has designed a National Climate Finance Strategy (2021) to address investment gaps in mitigation and adaptation. Ecuador uses debt-for-conservation swap mechanisms to finance projects such as the *Hermandad* marine reserve in Galápagos. To reverse its high levels of deforestation, Ecuador is implementing a REDD+ (Reducing Emissions from Deforestation and Forest Degradation) Action Plan 2016-2025 with financial support from the GCF.

	Key indicators – Ecuador					
	Ecuador		LAC		OECD	
Social						
	2016	**2020**	**2016**	**2020**	**2016**	**2020**
Extreme poverty	7.5	10.8	8.1	8.7	N/A	N/A
Poverty	24.3	30.6	25.9	26.3	N/A	N/A
Share of Internet users (% of population)	54.1	64.6	53.7	68.1	81.8	88.0
	2016	**2020**	**2016**	**2020**	**2016**	**2019**
Gini index	45.0	47.3	46.3	45.3	34.9	34.2
	2010	**2018**	**2009**	**2018**	**2009**	**2018**
Share of total population in informal households (%)	N/A	N/A	43.4	36.3	N/A	N/A
Share of total population in informal households, upper-income quintile (%)	N/A	N/A	24.5	13.6	N/A	N/A
Share of total population in informal households, lower-income quintile (%)	N/A	N/A	70.4	72.0	N/A	N/A
	2016	**2019**	**2016**	**2019**	**2016**	**2019**
Health expenditure (% of GDP)	7.3	7.8	6.5	6.8	8.7	8.8
SIGI index	N/A	28.9	N/A	25.4	N/A	17.5
	2015	**2018**	**2015**	**2018**	**2015**	**2018**
PISA score in science	N/A	N/A	411	407	489	487
Productivity and innovation						
	2016	**2021**	**2016**	**2021**	**2016**	**2021**
Labour productivity (% of the United States)	20.5	17.8	29.0	26.6	69.9	67.2
	2016	**2020**	**2016**	**2020**	**2016**	**2020**
High-tech exports (% of manufactured exports)	9.2	4.3	8.4	7.2	16.5	16.2
	2016	**2019**	**2016**	**2019**	**2016**	**2019**
R&D expenditures (% of GDP)	N/A	N/A	0.3	0.4	1.8	2.0
Citizens' perceptions and institutions						
	2016	**2021**	**2016**	**2021**	**2016**	**2021**
Share of population satisfied with efforts to preserve the environment (%)	67.9	45.0	46.2	42.0	55.3	51.2
Share of population with confidence in national government (%)	47.4	41.3	34.2	39.4	40.3	46.3
Share of population that thinks corruption is widespread throughout government (%)	64.6	69.7	74.5	70.0	59.0	54.9
Share of population satisfied with the education system (%)	77.6	54.9	64.9	54.3	67.0	66.8
Share of urban population satisfied with the availability of quality health care (%)	58.0	42.1	49.2	48.9	68.2	69.5
Environment and the green transition						
	2004-19		**2004-19**		**2004-19**	
Loss of natural and semi-natural vegetated land (%)	0.5		1.5		1.3	
	2016	**2019**	**2016**	**2019**	**2016**	**2019**
GHG emissions per capita excluding LUCF (t CO_2e)	4.4	4.2	6.3	6.3	9.6	9.1
Air pollution – exposure to PM2.5 (annual average exposure to more than 10 µg/m^3, % of population)	100	100	95.5	95.4	61.3	61.0
	2016	**2019**	**2016**	**2019**	**2016**	**2020**
Contribution of renewables to total primary energy supply (%)	13.4	17.5	34.2	33.4	19.7	22.9
	2016	**2021**	**2016**	**2021**	**2016**	**2021**
Marine protected areas (% of territorial waters)	13.1	13.3	2.5	7.3	16.5	18.6
Fiscal position						
	2016	**2020**	**2016**	**2020**	**2016**	**2020**
Environmentally related tax revenue (% of GDP)	0.4	0.3	1.1	1.0	2.4	2.1
Total tax revenues (% of GDP)	19.9	19.1	22.2	21.9	33.6	33.5
Share of VAT (% GDP)	6.0	5.5	5.8	5.6	6.6	6.7
Share of PIT (% GDP)	N/A	N/A	2.1	2.2	7.8	8.3
Share of CIT (% of GDP)	N/A	N/A	3.3	3.4	2.9	2.7
Perception of tax evasion (%)	N/A	27.9	N/A	27.3	N/A	N/A
Debt service (% of total tax revenue)	7.8	14.7	11.3	13.0	5.6	5.2
	2016	**2020**	**2016**	**2020**	**2016**	**2019**
Social expenditure (% of GDP)	8.8	11.7	11.5	13.6	20.0	19.9

Note: See the Reader's Guide for definitions and sources.

StatLink https://stat.link/6b21rc

EL SALVADOR

1. Recent trends

Poverty in El Salvador has decreased from 40.4% in 2016 to 30.7% in 2020, although it remained above the Latin America and the Caribbean (LAC) average of 26.3%. Extreme poverty also decreased in that period from 10.7% to 8.3%, below the LAC average (8.7%). The population living in completely informal households was 54.4% in 2018, above the LAC average (36.3%). Regarding environmental indicators, in 2019, greenhouse gas (GHG) emissions per capita were 2.0 tonnes of carbon dioxide equivalent (t CO_2e), lower than the averages for LAC (6.3) and countries belonging to the Organisation for Economic Co-operation and Development (OECD) (9.1). That year, the share of the population exposed to air pollution levels that pose risks to human health (PM2.5 at more than 10 $\mu g/m^3$) was 99.9%, higher than 95.4% for LAC and 61.0% for the OECD. The marine protected area of El Salvador accounted for just 0.7% of its territorial waters in 2021, compared to 7.3% for LAC and 18.6% for the OECD. On the fiscal side, environmentally related tax revenue was 0.5% of GDP in 2020, below LAC (1.0%) and the OECD (2.1%). Total tax revenue as a percentage of GDP in 2020 was 21.9%, on par with the LAC average (21.9%), but below the OECD average (33.5%).

2. Long-term development policies for a green transition

The impacts of climate change are already severely affecting El Salvador's livelihoods and economy. As a policy response, the National Climate Change Plan (2022-26) went through public consultation in 2022 and will serve as a strategic reference tool for public administration measures. In addition, the National Energy Policy 2020-2050 defines long-term strategic objectives for El Salvador's energy system, including the development of renewable energy sources to reduce dependence on imported fossil fuels.

Regarding **mitigation**, El Salvador is developing a Long-Term Climate Strategy (ED 2050), with support from EUROCLIMA+ and the International and Ibero-American Foundation for Administration and Public Policy of Spain. El Salvador participates in the Advancing a Regional Approach to e-mobility in Latin America initiative of the United Nations Environment Programme (UNEP) and the Green Climate Fund (GCF). **Adaptation** plays a crucial role in El Salvador. A project to elaborate a National Adaptation Plan, with support from the UNEP and the GCF, is expected to be launched before the end of 2022. Two large regional adaptation projects are currently underway. Project RECLIMA applies an integrated landscape approach in the country's dry corridor area, with support from the GCF and the Food and Agriculture Organization (FAO). To enhance the climate resilience of rural communities and ecosystems in the *Ahuachapán-Sur* region, El Salvador promotes forest landscape restoration, productive alternatives and the production of climate and hydrological information, with support from the United Nations Adaptation Fund and the United Nations Development Programme (UNDP).

El Salvador's **international partnerships** largely focus on environmental conservation and restoration. **Within** LAC, El Salvador co-operates with Costa Rica, Guatemala, Honduras, Nicaragua and Panama to promote ecosystem-based adaptation technologies in the Central American Dry Corridor, supported by the Central American Commission for Environment and Development, UNEP and the FAO. **Beyond** LAC, El Salvador is currently working with the Global Environment Facility (GEF) on the sustainable urban development of the San Salvador metropolitan area, supported by the UNDP (2019), and on integrated landscape management and restoration of land in the *El Imposible-Barra de Santiago* conservation area, supported by the World Bank (2021). Ecosystems in degraded areas of the *Complejo Jaltepeque* are restored, with financial support from the European Union and Germany (2021).

Regarding **green finance**, El Salvador established the *Mesa de Financiamiento Climático* for the implementation of a financial strategy to mobilise and track public and private funding for sectoral implementation plans to achieve its nationally determined contributions (NDCs). To reverse its high degree of environmental degradation, mainly due to agricultural activities and fires, El Salvador established a National REDD+ (Reducing Emissions from Deforestation and Forest Degradation) Strategy to access funding for the restoration of ecosystems and landscapes.

Key indicators – El Salvador	El Salvador		LAC		OECD	
Social						
	2016	**2020**	**2016**	**2020**	**2016**	**2020**
Extreme poverty	10.7	8.3	8.1	8.7	N/A	N/A
Poverty	40.4	30.7	25.9	26.3	N/A	N/A
Share of Internet users (% of population)	29.0	54.6	53.7	68.1	81.8	88.0
	2016	**2019**	**2016**	**2020**	**2016**	**2019**
Gini index	40.0	38.8	46.3	45.3	34.9	34.2
	2010	**2018**	**2009**	**2018**	**2009**	**2018**
Share of total population in informal households (%)	N/A	54.4	43.4	36.3	N/A	N/A
Share of total population in informal households, upper-income quintile (%)	N/A	24.4	24.5	13.6	N/A	N/A
Share of total population in informal households, lower-income quintile (%)	N/A	90.8	70.4	72.0	N/A	N/A
	2016	**2019**	**2016**	**2019**	**2016**	**2019**
Health expenditure (% of GDP)	7.7	7.2	6.5	6.8	8.7	8.8
SIGI index	N/A	22.9	N/A	25.4	N/A	17.5
	2015	**2018**	**2015**	**2018**	**2015**	**2018**
PISA score in science	N/A	N/A	411	407	489	487
Productivity and innovation						
	2016	**2021**	**2016**	**2021**	**2016**	**2021**
Labour productivity (% of the United States)	N/A	N/A	29.0	26.6	69.9	67.2
	2016	**2020**	**2016**	**2020**	**2016**	**2020**
High-tech exports (% of manufactured exports)	5.4	6.4	8.4	7.2	16.5	16.2
	2016	**2019**	**2016**	**2019**	**2016**	**2019**
R&D expenditures (% of GDP)	0.1	0.2	0.3	0.4	1.8	2.0
Citizens' perceptions and institutions						
	2016	**2020**	**2016**	**2021**	**2016**	**2021**
Share of population satisfied with efforts to preserve the environment (%)	44.8	58.1	46.2	42.0	55.3	51.2
Share of population with confidence in national government (%)	25.2	80.4	34.2	39.4	40.3	46.3
Share of population that thinks corruption is widespread throughout government (%)	70.0	39.4	74.5	70.0	59.0	54.9
Share of population satisfied with the education system (%)	62.3	69.9	64.9	54.3	67.0	66.8
Share of urban population satisfied with the availability of quality health care (%)	58.7	54.0	49.2	48.9	68.2	69.5
Environment and the green transition						
	2004-19		**2004-19**		**2004-19**	
Loss of natural and semi-natural vegetated land (%)	0.6		1.5		1.3	
	2016	**2019**	**2016**	**2019**	**2016**	**2019**
GHG emissions per capita excluding LUCF (t CO_2e)	1.9	2.0	6.3	6.3	9.6	9.1
Air pollution – exposure to PM2.5 (annual average exposure to more than 10 µg/m^3, % of population)	99.9	99.9	95.5	95.4	61.3	61.0
	2016	**2019**	**2016**	**2019**	**2016**	**2020**
Contribution of renewables to total primary energy supply (%)	45.8	42.9	34.2	33.4	19.7	22.9
	2016	**2021**	**2016**	**2021**	**2016**	**2021**
Marine protected areas (% of territorial waters)	0.7	0.7	2.5	7.3	16.5	18.6
Fiscal position						
	2016	**2020**	**2016**	**2020**	**2016**	**2020**
Environmentally related tax revenue (% of GDP)	0.6	0.5	1.1	1.0	2.4	2.1
	2016	2020	2016	2020	2016	2020
Total tax revenues (% of GDP)	20.5	21.9	22.2	21.9	33.6	33.5
Share of VAT (% GDP)	7.5	8.3	5.8	5.6	6.6	6.7
Share of PIT (% GDP)	3.0	3.3	2.1	2.2	7.8	8.3
Share of CIT (% of GDP)	3.2	3.8	3.3	3.4	2.9	2.7
Perception of tax evasion (%)	N/A	27.0	N/A	27.3	N/A	N/A
Debt service (% of total tax revenue)	**14.2**	**20.1**	**11.3**	**13.0**	**5.6**	**5.2**
	2016	2020	2016	2020	2016	2019
Social expenditure (% of GDP)	7.8	13.8	11.5	13.6	20.0	19.9

Note: See the Reader's Guide for definitions and sources.

StatLink https://stat.link/3am921

GUATEMALA

1. Recent trends

In 2019, total greenhouse gas (GHG) emissions per capita in Guatemala were 2.1 tonnes of carbon dioxide equivalent (t CO_2e), lower than the averages for Latin America and the Caribbean (LAC) (6.3) and countries belonging to the Organisation for Economic Co-operation and Development (OECD) (9.1). That same year, the share of the population exposed to air pollution levels that pose risks to human health (PM2.5 at more than 10 $\mu g/m^3$) was 100%, higher than 95.4% for LAC and 61.0% for the OECD. The marine protected area of Guatemala accounted for 0.8% of its territorial waters in 2021, compared to 7.3% for LAC and 18.6% for the OECD. On the fiscal side, environmentally related tax revenue was 0.8% of gross domestic product (GDP) in 2020, below the averages for LAC (1.0%) and the OECD (2.1%). Total tax revenue as a percentage of GDP (12.4%) in 2020 remained lower than the averages for LAC (21.9%) and the OECD (33.5%).

2. Long-term development policies for a green transition

Guatemala is among the ten countries most vulnerable to climate change in the world. Guatemala's policy response is backed by the National Development Policy 2032 (2015), the K'atun National Development Plan 2032 (2014) and the National Action Plan for Climate Change (2016), which set mitigation and adaptation objectives in environmental conservation, water resources, agriculture, spatial planning, solid waste, coastal marine systems and energy. Since 2009, Guatemala has a National Climate Change Policy in place. Also noteworthy, the Legislative Decree 7-2013 established a law to regulate the reduction of vulnerability, adaptation to climate change and mitigation of GHGs, being one of the first environmental laws to be approved in LAC. Later, the National Environmental Education Policy (2017), was created to strengthen responsibility for the environment.

Among Guatemala's **mitigation** efforts, the National Development Strategy with Low GHG Emissions (2018) promotes an efficient socio-economic model that decouples economic growth from increasing emission levels. The National Strategy for Sustainable Cattle Ranching (2018) regulates the sector's GHG emissions and prevents vulnerabilities. Regarding industry and services, Guatemala adopted a National Cleaner Production Policy (2010), which creates incentives, regulation, and simpler administrative procedures. The Energy Policy 2013-2027 aims to make use of Guatemala's great potential for renewable energy production but also promotes the exploration and exploitation of oil reserves. Guatemala has a System of Protected Areas, composed of 348 territories, that aims to conserve ecosystems and biodiversity, and is continuously monitored. Concerning **adaptation**, Guatemala developed a National Adaptation Plan (2018) for priority sectors, such as agriculture, water and sanitation, coastal zone management, and forestry. The National Action Plan for Climate Change has an adaptation strategy prioritising six sectors: health, marine and coastal areas, agriculture and food safety, forests and protected areas, infrastructure, and integrated water resource management. Guatemala also launched a project to promote policies for ecosystem-based adaptation practices in agricultural production and forestry. Guatemala's efforts to work with the private sector to strengthen climate resilience were recently analysed by the OECD, which noted promising approaches such as the provision of information, technologies and capacity building but also called for much stronger co-operation.

Guatemala's **international partnerships** emphasise environmental conservation. **Within** LAC, Guatemala promotes ecosystem-based adaptation technologies in the Central American Dry Corridor, in co-operation with Costa Rica, El Salvador, Honduras, Nicaragua and Panama. It also created a risk-sharing facility for small agricultural enterprises, with Mexico, the Green Climate Fund (GCF) and the Inter-American Development Bank (IDB). **Beyond** LAC, the ADAPTE project (2013), with the German Agency for International Cooperation (GIZ), supports climate change adaptation and rural development. Guatemala co-operates with the GCF on climate-resilient adaptation (project RELIVE) and watershed management. It also partners with the Rainforest Alliance and the GCF on risks mapping and strategic adaptation planning.

Regarding **green finance**, Guatemala developed the Action Plan of its National REDD+ (Reducing Emissions from Deforestation and Forest Degradation) Strategy (2020-50) to access funds from the Forest Carbon Partnership Facility and other sources for actions to reduce of GHG emissions. It also signed an Emissions Reductions Payment Agreement with the World Bank in 2021 to address deforestation and degradation and strengthen the management of protected areas, agroforestry systems and forest plantations.

Key indicators – Guatemala

	Guatemala		LAC		OECD	
Social						
	2016	**2020**	**2016**	**2020**	**2016**	**2020**
Extreme poverty	N/A	N/A	8.1	8.7	N/A	N/A
Poverty	N/A	N/A	25.9	26.3	N/A	N/A
Share of Internet users (% of population)	34.5	50.0	53.7	68.1	81.8	88.0
	2014	**2020**	**2016**	**2020**	**2016**	**2019**
Gini index	48.3	N/A	46.3	45.3	34.9	34.2
	2009	**2018**	**2009**	**2018**	**2009**	**2018**
Share of total population in informal households (%)	N/A	N/A	43.4	36.3	N/A	N/A
Share of total population in informal households, upper-income quintile (%)	N/A	N/A	24.5	13.6	N/A	N/A
Share of total population in informal households, lower-income quintile (%)	N/A	N/A	70.4	72.0	N/A	N/A
	2016	**2019**	**2016**	**2019**	**2016**	**2019**
Health expenditure (% of GDP)	6.1	6.2	6.5	6.8	8.7	8.8
SIGI index	N/A	28.6	N/A	25.4	N/A	17.5
	2015	**2018**	**2015**	**2018**	**2015**	**2018**
PISA score in science	N/A	N/A	411	407	489	487
Productivity and innovation						
	2016	**2021**	**2016**	**2021**	**2016**	**2021**
Labour productivity (% of the United States)	16.6	16.2	29.0	26.6	69.9	67.2
	2016	**2020**	**2016**	**2020**	**2016**	**2020**
High-tech exports (% of manufactured exports)	5.8	6.1	8.4	7.2	16.5	16.2
	2016	**2019**	**2016**	**2019**	**2016**	**2019**
R&D expenditures (% of GDP)	0.02	0.03	0.3	0.4	1.8	2.0
Citizens' perceptions and institutions						
	2016	**2019**	**2016**	**2021**	**2016**	**2021**
Share of population satisfied with efforts to preserve the environment (%)	54.2	57.0	46.2	42.0	55.3	51.2
Share of population with confidence in national government (%)	46.9	45.7	34.2	39.4	40.3	46.3
Share of population that thinks corruption is widespread throughout government (%)	69.5	73.2	74.5	70.0	59.0	54.9
Share of population satisfied with the education system (%)	67.0	72.7	64.9	54.3	67.0	66.8
Share of urban population satisfied with the availability of quality health care (%)	43.9	54.0	49.2	48.9	68.2	69.5
Environment and the green transition						
	2004-19		**2004-19**		**2004-19**	
Loss of natural and semi-natural vegetated land (%)	2.6		1.5		1.3	
	2016	**2019**	**2016**	**2019**	**2016**	**2019**
GHG emissions per capita excluding LUCF (t CO_2e)	2.0	2.1	6.3	6.3	9.6	9.1
Air pollution – exposure to PM2.5 (annual average exposure to more than 10 µg/m^3, % of population)	100	100	95.5	95.4	61.3	61.0
	2016	**2020**	**2016**	**2019**	**2016**	**2020**
Contribution of renewables to total primary energy supply (%)	63.4	62.1	34.2	33.4	19.7	22.9
	2016	**2021**	**2016**	**2021**	**2016**	**2021**
Marine protected areas (% of territorial waters)	0.8	0.8	2.5	7.3	16.5	18.6
Fiscal position						
	2016	**2020**	**2016**	**2020**	**2016**	**2020**
Environmentally related tax revenue (% of GDP)	0.9	0.8	1.1	1.0	2.4	2.1
Total tax revenues (% of GDP)	13.2	12.4	22.2	21.9	33.6	33.5
Share of VAT (% GDP)	4.8	4.8	5.8	5.6	6.6	6.7
Share of PIT (% GDP)	0.4	0.5	2.1	2.2	7.8	8.3
Share of CIT (% of GDP)	2.9	2.3	3.3	3.4	2.9	2.7
Perception of tax evasion (%)	N/A	22.3	N/A	27.3	N/A	N/A
Debt service (% of total tax revenue)	11.6	13.9	11.3	13.0	5.6	5.2
	2016	**2020**	**2016**	**2020**	**2016**	**2019**
Social expenditure (% of GDP)	7.2	9.5	11.5	13.6	20.0	19.9

Note: See the Reader's Guide for definitions and sources.

StatLink https://stat.link/z2uxct

MEXICO

1. Recent trends

Poverty in Mexico remained largely unchanged, at 37.4% in 2020, compared to 37.6% in 2016, above the Latin America and the Caribbean (LAC) average of 26.3%. Extreme poverty in that period rose from 8.4% to 9.2%, partly owing to the impact of the COVID-19 pandemic, above the LAC average (8.7%). The population living in completely informal households decreased from 41.6% in 2010 to 37.0% in 2018, in line with the LAC average of 36.3% in 2018. Regarding environmental indicators, in 2019, greenhouse gas (GHG) emissions per capita were 5.1 tonnes of carbon dioxide equivalent (t CO_2e), lower than the averages for LAC (6.3) and countries belonging to the Organisation for Economic Co-operation and Development (OECD) (9.1). That year, the share of the population exposed to air pollution levels that pose risks to human health (PM2.5 at more than 10 $\mu g/m^3$) was 99.5%, higher than 95.4% for LAC and 61.0% for the OECD. The marine protected area of Mexico accounted for 21.6% of its territorial waters in 2021, substantially higher than 7.3% for LAC and above 18.6% for the OECD. On the fiscal side, environmentally related tax revenue was 1.4% of gross domestic product (GDP) in 2020, above LAC (1.0%) but below the OECD (2.1%). Total tax revenue as a percentage of GDP (17.9%) remains lower than the averages for LAC (21.9%) and the OECD (33.5%).

2. Long-term development policies for a green transition

Mexico hosts 10-12% of the world's biodiversity and is geographically vulnerable to climate change risks. Climate change mitigation and adaptation were declared a public interest as early as 1988 (General Law of Ecological Equilibrium and Environmental Protection). In 2020, the Natural Resources Sectoral Programme 2020-2024 was launched. Mexico intends to submit improved 2030 NDC targets before the United Nations Climate Change Conference (COP27).

For its **mitigation** efforts, Mexico has published a Special Climate Change Program 2021-2024, which focuses on enhanced synergies with adaptation actions to generate environmental, social, and economic benefits. The National Development Plan 2019-2024 promotes sustainable development through inclusion of population and communities to produce energy from renewable sources. Current priorities are hydroelectric, solar and wind energy. Concerning carbon sinks, the General Law for Sustainable Forest Development and the National Forestry Program 2020-2024 promote sustainable forestry management, while the National Biodiversity Strategy of Mexico and the Action Plan 2016-2030 promotes biodiversity conservation. In 2019, Mexico launched the *Sembrando Vida* reforestation programme, which pays 420 000 farmers to plant trees.

Regarding **adaptation**, Mexico has focused its activities, through the Special Climate Change Program 2021-2024, on maintaining environmental services, reducing vulnerability to climate change, and increasing resilience of the communities. Regarding agriculture, Mexico published a Sectoral Programme for Agriculture and Rural Development 2019-2024, which resulted in 3 special and 11 institutional programmes, on fisheries and aquaculture, sugarcane, and national food security. The ADAPTUR project supports the adaptation of the tourism sector, with support from the German International Development Agency (GIZ). To monitor risks and adaptation progress, Mexico developed a National Environment and Natural Resource Information System, a National Risks Atlas, a National Atlas of Vulnerability to Climate Change, a National Program of Hydraulic Contingencies, and a National Water Program 2020-2024.

In its **international partnerships**, Mexico emphasises natural environmental conservation, agriculture, and energy. **Within** LAC, Mexico is fostering collaboration with Belize, Guatemala and GIZ to conserve the shared tropical forest region. **Beyond** LAC, Mexico collaborates on projects for adaptation technologies (2021) and adaptive river restoration (2021), with support from the Green Climate Fund (GCF). It is also creating a risk-sharing facility for agricultural businesses (2019) with the Inter-American Development Bank (IDB). Moreover, Mexico formed a policy-oriented Climate Change Alliance (2017) and an Energy Partnership (2017) both with Germany.

Regarding **green finance**, the Green Finance Advisory Council (2016) promotes a sustainable financial market. Exemplary results are Mexico's Green Bond Principles and several investor statements. The initiation of a national GHG emissions market was required by law in 2018, and an emissions trading scheme will be launched in 2023. Mexico also co-operates with the United States Agency for International Development (USAID) in the creation of carbon credits.

Key indicators – Mexico

	Mexico		LAC		OECD	
Social						
	2016	**2020**	**2016**	**2020**	**2016**	**2020**
Extreme poverty	8.4	9.2	8.1	8.7	N/A	N/A
Poverty	37.6	37.4	25.9	26.3	N/A	N/A
Share of Internet users (% of population)	59.5	72.0	53.7	68.1	81.8	88.0
	2016	**2020**	**2016**	**2020**	**2016**	**2019**
Gini index	47.7	45.4	46.3	45.3	34.9	34.2
	2010	**2018**	**2009**	**2018**	**2009**	**2018**
Share of total population in informal households (%)	41.6	37.0	43.4	36.3	N/A	N/A
Share of total population in informal households, upper-income quintile (%)	18.0	17.4	24.5	13.6	N/A	N/A
Share of total population in informal households, lower-income quintile (%)	78.0	70.9	70.4	72.0	N/A	N/A
	2016	**2019**	**2016**	**2019**	**2016**	**2019**
Health expenditure (% of GDP)	5.6	5.4	6.5	6.8	8.7	8.8
SIGI index	N/A	29.0	N/A	25.4	N/A	17.5
	2015	**2018**	**2015**	**2018**	**2015**	**2018**
PISA score in science	416	419	411	407	489	487
Productivity and innovation						
	2016	**2021**	**2016**	**2021**	**2016**	**2021**
Labour productivity (% of the United States)	37.7	32.8	29.0	26.6	69.9	67.2
	2016	**2020**	**2016**	**2020**	**2016**	**2020**
High-tech exports (% of manufactured exports)	20.6	21.5	8.4	7.2	16.5	16.2
	2016	**2020**	**2016**	**2019**	**2016**	**2019**
R&D expenditures (% of GDP)	0.4	0.3	0.3	0.4	1.8	2.0
Citizens' perceptions and institutions						
	2016	**2021**	**2016**	**2021**	**2016**	**2021**
Share of population satisfied with efforts to preserve the environment (%)	46.3	38.3	46.2	42.0	55.3	51.2
Share of population with confidence in national government (%)	27.8	48.3	34.2	39.4	40.3	46.3
Share of population that thinks corruption is widespread throughout government (%)	82.9	68.6	74.5	70.0	59.0	54.9
Share of population satisfied with the education system (%)	58.4	49.6	64.9	54.3	67.0	66.8
Share of urban population satisfied with the availability of quality health care (%)	50.9	47.4	49.2	48.9	68.2	69.5
Environment and the green transition						
	2004-19		**2004-19**		**2004-19**	
Loss of natural and semi-natural vegetated land (%)	0.9		1.5		1.3	
	2016	**2019**	**2016**	**2019**	**2016**	**2019**
GHG emissions per capita excluding LUCF (t CO_2e)	5.5	5.1	6.3	6.3	9.6	9.1
Air pollution – exposure to PM2.5 (annual average exposure to more than 10 µg/m^3, % of population)	99.5	99.5	95.5	95.4	61.3	61.0
	2016	**2020**	**2016**	**2019**	**2016**	**2020**
Contribution of renewables to total primary energy supply (%)	8.5	9.6	34.2	33.4	19.7	22.9
	2016	**2021**	**2016**	**2021**	**2016**	**2021**
Marine protected areas (% of territorial waters)	2.3	21.6	2.5	7.3	16.5	18.6
Fiscal position						
	2016	**2020**	**2016**	**2020**	**2016**	**2020**
Environmentally related tax revenue (% of GDP)	1.6	1.4	1.1	1.0	2.4	2.1
Total tax revenues (% of GDP)	16.6	17.9	22.2	21.9	33.6	33.5
Share of VAT (% GDP)	3.9	4.3	5.8	5.6	6.6	6.7
Share of PIT (% GDP)	3.4	3.8	2.1	2.2	7.8	8.3
Share of CIT (% of GDP)	3.5	3.6	3.3	3.4	2.9	2.7
Perception of tax evasion (%)	N/A	24.3	N/A	27.3	N/A	N/A
Debt service (% of total tax revenue)	18.8	21.8	11.3	13.0	5.6	5.2
	2016	**2020**	**2016**	**2020**	**2016**	**2019**
Social expenditure (% of GDP)	9.9	10.4	11.5	13.6	20.0	19.9

Note: See the Reader's Guide for definitions and sources.

StatLink https://stat.link/5w0719

PANAMA

1. Recent trends

Poverty in Panama decreased from 16.8% in 2016 to 15.0% in 2019, lower than the Latin America and the Caribbean (LAC) average for 2020 (26.3%). Extreme poverty decreased in that period from 7.4% to 6.8% and remains below the LAC average (8.7% in 2020). The Gini index decreased from 50.4 in 2016 to 49.8 in 2019, remaining above the LAC average (45.3 in 2019). Regarding environmental indicators, in 2019, greenhouse gas (GHG) emissions per capita were 4.9 tonnes of carbon dioxide equivalent (t CO_2e), lower than the averages for LAC (6.3) and countries belonging to the Organisation for Economic Co-operation and Development (OECD) (9.1). That year, the share of the population exposed to air pollution levels that pose risks to human health (PM2.5 at more than 10 $\mu g/m^3$) was 99.8%, higher than 95.4% for LAC and 61.0% for the OECD. The marine protected area of Panama accounted for 26.8% of its territorial waters in 2021, well above 7.3% for LAC and 18.6% for the OECD. On the fiscal side, environmentally related tax revenue was 0.4% of gross domestic product (GDP) in 2020, below LAC (1.0%) and the OECD (2.1%). Total tax revenue as a percentage of GDP (13.7%) remained lower than the averages for LAC (21.9%) and the OECD (33.5%).

2. Long-term development policies for a green transition

Panama is highly vulnerable to extreme weather events caused by climate change. In its National Strategy for Climate Change 2050 (2019), the country presents the vision for its green transition. The Executive Decree No. 100 (2020) regulates the mitigation chapter of the General Environmental Law and provides for the management and monitoring of low-carbon economic and social development.

Panama's **mitigation** efforts primarily address the energy transition. The National Energy Plan 2015-2050 establishes the goal of a 70% renewable energy mix. The government presented strategic guidelines (2020-2030) that focus on strengthening institutions, the energy sector and the hydrocarbon sector. Panama has developed National Strategies for Electric Mobility (2019), Distributed Generation (2021), Universal Energy Access (2022), Innovation for the National Interconnected System (2022), and the Rational and Efficient Use of Energy (2022). Additional priorities are set through an action plan for solar thermal energy (2021), a regulation on sustainable buildings (2019), the Panama Cooling Plan (2020), and the introduction of the Green Hydrogen Roadmap (2022). The *Reduce Tu Huella Corporativa* programme promotes voluntary private sector engagement by offering a standardised process to quantify of corporate carbon and water footprints. Panama has also launched a National Forest Restoration Programme 2021-25. Regarding **adaptation**, Panama recently launched a coastal adaptation project for gender-based planning instruments with the World Bank PROBLUE Fund (2022) and is preparing a nature-based adaptation project in *Bocas del Toro* and *Limon*, with support from the Adaptation Fund and the United Nations Environment Programme (UNEP). For monitoring purposes, Panama implemented a project to measure its adaptation progress, with support from the Adaptation Fund, and a platform (SIRED) that focuses on the losses and damages caused by climate change, with support from the World Bank. A National Climate Transparency Platform gathers all the information on Panama's climate actions.

In its **international partnerships, within** LAC, Panama is strengthening the adaptive capacity of coastal communities, in co-operation with Cuba, the Food and Agricultural Organization (FAO) and the Adaptation Fund, and promotes climate-resilient water management, with support from the Colombian *Fundación Natura*. **Beyond** LAC, Panama fosters sustainable land management and restoration, supported by the FAO and the Global Environment Facility (GEF). To increase urban resilience, Panama is part of the Nature4Cities Latam project, co-ordinated by UNEP and the Green Climate Fund (GCF). To incorporate gender considerations in green transition policies, Panama established a National Gender and Climate Change Plan in collaboration with the United Nations Development Programme (UNDP).

Regarding **green finance** Panama adopted a manual for labelling public spending projects according to climate change criteria in 2022. Panama's national stock exchange issued guidelines on sustainable investment. Panama is also in the process of structuring an exchange system for carbon credits and developing a national REDD+ (Reducing Emissions from Deforestation and Forest Degradation) strategy. Panama is part of the Working Group on Sustainable Finance Taxonomies in LAC to develop a common framework of sustainable finance taxonomies for the region.

Key indicators – Panama

	Panama		LAC		OECD	
Social						
	2016	**2019**	**2016**	**2020**	**2016**	**2020**
Extreme poverty	7.4	6.8	8.1	8.7	N/A	N/A
Poverty	16.8	15.0	25.9	26.3	N/A	N/A
Share of Internet users (% of population)	54.0	63.6	53.7	68.1	81.8	88.0
	2016	**2019**	**2016**	**2020**	**2016**	**2019**
Gini index	50.4	49.8	46.3	45.3	34.9	34.2
	2010	**2018**	**2009**	**2018**	**2009**	**2018**
Share of total population in informal households (%)	N/A	N/A	43.4	36.3	N/A	N/A
Share of total population in informal households, upper-income quintile (%)	N/A	N/A	24.5	13.6	N/A	N/A
Share of total population in informal households, lower-income quintile (%)	N/A	N/A	70.4	72.0	N/A	N/A
	2016	**2019**	**2016**	**2019**	**2016**	**2019**
Health expenditure (% of GDP)	7.3	7.6	6.5	6.8	8.7	8.8
SIGI index	N/A	N/A	N/A	25.4	N/A	17.5
	2015	**2018**	**2015**	**2018**	**2015**	**2018**
PISA score in science	N/A	365	411	407	489	487
Productivity and innovation						
	2016	**2021**	**2016**	**2021**	**2016**	**2021**
Labour productivity (% of the United States)	N/A	N/A	29.0	26.6	69.9	67.2
	2016	**2020**	**2016**	**2020**	**2016**	**2020**
High-tech exports (% of manufactured exports)	9.2	N/A	8.4	7.2	16.5	16.2
	2016	**2019**	**2016**	**2019**	**2016**	**2019**
R&D expenditures (% of GDP)	0.1	N/A	0.3	0.4	1.8	2.0
Citizens' perceptions and institutions						
	2016	**2019**	**2016**	**2021**	**2016**	**2021**
Share of population satisfied with efforts to preserve the environment (%)	48.8	50.3	46.2	42.0	55.3	51.2
Share of population with confidence in national government (%)	33.0	40.7	34.2	39.4	40.3	46.3
Share of population that thinks corruption is widespread throughout government (%)	77.5	82.9	74.5	70.0	59.0	54.9
Share of population satisfied with the education system (%)	69.2	64.9	64.9	54.3	67.0	66.8
Share of urban population satisfied with the availability of quality health care (%)	60.3	49.5	49.2	48.9	68.2	69.5
Environment and the green transition						
	2004-19		**2004-19**		**2004-19**	
Loss of natural and semi-natural vegetated land (%)	0.9		1.5		1.3	
	2016	**2019**	**2016**	**2019**	**2016**	**2019**
GHG emissions per capita excluding LUCF (t CO_2e)	4.4	4.9	6.3	6.3	9.6	9.1
Air pollution – exposure to PM2.5 (annual average exposure to more than 10 $\mu g/m^3$, % of population)	99.8	99.8	95.5	95.4	61.3	61.0
	2016	**2020**	**2016**	**2019**	**2016**	**2020**
Contribution of renewables to total primary energy supply (%)	22.4	17.2	34.2	33.4	19.7	22.9
	2016	**2021**	**2016**	**2021**	**2016**	**2021**
Marine protected areas (% of territorial waters)	1.7	26.8	2.5	7.3	16.5	18.6
Fiscal position						
	2016	**2020**	**2016**	**2020**	**2016**	**2020**
Environmentally related tax revenue (% of GDP)	0.7	0.4	1.1	1.0	2.4	2.1
Total tax revenues (% of GDP)	15.5	13.7	22.2	21.9	33.6	33.5
Share of VAT (% GDP)	2.6	1.8	5.8	5.6	6.6	6.7
Share of PIT (% GDP)	1.5	1.8	2.1	2.2	7.8	8.3
Share of CIT (% of GDP)	1.8	1.1	3.3	3.4	2.9	2.7
Perception of tax evasion (%)	N/A	29.8	N/A	27.3	N/A	N/A
Debt service (% of total tax revenue)	10.5	17.0	11.3	13.0	5.6	5.2
	2016	**2020**	**2016**	**2020**	**2016**	**2019**
Social expenditure (% of GDP)	8.3	N/A	11.5	13.6	20.0	19.9

Note: See the Reader's Guide for definitions and sources.

StatLink https://stat.link/epl7ni

PARAGUAY

1. Recent trends

Poverty in Paraguay decreased from 24.0% in 2016 to 22.3% in 2020 and remained below the Latin American and the Caribbean (LAC) average of 26.3%. These figures mask the effect of strong efforts to decrease poverty before the shock of the COVID-19 pandemic, which brought the figure down to 19.4% in 2019. Extreme poverty consistently decreased in that period, from 7.9% to 6.0%, below the LAC average (8.7%). The population living in completely informal households decreased from 66.4% in 2009 to 58.0% in 2018 higher than the LAC average of 36.3%. Regarding environmental indicators, in 2019, greenhouse gas (GHG) emissions per capita were 7.0 tonnes of carbon dioxide equivalent (t CO_2e), higher than the average for LAC (6.3) but below the average for countries belonging to the Organisation for Economic Co-operation and Development (OECD) (9.1). That year, the share of the population exposed to air pollution levels that pose risks to human health (PM2.5 at more than 10 μg/m^3) was 100%, higher than 95.4% for LAC and 61.0% for the OECD. On the fiscal side, environmentally related tax revenue was 0.9% of gross domestic product (GDP) in 2020, below LAC average (1.0%) and the OECD average (2.1%). Total tax revenue as a percentage of GDP in 2020 (13.4%) remained considerably lower than the averages for LAC (21.9%) and the OECD (33.5%).

2. Long-term development policies for a green transition

Paraguay depends economically on climatic conditions for income generation through agriculture, energy, and transport. The mitigation and adaptation efforts were made a national priority in 2011 through the National Climate Change Policy. Paraguay's National Development Plan 2014-2030 includes strategic objectives, such as the development of sustainable habitats, appreciation and protection of environmental capital, and promotion of productive diversification.

To achieve its **mitigation** goals, Paraguay adopted a National Plan in 2017, including programmes for integrated transport-sector management, clean cooking, replacement of liquefied petroleum (LGP) gas cookers, sustainable use of forests in the Chaco region, forest landscape restoration, waste management and sustainable architecture. A national electro-mobility strategy has been developed and the government is examining the potential to produce green hydrogen. Regarding **adaptation** policies, the National Adaptation Plan 2022-2030 was developed through a participatory process. It contains 25 objectives for resilient cities and communities, health and epidemiology, ecosystems and biodiversity, energy, agriculture and food security, forestry, water resources, and transport. As an example, the resilience of communities is fostered through the ecosystem-based adaptation *AbE Chaco* project, which aims to reduce the region's vulnerability to the impacts of climate change on food security.

In its **international partnerships within** LAC, Paraguay is engaged in a regional initiative to advance e-mobility, with support from the United Nations Environment Programme (UNEP) and the Green Climate Fund (GCF). For the post COVID-19 reactivation of sustainable tourism, the country is exchanging information with Costa Rica and Ecuador. **Beyond** the region, Paraguay is strengthening evidence-based actions and promoting private-sector involvement, with assistance from the Nationally Determined Contributions (NDC) Support Programme of the United Nations Development Programme (UNDP). Additional focal points of collaboration are sustainable agriculture and rural development (the *Samaeul Undong* project, with support from the Korea International Cooperation Agency), responsible agricultural production and consumption (in collaboration with the World Wildlife Fund [WWF]) and the strengthening of territorial governance capacities (with the Spanish Agency for International Development Cooperation [AECID]). Paraguay also supports plantation and reforestation efforts performed by 17 000 vulnerable families with the *PROEZA* project (2017), in collaboration with the GCF and the Food and Agriculture Organization (FAO).

In the area of **green finance**, Paraguay created the *Mesa de Finanzas Sostenibles*, a national platform for voluntary collaborations among financial actors to develop environmental and social risk-analysis systems and guidelines for sustainable financing of agricultural, livestock and agro-industrial activities. Paraguay also reviews environmental policies and fiscal incentives to stimulate the financing of sustainable agriculture, in collaboration with the United Nations' Good Growth Partnership initiative. Paraguay is also engaged in the voluntary REDD+ (Reducing Emissions from Deforestation and Forest Degradation) process to access results-based payments for sustainable forest management. To date, it has received payments from the GCF for emissions reduction of 23 Mt CO_2e achieved in 2015-17.

Key indicators – Paraguay

	Paraguay		LAC		OECD	
Social						
	2016	**2020**	**2016**	**2020**	**2016**	**2020**
Extreme poverty	7.9	6.0	8.1	8.7	N/A	N/A
Poverty	24.0	22.3	25.9	26.3	N/A	N/A
Share of Internet users (% of population)	53.4	74.0	53.7	68.1	81.8	88.0
	2016	**2020**	**2016**	**2020**	**2016**	**2019**
Gini index	47.9	43.5	46.3	45.3	34.9	34.2
	2009	**2018**	**2009**	**2018**	**2009**	**2018**
Share of total population in informal households (%)	66.4	58.0	43.4	36.3	N/A	N/A
Share of total population in informal households, upper-income quintile (%)	30.3	22.8	24.5	13.6	N/A	N/A
Share of total population in informal households, lower-income quintile (%)	99.0	95.3	70.4	72.0	N/A	N/A
	2016	**2019**	**2016**	**2019**	**2016**	**2019**
Health expenditure (% of GDP)	6.7	7.2	6.5	6.8	8.7	8.8
SIGI index	N/A	32.8	N/A	25.4	N/A	17.5
	2015	**2018**	**2015**	**2018**	**2015**	**2018**
PISA score in science	N/A	N/A	411	407	489	487
Productivity and innovation						
	2016	**2021**	**2016**	**2021**	**2016**	**2021**
Labour productivity (% of the United States)	21.1	20.0	29.0	26.6	69.9	67.2
	2016	**2020**	**2016**	**2020**	**2016**	**2020**
High-tech exports (% of manufactured exports)	7.6	5.1	8.4	7.2	16.5	16.2
	2016	**2019**	**2016**	**2019**	**2016**	**2019**
R&D expenditures (% of GDP)	0.1	0.1	0.3	0.4	1.8	2.0
Citizens' perceptions and institutions						
	2016	**2021**	**2016**	**2021**	**2016**	**2021**
Share of population satisfied with efforts to preserve the environment (%)	32.1	36.8	46.2	42.0	55.3	51.2
Share of population with confidence in national government (%)	28.5	29.0	34.2	39.4	40.3	46.3
Share of population that thinks corruption is widespread throughout government (%)	73.7	85.0	74.5	70.0	59.0	54.9
Share of population satisfied with the education system (%)	68.7	60.8	64.9	54.3	67.0	66.8
Share of urban population satisfied with the availability of quality health care (%)	49.7	52.2	49.2	48.9	68.2	69.5
Environment and the green transition						
	2004-19		**2004-19**		**2004-19**	
Loss of natural and semi-natural vegetated land (%)	3.5		1.5		1.3	
	2016	**2019**	**2016**	**2019**	**2016**	**2019**
GHG emissions per capita excluding LUCF (t CO_2e)	7.0	7.0	6.3	6.3	9.6	9.1
Air pollution – exposure to PM2.5 (annual average exposure to more than 10 µg/m³, % of population)	100	100	95.5	95.4	61.3	61.0
	2016	**2020**	**2016**	**2019**	**2016**	**2020**
Contribution of renewables to total primary energy supply (%)	127.9	101.5	34.2	33.4	19.7	22.9
	2016	**2021**	**2016**	**2021**	**2016**	**2021**
Marine protected areas (% of territorial waters)	N/A	N/A	2.5	7.3	16.5	18.6
Fiscal position						
	2016	**2020**	**2016**	**2020**	**2016**	**2020**
Environmentally related tax revenue (% of GDP)	1.0	0.9	1.1	1.0	2.4	2.1
Total tax revenues (% of GDP)	13.4	13.4	22.2	21.9	33.6	33.5
Share of VAT (% GDP)	5.1	4.9	5.8	5.6	6.6	6.7
Share of PIT (% GDP)	0.2	0.2	2.1	2.2	7.8	8.3
Share of CIT (% of GDP)	2.0	2.3	3.3	3.4	2.9	2.7
Perception of tax evasion (%)	N/A	27.0	N/A	27.3	N/A	N/A
Debt service (% of total tax revenue)	6.6	10.5	11.3	13.0	5.6	5.2
	2016	**2020**	**2016**	**2020**	**2016**	**2019**
Social expenditure (% of GDP)	8.5	11.4	11.5	13.6	20.0	19.9

Note: See the Reader's Guide for definitions and sources.

StatLink https://stat.link/90se86

PERU

1. Recent trends

Poverty in Peru increased in recent years, from 19.1% in 2016 to 28.4% in 2020, partly owing to the impact of the COVID-19 pandemic. As a result, it is now above the Latin America and the Caribbean (LAC) average of 26.3%. Extreme poverty also increased in that period from 5.2% to 8.6% and was on par with the LAC average (8.7%). The population living in completely informal households decreased from 67.9% in 2010 to 51.6% in 2018 but remained well above the LAC average of 36.3% in 2018. Regarding environmental indicators, in 2019, greenhouse gas (GHG) emissions per capita were 3.1 tonnes of carbon dioxide equivalent (t CO_2e), lower than the averages for LAC (6.3) and for countries belonging to the Organisation for Economic Co-operation and Development (OECD) (9.1). That year, the share of the population exposed to air pollution levels that pose risks to human health (PM2.5 at more than 10 $\mu g/m^3$) was 98.4%, higher than 95.4% for LAC and 61.0% for the OECD. The marine protected area of Peru accounted for just 0.5% of its territorial waters in 2021, lower than averages for LAC (7.3%) and the OECD (18.6%). In June 2021, the government signed a decree that establishes the protected marine area of the *Nazca Ridge National Reserve* to 62 392 km^2, thereby increasing the national marine protected area to almost 8% of territorial waters. On the fiscal side, environmentally related tax revenue was 0.5% of gross domestic product (GDP) in 2020, below LAC (1.0%) and the OECD (2.1%). Total tax revenues as a percentage of GDP in 2020 (15.2%) remained lower than the averages for LAC (21.9%) and the OECD (33.5%).

2. Long-term development policies for a green transition

Peru's National Strategy on Climate Change to 2050, recently updated, defines action on climate change in an integrated, cross-cutting and multi-sectoral manner. The National Environmental Policy 2030 aims to decrease the vulnerability of ecosystems, preserve biodiversity, contribute to the sustainable management of nature, and achieve a 64% renewable energy matrix by 2030. The National Plan of Competitiveness and Productivity 2019-2030 promotes sustainable economic growth and international trade, prioritising environmental sustainability.

Within Peru's **mitigation** efforts, the Ministry of Environment (MINAM) is executing the *Construyendo bienestar humano y resiliencia en bosques amazónicos* project (2019), financed by the Global Environment Facility (GEF), to preserve Amazon ecosystems. The MINAM is also leading a project with the NAMA (Nationally Appropriate Mitigation Actions) Facility for management of organic wastes. The *Mejora de la sostenibilidad en la movilidad eléctrica para el transporte urbano* project aims to enhance electric transport initiatives to reduce pollutants and improve air quality. Regarding Peru's **adaptation** measures, a National Adaptation Plan has been adopted in 2021.

Regarding **international partnerships, within** the region, Brazil, Colombia, Ecuador and Peru are collaborating on the integrated management of the Putumayo-Ica river basin, with the World Bank as implementer. The Pacific Alliance leads a project to promote the circular economy and sustainable management of plastics in Chile, Colombia, Mexico and Peru. The AICCA-Peru (Adapting irrigation to climate change) project, in partnership with Bolivia, Colombia and Ecuador, aims to strengthen adaptation measures in irrigation system projects. **Beyond** LAC, Peru has various projects with the GEF to preserve the Amazon including: Sustainable Productive Landscapes in the Peruvian Amazon; Securing the Future of Peru's Natural Protected Areas; Sustainable Industrial Zone Development in Peru; and Deforestation Free Commodity Supply Chains in the Peruvian Amazon. As a member of the Climate and Clean Air Coalition, Peru received technical assistance to scale up municipal projects in solid wastes. In 2022, the Amazon Cooperation Treaty Organization began creating a regional platform for indigenous people in the Amazon as part of *Pacto de Leticia por la Amazonía*, implemented with support from EUROCLIMA+.

Concerning **green finance**, the Green Finance Roadmap incorporates environmental guidelines in the management of the financial system, insurance sector and the stock market to increase access to green funds. *Iniciativa Patrimonio Natural del Peru* uses a financial model that grants resources to projects that meet environmental and social criteria for the conservation of protected natural areas. The Development Bank of Peru issued green bonds to finance renewable energies. Peru has two public-private initiatives to promote sustainable construction: *Bono Buen Pagador Sostenible*, a bonus granted by the government to reduce the mortgage loan of households considered eco-friendly; and *Bono Sostenible*, granted to residential projects with the sustainable household certificate issued by the Ministry of Housing.

Key indicators – Peru	Peru		LAC		OECD	
Social						
	2016	**2020**	**2016**	**2020**	**2016**	**2020**
Extreme poverty	5.2	8.6	8.1	8.7	N/A	N/A
Poverty	19.1	28.4	25.9	26.3	N/A	N/A
Share of Internet users (% of population)	45.5	65.3	53.7	68.1	81.8	88.0
	2016	**2020**	**2016**	**2020**	**2016**	**2019**
Gini index	43.6	43.8	46.3	45.3	34.9	34.2
	2010	**2018**	**2009**	**2018**	**2009**	**2018**
Share of total population in informal households (%)	67.9	51.6	43.4	36.3	N/A	N/A
Share of total population in informal households, upper-income quintile (%)	30.5	18.2	24.5	13.6	N/A	N/A
Share of total population in informal households, lower-income quintile (%)	94.5	89.2	70.4	72.0	N/A	N/A
	2016	**2019**	**2016**	**2019**	**2016**	**2019**
Health expenditure (% of GDP)	5.0	5.2	6.5	6.8	8.7	8.8
SIGI index	N/A	24.5	N/A	25.4	N/A	17.5
	2015	**2018**	**2015**	**2018**	**2015**	**2018**
PISA score in science	397	404	411	407	489	487
Productivity and innovation						
	2016	**2021**	**2016**	**2021**	**2016**	**2021**
Labour productivity (% of the United States)	18.9	17.8	29.0	26.6	69.9	67.2
	2016	**2020**	**2016**	**2020**	**2016**	**2020**
High-tech exports (% of manufactured exports)	4.8	4.8	8.4	7.2	16.5	16.2
	2016	**2019**	**2016**	**2019**	**2016**	**2019**
R&D expenditures (% of GDP)	0.1	0.2	0.3	0.4	1.8	2.0
Citizens' perceptions and institutions						
	2016	**2021**	**2016**	**2021**	**2016**	**2021**
Share of population satisfied with efforts to preserve the environment (%)	49.6	36.5	46.2	42.0	55.3	51.2
Share of population with confidence in national government (%)	28.2	23.5	34.2	39.4	40.3	46.3
Share of population that thinks corruption is widespread throughout government (%)	82.7	88.4	74.5	70.0	59.0	54.9
Share of population satisfied with the education system (%)	58.4	30.4	64.9	54.3	67.0	66.8
Share of urban population satisfied with the availability of quality health care (%)	44.1	28.9	49.2	48.9	68.2	69.5
Environment and the green transition						
	2004-19		**2004-19**		**2004-19**	
Loss of natural and semi-natural vegetated land (%)	0.3		1.5		1.3	
	2016	**2019**	**2016**	**2019**	**2016**	**2019**
GHG emissions per capita excluding LUCF (t CO_2e)	3.2	3.1	6.3	6.3	9.6	9.1
Air pollution – exposure to PM2.5 (annual average exposure to more than 10 µg/m^3, % of population)	98.4	98.4	95.5	95.4	61.3	61.0
	2016	**2020**	**2016**	**2019**	**2016**	**2020**
Contribution of renewables to total primary energy supply (%)	24.5	24.5	34.2	33.4	19.7	22.9
	2016	**2021**	**2016**	**2021**	**2016**	**2021**
Marine protected areas (% of territorial waters)	0.5	0.5	2.5	7.3	16.5	18.6
Fiscal position						
	2016	**2020**	**2016**	**2020**	**2016**	**2020**
Environmentally related tax revenue (% of GDP)	0.5	0.5	1.1	1.0	2.4	2.1
Total tax revenues (% of GDP)	16.2	15.2	22.2	21.9	33.6	33.5
Share of VAT (% GDP)	6.0	5.9	5.8	5.6	6.6	6.7
Share of PIT (% GDP)	1.8	1.8	2.1	2.2	7.8	8.3
Share of CIT (% of GDP)	3.8	3.3	3.3	3.4	2.9	2.7
Perception of tax evasion (%)	N/A	34.8	N/A	27.3	N/A	N/A
Debt service (% of total tax revenue)	5.8	9.2	11.3	13.0	5.6	5.2
	2016	**2020**	**2016**	**2020**	**2016**	**2019**
Social expenditure (% of GDP)	N/A	N/A	11.5	13.6	20.0	19.9

Note: See the Reader's Guide for definitions and sources.

StatLink https://stat.link/3os2fx

URUGUAY

1. Recent trends

Poverty in Uruguay increased from 3.6% in 2016 to 5.2% in 2020, partly owing to the impact of the COVID-19 pandemic, but remains substantially lower than the Latin America and the Caribbean (LAC) average of 26.3%. Extreme poverty increased in that period from 0.2% to 0.3%, also far below the LAC average (8.7%). The population living in completely informal households decreased by almost half – from 28.5% in 2008 to 16.3% in 2018 – bringing it far below the LAC average of 36.3% in 2018. Regarding environmental indicators, in 2019, greenhouse gas (GHG) emissions per capita were 10.4 tonnes of carbon dioxide equivalent (t CO_2e), higher than the averages for LAC (6.3) and countries belonging to the Organisation for Economic Co-operation and Development (OECD) (9.1). That year, the share of the population exposed to air pollution levels that pose risks to human health (PM2.5 at more than 10 $\mu g/m^3$) was 26.5%, substantially lower than 95.4% for LAC and 61.0% for the OECD. The marine protected area of Uruguay accounted for 0.75% of its territorial waters in 2021, far below 7.3% for LAC and 18.6% for the OECD. On the fiscal side, environmentally related tax revenue was 1.8% of gross domestic product (GDP) in 2020, above LAC (1.0%) but slightly below the OECD (2.1%). Total tax revenues as a percentage of GDP in 2020 (26.6%) remained higher than the average for LAC (21.9%) but below the OECD (33.5%).

2. Long-term development policies for a green transition

At the institutional level, the Ministry of Environment was created in 2020 as the first state secretariat to deal exclusively with environmental and institutional concerns. The National Climate Change Plan (2017) and the Long-Term Climate Strategy (2021) lay the foundations of the national environmental policy.

With respect to its **mitigation** measures, Uruguay places a specific focus on the energy, agricultural, transport and industrial sectors. In transport, Uruguay relies on legal incentives, for example by subsidising investments in electric public transport buses (Budget Law No. 19.670), promoting investments in electric mobility (Decree 268/020), and reducing taxes for hybrids and electric vehicles (Decree 321/021). In the agricultural sector, Uruguay aims to reduce emissions through productivity gains thanks to investments and the adoption of new technologies. In terms of energy production, Uruguay has managed to generate over 98% of its electricity from renewable sources (mainly hydropower, wind, solar, biofuels and biomass) and has launched a Green Hydrogen Roadmap in 2022. Among its **adaptation** policies, Uruguay has developed three national adaptation plans (NAPs) focused on agriculture (2019), cities and infrastructure (2021), and coastal adaptation (2021). For the protection against negative impacts of climate change, Uruguay established a National Emergency System (SINAE) for integrated disaster risk management.

In its **international partnerships within** LAC, Uruguay attaches particular importance to the transport sector. It developed the VERNE project in 2020, in partnership with the Inter-American Development Bank (IDB), to support creation of a hydrogen ecosystem to decarbonise the transport sector. In collaboration with the United Nations Development Programme (UNDP), Uruguay has created a sustainable and efficient urban mobility system (*Movés*). The country among other regional partnerships also launched a joint programme with Honduras in capacity building for electric mobility. **Beyond** LAC, Uruguay closely collaborates with the European Union (EU) through the EUROCLIMA+ programme which runs 15 mitigation and adaptation projects in the country, the EU AL-INVEST Verde programme which promotes the green transition, and a bilateral association that strengthens strategic dialogue. Lastly, the Partnership for Action on Green Economy, led by five United Nations agencies, supports Uruguay's efforts to incorporate the green economy in public policies of key economic sectors.

Regarding **green finance**, Uruguay's central bank joined the Network for the Greening of Financial Systems. The Economy and Finance Ministry joined the Coalition of Finance Ministers for Climate Action and adopted the Helsinki Principles for climate action. Uruguay is a member of the United Nations Joint Sustainable Development Goal Fund and is the only LAC country selected to receive funding from it for its decarbonisation efforts and energy sector transition. The programme will establish a Renewable Energy Innovation Fund to support Uruguay's second energy transition. Uruguay is also planning to issue an innovative sovereign bond that will tie its interest rate to environmental goals.

Key indicators – Uruguay

	Uruguay		LAC		OECD	
Social						
	2016	**2020**	**2016**	**2020**	**2016**	**2020**
Extreme poverty	0.2	0.3	8.1	8.7	N/A	N/A
Poverty	3.6	5.2	25.9	26.3	N/A	N/A
Share of Internet users (% of population)	66.4	86.1	53.7	68.1	81.8	88.0
	2016	**2020**	**2016**	**2020**	**2016**	**2019**
Gini index	39.7	40.2	46.3	45.3	34.9	34.2
	2008	**2018**	**2009**	**2018**	**2009**	**2018**
Share of total population in informal households (%)	28.5	16.3	43.4	36.3	N/A	N/A
Share of total population in informal households, upper-income quintile (%)	17.1	2.2	24.5	13.6	N/A	N/A
Share of total population in informal households, lower-income quintile (%)	56.5	48.0	70.4	72.0	N/A	N/A
	2016	**2019**	**2016**	**2019**	**2016**	**2019**
Health expenditure (% of GDP)	8.7	9.3	6.5	6.8	8.7	8.8
SIGI index	N/A	22.2	N/A	25.4	N/A	17.5
	2015	**2018**	**2015**	**2018**	**2015**	**2018**
PISA score in science	435	426	411	407	489	487
Productivity and innovation						
	2016	**2021**	**2016**	**2021**	**2016**	**2021**
Labour productivity (% of the United States)	38.9	38.0	29.0	26.6	69.9	67.2
	2016	**2020**	**2016**	**2020**	**2016**	**2020**
High-tech exports (% of manufactured exports)	10.6	9.8	8.4	7.2	16.5	16.2
	2016	**2019**	**2016**	**2019**	**2016**	**2019**
R&D expenditures (% of GDP)	0.4	0.5	0.3	0.4	1.8	2.0
Citizens' perceptions and institutions						
	2016	**2021**	**2016**	**2021**	**2016**	**2021**
Share of population satisfied with efforts to preserve the environment (%)	57.9	48.8	46.2	42.0	55.3	51.2
Share of population with confidence in national government (%)	42.3	48.6	34.2	39.4	40.3	46.3
Share of population that thinks corruption is widespread throughout government (%)	54.3	50.7	74.5	70.0	59.0	54.9
Share of population satisfied with the education system (%)	58.0	68.8	64.9	54.3	67.0	66.8
Share of urban population satisfied with the availability of quality health care (%)	70.9	69.4	49.2	48.9	68.2	69.5
Environment and the green transition						
	2004-19		**2004-19**		**2004-19**	
Loss of natural and semi-natural vegetated land (%)	0.5		1.5		1.3	
	2016	**2019**	**2016**	**2019**	**2016**	**2019**
GHG emissions per capita excluding LUCF (t CO_2e)	10.8	10.4	6.3	6.3	9.6	9.1
Air pollution – exposure to PM2.5 (annual average exposure to more than 10 μg/m^3, % of population)	33.2	26.5	95.5	95.4	61.3	61.0
	2016	**2020**	**2016**	**2019**	**2016**	**2020**
Contribution of renewables to total primary energy supply (%)	58.8	58.6	34.2	33.4	19.7	22.9
	2016	**2021**	**2016**	**2021**	**2016**	**2021**
Marine protected areas (% of territorial waters)	0.72	0.75	2.5	7.3	16.5	18.6
Fiscal position						
	2016	**2020**	**2016**	**2020**	**2016**	**2020**
Environmentally related tax revenue (% of GDP)	1.7	1.8	1.1	1.0	2.4	2.1
Total tax revenues (% of GDP)	25.6	26.6	22.2	21.9	33.6	33.5
Share of VAT (% GDP)	7.1	7.0	5.8	5.6	6.6	6.7
Share of PIT (% GDP)	3.0	4.1	2.1	2.2	7.8	8.3
Share of CIT (% of GDP)	2.6	2.7	3.3	3.4	2.9	2.7
Perception of tax evasion (%)	N/A	32.6	N/A	27.3	N/A	N/A
Debt service (% of total tax revenue)	9.5	9.6	11.3	13.0	5.6	5.2
	2016	**2020**	**2016**	**2020**	**2016**	**2019**
Social expenditure (% of GDP)	14.8	16.7	11.5	13.6	20.0	19.9

Note: See the Reader's Guide for definitions and sources.

StatLink https://stat.link/5wevlx

www.ingramcontent.com/pod-product-compliance
Lightning Source LLC
Chambersburg PA
CBHW061159240926
51846CB00006B/1059